Mineral Resource Estimation Conference 2025

6–7 May 2025
Perth, Australia

The Australasian Institute of Mining and Metallurgy
Publication Series No 2/2025

AusIMM

Published by:
The Australasian Institute of Mining and Metallurgy
Ground Floor, 204 Lygon Street, Carlton Victoria 3053, Australia

ISBN 978-1-922395-47-4

Advisory Committee

René Sterk
FAusIMM(CP)
Conference Advisory Committee Chair

Isobel Clark

Danny Kentwell
FAusIMM

Leigh Slomp
FAusIMM(CP)

Kathleen Body

Ilnur Minniakhmetov
MAusIMM

Jacqui Coombes

Dhaniel Carvalho
MAusIMM(CP)

Lynn Olssen

Scott Dunham

Mike Stewart
MAusIMM

Jeff Boisvert

Clint Ward
MAusIMM

Xavier Emery

AusIMM

Julie Allen
Head of Events

Kathryn Laslett
Manager, Events

Melissa Ogier
Program Manager

Reviewers

We would like to thank the following people for their contribution towards enhancing the quality of the papers included in this volume:

Olivier Bertoli

Kathleen Body

Jeff Boisvert

Dhaniel Carvalho

Isobel Clark

Jacqui Coombes

Scott Dunham

Xavier Emery

Danny Kentwell

Ilnur Minniakhmetov

Lynn Olssen

Oscar Rondon

Leigh Slomp

René Sterk

Mike Stewart

Clint Ward

Foreword

The support for the inaugural MREC23 conference was surprisingly strong, with more than 500 delegates attending.

Maybe that's just down to good old marketing. Maybe it's down to a genuine and strong desire to come together and discuss the issues of our day in one room.

Whatever it is, based on the feedback received from delegates during and after the event, it was clear that continuing this conference series is something people were looking for. The decision to organise the next one was therefore easy!

One thing that the Organising Committee is particularly keen to continue to drive, is a move away from what has turned into convention over the years – back-to-back presentations in blocks of five, sometimes even in parallel sessions, from early morning to late afternoon. Yes, that does drive ticket sales, but it does not stimulate the conversation and discussion that we desperately need to solve the questions of our time.

Our strong focus therefore has been to pick the best 22 out of 80 submitted papers and provide these the stage to present their research, as well as ample time to discuss, challenge, and be challenged.

The final papers were not chosen lightly. A detailed and multi-stage review process lay at the foundation of the final selection, with some papers even being reviewed by ten experts. As practitioners for whom the word 'bias' is equivalent to a wooden stake through the heart, care was taken to make this a fair process across the board.

We are hoping to bring back healthy debate. Not hiding behind online pseudonyms or linkedIN accounts to comment on a paper or a novel thought, but back to what conferences used to be – a place to listen to the latest research and to respectfully challenge each other to keep standards in our industry at the highest level.

The discussion forums two years ago at MREC23 worked well, but we've taken our learnings from the feedback and hope to be presenting an even better experience for conference goers this year to achieve exactly that.

Thank you to our delegates for taking busy time out of your lives to attend, and thanks for helping to make this conference one of the standout experiences this year.

Yours faithfully,

René Sterk

Mineral Resource Estimation Conference 2025 Advisory Committee Chair

Sponsors

Welcome Reception Sponsor

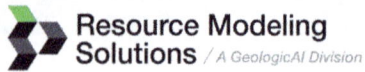

Resource Modeling Solutions / A GeologicAI Division

Lunch Sponsor

CAPITAL

Parker Challenge Sponsor

BARRICK

Name Badge and Lanyard Sponsor

entech.
Mineral Resources | Mentoring | Mining

Supporting Partners

Australian Institute of Geoscientists

BUSINESS EVENTS PERTH

Contents

Classification, RPEEE and reporting

Improving transparency in mineral resource classification using multi criteria decision analysis   3
  C Barton and J Dalrymple

Application of Quantile Regression Forest for resource classification   19
  K Maxwell

Culture, failures/learnings, case studies

Peer accountability and communication through a critical checkpoint framework – driving a   35
culture of critical thinking
  J M Clark, C Fiddes and M Roux

Rethinking the world's largest gold producing complex through geologic improvements to   45
mineral resource estimates
  J M Clark, M Samson, C Le Cornu, C Fiddes, M Vandelle, A Williams and M Roux

You're so vein – cognitive bias in the resource estimation process   57
  D A Sims

Data science, input data, machine learning and AI

Dare to transform – leveraging new orebody knowledge for next-generation resource   75
estimation
  J Coombes, P Hodkiewicz and M Mattingley-Scott

Use of machine learning to predict recovery of alumina involving multiple analytical methods   87
at Worsley Alumina
  J Dangin, P Soodi Shoar, T Wilson, J Harvey and C Cavelius

Building on the shoulders of giants – geostatistics and geodata science   101
  G Nwaila and S Zhang

A practical use of machine learning to aid gold recovery studies at Liberty Gold's Black Pine   119
Oxide Gold Project in Southern Idaho
  V Wilson, C Gomes and A Barrios

Domaining and geological modelling

The importance of geologically driven estimation domaining on resource estimation – an   133
example from the Turquoise Ridge Gold Mine within the Getchell Trend in Nevada
  C Le Cornu, J M Clark, L Snider and J Pari

Advances in geologic and resource modelling methods for locally-varying controls on   147
orogenic gold mineralisation
  J A McDivitt, D Greene, E Hart, B Wilson and M Hadavand

Estimation

DIY – implementing customised Python workflows for flexible and effective mineral resource estimation   171
L Bertossi and D Carvalho

Spatial estimation with machine learning of multiple-fluid phase veining in shear-hosted intrusion-related deposits using geospatial features – a case study from Pogo Mine, Alaska   185
E Ramos, S Sen, M Matthews and J Machukera

Estimating recoverable resources – is it still hopeless?   209
O Rondón and W Assibey-Bonsu

Preg-robbing gold ores – an example of a non-linear estimation workflow when dealing with non-additive geometallurgical variables   227
M Samson, G Benthen, J M Clark, C Le Cornu and D Conn

Exploring tailings reprocessing options – from characterisation to preliminary block modelling of the Mount Isa Mines tailings storage facility   243
L Stefoni, A Wilhelmsen and D Ambach

Quantifying, integrating and modelling

Assessment of revenue uncertainty in life-of-mine planning at George Fisher deposit – Queensland, Australia   269
B C Afonseca

Combining geological and grade uncertainties in drill hole spacing studies   289
C Gomes

Onto copper-gold deposit, Indonesia – evaluation of geological and grade uncertainty in a block cave operation   303
R C Hague

Consuming *in situ* uncertainty into mine planning processes – from reserve calculation to dynamic decision-making   319
I Minniakhmetov and J Serjeantson

In situ uncertainty quantification and implications for risk mitigation – a case study from WAIO Newman Hub   333
C Williams and I Minniakhmetov

Novel stratigraphy simulation – a case study for ESG and mine planning application   357
B Wilson, S Sanchez, D Carvalho and S Horan

Author index   373

Classification, RPEEE
and reporting

Improving transparency in mineral resource classification using multi-criteria decision analysis

C Barton[1] and J Dalrymple[2]

1. Senior Resource Geologist, IGO, South Perth WA 6151. Email: cathy.barton@igo.com.au
2. MAusIMM(CP), Senior Resource Geologist, IGO, South Perth WA 6151. Email: jenny.dalrymple@igo.com.au

ABSTRACT

This study proposes an innovative framework to enhance JORC Code-compliant Mineral Resource classification by providing a versatile and transparent method to quantify risks in Mineral Resource estimation (MRE). The framework combines multi-criteria decision analysis (MCDA) with the analytical hierarchy process (AHP). These methods enable the systematic evaluation of complex, multi-faceted criteria by breaking down the decision-making process into a hierarchy of simpler components. It integrates key parameters that can influence MRE classification, such as geological interpretation uncertainty, grade estimation confidence, data integrity and quality, data spacing and geostatistical continuity, and confidence in *in situ* dry density measurements. The approach is further adaptable to incorporate multiple modifying factors, such as those requiring assessment for JORC Code 2012 reporting and associated determination of reasonable prospects of eventual economic extraction (RP3E).

At its core, the framework generates a unified confidence score (UCS) for each MRE block, which can be used in the classification of Mineral Resources into appropriate JORC Code 2012 risk categories. The UCS is established by using a graphical tool, known as the confidence decision tree (CDT), which provides stakeholders with a clear, visual representation of classification criteria and decisions. The framework was applied to two magmatic nickel sulfide deposits at IGO's Cosmos Project in Western Australia to test and demonstrate its effectiveness and flexibility for both open pit and underground MREs. The study concludes that this approach offers a robust and user-friendly method for MRE classification, offering the clarity and consistency in reporting practices essential for informed decision-making and stakeholder confidence. Furthermore, the development and presentation of a CDT in supporting MRE documentation advances the JORC Code's principles of transparency and materiality.

INTRODUCTION

The estimation of Mineral Resources involves inherent uncertainty, as it relies on interpreting geological, statistical, and spatial data, each with intrinsic limitations and potential errors. This uncertainty can propagate through the evaluation process, impacting critical decisions on MRE grade or quality, quantity, and continuity (Lindi *et al*, 2024). If inadequately managed or communicated, these uncertainties can lead to severe consequences, including premature mine closure, financial losses, operational setbacks, environmental harm, and erosion of stakeholder trust (Laurence, 2006). Consistent communication of MRE uncertainty is therefore essential to ensure stakeholders have the information for balanced decision-making and informed risk management. To address this, several internationally accepted codes have been established to guide MRE practitioners in the public disclosure of Mineral Resources, by setting minimum reporting standards. In Australasia, this code is governed by the Joint Ore Reserves Committee (JORC). The JORC Code 2012 standardises public communication of uncertainty in MREs by classifying them into three categories – Inferred Resources, Indicated Resources, and Measured Resources – based on increasing levels of geological confidence (JORC, 2012). Three fundamental principles must be adhered to when assigning these categories:

- **Transparency:** Information must be clearly and explicitly presented.

- **Materiality:** All relevant information must be disclosed.

- **Competence:** A suitably qualified and experienced person (the Competent Person) must be responsible for classification (JORC, 2012).

While the JORC Code 2012 sets clear principles and some guidelines, it leaves room for interpretation in classification methodologies. The Competent Person can choose a classification method that aligns with the unique geological and operational characteristics of each deposit. This flexibility is necessary to accommodate diverse circumstances, yet it often leads to inconsistencies in MRE classification approaches. Most classification methods rely on geometric criteria, such as drill hole spacing analysis (DHSA) or search neighbourhood, which evaluate the amount and distribution of data available for estimation (Owusu and Dagdelen, 2019; Silva and Boisvert, 2014). However, they do not account for grade continuity and data redundancy and should be supported by geostatistical methods to address these gaps (Deutsch, Leuangthong and Ortiz, 2006). Geostatistical methods are becoming more common, including quantitative metrics that assess grade estimation precision relative to data spacing, geostatistical simulation, and probabilistic methods (Lindi *et al*, 2024). They can be used to guide classification using confidence intervals, such as 90 per cent probability that future production volumes will be within 15 per cent of the estimated grade and tonnage (Verly, Postolski and Parker, 2014). However, these commonly used approaches often overlook equally critical factors, such as data quality, data acquisition processes, *in situ* dry density, and confidence in the geological interpretation of the model. Scorecard approaches have the flexibility to include all critical factors and are well established in the mining industry (Rocha and Bassani, 2023). However, they frequently involve subjective assessments, with little standardisation in criteria prioritisation or uncertainty thresholds. This lack of uniformity can make it difficult for stakeholders to understand how uncertainty is classified into the JORC Code categories.

To address the challenges posed by inconsistent classification methods, overlooked uncertainty factors, and clear communication of subjective decisions, this study introduces a systematic framework that prioritises, stores, integrates, and communicates risk factors in MRE classification. The proposed approach complements existing methodologies while improving the clarity, consistency, and transparency essential for effective MRE public reporting and decision-making.

BACKGROUND

In June 2022, IGO acquired the Cosmos Project (Cosmos), which included MREs for several magmatic nickel sulfide deposits (IGO, 2022). The deposits are clustered within or adjacent to a 2.7-billion-year-old metamorphosed komatiite, part of the Agnew-Wiluna Greenstone Belt in the Kalgoorlie Terrane of the Eastern Yilgarn Craton. Despite their proximity, each deposit presents a unique combination of risk factors from geological complexity and development history. The deposits range from discrete remobilised massive sulfide lenses to large, disseminated Type 2 (Mount Keith-style) bodies, all of which have undergone a complex deformational history (IGO, 2024b). Localised nickel-barren pegmatite dykes significantly affect the precision of some MREs, while others experience minimal impact from occasional narrow pegmatite dykes and sills (Barton and Murphy, 2023). The deposits also vary in-depth, from near surface occurrences to approximately 1.5 km below the surface and exhibit distinct mineral assemblages and orientations. Adding to the complexity, fragmented ownership histories and inconsistent data collection techniques have led to variability in the methods used for geochemical analysis, *in situ* dry density measurements, and determining sample locations. Some deposits rely primarily on diamond drill hole data, while others incorporate a broader range of drilling methods. Poor-quality drill hole data has the potential to undermine confidence in geological interpretations, compromising the accuracy and reliability of resulting models.

Following the acquisition, IGO recognised the need to standardise Cosmos' MREs to address these challenges. Existing classification methods were inconsistent and lacked the flexibility to account for the deposits' varied characteristics without significantly modifying or excluding locally relevant risk factors. Additionally, volatile commodity market conditions underscored the importance of robust evaluations to ensure project viability (IGO, 2024a). To tackle these challenges, IGO investigated MCDA combined with AHP. These methodologies were selected for their ability to evaluate complex, multi-dimensional problems systematically. The MCDA-AHP framework was tested on two contrasting magmatic nickel sulfide deposits within Cosmos. Mt Goode represents a near-surface, unmined, low-grade, disseminated deposit, while Odysseus has been partially mined from underground and contains both disseminated and massive nickel sulfides. Together, these deposits illustrate the spectrum of geological and operational diversity inherent to Cosmos.

MULTI-CRITERIA DECISION ANALYSIS

MCDA is a decision-making approach that is designed to address challenges involving multiple, complex criteria with conflicting relationships and has been successfully applied and adapted to healthcare, renewable energy, natural resource management, and engineering applications (Więckowski et al, 2023). AHP is a subset of MCDA developed by Saaty (1988), whereby a hierarchy of components is created by assigning numerical weights to the criteria through expert judgement guided pairwise comparisons. This approach can integrate quantitative and qualitative criteria, accommodating metrics with differing or no measurement value. AHP's structured methodology prioritises the most relevant criteria while resolving conflicting viewpoints through a consensus-based process. Further details about AHP and a discussion of its limitations can be found in Chaube et al (2024) and Munier and Hontoria (2021). MCDA with AHP was selected for its ability to account for conflicting relationships among the multiple risk factors identified at Cosmos. The objective was to develop a UCS for each parent block in the block model, integrating locally relevant risk factors to inform MRE classification. To ensure consistency between deposits, the approach was structured around primary criteria deemed integral to understanding MRE uncertainty, under which deposit-specific risk factors were organised.

The workflow for implementing the MCDA-AHP framework to Cosmos MREs included these steps:

1. **Define the goal:** to develop a UCS for each parent block in the block model to guide MRE classification.

2. **Identify the criteria:** isolate the key parameters (primary criteria) essential for understanding uncertainties in MRE work.

3. **Construct the hierarchy:** develop a decision tree below the primary objective (goal), followed by primary criteria and sub-criteria tailored to locally relevant risk factors.

4. **Weight the criteria:** rank the relative importance of each primary criterion using pairwise comparisons. Consensus weights were derived from the geometric mean of the comparison matrices. Where necessary, sub-criteria were weighted using the same process.

5. **Data value rating:** rate the data values, representing the lowest level of the hierarchy, from zero (least confidence) to one (highest confidence).

6. **Database development:** source data values from drill hole records, 3D volumes of geological interpretations, and block model fields, to convert into formats compatible with appropriate Mineral Resource modelling software.

7. **Build the model:** link sub-criteria to the rating scale and combine using their weightings to generate fields for primary criteria. Aggregate these to calculate a UCS for each parent block.

8. **Define classification ranges:** categorise JORC Code 2012 classification levels using the UCS, with thresholds defined by primary criteria weights. Apply restrictions to meet conditions for RP3E.

This approach ensures consistency in MRE classification by structuring the process around key parameters affecting MRE work, supporting the consideration of all material factors influencing uncertainty. Transparency is provided by presenting the resulting hierarchy as a decision tree and categorising uncertainty in alignment with JORC Code requirements using the UCS as a guide. Furthermore, collective stakeholder buy-in is achieved by synthesising expert input into a single, consensus weight for each criterion through a transparent and inclusive ranking process. This method fosters a shared commitment to the outcomes and promotes effective organisational standardisation, ensuring robust and credible MRE reporting of uncertainty in the JORC Code classifications applied.

THE CONFIDENCE DECISION TREE

The hierarchy constructed to establish the proposed UCS – the CDT – is composed of primary criteria central to understanding uncertainty in MRE work, adaptable sub-criteria that incorporate locally relevant risk factors at each deposit, and optional restrictions (Figure 1).

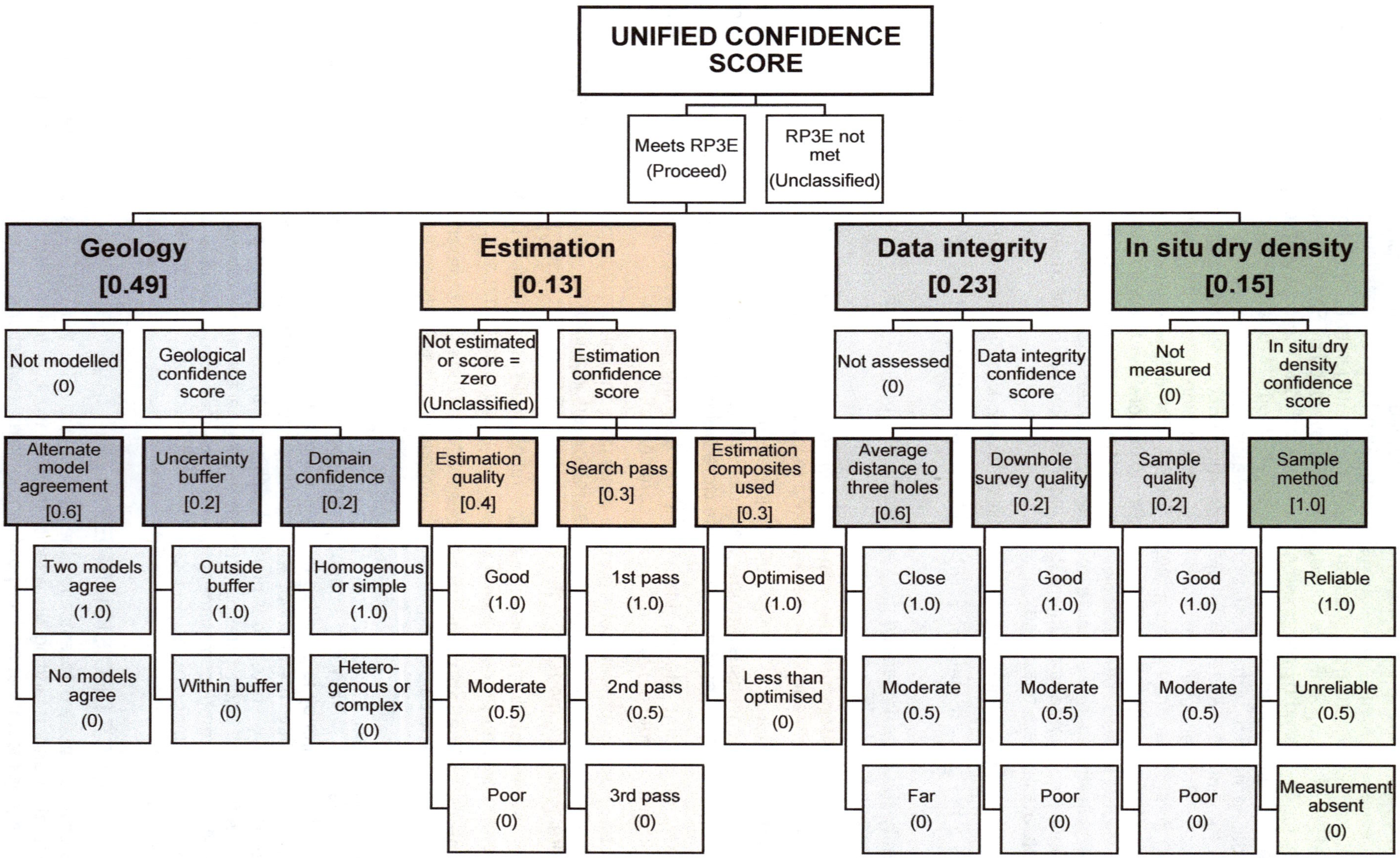

FIG 1 – The decision tree was developed to obtain a unified confidence score for each parent block in the block model, incorporating the criteria identified as relevant to understanding uncertainty in a Mineral Resource estimate. Square brackets indicate the weighting of each criterion, while parentheses specify data value ratings. Data value ratings range from zero (least confidence) to one (highest confidence).

IGO resource geologists identified the following primary criteria as essential:

- **Geology:** confidence in the geological interpretation.
- **Estimation:** confidence in the grade or quality estimation.
- **Data integrity:** an evaluation of data spacing and confidence in the integrity of the data used.
- ***In situ* dry density:** confidence in the quantity or tonnage estimation as a function of the integrity of *in situ* dry density measurements.

The relative importance of these criteria was determined using pairwise comparisons with Saaty's priority scale, as listed in Table 1. The criteria were ranked by considering the intensity of importance between each criterion, producing a comparison matrix for each geologist. The geometric mean of the comparison matrices was then calculated to derive a consensus weight for each primary criterion, integrating diverse expert viewpoints and resolving potential conflicts in criteria priority. The consensus results were scaled to one for simplification. Scores for the primary criteria were derived by identifying locally relevant risk factors, termed sub-criteria. These sub-criteria are flexible, allowing them to be tailored to the unique characteristics of the deposit under consideration. If multiple sub-criteria are identified for a primary criterion, they can be weighted using the same pairwise comparison process. Data value ratings for each sub-criterion should be determined by the Competent Person, which may include subjective decisions based on experience and knowledge of the deposit under consideration, or decisions quantified by specific studies of the data, such as DHSA. Subjective and objective data value ratings are displayed transparently in the CDT, enhancing transparency of the decision-making process. The weights for each primary and sub-criterion as depicted in Figure 1 are shown in square brackets, while data value ratings are presented in parentheses. Examples of weights and data value ratings are illustrated in Figure 1 for demonstration purposes.

TABLE 1

The priority scale developed by Saaty (1988) used in the pairwise comparison.

Intensity of importance	Description
1	With equal importance, two criteria contribute equally to the objective.
3	Moderate importance, judgement slightly favours one criterion over the other.
5	Strong importance, judgement strongly favours one criterion over the other.
7	Very strong importance, the evidence favouring one criterion over the other is demonstrated in practice.
9	Extreme importance, the evidence favouring one criterion over the other is unequivocal.
2, 4, 6, 8	Used to express intermediate values.

The sub-criteria suggested in Figure 1 involve assumptions about the techniques used for data collection, geological interpretation, and statistical evaluation. The Competent Person must ensure the selected sub-criteria align with the practices and data employed in the MRE. For example, it is assumed that an alternative geological model is produced to determine a geological confidence score, which may not always be practical. Likewise, the estimation confidence sub-criteria in Figure 1 are influenced by the assumption that most grade estimates use a kriging method, which may not apply to all deposits. Data value ratings must also align with the characteristics of each deposit. For instance, in determining a data integrity confidence score, the sub-criterion with the highest priority was data spacing, measured by the average distance of a parent block centroid to the nearest sample in three drill holes. A high data value rating is given for distances considered close, while lower ratings are given for blocks where samples are further away. The distance ranges for each value must be determined by the Competent Person after consideration of the circumstances at the deposit under evaluation. In some complex deposits, an optimised distance range might be from zero to 15 m, while in others, distances from zero to 60 m may receive an equivalent rating.

In the CDT provided in Figure 1, two levels of optional restrictions are included. The first ensures that parent blocks meet specified requirements for RP3E, where they can be attributed to individual blocks. For example, this may include economic, metallurgical, technical, and geometric constraints, assisting with the Competent Person's preliminary assessment of economic viability for a MRE. The second application guarantees that parent blocks with an estimation confidence score of zero are not classified in the MRE. The use of optional restrictions in the CDT is beneficial when combining primary criteria scores by allowing the decision process to become more adaptable to unique geological, technical, or economic characteristics. The CDT can be visually represented in MRE reporting, succinctly outlining the risk factors considered, criteria prioritised, restrictions applied, and systematic decisions made during the classification process. While the Competent Person must account for the risk factors specific to each deposit, the differences in decisions are displayed in one graphic, enhancing transparency in MRE documentation.

CLASSIFICATION USING THE UNIFIED CONFIDENCE SCORE

The UCS serves as a versatile metric for internal and external purposes, by combining the material factors influencing uncertainty in MRE work into a single confidence score for each parent block. Internally, it allows for the transparent communication of risks related to multiple factors influencing MRE uncertainty. Externally, it can guide the public disclosure of MRE classification under the JORC Code by defining a method that uses UCS scores to inform categorisation. However, the methodology developed must be established with strong justification, as the prioritisation of primary criteria may vary depending on the specific context of a project. An example of using the UCS to assign JORC Code categories is provided in Table 2.

TABLE 2

An example of how the UCS can guide Mineral Resource classification into the JORC Code classification categories. The UCS ranges corresponding to each JORC category are determined by considering the priority of primary criteria used to establish the UCS.

UCS from	UCS to	JORC Code 2012 classification assignment	Justification
0	0.51	Unclassified	Unclassified if the estimation confidence score or the geological confidence score is zero. Unclassified if the MRE block does not meet specified RP3E requirements.
>0.51	0.75	Inferred	The estimation confidence score must be greater than zero. Some geological confidence is required along with some confidence in other criteria.
>0.75	<1.00	Indicated	High geological confidence is required with high confidence in at least two other criteria.
1.00	1.00	Measured	All criteria must achieve a high confidence score.

In this example, UCS ranges associated with JORC Code categories are determined by considering the weights of the primary criteria and restrictions outlined in the CDT in Figure 1. Geological confidence – the primary criterion with the highest weight (0.49) – is prioritised. A UCS with a geological confidence score of zero (no confidence) cannot result in classification, regardless of high scores in other criteria, because the combined weight of the other primary criteria is 0.51. Furthermore, the optional restrictions ensure that if the estimation confidence score is zero or the MRE block does not meet specified RP3E requirements, the block cannot be classified. According to JORC Code classification logic, the Inferred Resource category has the lowest level of confidence with limited geological evidence and sampling (JORC, 2012, Section 21). The corresponding UCS range can comprise some geological confidence, must have some confidence in grade estimation, and some confidence in other criteria. The Indicated Resource category demands sufficient geological evidence to assume grade and geological continuity between sample points (JORC, 2012, Section 22). The UCS range related to Indicated must include high geological confidence alongside high confidence in at least two other criteria. The Measured Resource category requires enough geological evidence to confirm grade and geological continuity between sample locations (JORC, 2012, Section 23). Therefore, the associated UCS requires a high score in all primary criteria. Combining a CDT and UCS in MRE classification in this structured approach ensures that classification under the JORC Code remains transparent, well-informed, and consistent while prioritising the geological evidence necessary to underpin resource estimations.

CASE STUDIES

IGO's review of MREs at Cosmos began with analysing the Odysseus and Mt Goode deposits, with each deposit demonstrating unique geological challenges. Odysseus comprises two domains of disseminated to net-textured pentlandite-pyrrhotite with minor pyrite, alongside structurally remobilised massive nickel sulfide lenses. The disseminated domains are shallow-dipping, Mount Keith-style Type 2 magmatic nickel sulfide deposits. They have been heavily deformed and disrupted by nickel-barren pegmatite dykes intruding along brittle faults, contributing to uncertainty in the MRE (IGO, 2024b). Discovered by Xstrata Nickel Australasia (XNA) between 1.0 and 1.3 km below the surface, Odysseus was initially defined through surface diamond drilling to 40 m spacing before the Cosmos Project was placed into care and maintenance in 2012. Subsequent owners Western Areas Ltd (WSA) publicly reported the first JORC Code 2012 reportable MRE and initiated underground development. After the acquisition by IGO in 2022, further development and infill drilling from underground occurred before the project was again placed into care and maintenance in 2024. The Mt Goode deposit, discovered earlier than Odysseus by Lachlan Resources NL, is a steeply dipping, low-grade disseminated magmatic nickel sulfide domain located on an adjacent tenement. Striking north–south and extending from approximately 100 to 600 m below the surface, Mt Goode is characterised by pentlandite-pyrrhotite-heazlewoodite mineralisation, with elevated nickel grades closer to the eastern margin. Boundaries are defined by a felsic porphyry intrusion to the east and a gradational western margin above the basal komatiite contact. Over decades of exploration, various data collection methods were employed, including multiple drilling techniques from the surface. Drilling orientations have been primarily from east to west at an approximate spacing of 40 m. Despite periodic updates to the MRE by several prior owners, no JORC Code reportable Ore Reserve has been reported, and IGO ceased reporting an MRE for Mt Goode as of 2024 (IGO, 2024b). Regardless of the Cosmos Project's status, a comparison of UCS and CDT applications at both deposits simultaneously demonstrates the versatility of the approach, and the consistency required to inform decision-makers of material factors influencing uncertainty. Furthermore, the comparison provides the opportunity to discuss the uncertainty criteria in further detail using examples from each deposit.

Geological confidence

Differences in geological uncertainty across the deposits stem from specific factors. At Mt Goode, challenges include the gradational western margin, reliance on a nickel grade cut-off for mineralisation domaining, and inconsistent data informing the weathering model. The ambiguity surrounding a major fault intersecting the deposit further reduces confidence. In contrast, Odysseus' uncertainties centre on interpreting nickel-barren pegmatite dykes and remobilised massive sulfide lenses, with their variable thickness and complex deformation history requiring careful sub-criteria adjustments.

Alternate model agreement

Geological models for both deposits were created using implicit geological modelling tools. These were exported as 3D volumes and used to code the geological interpretation field in the block model. Alternative models were also developed using machine learning (ML) software and used to expose areas in the block model where the geological interpretation differed at the parent block scale. Higher confidence was assigned where models agreed, while disagreements resulted in lower scores.

Uncertainty buffer

Buffer distances were applied to critical geological features to account for interpretative uncertainty. At Mt Goode, a 20 m buffer around a major fault, modelled as a planar feature and expanded to produce a closed volume, reduced confidence scores for proximate blocks. The decision to use a 20 m buffer reflects significant uncertainty in the fault's location and displacement. Similarly, a 5 m buffer around pegmatite dyke interpretations at Odysseus accounted for their irregular contacts and variability. This low confidence buffer reinforced cautious scoring of the pegmatite dyke interpretation by the alternate model agreement.

Domain confidence

Estimation domains at Mt Goode and Odysseus are derived by intersecting geological and nickel grade interpretation 3D volumes. Confidence in the interpretation of these domains vary based on geological complexity and a domain confidence score can be applied to reflect uncertainty which may not be accounted for in the alternate model agreement score. For example, inconsistent weathering data reduced confidence in the transitional and oxidised domain interpretation at Mt Goode. As such, the fresh rock domain at Mt Goode was assigned a higher confidence score than weathered domains. At Odysseus, disseminated nickel sulfide domains were assigned higher scores due to their relative homogeneity, while remobilised massive sulfide lenses, being discrete and challenging to interpret, received lower scores.

Estimation confidence

The mineralised domains in both case studies had nickel grades and density estimated using Ordinary Block Kriging (OBK). A variety of neighbourhood parameters were tested to optimise the efficiency of the interpolant while considering smoothing and the influence of negative weights. For blocks that are well supported by data we expect the grade to be interpolated in the first search volume, using an optimised number of samples. For blocks that are not so well supported, additional search volumes may be required and less samples may be available for the estimate. The confidence in the estimation of the nickel grade and density for the two case studies was evaluated based on key estimation metrics, the search volume, and the number of composites used in the estimate.

Estimation quality

Various metrics can be used to evaluate the quality of the grade estimate by OBK. For these case studies, the kriging theoretical Slope of Regression (SOR) was populated when preparing the estimate. The SOR metric quantifies the ratio of variance between the theoretical true (Z) distribution of block grades and the variance of the estimated value (Z^*) is used as an indicator of conditional bias (Deutsch, 2007).

Search pass

The OBK estimate was run using a search neighbourhood optimised by selecting a search distance and number of samples to maximise the Kriging Efficiency (KE), while considering negative weights and the variogram range. A three-pass search approach was used whereby to estimate blocks that were not estimated in the first search volume, a second and third search volume were run with an increase in the search distance and a decrease in the allowable minimum number of estimation composites. Blocks populated in the first search volume were assigned a higher confidence score than those populated by the second and third search volumes.

Estimation composites used

Blocks that were estimated with less than the optimised number of samples were flagged as having lower confidence. Lower data support for these blocks means that less samples are available inside the search neighbourhood for the estimate. Insufficient sample data may conceal spatial patterns, which could produce an unreliable estimate at unsampled locations.

Data integrity confidence

Data spacing significantly affects confidence in geological interpretations and statistical evaluations and should be used to support an assessment of uncertainty (Deutsch, Leuangthong and Ortiz, 2006). However, the age and provenance of each deposit heavily influence confidence in the integrity of the data. Mt Goode exhibits variability in the techniques used for geochemical analysis and determining sample locations, as well as inconsistent sampling frequencies. Although Odysseus is a more recent discovery than Mt Goode, its initial surface-based drill-out required diamond drill holes exceeding 1 km in length. This introduced uncertainty in final sample locations which has been partially mitigated by subsequent infill drilling from underground. However, some areas of the Odysseus MRE still rely on data from long and less precisely located surface holes. Therefore, an evaluation of data spacing was augmented by incorporating factors influencing confidence in sample location and quality.

Average distance to three drill holes

Data spacing was evaluated by calculating the average distance from the parent block centroid to the nearest sample in three drill holes. An inverse distance weighting (to the power of one) interpolation of nickel grades was run with a wide search radius limited to a minimum and maximum of three composites to obtain the average distance from the block centroid to the samples, recorded as an output of the interpolation. At Mt Goode, a high confidence score was applied to estimates where data distances to block centroids were in the range zero to 25 m, while distances exceeding 40 m were given a low confidence score. Due to the added geological complexity from pegmatite dykes and remobilised massive sulfides at Odysseus, slightly stricter data distance criteria ranges were applied. These data spacing distances were selected based on (DHSA) using estimated quarterly and annual volumes. A high confidence score was given for data spacing of zero to 23 m, and a low confidence score for distances exceeding 37 m.

Downhole survey quality

The methods used to record downhole sample locations vary across drill holes. While most diamond drill holes at Cosmos were continuously surveyed with north-seeking gyroscopic instruments, technical difficulties occasionally necessitated alternative methods. In such cases, single survey recordings or planned hole paths were used as substitutes, introducing greater uncertainty. Precision issues were particularly pronounced in long (>1 km) surface-based diamond drill holes at Odysseus, where up to 15 m discrepancies between planned and actual intersections were observed during underground development. As such, hole depth was used as a metric for downhole survey quality. At Odysseus, drill holes collared at the surface were given a lower confidence score than the shorter holes collared from underground workings. At Mt Goode, hole depths of less than 400 m were given high confidence scores, while depths surpassing 800 m received low confidence. Holes with incomplete or missing gyroscopic surveys were automatically assigned a low confidence score, irrespective of depth.

Sample quality

The varying age of data collected across Cosmos has led to inconsistencies in geochemical assay techniques. At Odysseus, the predominant technique employed was a four-acid digest followed by inductively coupled plasma atomic emission spectroscopy (ICP-AES) (IGO, 2024b). At Mt Goode, older drill holes often employed less consistent methods, and in some cases, the assay technique is unknown (IGO, 2022). To quantify uncertainty in the sample quality at both deposits, the age of the data was used as a proxy. Holes drilled before 2004 were assigned a low confidence score, while holes drilled after 2012 received higher confidence.

In situ dry density confidence

Density is critical in Mineral Resource estimation, as volume estimates must be converted to tonnages using *in situ* dry density values. Poor-quality *in situ* dry density measurements can introduce significant errors in tonnage estimations, ultimately affecting the contained product estimate (Lipton and Horton, 2014). Historical MREs for the Mt Goode deposit relied on a regression between *in situ* dry density and nickel grades to address gaps in measurement data (IGO, 2022).

However, IGO's recent review found only very low correlation between nickel grades and *in situ* dry density. This lack of correlation was attributed to errors arising from the field methods used to obtain *in situ* dry density measurements, which may have obscured any potential relationship, particularly at low nickel grades. *In situ* dry density measurement methods of fresh rock included a pycnometer of pulverised samples, water displacement of competent drill core using Archimedes' principle, and historical methods listed as unknown. In some drill holes, measurements were collected for every tenth geochemical sample. Fresh rock samples using a pycnometer demonstrated the highest precision and were assigned a high confidence score, as there is typically low porosity in the fresh rock samples. Water displacement and unknown methods showed lower precision and were awarded a moderate confidence score due to the variation in field measuring approach. A low confidence score was given to samples without any recorded *in situ* dry density measurements. At Odysseus, the correlation between *in situ* dry density and nickel grades is evident. However, pycnometer measurements were less common, and the water displacement method is predominant, particularly on samples from the underground infill drilling. As *in situ* dry density is interpolated into the block model using OBK, the main cause of uncertainty in *in situ* dry density estimation at Odysseus stems from missing values. A high confidence score was assigned to samples with measurements and a low confidence score was given to samples with absent values. Assigning confidence scores to *in situ* dry density values ensures that the variable reliability of density data is accounted for in the estimation, reducing the risk of overstating the tonnage.

Reasonable prospects for eventual economic extraction

The JORC Code 2012 states that a Mineral Resource must have RP3E (JORC, 2012, Section 20). The Competent Person is required to conduct a preliminary assessment of the technical and economic assumptions underlying the expected mining parameters, and portions of the MRE that do not have RP3E cannot be publicly reported. Prior to Mt Goode becoming declassified in 2024, the preliminary assessment by the Competent Person included a nickel grade cut-off and evaluation of the block model against an open pit shell design, among other technical and economic considerations. These restrictions were applied after a UCS had been obtained to ensure blocks with a high UCS did not result in classification where RP3E requirements were not met. At Odysseus, the assessment is ongoing following the transition of Cosmos into care and maintenance in 2024.

Results

The confidence scores for primary criteria and resulting UCS for Mt Goode and Odysseus are depicted as colour-coded block centroids in Figures 2 and 3 respectively, with hot colours indicating lower confidence than cool colours. The UCS developed for Mt Goode identified areas of low geological confidence around the edges of the interpreted mineralised domain and at depth where sample spacing is greater. Low confidence in the nickel grade estimation was highlighted on the western margin, where estimation quality dropped, likely due to sample concentration on higher nickel grades in the east. Confidence in the data integrity highlighted areas where wider drill spacing was compounded by older, poor-quality data. A zebra striped pattern was dominant due to the chosen data spacing ranges. The evaluation of *in situ* dry density revealed significant areas of low confidence, particularly where measurements were absent and nearby drill holes only had measurements for every tenth sample. Understanding of *in situ* dry density in the weathered zones is poor, and the existing measurement methods are unlikely to account for porosity. The final UCS result for Mt Goode was characterised by a zebra striped pattern along the eastern side of the domain, lower confidence at depth, along the western boundary, and within weathered domains. This pattern indicates that infill drilling is required to reduce uncertainty, additional samples are needed in the western portion of the domain, and a targeted approach to understanding density in the weathered zones is required if development at the deposit progresses in future.

At Odysseus, areas of low geological confidence concentrated along pegmatite dyke boundaries, around the margins of the disseminated domains, and in the interpretation of the massive domains. Small areas of low estimation confidence were highlighted between the disseminated domains, and in the south-east extension of the southern disseminated domain, where there is less data informing the estimate. Evaluation of the data integrity exposed areas of low data frequency, compounded by a reliance on longholes drilled from the surface. Confidence in *in situ* dry density measurements was reduced in isolated portions of the deposit with missing density data, particularly in the southern disseminated domain. The UCS result for Odysseus identified locations where low geological confidence was exacerbated by challenges with estimation confidence, low confidence in the data integrity, and absent *in situ* dry density measurements.

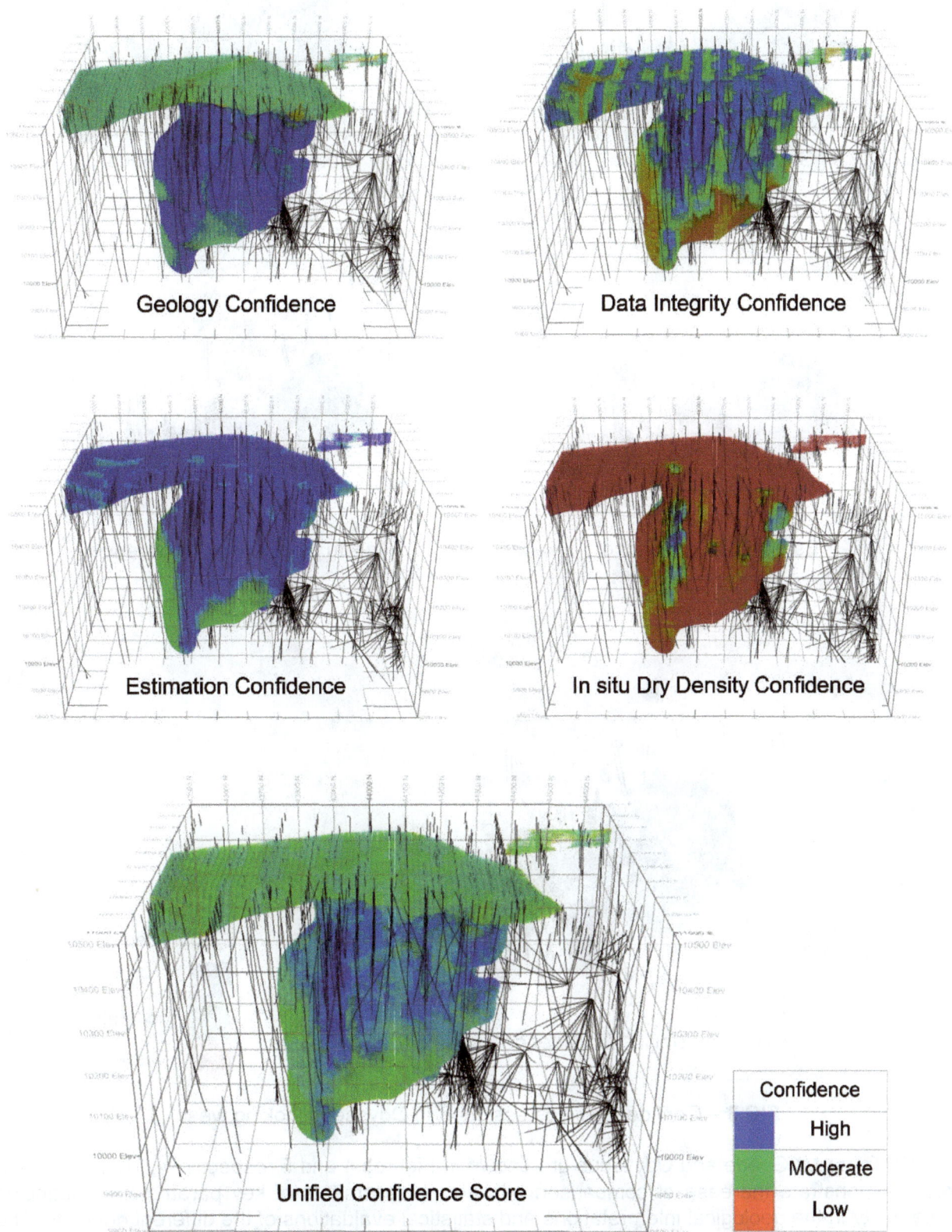

FIG 2 – Confidence fields assigned to Mt Goode looking west.

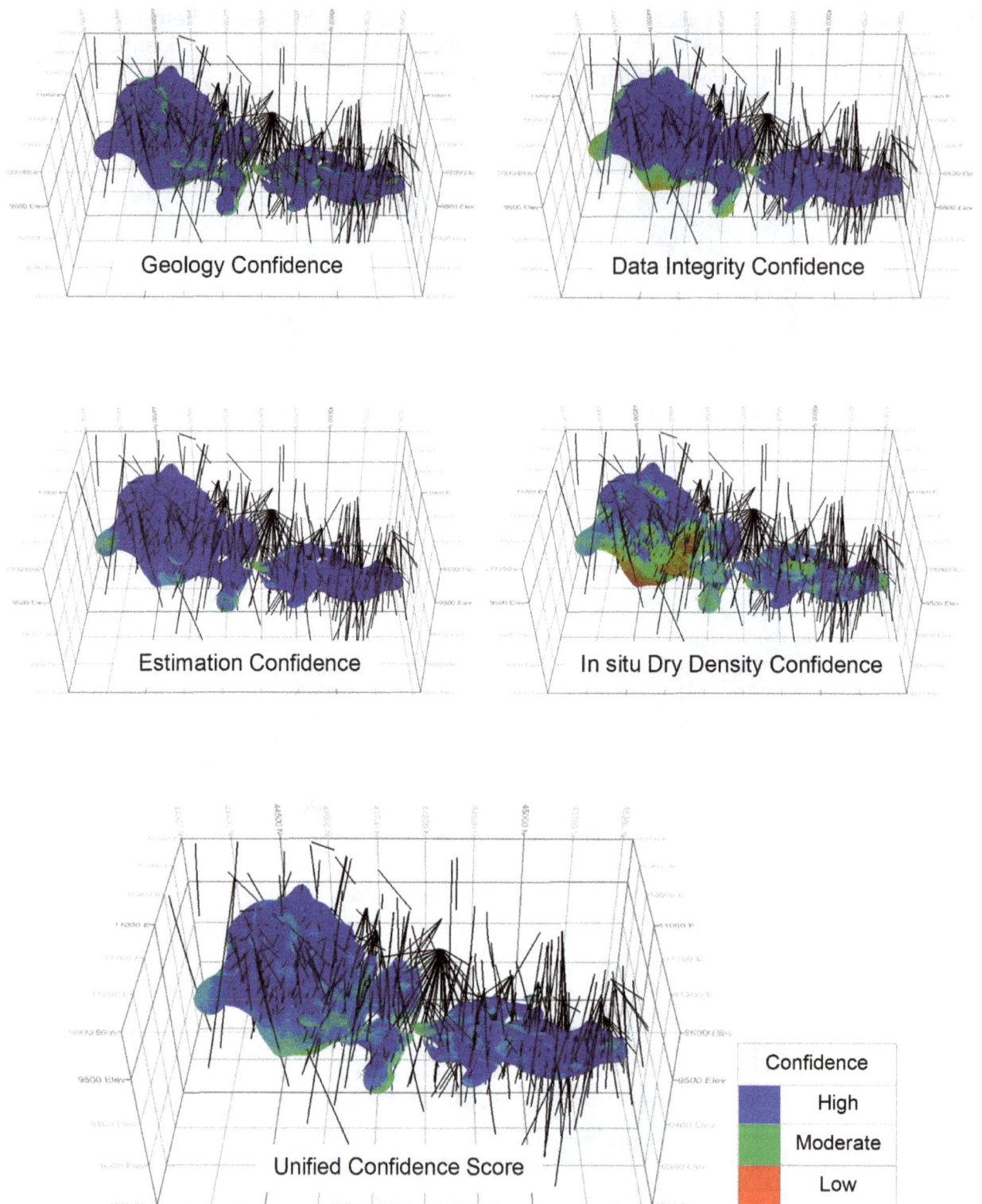

FIG 3 – Confidence fields assigned to Odysseus looking west.

The CDTs for Mt Goode and Odysseus are shown in Figures 4 and 5 respectively. These decision trees demonstrate the ease of comparison between assessments of key parameters influencing uncertainty in the geological interpretations and statistical evaluations of the different deposits. The primary criteria of geological confidence, estimation confidence, data integrity confidence, *in situ* dry density confidence, and their weightings remain consistent. However, the sub-criteria established to measure the primary criteria are modified slightly to ensure they remain relevant to the localised conditions. As the Mt Goode deposit is no longer reported as a Mineral Resource and the assessment of RP3E at Odysseus is ongoing, JORC Code classification using the UCS as a guide could not be completed for this study.

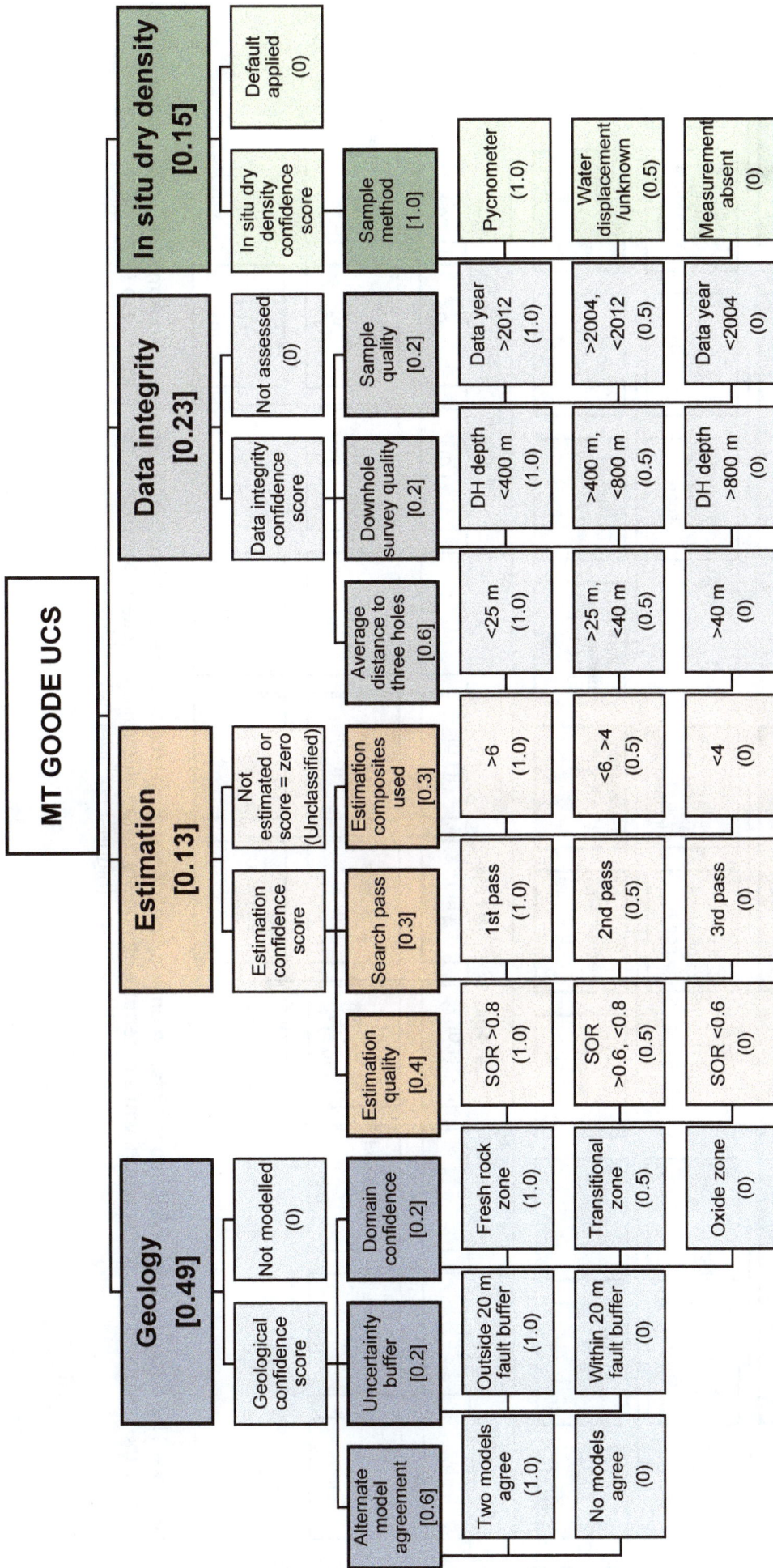

FIG 4 – The decision tree developed to obtain a unified confidence score for each parent block in the Mt Goode block model. Square brackets indicate the weighting of each criterion, while parentheses specify data value ratings. Data value ratings range from zero (least confidence) to one (highest confidence).

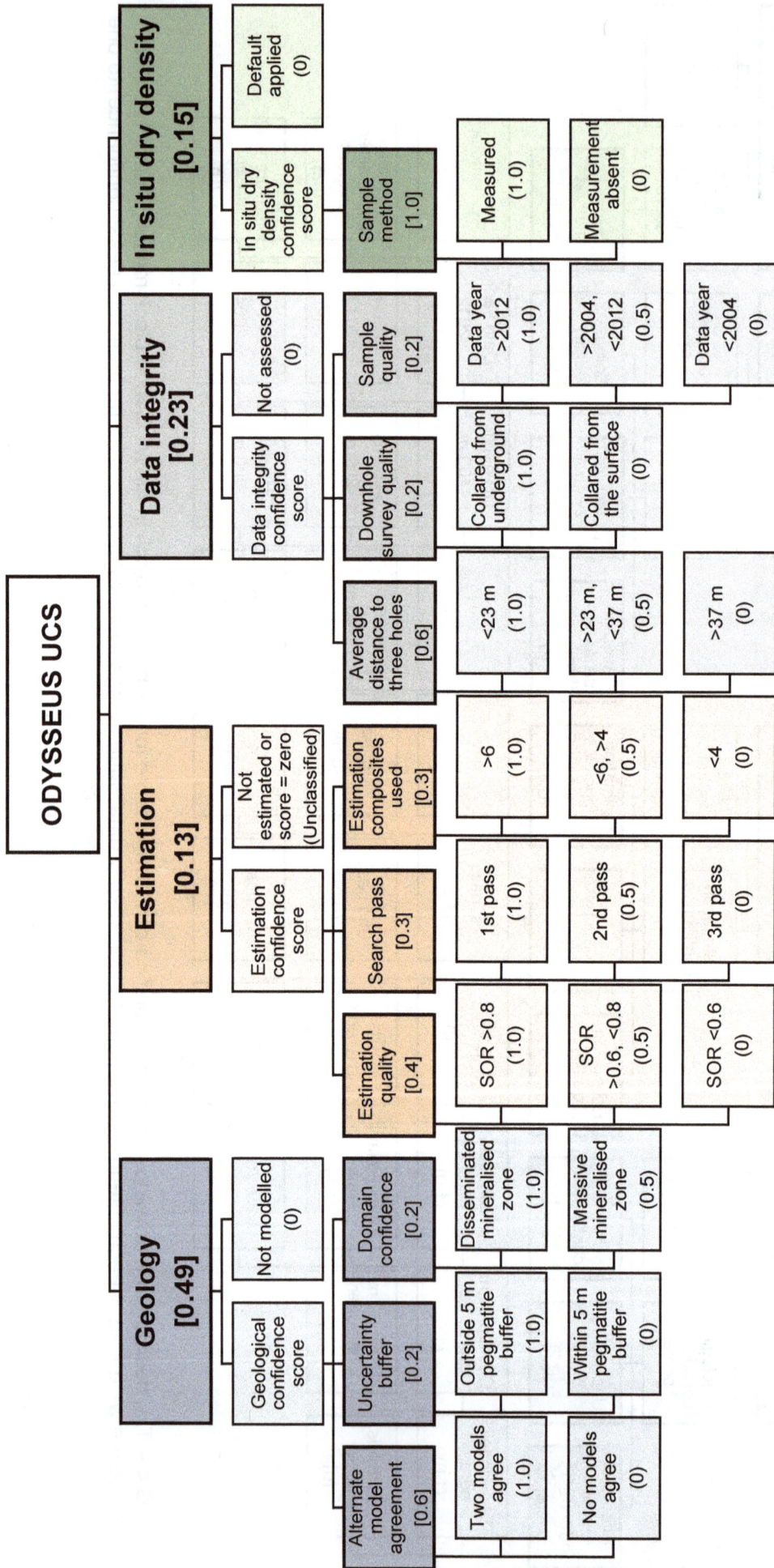

FIG 5 – The decision tree developed to obtain a unified confidence score for each parent block in the Odysseus block model. Square brackets indicate the weighting of each criterion, while parentheses specify data value ratings. Data value ratings range from zero (least confidence) to one (highest confidence).

CONCLUSION

A comprehensive review of MREs at Cosmos exemplifies the importance of systematic evaluation in managing geological uncertainty. Each deposit's unique geological characteristics and data histories necessitate a flexible approach, yet public reporting requirements demand consistent and transparent communication of material factors that influence MRE confidence and JORC Code classification. The key parameters considered integral in MRE classification include confidence in the interpretation of the geological model, grade estimation confidence, data integrity and quality, data spacing and geostatistical continuity, and confidence in *in situ* dry density measurements. The application of MCDA-AHP delivers the necessary structure required to evaluate these criteria and can be adapted to integrate existing best practice methodologies for quantifying uncertainty. This framework incorporates the complex and interrelated factors that influence confidence in MREs in a UCS for each parent block. A JORC Code Competent Person can then incorporate the assessment of locally relevant risk factors using this structure, and the inclusive hierarchy process can standardise the evaluation of MRE uncertainty across organisations. The resulting CDT can be displayed in MRE reporting to concisely demonstrate the material risk factors considered and their relative importance in systematic decisions underpinning confidence evaluations. This approach provides Competent Persons with a defensible, straightforward framework for assigning JORC Code classifications and justifying their decisions regarding acceptable levels of uncertainty in the Mineral Resource estimate.

The authors invite Mineral Resource practitioners to test the CDT and UCS method on other deposits. Viability of the method in other commodities, comparative studies against alternative classification methods, and incorporation of untested criteria will be necessary further work to critically review the method's effectiveness.

ACKNOWLEDGEMENTS

The authors would like to sincerely thank IGO for providing permission to publicly release the technical information included in this paper and Mark Murphy for providing his guidance and review of this work.

REFERENCES

Barton, C and Murphy, M, 2023. Testing and quantifying geological uncertainty (or a gram of drill data is worth a kilogram of geological interpretation), in *Proceedings of the Mineral Resource Estimation Conference* (ed: R Sterk), pp 96–110 (The Australasian Institute of Mining and Metallurgy: Melbourne).

Chaube, S, Pant, S, Kumar, A, Uniyal, S, Singh, M K, Kotecha, K and Kumar, A, 2024. An Overview of Multi-Criteria Decision Analysis and the Applications of AHP and TOPSIS Methods, *International Journal of Mathematical, Engineering and Management Sciences*, 9(3):581–615. https://doi.org/10.33889/IJMEMS.2024.9.3.030

Deutsch, C V, 2007. The Slope of Regression for Kriging Estimators. Available from: <https://ccg-server.engineering.ualberta.ca/CCG%20Publications/CCG%20Annual%20Reports/Report%209%20-%202007/311%20Slope%20of%20Regression.pdf>

Deutsch, C V, Leuangthong, O and Ortiz, J, 2006. A Case for Geometric Criteria in Resources and Reserves Classification. Available from: <https://ccg-server.engineering.ualberta.ca/CCG%20Publications/CCG%20Annual%20Reports/Report%208%20-%202006/301-Prob%20Classification.pdf>

IGO, 2022. FY22 Cosmos and Forrestania Resources and Reserves, *ASX Release (30): August 2022* (IGO Limited). Available from: <https://www.igo.com.au/site/PDF/549915a8-ff6e-411d-9211-1e85fdf2e853/FY22CosmosandForrestaniaResourcesandReserves>

IGO, 2024a. Cosmos Project Update, *ASX Release (31): January 2024* (IGO Limited). Available from: <https://www.igo.com.au/site/pdf/def9f6cc-fe43-4acb-9bc0-0428004ee14f/Cosmos-Project-Update.pdf>

IGO, 2024b. FY24 Mineral Resources and Ore Reserves Report, *ASX Release (29): August 2024* (IGO Limited). Available from: <https://www.igo.com.au/site/investor-center/resources-and-reserves1>

JORC, 2012. Australasian Code for Reporting of Exploration Results, Mineral Resources and Ore Reserves (The JORC Code) [online]. Available from: <http://www.jorc.org> (The Joint Ore Reserves Committee of The Australasian Institute of Mining and Metallurgy, Australian Institute of Geoscientists and Minerals Council of Australia).

Laurence, D C, 2006. Why do mines close?, in *Proceedings of Mine Closure 2006* (eds: A Fourie and M Tibbett), pp 83–94 (Australian Centre for Geomechanics: Perth).

Lindi, O T, Aladejare, A E, Ozoji, T M and Ranta, J P, 2024. Uncertainty Quantification in Mineral Resource Estimation, *Natural Resources Research*, 33:2503–2526. https://doi.org/10.1007/s11053-024-10394-6

Lipton, I T and Horton, J A, 2014. Measurement of Bulk Density for Resource Estimation – Methods, Guidelines and Quality Control, in *Mineral Resource and Ore Reserve Estimation: The AusIMM Guide to Good Practice* (2nd ed), pp 97–108.

Munier, N and Hontoria, E, 2021. *Uses and Limitations of the AHP Method: A Non-Mathematical and Rational Analysis*, (Springer Nature: Switzerland). http://www.springer.com/series/10101

Owusu, S K A and Dagdelen, K, 2019. Critical review of mineral resource classification techniques in the gold mining industry, in *Mining Goes Digital* (ed: E Mueller), pp 201–209 (CRC Press / Taylor and Francis Group).

Rocha, V and Bassani, M A A, 2023. Practical application of a multi-layer scorecard workflow (MLSW) for comprehensive mineral resource classification, *Applied Earth Science: Transactions of the Institute of Mining and Metallurgy*, 132(3–4):216–226. https://doi.org/10.1080/25726838.2023.2244775

Saaty, T, 1988. *The analytic hierarchy process: planning, priority setting, resource allocation* (2nd ed), University of Pittsburgh.

Silva, D S F and Boisvert, J B, 2014. Mineral resource classification: a comparison of new and existing techniques, *The Journal of The Southern African Institute of Mining and Metallurgy*, 114:265–273.

Verly, G, Postolski, T and Parker, H M, 2014. Assessing Uncertainty with Drill Hole Spacing Studies – Applications to Mineral Resources, in *Proceedings of Orebody Modelling and Strategic Mine Planning Symposium*, pp 109–118.

Więckowski, J, Sałabun, W, Kizielewicz, B, Bączkiewicz, A, Shekhovtsov, A, Paradowski, B and Wątróbski, J, 2023. Recent advances in multi-criteria decision analysis: A comprehensive review of applications and trends, *International Journal of Knowledge-Based and Intelligent Engineering Systems*, 27(4):367–393. https://doi.org/10.3233/KES-230487

Application of Quantile Regression Forest for resource classification

K Maxwell[1]

1. Director and Principal Consultant, Matrix Geoscience, Tas 7255.
   Email: kane.maxwell@matrixgeoscience.com

ABSTRACT

This study evaluates Geographic Quantile Regression Forest (GQRF), a machine learning-based spatial prediction method, as an alternative to Sequential Gaussian Simulation (SGS) for uncertainty quantification in coal resource classification. The ASTM D8215-21 standard (2021), commonly applied with SGS, was used to estimate coal tonnage and classify resources into measured, indicated, and inferred categories at an active mine in the Bowen Basin, Queensland, Australia.

While SGS is widely used, its reliance on variogram modelling and significant user input makes it resource-intensive and complex, particularly for heterogeneous data sets. GQRF offers advantages, including the ability to integrate multiple covariates, account for non-linear relationships, and reduce user dependency by eliminating variogram modelling.

The performance of both methods was assessed through leave-one-out cross-validation. Metrics such as mean absolute error (MAE), root mean squared error (RMSE), and prediction interval coverage were calculated to evaluate predictive accuracy and uncertainty estimation. Results showed that GQRF outperformed SGS in predictive accuracy, producing lower errors and higher explanatory power. However, SGS demonstrated slightly better uncertainty estimates, with higher coverage probabilities.

Despite these differences, GQRF provided comparable resource classifications to SGS while significantly simplifying the modelling process. These findings highlight the potential of GQRF to enhance resource modelling efficiency and accuracy, particularly in projects with limited resources or complex spatial heterogeneity. Future work should aim to optimise GQRF's uncertainty estimation to fully match the robustness of SGS while retaining its practical and computational advantages.

This research underscores the growing role of machine learning techniques in resource classification, offering innovative solutions to longstanding challenges in spatial uncertainty modelling.

INTRODUCTION

The 'Standard Practice for Statistical Modelling of Uncertainty in Assessment of In-Place Coal Resources' (ASTM D8215-21, 2021) is a recent standard outlining a methodology for determining uncertainty in coal Resource estimates. The method uses conditional simulation (CS) to derive uncertainty of kriging estimates which are used to categorise Resource estimates into measured, indicated and inferred categories. This method and its derivatives have been promoted as the preferred methodology for coal resource classification by numerous authors (Scott, Dimitrakopoulos and Shuxing, 2005; Emery, Ortiz and Rodriguez, 2004; Olea, 2013).

However, a criticism of the CS method for Resource classification is that it requires significant amounts of user input and validation (such as variogram interpretation and modelling, transform etc) and can be computationally intensive (Maxwell, Rajabi and Esterle, 2021a; Hengl *et al*, 2018; Georganos *et al*, 2019). In addition, the method becomes even more complex if there are numerous domains that need to be modelled, and if additional co-correlated variable is incorporated (Maxwell, Rajabi and Esterle, 2021a; Maxwell *et al*, 2024).

Quantile regression forest (QRF), which is a machine learning method, has recently been demonstrated to provide uncertainty estimates of spatial predictions which are as accurate as CS (Szatmári and Pásztor, 2019). When QRF is extended account for spatial data, QRF has also proven to have equivalent or more accurate predictions than kriging and regression kriging methods and has numerous advantages over these methods including; QRF can easily incorporate numerous co-variates and account for non-linear relationships, does not require variogram modelling, requires little user input (Hengl *et al*, 2018; Georganos *et al*, 2019; Maxwell, Rajabi and Esterle, 2021a;

Maxwell *et al,* 2024). In this study the ASTM D8215-21 methodology (2021) is followed but in addition to CS, QRF is used to determine uncertainty. The primary aim of the study is to determine if QRF can be used as an alternative to CS for determining uncertainty of resource estimates to derive resource classification. The accuracy of the predictions of CS and QRF will also be compared.

To compare methods, the study uses data from an active coalmine site in the Bowen Basin Queensland Australia. To evaluate the predictive performance of each of the methods (CS and QRF), leave one out cross validation will be used to derive several performance measures including mean absolute error (MAE), mean absolute percentage error (MAPE), mean squared error (MSE), root mean squared error (RMSE) and the coefficient of determination (R^2). To evaluate the uncertainty of each method, metrics including coverage probability, prediction interval width, the g-statistic (Goovaerts, 1997), and the goodness statistic (Deutsch,1997) were calculated.

CASE SITE

Geology

The study area is in the eastern Bowen Basin, Queensland, Australia (Figure 1). The Permo-Triassic Bowen Basin is globally recognised as a significant coal-bearing province with a complex tectonic and structural history (Green *et al,* 1997; Salmachi *et al,* 2021; Sliwa *et al,* 2017). At the local scale, the mine site is situated within a synclinal structure, with prominent structural features including north-north-west trending normal and thrust faults (Figure 2) (Sliwa *et al,* 2017).

The primary coal seam of interest at the site is the Permian-aged Leichhardt coal seam, comprising multiple sub-plies with variable thickness and lateral continuity (Figure 3; Maxwell, Rajabi and Esterle, 2022). Resource geologists have defined three geological domains within the site—CP1, CP2, and EP—which correspond to existing open cut mining pits (Figures 2 and 3). These domains are delineated by coal crop lines and mining lease boundaries. Domain EP is distinct due to the occurrence of Cretaceous-aged lamprophyre and dolerite intrusions (Ritchie, 2010). Coal proximal to these intrusions exhibits elevated density, reduced volatile matter, and a marginal increase in ash content relative to unaltered coal (Maxwell, Rajabi and Esterle, 2019).

Domain CP1, which encompasses the largest area, is characterised by coal ply splitting, resulting in reduced mineable coal seam thickness. The five-year forward mine plan incorporates all geological domains, with primary activities concentrated in CP2, where coal is largely unaffected by intrusions (Table 1). Additional information regarding the regional and local geology of the site can be found in Maxwell, Rajabi and Esterle (2019) and Ritchie (2010).

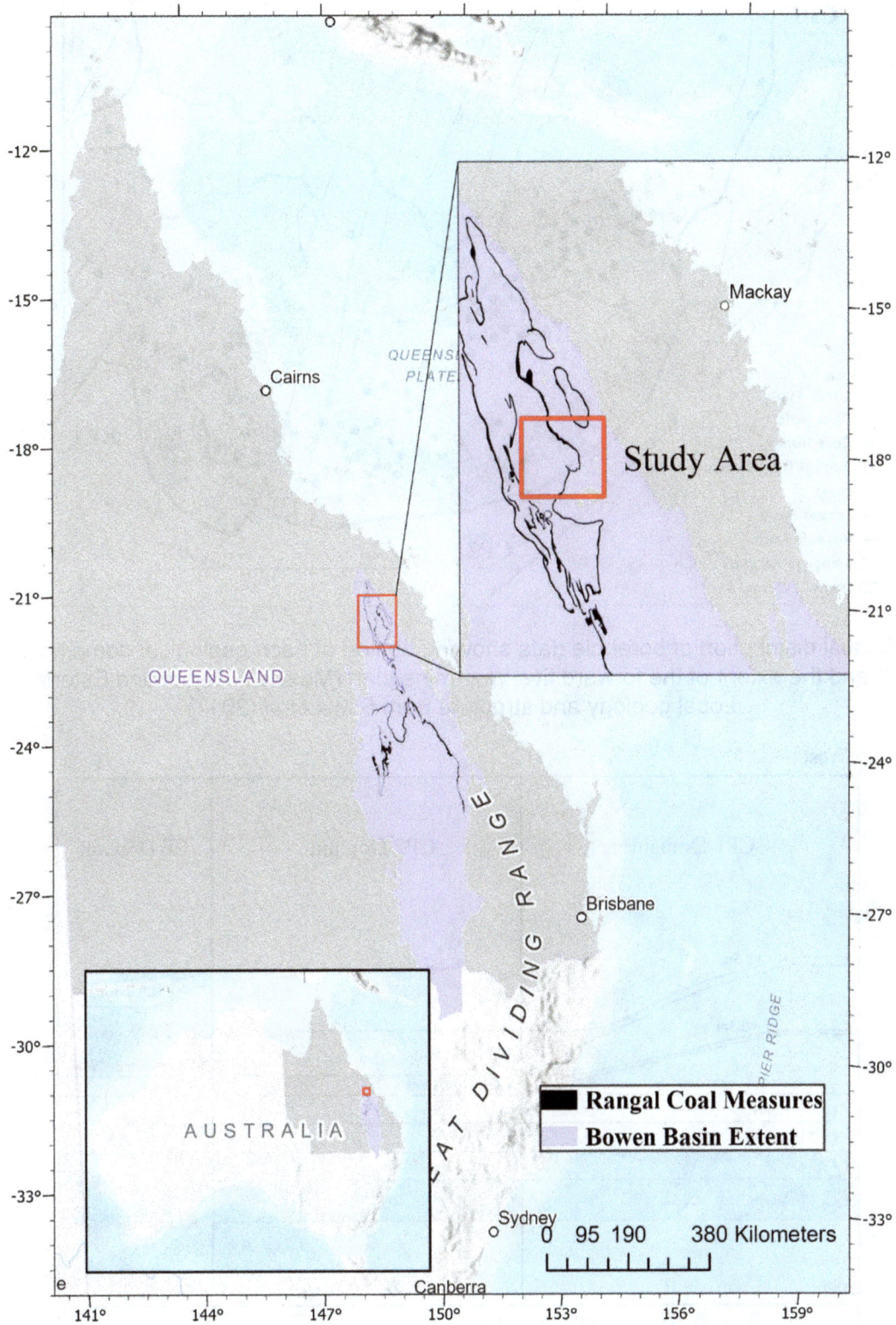

FIG 1 – Approximate location of the study area (Maxwell, Rajabi and Esterle, 2022).

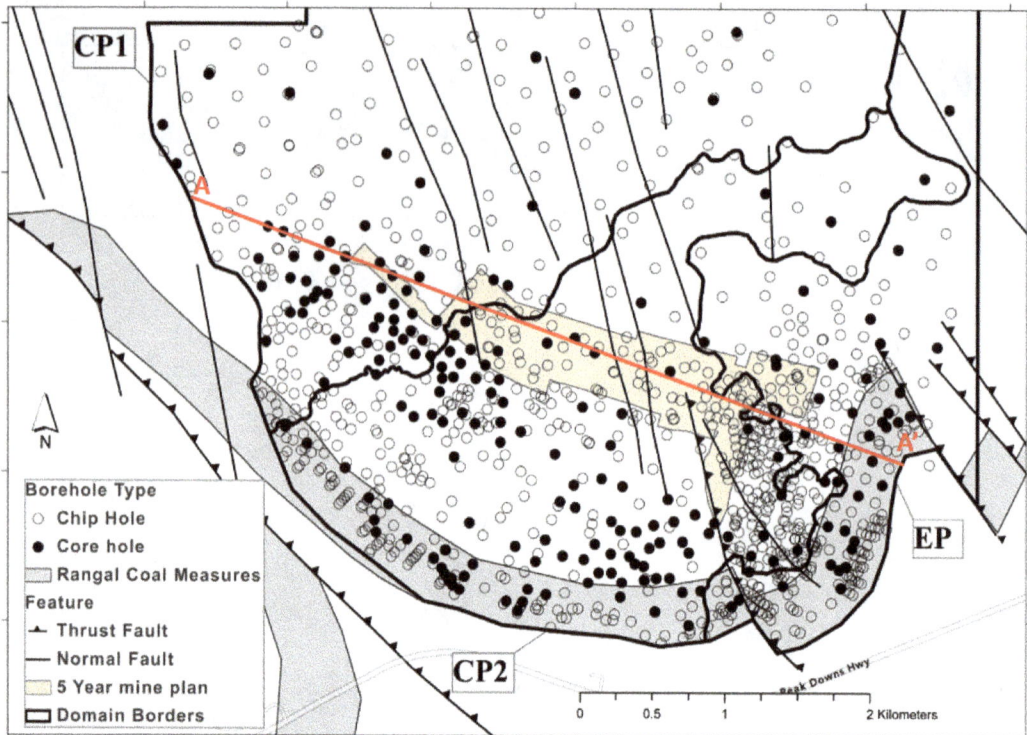

FIG 2 – Spatial distribution of borehole data showing outline of each geological domain (CP1, CP2 and EP) and the extent of the forward five-year mine plan (Maxwell, Rajabi and Esterle, 2022). Local geology and structure from Sliwa *et al* (2017).

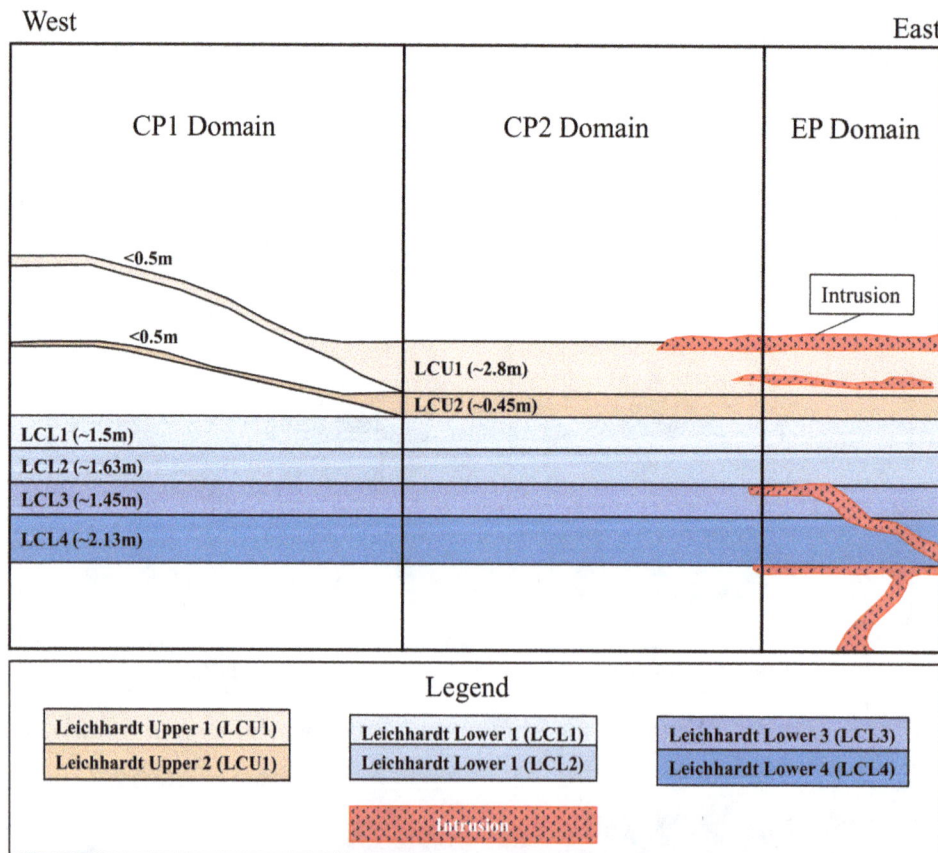

FIG 3 – Stratigraphic cross-section through A-A' in Figure 2) showing the approximate thickness and extent of individual plies at the mine site (Maxwell, Rajabi and Esterle, 2022). CP1 domain is characterised by the thinning and splitting of LCU1 and LCU2 while EP domain is heavily intruded.

TABLE 1

In situ density statistics for all coal plies in each geological domain.

Domain	N	Min	Max	Mean	Sd	Var	Range	Skewness	Kurtosis
CP1	179	1.28	1.92	1.43	0.06	0.00	1.28	3.65	30.36
CP2	511	1.30	1.89	1.44	0.10	0.01	1.30	1.96	3.50
EP	101	1.34	2.00	1.64	0.21	0.04	1.34	0.13	-1.57

Data

This study utilised a data set previously prepared and analysed by Maxwell, Rajabi and Esterle (2022), which provided a comprehensive foundation for the analysis. The data set comprised 246 core holes and 878 chip holes. Core holes were accompanied by laboratory analyses, particularly measurements of relative density, which were subsequently converted to *in situ* density, as required for implementing the ASTM D8215-21 methodology (2021). Laboratory testing of relative density was conducted on core hole samples by NATA-accredited laboratories. Additionally, all boreholes were geophysically logged, providing detailed data on natural gamma, density (short, long, and compensated), and calliper measurements. These data facilitated precise determination of coal ply thickness.

Calculation of in situ density

In addition to thickness, calculation of coal resource tonnage requires knowledge of *in situ* density (reported in g/cm³). Calculation of *in situ* density can be derived from the lab analysed relative density using the method outlined by Preston and Sanders, (1993):

$$RD_{is} = \frac{RD_{ad} * (100 - M_{ad})}{100 + RD_{ad} * (M_{is} - M_{ad}) - M_{is}}$$

where:

RD_{ad} is relative density (air-dry basis)

M_{ad} is moisture (air-dry basis)

M_{is} is *in situ* moisture

In this study a default *in situ* moisture value of 4.5 per cent was used. This default value was provided by the mine site and was based on an internal report that evaluated various methods for predicting *in situ* moisture according to Fletcher and Sanders (2003) and Meyers and Clarkson (2004).

Descriptive statistics

Figure 4 and Table 1 present the distribution and descriptive statistics of calculated *in situ* density for each geological domain. The EP domain exhibits the highest variability and widest range of *in situ* density values, with its data following a broad uniform distribution, indicative of the influence of intrusions in this region. In contrast, the CP1 domain shows a normal distribution with the narrowest range and lowest variability among the domains. The CP2 domain, which contains the largest number of samples, is moderately right-skewed, potentially due to intrusion effects increasing *in situ* density near the intrusion boundary. Additionally, an outlier in CP1 with a high *in situ* density value (>1.8 g/cm³) suggests that the impact of lamprophyre and dolerite intrusions extends beyond the EP domain.

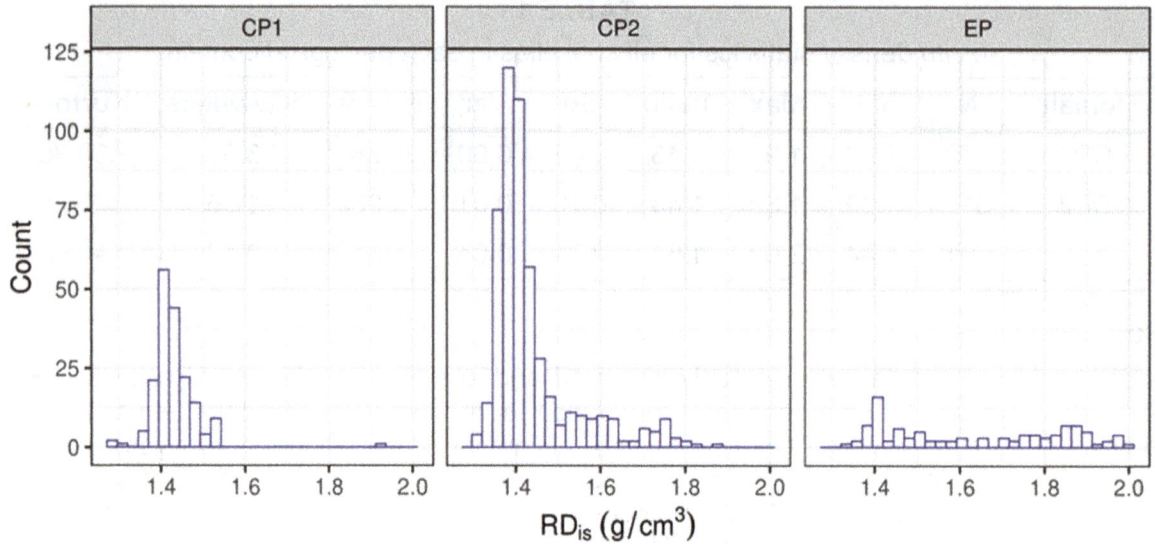

FIG 4 – Distribution of *in situ* density for all coal plies across each geological domain.

THEORY

Sequential Gaussian Simulation (SGS)

Sequential Gaussian Simulation (SGS) is a geostatistical CS method that generates multiple realisations of a spatial property, preserving its variability while honouring observed data and the spatial correlation structure defined by a variogram (Goovaerts, 1997). The process starts by transforming the data into a standard normal distribution to meet the Gaussian assumptions required for simulation. The simulation proceeds sequentially, visiting unsampled locations in random order. At each location, the conditional distribution is determined:

$$Z * (x_i) \sim N(m_i, \sigma_i^2)$$

where m_i is the kriging mean and σ_i^2 is the kriging variance, calculated using the variogram model. The simulated value is drawn as:

$$Z * (x\_i) = m\_i + \varepsilon \sqrt{\sigma\_i\textasciicircum2}$$

where $\varepsilon \sim N(0, 1)$. Each simulated value is added to the conditioning data set to maintain the spatial structure for subsequent simulations. After all locations are simulated, the values are back-transformed to their original distribution. Repeating this process produces multiple realisations, each with their own uncertainty (kriging variance).

Geographical Quantile Regression Forest

Quantile Regression Forest (QRF) is an extension of random forests introduced by Meinshausen, (2006) that is designed to estimate the conditional distribution of a response variable Y given predictor variables X. Unlike traditional random forests, which provide a single mean prediction, QRF predicts conditional quantiles, such as the median or other percentiles, allowing for a richer understanding of uncertainty and variability in the predictions. The algorithm works by constructing regression trees using bootstrapped samples of the data. For a given query point x QRF collects all training samples that fall into the same leaf nodes across all trees. These pooled samples form an empirical distribution from which the desired quantiles are calculated.

Mathematically:

For each tree t:

$L_t(x)$: Leaf node of tree t corresponding to x.

$y_{t,i}$: Response values of the data points in $L\_t(x)$

The quantile q is estimated as:

$$Q_q(x) = Quantile_q\left(\bigcup_{t=1}^{T} y_{t,i}\right)$$

Where $Quantile_q$ calculates the $q-th$ quantile of all predictions across the trees.

In this way, any number of quantiles can be predicted, and those quantiles can be used to represent 'estimation error'. For example, in addition to using the algorithm to predict the 0.5 quantile, by also choosing to predict the 0.05, 0.95 quantiles, these may be used as the upper and lower bounds of error.

However, a disadvantage of the QRF algorithm is that is does not account for spatial data. To overcome this issue, Maxwell, Rajabi and Esterle (2021a) proposed a method termed geographical quantile regression forest (GQRF) that incorporates spatial data by using locally constrained QRF models and distance weightings. Maxwell, Rajabi and Esterle (2021a, 2021b) demonstrated that GQRF can outperform inverse distance weighting (IDW), ordinary kriging (OK), and random forest Kriging (RFK) in predicting coal relative density values. Further, Maxwell, Rajabi and Esterle (2021a, 2021b) described numerous advantages of the method over traditional geostatistical methods including that: it does not require strict assumptions about data normality or stationarity, variogram modelling is not required, it is robust to outliers and skewed data distributions, it can more easily handle co-variates, and that a single model can be used for multiple domains.

The method (fully described in Maxwell, Rajabi and Esterle, 2021a) involves a systematic set of steps including:

1. Calculate the distance between $Z * (x0)$ and $Z * (xi)$, $i = 1.., k$ (number of observed data).

2. Scale the distances calculated in the previous step between 0 and 1. These will be used as weighting values in the following step.

3. Train a quantile regression forest algorithm using the points from Step 1 and use the weights from Step 2 as weighting values in the quantile regression forest algorithm.

4. Use the trained quantile regression forest to predict the value at $Z * (x0)$. Note that at this step, an uncertainty (variance) estimate is produced based on user-defined quantiles.

5. Repeat this process for all locations.

In the context of this study, the user defined quantiles are set as 0.05, 0.5 and 0.95. The 0.05 and 0.95 quantiles are utilised as estimation error to so that this variance may directly be used to produce a cumulative cell (grid) tonnage plot in accordance with the ASTM D8215-21 standard (2021). In addition, these quantiles are used to assess the 'goodness' of the estimation error as compared to the kriging errors produced by SGS.

METHOD

Materials

Data preparation and spatial modelling were performed using the R programming language (R Core Team, 2017). Estimates for SGS were generated with the R library gstat (Gräler *et al*, 2016), while the library gqrf (Maxwell, Rajabi and Esterle, 2021a) was utilised for producing GQRF estimates. Evaluation metrics were calculated, and figures were generated in R with the support of various R libraries.

ASTM D8215-21

The ASTM standard (2021) defines a procedure to classify Resources based on coal tonnage uncertainty produced from SGS (or alternative CS methods). The primary outputs of the procedure are a total tonnage uncertainty histogram and cumulative tonnage distribution plot (that is used for Resource classification). The procedure for using SGS is summarised as:

1. Use SGS to produce a minimum of 100 coal thickness and density realisations for each seam of interest across a regularised grid (constrained by geological and available data).

2. Derive coal tonnage realisations by multiplying the thickness and density realisations.

3. Produce a total tonnage uncertainty distribution plot and associated statistics.

4. Produce a cumulative cell (grid) tonnage plot by ordering the tonnage realisations by increasing uncertainty. For simplicity, only plot the spread of realisations that fall between the 5th and 95th percentile.

5. Using the cumulative tonnage plot, determine Resource categories by using nominal quantile cut-offs for measured, indicated and inferred.

The procedure also outlines minimum sampling and data preparation steps and provides annexes for specific geological scenarios (such as discontinuous and faulted coal seams).

For GQRF the procedure is modified to the following:

1. Use GQRF to predict coal thickness and density across all grid nodes using easting, northing, ply code and domain as input data. Select 0.05, 0.50 and 0.95 as the quantiles to predict. The predicted value is taken as the output of the 0.50 quantile, while the upper and lower error limits were taken as 0.05 and 0.95. Calculate the error uncertainty by computing the variance between the upper (0.95) and lower quantiles (0.05).

2. Derive coal tonnage and associated uncertainty by multiplying the thickness and density results multiplied by the grid cell area.

3. Produce a cumulative cell (grid) tonnage plot by ordering the tonnage realisations by increasing uncertainty.

4. Using the cumulative tonnage plot, determine Resource categories by using nominal quantile cut-offs for measured, indicated and inferred.

Model implementation

For both SGS and GQRF, the grid (cell) spacing was set to 20 and was cropped to the geological domains (CP1, CP2 and EP). The cell spacing represents a spacing the is approximately one fourth the average distance to the closest drill hole as recommended in the ASTM standard. In addition, the mined-out area was removed from the grid.

SGS required two dimensional variogram modelling for each ply, domain and coal parameter (thickness and density). Anisotropic ellipses were defined in four directions (0, 45, 90 and 135) to determine the most appropriate variograms. Variogram models were then carefully fit by eye (variogram output and settings available on request). In all, this amounted to 144 variogram models. Following variogram modelling, SGS predictions were conducted for thickness and density for each coal ply and domain across all gid nodes. The number of simulations was set as 100 (as the minimum recommended in the standard), and the neighbours were set to 12. For each grid node, the predicted value was taken as the mean of the 100 simulated values, while the error range was constrained within the 5 and 95 per cent quantiles of the kriging variance. The grid node predictions and associated kriging variance for thickness and density were multiplied together for later evaluation. Finally, the thickness and density product results were multiplied by the grid cell area to produce tonnage estimates. These tonnage estimates were aggregated for all plies to produce a cumulative cell tonnage plot, that form the basis of Resource categorisation.

GQRF was used to predict coal thickness and density across all grid nodes using easting, northing, ply code and domain as input data. The quantiles selected in the algorithm were 0.05, 0.50 and 0.95. The predicted value is taken as the output of the 0.50 quantile, while the upper and lower error limits were taken as 0.05 and 0.95. The number of neighbours was set to the same as SGS for consistency. The procedure to produce tonnage estimates (multiplication of density thickness and cell area) was then conducted to produce a cumulative cell tonnage plot.

Finally, for both SGS and GQRF to estimate the amount of measured, indicated and inferred tonnes, 25 per cent and 75 per cent cut-off values were used on the cumulative cell tonnage plots. That is, tonnages between 5 and 25 per cent variance were classed as measured, tonnes between 25 and

75 per cent variance were considered indicated, and tonnages >75 per cent variance were considered inferred.

Key model settings are summarised in Table 2.

TABLE 2

Key model settings.

Setting	Description
Grid extents and setting	Easting limits: 646 000 m, 653 000 m
	Northing limits: 7 580 000, 7 586 000 m
	Spacing: 20 m
	Constraints: Coal crop line, CP1, CP2, EP domain, mined out area
SGS algorithm settings	Number of simulations: 100
	Number of neighbours: 12
GQRF algorithm settings	Number of neighbours: 12
	Predictors: Easting, northing, ply, domain
	Number of trees: 50
	Number of variables to possibly split at in each node: 2

Model evaluation

To summarise the results, leave one out cross validation was conducted on thickness multiplied by density for all plies for both SGS and GQRF models. Evaluation metrics were computed to assess the accuracy of predictions and the accuracy or 'goodness' of the estimation error produced from SGS and GQRF.

In this context, for SGS the upper and lower bounds of the estimation error correspond to the lowest and highest values of the kriging variance from the 100 simulations (within the 5 per cent and 95 per cent confidence interval). For GQRF, the 5 per cent and 95 per cent quantiles were used.

To evaluate the accuracy of predictions, mean absolute error (MAE), mean absolute percentage error (MAPE), mean squared error (MSE), root mean squared error (RMSE) and the coefficient of determination (R^2) were computed. Each of these metrics serves a different purpose. MAE and MAPE focus on the magnitude of errors, while MSE and RMSE emphasise penalising larger errors. R^2 provides an overarching view of the model's ability to explain the variance in the data. An ideal model will have an MAE, MAPE, MSE and RMSE of zero, and an R^2 of 1.

To assess the error estimates generated by SGS and GQRF, metrics including coverage probability, prediction interval width, the g-statistic, and the goodness statistic were calculated. As these metrics are less commonly used, detailed explanations are provided below.

Coverage probability is the proportion of observed values that fall within the upper and lower prediction intervals (5 per cent and 95 per cent). Higher coverage probability indicates well calibrated model.

Prediction interval width is the average width of the prediction intervals. Narrower intervals indicate more precise predictions, but they must still achieve adequate coverage probability.

G-statistic (g) (Goovaerts, 1997) assesses whether observed values fall within the predicted intervals and penalises those that fall outside.

The g-statistic is defined as:

$$g = \frac{1}{n} \sum_{i=1}^{n} \left[\frac{1}{\alpha} \max(0, \text{Lower}_i - \text{Observed}_i) + \frac{1}{\alpha} \max(0, \text{Observed}_i - \text{Upper}_i) \right]$$

where:

n	the number of observations
α	the nominal level of the prediction interval (eg α = 0.05 for a 95 per cent confidence interval)
$Lower_i$	the lower bound of the prediction interval for observation i
$Upper_i$	the upper bound of the prediction interval for observation i
$Observed_i$	the observed value

An ideal value (g = 0) indicates that all observed values are within the prediction intervals. Higher values Indicate poorer quality of prediction intervals, either due to systematic bias or overly narrow intervals. The g-statistic is useful for comparing the quality of prediction intervals across different models or methods.

Goodness statistic (G), introduced by Deutsch (1997) is a measure used to evaluate the agreement between the expected coverage of prediction intervals and the actual coverage observed in the data.

The goodness statistic is defined as:

$$G = 1 - \frac{|P - P_{nominal}|}{P_{nominal}}$$

where:

P	the observed coverage probability (proportion of observed values within the prediction intervals)
$P_{nominal}$	the nominal coverage probability (eg 0.95 for a 95 per cent prediction interval)

G ranges between 0 and 1, where 1 indicates that all observed values fall within the estimation error intervals, and 0 indicates observed coverages is completely outside of the estimation error range.

RESULTS

Table 3 contains the Resource estimates for each model method produced form the cumulative cell tonnage plots (Figure 5). Resource estimates between models are very similar, with SGS containing slightly more (between 3.41 and 4.98 per cent) measured, indicated (4.41 per cent) and inferred (4.98 per cent) tonnes than GQRF. While the cumulative cell tonnage plots appear to show significant variance between models, the intersection of the 5 to 95 spread for each Resource category is similar.

TABLE 3

Resource estimates produced from each model following the ASTM D8215–21 (2021).

Model	Measured (Mt)	Indicated (Mt)	Inferred (Mt)	Total (Mt)
GQRF	17.0	34.7	19.1	70.8
SGS	17.6	36.3	20.1	74
% Difference	3.41	4.41	4.98	4.32

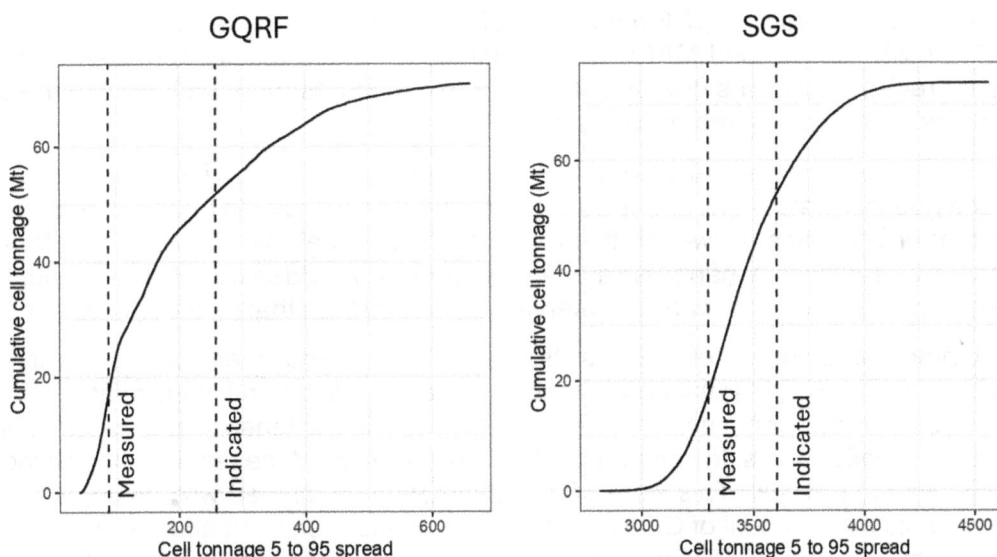

FIG 5 – Cumulative cell (grid) tonnage plot for GQRF and SGS models (total tonnage, all plies). Vertical lines intersect each plot at 25 per cent and 75 per cent confidence intervals, representing the measured and indicated Resource category cut-offs.

Tables 4 and 5 contain the evaluation metrics for the predictions and estimates of error for both SGS and GQRF models. Across all evaluation metrics, GQRF has slightly better prediction metrics compared to SGS. Specifically, GQRF has lower MAE, MSE, RMSE and MAPE and higher R^2 than SGS, indicating that GQRF has produced a more accurate model than SGS. Lower MAE reflects closer average predictions, while lower MSE and RMSE indicate fewer large errors and better overall performance. A lower MAPE shows greater accuracy relative to the magnitude of the target values. A higher R^2 signifies that the model explains more of the variability in the data, making it a better fit.

TABLE 4

Evaluation metrics for the leave one out predictions for SGS and GQRF models.

Model	MAE	MSE	RMSE	R^2	MAPE
GQRF	4.42	36.73	6.06	0.02	110.37
SGS	4.81	39.54	6.29	-0.05	113.45

TABLE 5

Evaluation metrics for the leave one out estimates of error for SGS and GQRF models.

Model	Coverage probability	Average interval width	G-statistic (g)	Goodness statistic (G)
GQRF	0.71	11.79	15.74	0.75
SGS	0.82	16.69	13.46	0.86

Conversely, SGS has better evaluation metrics (higher coverage probability, goodness statistic and lower g-statistic values) compared to GQRF for estimates of error. However, GQRF has a lower average interval with.

DISCUSSION

Resource estimates derived from cumulative cell tonnage plots indicate that the SGS model classifies slightly higher tonnage across measured, indicated, and inferred categories, with differences ranging from 3.41 per cent to 4.98 per cent. However, despite visual variances in the cumulative tonnage plots, the overlapping 5–95 per cent confidence intervals suggest that GQRF provides broadly consistent Resource classifications to SGS.

In terms of predictive accuracy, GQRF outperforms SGS across all evaluation metrics, as evidenced by lower MAE, MSE, RMSE, and MAPE, and a higher R^2. These results demonstrate that GQRF delivers more precise predictions with smaller errors and captures more variability in the data, thus producing 'better' Resource estimates in this case.

Conversely, SGS demonstrates better metrics for uncertainty estimation, including higher coverage probability and goodness statistics, and a lower g-statistic. These indicate that SGS produces slightly more reliable error estimates. However, this robustness comes at the cost of a broader prediction interval, as GQRF achieves a narrower average interval width. Ideally, a model should maintain robust uncertainty estimates with a narrow interval width, a balance that neither model achieves fully.

A significant practical limitation of SGS is its computational and interpretative complexity. The method required numerous (144) variogram models to be developed and interpreted across multiple coal plies, variables, and domains. This process is both time-intensive and requires significant expertise to ensure valid results. In contrast, GQRF was able to produce results using a single model, greatly simplifying the process. This ease of implementation, combined with its superior predictive accuracy, highlights the potential of GQRF as a viable and more efficient alternative to SGS. Future work could explore optimising GQRF's uncertainty estimation capabilities, particularly by addressing its slightly lower performance in robustness metrics, to further enhance its utility in resource classification.

CONCLUSION

This study demonstrates the potential of Geographical Quantile Regression Forest (GQRF) as a viable alternative to Sequential Gaussian Simulation (SGS) for resource classification in coal deposits. While both methods produce broadly similar resource estimates, GQRF exhibits superior predictive accuracy with lower errors and higher explained variance compared to SGS. GQRF's ability to efficiently handle non-linear relationships, multiple covariates, and spatial variability with minimal user input positions it as a robust and practical modelling tool.

Conversely, SGS provides slightly more reliable uncertainty estimates, with higher coverage probability and goodness statistics, but at the cost of broader prediction intervals and significant computational and interpretive complexity. The requirement to develop and interpret numerous variogram models across multiple coal plies, variables, and domains makes SGS both resource-intensive and reliant on specialised expertise.

GQRF's ability to operate with a single model while achieving comparable or superior results underscores its potential as a more efficient and accessible method for resource classification. Future research should focus on refining GQRF's approach to uncertainty estimation to match or surpass SGS in robustness while maintaining its practical advantages. This work highlights the increasing relevance of machine learning techniques like GQRF in enhancing the efficiency and accuracy of spatial resource modelling.

ACKNOWLEDGEMENTS

The author would like to thank Peabody Energy for providing data for this study and for granting permission to publish the results of this study.

REFERENCES

ASTM International, 2021. ASTM D8215-21: Standard Practice for Statistical Modeling of Uncertainty in Assessment of In-place Coal Resources, West Conshohocken, PA: ASTM International. https://doi.org/10.1520/D8215-21

Deutsch, C V, 1997. Direct assessment of local accuracy and precision, *Geostatistics Wollongong*, 96(1):115–125.

Emery, X, Ortiz, J C and Rodriguez, J J, 2004. Quantifying uncertainty in mineral resources with classification schemes and conditional simulations, Center for Computational Geostatistics Annual Report Papers, pp 1–13. papers2://publication/uuid/DDFCD6B7-9AA0-4E38-8372-345272E69FA8

Fletcher, I S and Sanders, R H, 2003. Estimation of *in situ* moisture of coal seams and product total moisture: Final Report for ACARP Project C10041.

Georganos, S, Grippa, T, Niang Gadiaga, A, Linard, C, Lennert, M, Vanhuysse, S, Mboga, N, Wolff, E and Kalogirou, S, 2019. Geographical random forests: a spatial extension of the random forest algorithm to address spatial heterogeneity in remote sensing and population modelling, *Geocarto International*, 1(12). https://doi.org/10.1080/10106049.2019.1595177

Goovaerts, P, 1997. *Geostatistics for Natural Resources Evaluation* (Oxford University Press).

Gräler, B, Pebesma, E and Heuvelink, G, 2016. Spatio-temporal interpolation using gstat, *The R Journal*, 8(1):204–218.

Green, P M, Carmichael, D C, Brain, T J, Murray, C G, McKellar, J L, Beeston, J W, Gray, A R G and Green, P, 1997. Lithostratigraphic units in the Bowen and Surat basins, Queensland, The Surat and Bowen Basins, South-East Queensland, 1:41–108.

Hengl, T, Nussbaum, M, Wright, M N, Heuvelink, G B and Gräler, B, 2018. Random forest as a generic framework for predictive modeling of spatial and spatio-temporal variables, *PeerJ*, 6:e5518.

Maxwell, K, Rajabi, M and Esterle, J, 2019. Automated classification of metamorphosed coal from geophysical log data using supervised machine learning techniques, *International Journal of Coal Geology*, 214:103284. https://doi.org/10.1016/j.coal.2019.103284

Maxwell, K, Rajabi, M and Esterle, J, 2021a. Spatial interpolation of coal properties using geographic quantile regression forest, *International Journal of Coal Geology*, 248:103869. https://doi.org/10.1016/j.coal.2021.103869

Maxwell, K, Rajabi, M and Esterle, J, 2021b, 14 September. Geographic quantile regression forest: a new method for spatial modelling of mineral commodities, in 3rd AEGC: Geosciences for a Sustainable World.

Maxwell, K, Rajabi, M and Esterle, J, 2022. Impact of *In situ* Density Spatial Model Methods on Resource Tonnages in Highly Intruded Coal Deposits, *Natural Resources Research*, 31:499–515.

Maxwell, K, Rajabi, M, Esterle, J, Tivane, M and Travassos, D, 2024. Spatial modelling and classification of altered coal using random forest-based methods at Moatize Basin, Mozambique, *Journal of African Earth Sciences*, 215:105279.

Meinshausen, N, 2006. Quantile regression forests, *Journal of Machine Learning Research*, 7:983–999.

Meyers, A and Clarkson, C, 2004. Estimation of *In situ* Density from Apparent Relative Density and Relative Density Analyses: Final Report for ACARP Project C10042.

Olea, R A, 2013. Special issue on geostatistical and spatiotemporal modeling of coal resources, *International Journal of Coal Geology*, 112:1. https://doi.org/10.1016/j.coal.2013.01.010

Preston, K B and Sanders, R H, 1993. Estimating the *in situ* relative density of coal, *Australian Coal Geology*, 9:22–26.

R Core Team, 2017. R: A language and environment for statistical computing. https://www.R-project.org/

Ritchie, C, 2010. Lamprophyric intrusions in the Rangal Coal Measures, Bowen Basin : classification, geochemistry and tectonic significance, Honours Thesis, The University of Queensland, UQ eSpace. https://search.library.uq.edu.au/permalink/f/18av8c1/61UQ_eSpace734827

Salmachi, A, Rajabi, M, Wainman, C, Mackie, S, McCabe, P, Camac, B and Clarkson, C, 2021. History, Geology, In Situ Stress Pattern, Gas Content and Permeability of Coal Seam Gas Basins in Australia: A Review, *Energies*, 14(9). https://doi.org/10.3390/en14092651

Scott, J, Dimitrakopoulos, R and Shuxing, L, 2005. Quantification of geological uncertainty and risk assessment in resource/reserve classification, vol 1, C11042, Australian Coal Association Research Program (ACARP).

Sliwa, R, Esterle, J, Phillips, L and Wilson, S, 2017. Rangal supermodel 2015: The Rangal-Baralaba-Bandanna Coal Measures in the Bowen and Galilee Basins, Final Report ACARP Project C22028, ACARP (Australian Coal Industry Research Program).

Szatmári, G and Pásztor, L, 2019. Comparison of various uncertainty modelling approaches based on geostatistics and machine learning algorithms, *Geoderma*, 337:1329–1340.

Culture, failures/learnings, case studies

Peer accountability and communication through a critical checkpoint framework – driving a culture of critical thinking

J M Clark[1], M Roux[2] and C Fiddes[3]

1. Regional Chief Geologist, Nevada Gold Mines, Elko NV 89801, USA.
   Email: jesse.clark@nevadagoldmines.com
2. Regional Lead, Resource Geology, Barrick Gold, Elko NV 89801, USA.
   Email: mark.roux@barrick.com
3. Lead, R&R Governance, Nevada Gold Mines, Elko NV 89801, USA.
   Email: craig.fiddes@nevadagoldmines.com

ABSTRACT

Nevada Gold Mines operates over 15 active mining operations, a large network of stockpiles, and half a dozen projects at various stages of study. These result in over 20 project updates converging within a four-month window each year. Comprehensive peer review is essential at key milestones to ensure compliance to standards without surprises but is challenging to manage.

Four critical checkpoints were initially developed to track resource model deliverables. Approval was required at each milestone (eg database validation, domain updates, estimation set-up etc) before a final estimation review which applied a standard suite of validations. The responsible resource geologist populated these validations and findings using a template. This resulted in mass production of plots and images without conclusively presenting the material validation themes. The result was a linear and inefficient transactional process which we were trying to eradicate.

This provided the catalyst for a revised framework of a 'Critical Checkpoint' system extending across the mining value chain and covering all stakeholder inputs that feed the final resource model, Resource and Reserve inputs, and life-of-mine planning assumptions. Geologists, mining engineers, metallurgists and key management are all involved across ten stages. An auditable online system integrated within Barrick's existing systems was adopted with electronic approvals and a document repository.

A Critical Checkpoint *scorecard* was also implemented, which is a series of 21 focused questions requiring the project owner to describe, explain and demonstrate the logic behind key decision points of the estimation workflow. While daunting at first, the questions are static and freely available to the team for review. The questionnaire is finalised during the final estimation review where Qualified Persons and management (a minimum of three) can score each answer using a pre-defined three-point scorecard. These scores are compiled and weighted 30 per cent towards their end of year performance rating. The goal of this quantitative approach is to motivate and drive a culture of critical thinking.

As technical professionals, we problem solve non-trivial issues that have a spectrum of solutions and no discernible right or wrong answer. It is therefore critical that as Qualified Persons we weight the decision-making process higher than the physical final product. Streamlining processes through automation or scripting offers efficiencies but must be balanced with critical thinking.

INTRODUCTION

Critical thinking is the process of analysing information to make informed decisions. It is a core skill for mining professionals who deal with complexities across multiple data sets simultaneously. Yet, critical thinking is in alarming decline. It is endemic, not only within geoscience, but reflects on a broader societal trend of excess information. The physiological response of cognitive overload fosters bad habits like task-switching, reacting with urgency, and an unhealthy reliance on gradatim procedures (Rutkowski and Saunders, 2018), all of which are at the sacrifice of deep, reflective thinking and focused outcomes.

The absence of critical thinking hinders the ability to make sensible judgements, hundreds of which are required by a diverse team of professionals to construct mineral resource estimates. These estimates support detailed mining plans, and the compilation of Mineral Resources and Ore

Reserves, which are the largest asset of a commodity-focused mining company (Noppe, 2017). The accurate compilation of which is the responsibility of Qualified Persons (QP), in compliance to relevant national reporting codes and company-specific corporate governance guidelines.

Peer Review and Audits are the primary tools used to satisfy the QP that the resultant work is acceptable for public disclosure. Both tools focus on identifying material flaws of global estimates, yet the critical thinking rationale behind small decision points are rarely investigated in much detail assuming immateriality. Multiple incorrect small decisions can quickly materialise into sources of poor reconciliation, not necessarily fully resolved with infill drilling, leading to challenges during mining.

This paper presents an internal Peer Review system of Critical Checkpoints that occur during the mineral resource estimation process. Final approval of the estimate utilises a series of questions to understand the rationale behind decision points. Questions are transparently available to all resource geologists, intended to reinforce the importance of critical thinking over process. Each question is scored by a panel of subject matter experts, including relevant QPs and management, that is ultimately weighted against their end of year performance review to ensure accountability.

WHY DRIVE A CULTURE OF CRITICAL THINKING?

The traditional role of a mine or resource geologist focuses on enhancing the profitability of the orebodies we exploit (McKinstry, 1948). Value creation involves balancing organic growth with improving the quality of the material we seek to mine. Achieving these objectives requires a deep level of orebody knowledge, which can only be obtained through an iterative process of geological interpretation of mineralisation controls. Orebody knowledge is harvested over years of continuous exposure where every new geologic observation plants a new seed that is cultivated through deep learning. A profound sense of ownership across all aspects of mining geology, from target generation to model reconciliation, is unavoidable. It is an investment in intellectual property that compounds in value over time.

Challenging this conventional approach is an undercurrent of changes that are slowly transforming mine geology into a process-driven, transactional culture. Though well intentioned, the adoption of functionalised organisational structures accommodating flexible work options is at the sacrifice of orebody knowledge. Organised rotations are the best strategy to maximise exposure across the value chain obtaining a breadth of skills, though it still does not solve the dilution of orebody knowledge. A contributing factor, if not a root cause, is the undercutting of earth science degrees shifting the focus away from field skills to a dominantly analytical-based learning environment.

The consequence is a normalisation of mediocrity and the dilution of critical thinking leading to decision paralysis. A symptom is an obsession for technological 'improvements' that mostly result in value destructive 'busy work', increasingly reliant on procedures that absurdly complicates the original process. Another symptom is a constant perception of production pressure that drives an irrational sense of urgency inflating menial tasks of low to no value-add.

A critical thinking mindset cannot be mandated. It is instilled through a company's culture, measured by the values that we choose to act on. Though cultural change is a difficult and polarising process as humans naturally repulse change, defaulting to the stages of grief, it requires a slow process of awareness, a level of trust and autonomy balanced by guidelines and careful training, but ultimately, it is leading by example to empower teams to drive ownership of decisions in a safe environment without retribution. A culture of critical thinking can be measured by the level of orebody knowledge gained coupled with a profound sense of ownership of the orebodies we seek to characterise.

A litmus test for a critical thinking culture is listening to a model handoff meeting to peers. The quality of an estimate is a result of careful planning to ensure a fit-for-purpose product each time, as it reflects our best knowledge of the orebody at that moment in time. The validation of the final estimate should tell a story that explains the work undertaken to achieve its intended purpose. The focus is on continuous improvement complemented by a discussion on how these changes will optimise the mining plans or our growth targeting. A clear action plan should be captured on how to address outstanding items with a committed due date. If presented with a slew of geostatistical plots following a rigid template without any critical explanation of *why*, immediately fails the test and is simply the result of hitting 'go' on an automated workflow.

Challenging the improvement paradigm

Estimation workflows are increasingly complex in nature reflecting a self-driven fallacy under the guise of improvement. While automation eliminates repetitive processes allowing the user to focus on what adds the most value, it is the value opportunity that often loses focus. Geologists create value by identifying growth opportunities balanced by characterising resource risk in the mine plan, which again, requires a deep level of orebody knowledge. The obsession to improve workflow efficacies can lead to overlooked flaws by not understanding the geologic controls on mineralisation at appropriate scales to the mining method.

The toolkit available to resource practitioners today is immense, making it easier than ever to automate increasingly complex estimation workflows. The days of trolling through endless coding textbooks or online threads are now largely replaced with the introduction of artificial intelligence, which can improve seemingly infinite lines of codes within seconds. Automation not only extends to estimation, but also geologic domain interpretations using machine learning algorithms. With minimal effort, hundreds of steps that took countless weeks, can be replaced with a highly auditable notebook that can execute a full model within hours. But is it an improvement?

A potential explanation for the improvement paradigm is the shift to non-linear estimation workflows through the 2000s to 2010s. Adoption of multiple indicator kriging focused efforts to automate workflows with various scripts as it was otherwise too demanding a process to execute. Linked to the inherent workflow complexity was a desire to continually 'improve' the efficacy of the workflows through increased scripting and minor variations to the estimation set-up itself. The consequences include the sacrifice of foundational geology underpinning stationary estimation domains, and overly-complex and time-consuming workflows that are difficult to audit. The past decade has seen a correction back to simplified ordinary kriging workflows, yet the improvement paradigm is engrained.

Though auditability is perceptibly increased in well-organised notebooks or scripts, they are paradoxically still a black box. Its initial creation likely solved a complicated problem, but once static, often becomes the source of the problem itself, and difficult to update by different practitioners. An example is standardising basic validation plots intended for fast reviews looking for material flaws, quickly becomes the exclusive procedure for validation. Changes are rarely considered, if ever, no matter how irrelevant or 'off' the plot is, driven by a state of decision paralysis as they feel locked within the constraints of a corporate procedure.

Automation can be hugely positive in streamlining menial tasks or allowing for fast analysis of big data sets. The short-term gains must be balanced by a sustainable approach that does not interfere with the ability to critically think. An unintended consequence is a desire to continually improve the process and not the *product* of the process. This deflection of focus is sub-conscious yet detrimental to the foundation of the estimate rationalised by the improvement paradigm.

Critical thinking is important in estimation

Professor Charles Perrow in his 1984 book Normal Accidents presented a methodology to assess the inherent risk of any process at inducing a normal accident, which is equivalent to the scale of a meltdown in a nuclear reactor. It is measured by how tightly coupled or interdependent each process is against the linearity or control over the tasks necessary to execute the processes. A post office for example is loosely coupled and non-complex, meaning a lost letter does not induce a domino-style collapse of mail processing nor is it difficult to locate the source of how it came to be lost. Estimation is conversely a tightly coupled and complex process. Without critical thinking, the probability of making a mistake is high and the likelihood of material consequences is inevitable, yet it can be difficult to locate the source(s).

Estimations should serve a specific purpose for one or multiple stakeholders. Routine updates with fast turnaround times force practitioners to surrender to process over purpose. Process is often partnered with the common occurrence in functionalised teams to have a specialist geologic modeller who provides domains, often in isolation without any training in geostatistical fundamentals, displacing the traditional ownership away from the resource geologist. Uncertainty arises when there is an impasse on the geologic rationalisation for the resultant geometries, or worse, blind acceptance of the geologic interpretation, displacing critical decision points.

The accuracy of a resource model materially relies on a correct interpretation of the mineralisation geometry (the modelled domain) and grade distribution (the estimate). The hardest and most avoided decisions are those that are qualitative, not easily reproducible by a geostatistical tool, often requiring a strong dose of gumption once all the details are carefully considered. Estimation choices must complement the geologic interpretations driving an iterative feedback loop to balance geostatistical inferences. Field validation and an intense scrutiny of products in three-dimensional software is essential to decoupling the layered uncertainty of increasingly complex interpretation choices. The devil is always in the detail.

Peer Review and Audits are the primary tools used to satisfy the QPs that the resultant work is acceptable for public disclosure. Audits are typically conducted by independent parties to verify regulatory and reporting code compliance. Investigative review of provided files focuses on identifying material flaws in estimation inputs and resource and reserve assumptions. Peer Review is an internal process that is typically governed by corporate standards and guidelines, often written by the QP, if they exist at all. Common industry practice uses a 'tick and flick' checklist that is discerned by QPs during site visits or meetings. The focus is on the result, rarely considering the critical thinking rationale behind each decision and how that judgement was made.

QUALITY ASSURANCE

The model planning process is designed to maintain communication across the project team (Figure 1). Setting deadlines and deliverables early clarifies accountability for on-time delivery of fit-for-purpose resource models. For annual model delivery supporting public disclosure and life-of-mine planning, a rolling 18-month model delivery timeline is maintained for 22 active models across NGM. Additionally, an action register tracking key items for continuous improvement is connected to the model delivery timeline with its own priority ranking with committed due dates. Actions must be reviewed and finalised before the model time frame begins.

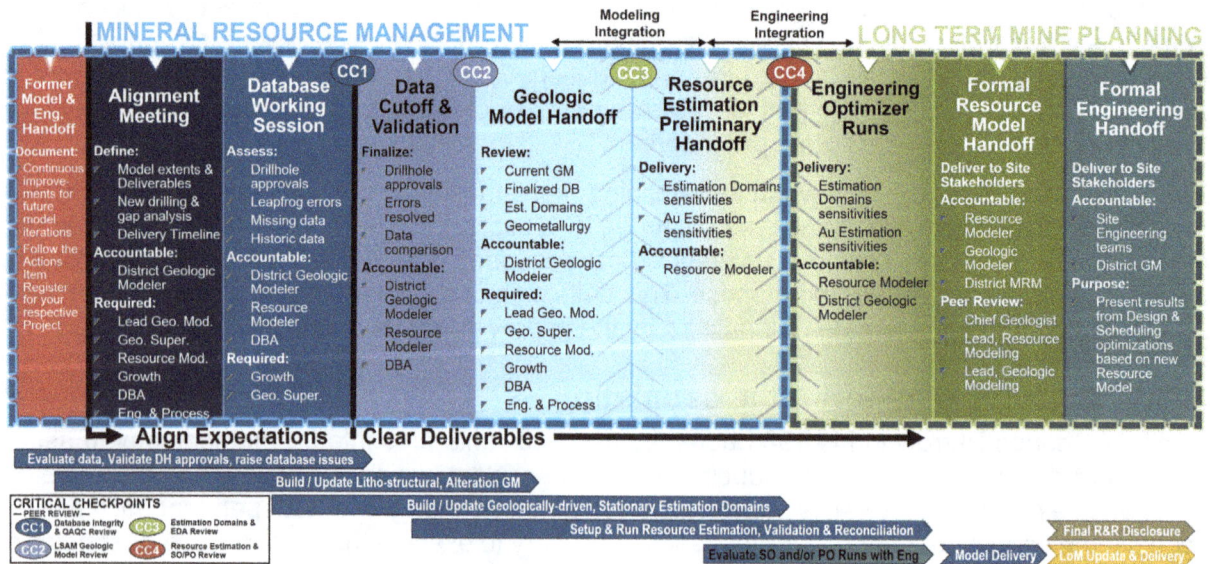

FIG 1 – Simplified flow chart of the Model Planning Process at NGM.

A model Alignment Meeting is required prior to commencement of work on any project. The purpose of this is to clarify across the project team the purpose and intent of the model update. Additional considerations to factor are areas where we expect to potentially see material changes including, but not limited to, disappointing conversion rates or exploration results, metallurgical recoveries or other parameter changes, significant changes to the geologic interpretation, changes to geotechnical zones or deviation to geologic model.

Internal peer review plays a crucial role in maintaining model quality. Resource geology superintendents work alongside their respective teams, assisting with decision-making during model construction. This collaboration not only fosters skill development within the team but also helps identify potential issues early, preferably before they arise. Routine check-ins review detailed estimation decisions to ensure that the model is progressing as planned, is fit-for-purpose and at an acceptable quality, while ensuring compliance to internal corporate standards. Any concerns are addressed before the model proceeds to the subject matter expert review of critical checkpoints.

Once the model reaches the signoff stage, it ensures that all stakeholders are informed of any updates or changes made to the model. Effective communication of these changes helps prevent the inadvertent use of incorrect model files in geology and mine planning work, thereby reducing the risk of errors. Finally, an external review and audit are conducted approximately every three years to independently assess the model. This review supports the disclosure of Reserves and Resources and may be combined with the review of other aspects of Reserve and Resource Reporting, further strengthening the model's accuracy and reliability.

CRITICAL CHECKPOINT SYSTEM

A ten-point critical checkpoint system was established in 2023 to help standardised our approach to internal peer review emphasizing a dynamic approach as the estimation progresses. Due to the scale of NGM operations, with over 22 model updates converging within a four-month window, the likelihood for three subject matter experts to complete detailed peer reviews was near-impossible. While detailed review is not always necessary, several significant changes in our internal standard practices resulted in material changes to workflows requiring assessment (Clark et al, 2025).

The benefit of being dynamic is that it compartmentalises the workload as different projects and people progress at differing rates to accommodate a staggered approach to peer review. The idea is to encourage 100 small peer conversations during critical moments of the resource estimation process, encouraging deep reflection, and providing near-live feedback. By keeping a finger on the pulse throughout model development allows streamlining of the approval processes while engaging in one-on-one coaching and mentorship without the pressure and intensity of a multi-day deep-dive peer review of a completed model that usually occurs adjacent to a final deadline.

The framework covers the end-to-end resources and reserve cycle, with five critical checkpoints for both Mineral Resource Management and Long-Term Planning teams. Beyond the review efficiencies, the primary philosophy is to instil a culture of critical thinking by empowering the teams with a level of autonomy to control their estimation and validation choices. Through a dynamic approach of review as the estimate progresses, it not only encourages a series of estimation sensitivities to rationalise choices, but it also catches potentially material changes and/or flaws early. This allows enough time to support any training or guidance necessary to close the action item out, formulate stakeholder communication, and/or a revised estimation strategy.

Each critical checkpoint was initially introduced and managed using a series of trigger points within a model documentation template. The downsides to managing it using this template was that it ultimately drove the opposite effect of its intended purpose. Rather than encourage critical thinking, it became a complacent exercise to populate hundreds of slides with a slew of geostatistical plots and data table summaries, of which, only a few plots revealed something interesting but could rarely be explained adequately.

Classic approaches to internal peer review are developing a tick-and-flick style checklist that constrains the reviewer to a few, focused items. After the initial enthusiasm wears off, complacency quickly kicks in and the success of that system to protect the business diminishes rapidly. There are benefits though, firstly, it builds an auditable trail of consistent checks and balances of assumingly material items. And it avoids a second example of complacency, which is no system at all. A minimum checklist of critical items that must be reviewed is an important aspect to help govern and guide critical checkpoint reviews and ultimately provides a consistent trail of supporting documentation that basic checks and balances were conducted.

A review of the process drove two fundamental improvements, the introduction of a critical thinking scorecard, and to standardise the critical checkpoint framework by creating a web-based tool embedded in Barrick's intranet with electronic approvals (Figure 2). Project team members are assigned to a designated role, which determines automated email notifications triggering required approvals at various stages.

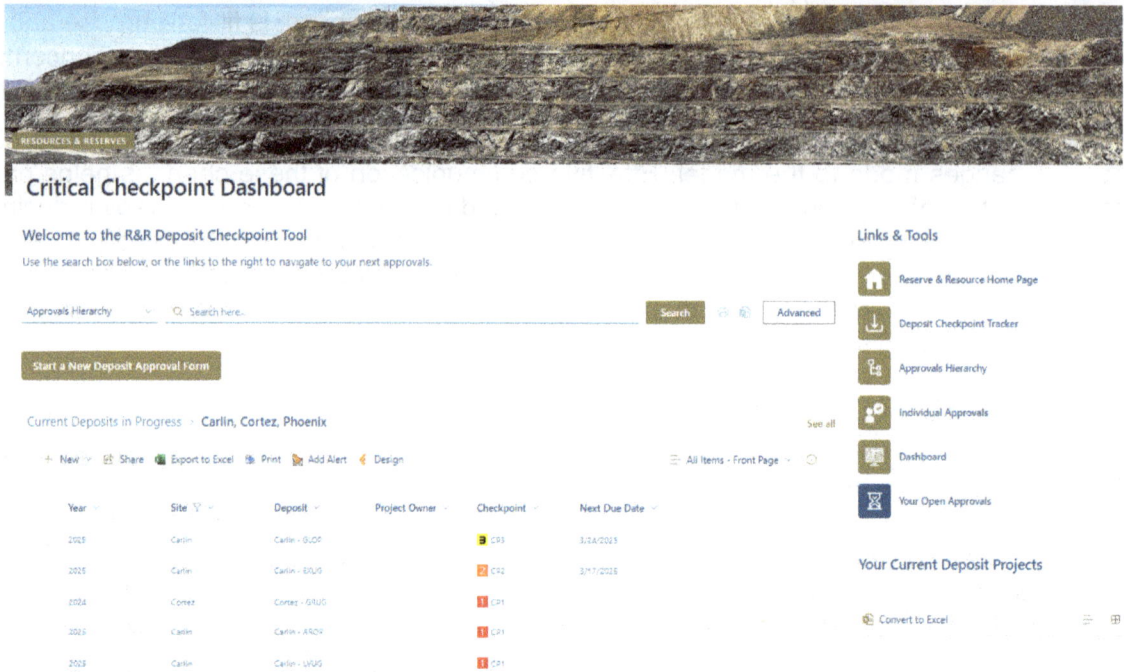

FIG 2 – The Critical Checkpoint Dashboard online tool embedded within Barrick's web-based server.

Figure 3 is an example of the designated role summaries with their respective approval responsibilities against each stage. Approvers are not required for each stage but receive an email notification as each stage progresses to remain informed. Management receives a high-level email summarising approvals to date with direct web-links to all relevant documentation.

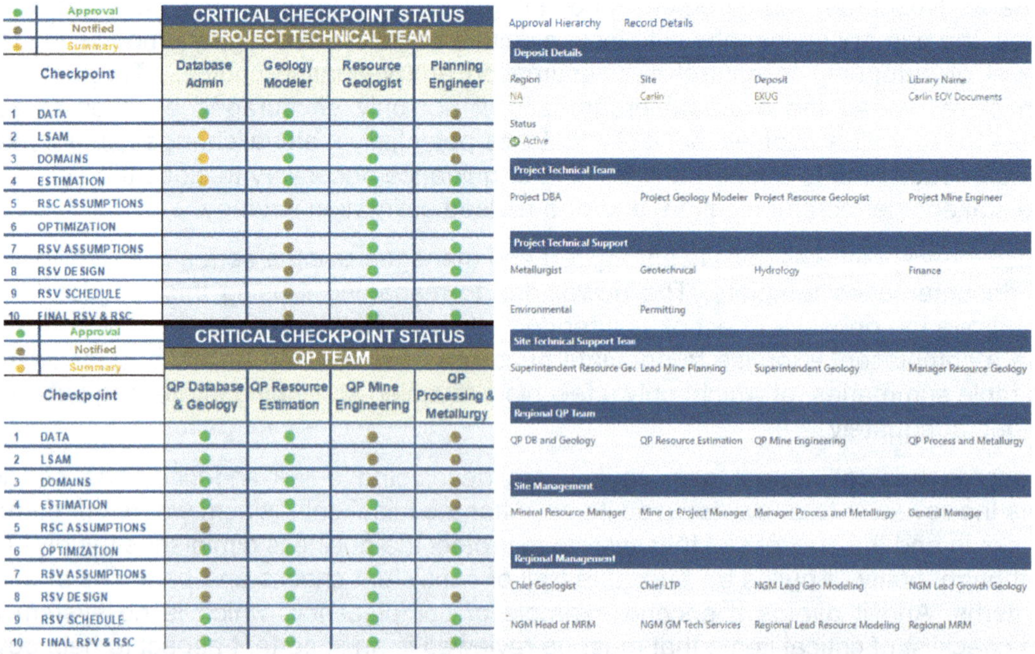

FIG 3 – Example of the Designated Roles and their respective responsibility against each stage with the final Approval Hierarchy that usually has an assigned user underneath each designation role.

Documentation must be uploaded to the system to progress each checkpoint to an approval stage. Once uploaded, documents are automatically stored on the Barrick server in a pre-determined location. Documentation includes written reports or summaries, meeting minutes and action registers, as well as any other relevant communications. Basic templates are provided per our internal Standards and Guidelines reflecting the minimum requirement. This enables an auditable trail for each critical checkpoint review and approval.

The critical thinking scorecard

The initial template-based system was too often ignored and the requests for decision rationalisation met with short, non-answers reinforcing a transactional culture of low accountability. The buy-in was not there. Thinking of the adage of when something is free, it is often ignored and considered someone's junk, but when you put a price on it, it is now considered something of value, and demand follows. To incentivise the process, a three-point scorecard was established to rate a series of questions generated for each critical checkpoint (Figures 4 and 5).

The scorecard is transparently accessible to the team and the questions are static. It is a process of demonstrating critical thinking and decision-making rationale. Every model update serves a different purpose, with various actions or improvements, meaning the questions can remain static as the focus shifts each time underpinned by different decision points. A minimum of three people comprise the panel that is scoring the project team at each critical checkpoint representing a combination of regional or site management and subject matter experts or QPs. The final score is an aggregate of the panel's score that is used towards individual's end of year performance rating weighted at 30 per cent.

The response was positive as it serves as an excellent tool to structure individual development plans and personalised coaching. If certain areas were more difficult to rationalise decisions, there is an opportunity to revise the score if you can take on the feedback and demonstrate it during a follow-up session, or in future updates after more structured training has been completed. As resource geologists are often introverted personality types, they may struggle to recognise or be self-aware of their performance, usually casting a negative light on themselves. This process allows full transparency without any surprises during performance discussions regarding technical capability on model delivery.

SCORECARD METRICS:
1. Presentation of minimum baseline standard list of items will score a 0/10 (it is the minimum requirement).
o Failure to show the full minimum requirements will see a negative or deduction in score.
2. A guide to scoring:
o [0] **Not Satisfactory.** Could not explain the key concepts of the requested item. E.g., cannot explain or justify a recommended approach to boundary conditions despite presenting the contact analysis plots or relating such plots to the geology.
o [3] **Satisfactory.** Can explain critical concepts of the requested item beyond the general components that is presented. E.g., can justify a recommended approach to boundary conditions but is not entirely sure how it relates back to the geology; however, is supported by the presented contact analysis plot.
o [5] **Exemplary.** Can explain critical concepts of the requested item with demonstrable confidence and critical thinking preferably with an introspection on the chosen approach and what we can do in the future to improve. E.g., while the contact analysis plot suggest a hard-boundary, on relating it back to the geologic controls it is obvious that Domain 1 and 2 show a diffusive relationship and the contact plot is an artefact from the domaining approach. However, Domain 3 is truly a hard-boundary due to x,y,z reasons. Backed up with 3D model images or live demonstration.

SCORE ANALYSIS:	
▪ 0 - 57	**= FAIL**
▪ 58 – 75	**= MODERATE PERFORMANCE**
▪ 75 – 100	**= GOOD PERFORMANCE**
▪ 100 – 115	**= EXEMPLARY PERFORMANCE**

FIG 4 – Critical thinking scorecard criteria.

MRM RESOURCE MODEL REVIEW ITEM CHECKLIST

NEVADA GOLD MINES OPERATED BY BARRICK

	Critical Checkpoint #1	Critical Checkpoint #2	Critical Checkpoint #3	Critical Checkpoint #4
Model:				
Resource Geologist:				
Reviewer/s:				
Date/s:				

#	ITEM	SATISFACTORY? Y/N	Score — If NO, can be mitigated by (date)?	REVIEWER COMMENTS
	CC #1: Database Integrity and QAQC			
1	Describe data collection methods (collar, DHS, assay) are appropriate and analyses are representative for all critical variables (e.g. econ metal, geomet, density, alteration)			
2	Describe historical data risks and our level of confidence related to our approved resource extraction is appropriate for use			
3	Present the spatial extents of approved resource drillholes colored by new vs old data. Describe how these new holes impact the LSAM controls, domaining and estimation.			
	CC #2: Geologic Model and LSAM Controls Review			
4	Describe the LSAM controls of the project, how it has evolved since the previous model update, and present the material changes to the geologic model that impact the estimation			
5	Present visual section comparisons of material domains comparing old model vs new model. Describe the observations against geological context and discuss the potential implications.			
	CC #3 Estimation Domains and Exploratory & Spatial Data Analysis			
6	Describe the approach to estimation domaining for all variables. For economic metals, describe how the domains represent LSAM controls, growth upside representation.			
7	Describe the basis for geometallurgical domains, how they relate to the LSAM controls or NOT and reasoning, the robustness of the mineralogical inputs and how they link to the metallurgical recovery and process routing allocations (Include all related variables)			
8	Describe how the density domains are generated and if density is assigned or estimated and what is it linked to (LSAM). Include areas for future improvement.			
9	Full Estimation Domain statistics review describe how estimation domains are derived, then show both raw and composited data and including mean, variance (Std Dev) and associated CV. Describe compositing decision and methodology. Provide full discussion within context of estimation domain materiality (focus on main ore domains and less focus on waste domains). Present contact analysis and interpretations for how estimation domains relate to one another.			
10	Plot and describe all Histogram and Cumulative distribution plots AND any relevant additional supporting plots. Include full discussion of the observed distributions and how Top Cuts and High Yields values have been determined (if mirroring previous values comment on their appropriateness for the updated distributions). Include Geological context for ALL boundary, yielding and top cutting decisions. (Include all variables)			
11	Present variography analysis and model for each Domain. Include motivation for chosen orientations and provide geological context for the modelled sills, ranges and the orientations. (Include all variables)			
12	Describe Dynamic Anisotropy methodology and controls. Relate back to geological context (include all variables). If not applied motivate to support decision.			
	CC #4: Final Estimation Review			
13	Describe the applied Parent Block size. Include any analysis or sensitivity informing the decision (e.g. KNA). If related to the SMU, discuss the impact on smoothing / resolution of the estimate. Describe the sub-cell approach and wireframe volume representation (block volume vs. estimation domain volumes).			
14	Describe the applied Search Neighborhood. Include any analysis or sensitivity informing decisions (e.g. KNA).			
15	Describe Discretization approach and include analysis informing decision (e.g. KNA).			
16	Describe Estimation technique/s and provide motivation for selection. Include support for applying different techniques for different domains (and/or variables). Compare to previous approach and if no change then demonstrate that previous techniques are stll relevant for current estimation domains and data distributions.			
17	Present Classification approach. Describe how geological and statistical behavior have been incorporated (Domain resolution)			
	Model Validation			
18	Tabulate block model against declustered composite data at Domain resolution. Describe observations.			
19	Present visual section comparisons for material domains (Block model versus composite data). Illustrate the behavior of the high- and low-grade estimate portions and describe the observations against geological context.			
20	Plot Domain declustered plots and histograms. Explain observations. Describe departures of the estimate against the produced DGM.			
21	Produce Swath plots at domain resolution. Describe observations and explain any observed bias.			
	Model Reconciliation			
22	At Domain resolution, compare Grade, Tonnes and Ounces of produced model against previous model (G-T curve). Also, tabulate model comparisons by Resource Categorization. Fully describe differences against new model inputs.			
23	Compare new model against production performance (data/ model) for the past 6 / 12 months. If in a multi-feed process plant change only the 1 model and maintain all other feeds and show the impact/change on the production reconciliation. Describe differences and observations.			
	FINAL PERFORMANCE RATING SCORE	0		Maximum possible score = 115

FIG 5 – Critical thinking scorecard question sheet.

CC#1 → 4 – integrity of the mineral resource estimate

The first four critical checkpoints are intended to dynamically peer review critical inputs that will inform the mineral resource estimate (Figure 6).

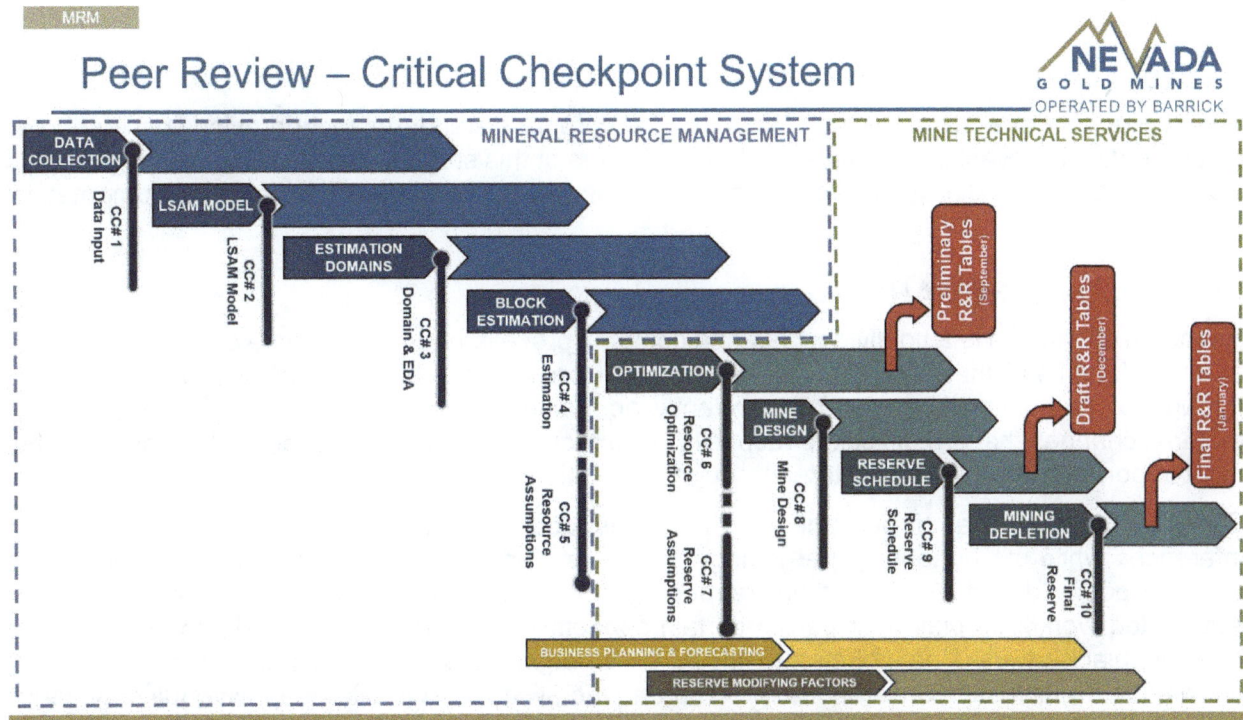

FIG 6 – Overview of the Critical Checkpoint Framework from 1 through 10.

The resource geologist will conduct a review of the resource database extraction, including quality control checks based on internal standards in place for spatial and assay validation, and if satisfied, request a critical checkpoint #1 to initiate. The QP Geology and Database is the responsible approver and will coordinate a time to conduct the required due diligence.

The work continues once the first checkpoint is approved. The next two critical checkpoints focus on the geologic model updates, with #3 aimed exclusively at reviewing the estimation domain updates, changes and rationalisation behind the geologic interpretations. Part of this will also be a review of the exploratory data analysis and variography as it is intimately tied into the decision-making process during domain construction.

The final critical checkpoint is to review the estimation parameter and kriging set-up in contrast to previous models, the orebody knowledge and drilling garnered since that last model, and if there were fundamentally changes in the geological model or during review of the geostatistics. If there are material changes to the resource estimate, a detailed model review report is prepared by the QP prompting a comprehensive independent peer review. Resultant actions are shared once the review is completed, and if anything is material, the model delivery will wait until rectified.

Once on-track for approval, we will finalise the remaining questions on the critical thinking scorecard to calculate a final score and provide final feedback to the resource geologist. If necessary, an update to their individual development plan may be an action to close out any identified gaps discovered during the process. The next stages are preliminary optimisations using resource assumptions to understand the potential impacts that will be realised in either the resource and/or life-of-mine plan. This is an iterative process where the model is in a preliminary state until final optimisations have been reviewed and the model deemed fit for purpose.

CC#5 → 9 – resource and reserve assumptions and optimisations

Critical checkpoints 5 through to 9 are primarily reviews set-up by the Reserve QP to first review the resource assumptions (5), of which we will use to inform a preliminary tabulation of expected resources and reserves, the resource and reserve optimisation shapes (6) alongside reserve assumptions (7). At this point, we have a draft resource and reserve tabulation to begin deeper-dive reviews and compile reconciliations. These steps also complement the final approval of the model, which once approved, will be used to construct an updated life-of-mine plan, which once completed is the product to review in (8). Checkpoint (9) is the final wrap up the reserves before final declaration and public disclosure.

CC#10 – final approval of outputs

This process is more closely aligned to our internal corporate standards for approving final resources and reserves. Relevant Barrick QPs are key approvers at this stage, which involves working through a series of actions and requests from external auditors. As well as review if technical documentation, if relevant.

CONCLUDING REMARKS

Resource estimation is a tightly coupled and complex process often relying on a diversity of experts to help inform its ultimate use. Practitioners are required to make hundreds of sensible judgements throughout its construction that independently appear harmless but can quickly compound into material changes that are difficult to reconcile. This requires a deep level of orebody knowledge but most importantly, ownership, fostered through a critical thinking mindset.

The qualitative nature of geologic interpretations coupled with the mashing of geostatistical inferences while still honouring the geology is no easy feat, especially in complicated geometries that are poorly drilled. Avoid getting trapped by the dogma that is the improvement paradigm. Automated workflows play their part in certain aspects of estimation, but not all, so always avoid creating black-boxes. The point is to focus on what matters, which is building deep orebody knowledge, to ultimately drive an interpretation an inch-wide, and a mile-deep, with full confidence of the final products, every time, slowly improving the entire deposit over time.

The evolution of national reporting Codes motivates commodity-focused mining companies to continually improve corporate governance practices, not only for regulatory compliance but to protect its primary assets of Mineral Resources and Ore Reserves. It is, however, a largely self-regulating system, often determined exclusively by the QP, who is responsible for the accurate and transparent compilation of such. By implementing a structured peer review process through the critical checkpoint framework, it takes the bias away and introduces focused feedback and review.

The ongoing challenge is to continue to instil a culture of critical thinking across all teams. It is a core skill and requirement for mining professionals who are paid to solve complex problems through their expertise and experience. We have adopted a back-to-basics approach to not only estimation but also a commonsense approach to peer review ensuring adequate due diligence. It is ultimately underpinned by the geologic interpretations that support estimation domaining. Through these efforts, we will destroy the decision paralysis and rebuild confidence in teams as they are more empowered to make and rationalise decisions with a strong support network by their side.

REFERENCES

Clark, J M, Samson, M, Le Cornu, C, Fiddes, C, Vandelle, M, Williams, A and Roux, M, 2025. Rethinking the world's largest gold producing complex through geologic improvements to mineral resource estimates, in *Proceedings of the Mineral Resource Estimation Conference 2025*, pp 45–56 (The Australasian Institute of Mining and Metallurgy: Melbourne).

McKinstry, H, 1948. *Mining Geology* (Prentice-Hall: Englewood Cliffs).

Noppe, M, 2017. Improving Assurance for Mineral Resource and Ore Reserve Estimates and Reporting, in Proceedings of the Mining Geology Conference 2017 (The Australasian Institute of Mining and Metallurgy: Melbourne).

Perrow, C, 1984. *Normal Accidents: Living with High-Risk Technologies* (Basic Books: New York).

Rutkowski, A-F and Saunders, C, 2018. *Emotional and Cognitive Overload: The Dark Side of Information Technology,* 1st edition (Routledge). https://doi.org/10.4324/9781315167275

Rethinking the world's largest gold producing complex through geologic improvements to mineral resource estimates

J M Clark[1], M Samson[2], C Le Cornu[3], C Fiddes[4], M Vandelle[5], A Williams[6] and M Roux[7]

1. Regional Chief Geologist, Nevada Gold Mines, Elko NV 89801, USA.
   Email: jesse.clark@nevadagoldmines.com
2. Principal Geostatician, Nevada Gold Mines, Elko NV 89801, USA.
   Email: matthew.samson@nevadagoldmines.com
3. Manager, Resource Geology, Nevada Gold Mines, Elko NV 89801, USA.
   Email: christopher.lecornu@nevadagoldmines.com
4. Lead, R&R Governance, Nevada Gold Mines, Elko NV 89801, USA.
   Email: craig.fiddes@nevadagoldmines.com
5. Group Resource Geologist, Barrick Gold, Africa and Middle East.
   Email: mathias.vandelle@barrick.com
6. Mineral Resource Manager, Carlin Complex, Nevada Gold Mines, Elko NV 89801, USA.
   Email: adrian.williams@nevadagoldmines.com
7. Regional Lead, Resource Geology, Barrick Gold, Elko NV 89801, USA.
   Email: mark.roux@barrick.com

ABSTRACT

Nevada Gold Mines was established in 2019 as a joint venture between Barrick Gold (61.5 per cent), the operating partner, and Newmont Corporation (38.5 per cent), consolidating over 15 operations that produce around 3 Moz of poured gold per annum, supplanting Nevada Gold Mines as the largest gold producing complex in the world. The prolific Carlin Trend has produced over 100 Moz of poured gold since discovery of the archetype carlin deposit in 1961. With about 100 Moz of Mineral Resources, and several projects awaiting various stage-gates, this complex offers significant organic growth opportunities.

The technical challenges of bringing this venture together were immense. The amalgamation of three company cultures within a short period of time introduced significant change to a team of over 180 geologists. Though, the consolidation of Trends offered an immense opportunity to unify the geologic understanding of Carlin-type deposits. Recognising this, a concerted effort in geologic modelling in conjunction with aggressive organic growth drilling has enabled a paradigm shift in the structural understanding of mineralisation controls. This led to a shift in understanding the complex tectonic evolution of Nevada that dictates the local orebody geometries observed. The principal controls on mineralisation are intersection lineations at the confluence of favourable lithology and/or fold hinges.

Value is not realised until it is built into a mine plan. Armed with a new geologic foundation, the focus shifts to integrating detailed orebody knowledge into resource models to improve mining plans. With sustainability in mind, simple Ordinary Kriging workflows were preferred as the minimum baseline standard. The focus is always on geologically driven estimation domains and trend interpretations to support dynamic anisotropy. After 18-months, all workflows have been completely overhauled compliant to regional standards.

This paper will discuss the technical and cultural challenges in pursuing the significant changes, as well as the methodologies to workflows adopted to reinforce the geo-centric mindset that we pursue.

INTRODUCTION

The strength of globally mature tier one complexes are also its biggest weakness, that is, large tonnages of high-grade material often hide the sins of complacent resource estimation and mine planning practices. Technical innovations necessary to bring these deposits into production likely occurred decades ago. The consequences of which are often reflected in process-driven mining practices where teams become trapped by institutional dogma.

While the giant carlin-style gold deposits in north-east Nevada are not immune to tier one afflictions, Randgold Resources identified the obvious synergies of unifying these deposits when they merged with Barrick Gold in 2018, and in July 2019, Nevada Gold Mines LLC (NGM) was born from a joint venture deal between Barrick Gold Corporation (61.5 per cent), the operating partner, and Newmont Corporation (38.5 per cent). This consolidated the tier one Carlin, Battle Mountain and Getchell gold districts, supplanting NGM as the world's largest gold producing complex, with over 15 open pit and underground operations and half a dozen projects awaiting stage-gates.

The initial focus beyond obvious infrastructure, mine and processing synergies, was unifying the holistic geologic understanding of carlin-style deposits. A back-to-basics approach focused on the deposit-scale lithology, structure and alteration characteristics was undertaken to explain the primary controls on mineralisation and challenge existing interpretations. Geological models for each deposit were created utilising all data sets across the region, serving as a foundation for organic growth upside as well as a framework to refine local controls within grade control models at an appropriate resolution to the respective mining methods.

Significant advancements in orebody knowledge introduced potentially material changes to orebody geometry and grade distributions that were not reflected in the resource models. An 'inch-wide, mile-deep' approach to reviewing 20 projects proved difficult as there was no standard approach to estimation with wide time gaps between documentation or internal peer review. The findings ranged between probably okay to potentially troublesome signalling a similar approach to overhauling the geologic models was required.

This paper documents the challenges and opportunities of rethinking the world's largest gold producing complex. A set of regional minimum standard practices were established adopting a sustainable workflow flexible for most epigenetic ore deposits. An extensive training program was developed to refocus teams on geologically driven estimation domaining, accurate mineralisation trends to support dynamic anisotropy and simple ordinary kriging workflows underpinning a practical, fit-for-purpose product focused on quality over complexity (Figure 1).

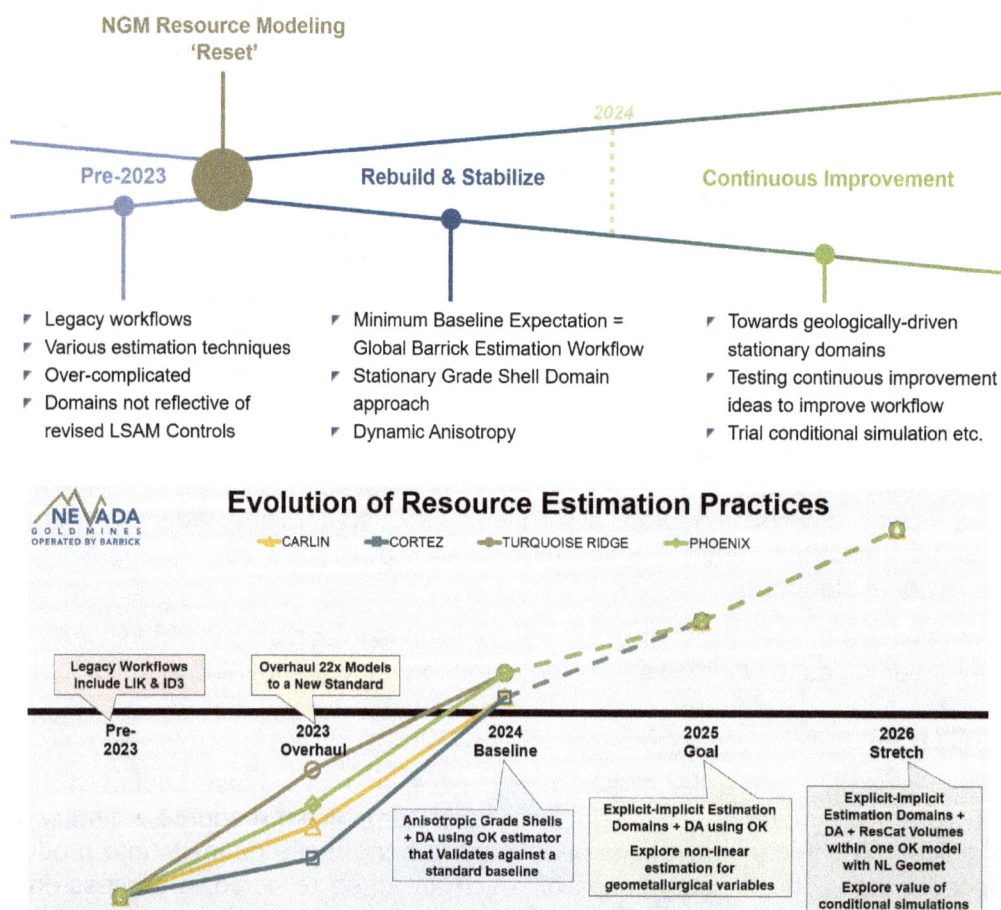

FIG 1 – Vision and strategy presented to the NGM team in 2023 on how to forge a pathway towards a newly defined regional standard versus how we are tracking today for each Tier One District.

A geologically driven asset management culture, together with accurate models and iteratively testing new growth concepts, underpin our successful reserve replacement. Cumulative gold production between 2019 and 2023 was about 20 Moz of gold poured (Barrick Gold Corporation, 2020–2023), of which every ounce mined was replenished securing our Mineral Reserve base, also increasing total Mineral Resources to over 80 Moz (Barrick Gold Corporation, 2020–2023).

GEOLOGICALLY DRIVEN ASSET MANAGEMENT

Mineral resource management is an organisational structure that underpins Barrick's core focus on geologically driven asset management, central to achieving the reserve replacement record but also to unlocking the organic growth portfolio. A unified team fostering a geo-centric culture of critical thinking empowers geologists to be true custodians of the orebody, integrated across the full mining value chain. The role is to optimise the *quality* of the Mineral Reserves we mine, balanced with a growth pipeline to support the Mineral Resources we define.

Geologists are employed to predict uncertainties through the creation of three-dimensional interpretations of orebody geometries that explain internal grade distributions of varying geologic confidence. These are represented within a resource model. The quality, accuracy, and relevancy to the mining method is the responsibility of every geologist on that project, no matter how perceptibly big or small their contribution is to the final inputs. To achieve this requires a deep level of orebody knowledge that is only realised through an iterative process of scaled geologic interpretations of modelled mineralisation controls.

Deep orebody knowledge is as much an art as it is a science, only achieved through continuity of project ownership over a significant period, with a paranoid focus on the nuances of hundreds of finer details. It is a significant investment and arguably the most valued asset of a commodity-focused mining company—intellectual property. It requires a fine balance of characterising geologic features in enough detail to be able to quickly filter what is most practically relevant to optimising the mining plans or reserve modifying factors as variances arise. In doing so develops a deep sense of ownership of the model, and the orebody itself, further driving a positive feedback loop of continuous improvement, as no model should ever be static.

The LSAM approach

The complexity of NGM is that deposits are close enough for mining, processing and infrastructure synergies yet too far between deposits warranting isolated technical services teams to support each mine. A concerted effort to consolidate the geologic understanding of carlin-style gold deposits in 2020 revealed vast orebody knowledge gaps across each districts reflecting a vast network of sub-cultures. Each deposit had developed their own modified set of standards and practices independent to whatever company policy may have existed in previous decades.

Adding to these challenges is a significant addition of new data since the creation of NGM about five-years ago. Over 1500 km of new drilling data, and at least 3000 km of open pit and underground mapping observations were captured. This volume of data drove several paradigm shifts in geologic thinking and subsequent interpretations that required several iterations before fully understanding the impacts.

A back-to-basics strategy to drive a geo-centric mining business at NGM led to the development of the Lithology – Structure – Alteration – Mineralisation (LSAM) approach. This remains a core function of NGM, successfully embedding geology through the mining value chain with LSAM at the heart of every discussion with all stakeholders. A simple yet effective approach to reinforce the essential components controlling mineralisation, precipitating an overwhelming cultural change impacting well over 180 geologists.

A geologic summary of carlin-type deposits

Carlin-style mineral systems form some of the largest gold deposits on earth offering a unique combination of hydrothermal characteristics dissimilar to most epigenetic deposits. The deposit class has been extensively researched with detailed geologic descriptions provided by Cline *et al* (2005) and Muntean *et al* (2011). Rhys *et al* (2015) provide an eloquent review of the structural evolution of Nevada in context of local mineralisation controls and orebody geometries.

Research efforts in the 1990s through 2000s focused on understanding the Ordovician to Devonian carbonate host rocks that were deposited along a passive margin on the western margin of the North American craton. Cook and Corby (2004) identified shelf- and platform-facies rocks as important host rocks characterised by highly permeable, mixed carbonate-siliciclastic compositions. Efforts paid off with several discoveries targeting along-strike interpretations of favourable host rocks, most notably the tier one Goldrush-Fourmile deposit (Bradley and Eck, 2015).

Deposit characteristics are too often simplified to high-angle faults with oblate mineralisation zoning out into favourable sub-units (Robert *et al*, 2007). Though the first-order controls of carlin mineral systems is the underlying tectonic architecture that controlled the depositional environment of sedimentary host rocks with high-grade mineralisation locally associated with imbricate sets of thrust-propagated inclined to recumbent anticlinal folds or thrust-ramp geometries reflecting over 200 million years of protracted deformation (Rhys *et al*, 2015).

Carlin-style deposits are characterised by two fundamental scales of mineralisation observed:

1. Discrete, prolate zones of hyper-focused fluid-rock interaction generating intensely decalcified rocks and dissolution breccia at favourable structural intersections eg intersection lineation of an axial plane of a recumbent fold and a favourable sub-unit contact.

2. Laterally extensive, oblate mineralisation that is typically lower grade than (1.) and is dominantly controlled by favourable stratigraphy nearby a feeder fault of some description, and may be disseminated eg it is common to observe over 6 miles (10+ km) of mineralisation continuity above 3 g/t in all districts.

Geologic modelling

Geological interpretation and modelling is based on internal NGM Guidelines that outline standard practices for the construction, maintenance and version control of a three-dimensional geological model using the Seequent Leapfrog Geo software package. Each model adheres to a detailed review process performed by NGM Subject Matter Experts to ensure geological integrity and compliance to Guidelines.

NGM uses a server-hosted database, to store, share and review all geological models that have been constructed in 3D modelling software. Each project retains a chain of custody version control showing each published change, the user, project status (Draft, Ready for Review or Peer Review) and the date it was uploaded. Server-hosted projects enable multiple users to work on a model at one time while retaining model integrity. Subject Matter Experts are regional administrators that can grant or refuse access to any user at any time, as well as change project permissions including read or write access.

Construction and maintenance of representative geologic cross-sections is required at each deposit that adequately describes the underlying geologic framework and controls on mineralisation. Interpretations are supported by geologic observations (both drill holes and mapping), multi-elemental assay data, and field verification. Once approved, representative sections are expanded into a three-dimensional lithostructural geological model.

A comprehensive visual review of the updated drill hole and/or mapping database, including a comparison to previous extractions, is required prior to its incorporation or use in the geological model. A focus on the spatial validity and verification of collar and downhole survey data is paramount to ensure data integrity. Any issues are raised to the database teams using a Digital Service Request system for further investigation to include or exclude from future extractions.

Mature districts like the Carlin Trend often have inconsistent logging databases as hundreds of geologists have contributed data reflecting periods of changing geological thinking or priorities. Initial lithological models were built utilising the interval selection tool in Geo as it enables the user to modify or reinterpret the lithology logs onto a new table thus preserving the original data. Additionally, if an existing model from a different software or external source exists, it can be evaluated on to the new lithology field.

In 2021, each deposit underwent this process to produce a foundational litho-structural model using the reinterpreted selection field. Geological Models were constructed mostly using the 'deposit' interpolation method in Geo, locally manipulating contact geometries using explicit points to honour the geological interpretations in areas of low data support. Deposit-scale structures were modelled and those with discernible offsets were activated.

Deposits with active operations are prioritised for further refinements of the geological model to honour local geologic controls on mineralisation closer to the scale of the selective mining unit. This involves translating the lithology selection table into a drill hole correlation table to allow metre-scale refinements using lithogeochemical signatures. Local-scale structures gleaned from open pit or underground mapping are modelled as a separate structural model to appropriately restrict the strike continuity to relevant geological regions and minimise the number of active fault blocks.

Large deposits with many sub-regions manage each refined model as a separate branch so multiple geologists can manage the workload. These models undergo the same rigorous model reviews as the foundational models. As models are further refined, the more explicit functions are used to control the modelled outputs to the geological interpretation.

Extensive model validation is conducted annually to ensure model integrity and general representativeness to the input data sets. Standard practices include proportional volume comparisons of lithology units using a nearest neighbour estimate as a baseline for comparison against modelled outputs. Additionally, swath plots in northing, easting and depth at regular increments are also completed to understand major changes spatially. This can be easily visualised using the Combined Model function in Geo that creates a common, gain and loss volume in reference to the old model.

Mineralisation trend analysis

A common misnomer is that estimation domains are interchangeable with geologic domains, yet for most epigenetic hydrothermal deposits, especially carlin-style, it could not be furthest from the reality. (Le Cornu *et al*, 2025). A simple tool to test this theory is looking at enhance values in Geo, otherwise known as maximum intensity projection, or introduced by Cowan (2014) as X-Ray plunge projection. This allows the user to highlight the third quantile of the global data distribution through the sea of lower-grade values essentially lifting it out of the page. The simple exercise we workshop regularly with teams is to interpret structural trends using discs in Geo when aligned to a down-plunge view of a local control or trend. The result is typically <20 trend surfaces that you can load against modelled geologic features to see how many are explained; it is typically <5 out of the 20.

The concentration of hydrothermal mineralisation is the result of multiple geologic occurrences confluencing. A simplistic perspective is to consider a mineral systems approach in the context of geometric controls, which in the case of carlin-style deposits, represents a feeder-style structure that transports fluid from an unknown source upwards until reacting with a favourable host-unit. The presence of ferrous iron triggers a sulfidation reaction of the fluid ultimately trapping gold within the crystal lattice of hydrothermal pyrites. Sealing this system are non-reactive rocks, typically within the thrusted Roberts Mountain Allochthon (Muntean *et al*, 2011).

Individually, each LSAM category offers a unique set of ingredients that must react with one or all the others forming a Venn diagram of controls (Figure 2). The approach informs that one or several sub-facies are the best host but is not uniformly mineralised, as it must have ferrous iron to react with. Adding to this is the proportion of carbonate to siliciclastic components where it requires a proportional balance, otherwise the mildly acidic fluid is buffered by too much carbonate or does not propagate into the rock when too many siliciclastics. Variably orientated structures are a focusing point to the favourable stratigraphy but can be mineralised themselves as they may also host pre-mineral dyke swarms with reactive ferrous iron. Lastly is the intensity of decalcification which is intimately linked to the former two categories but may also develop intense silicified brecciation within the core hosting the highest grades.

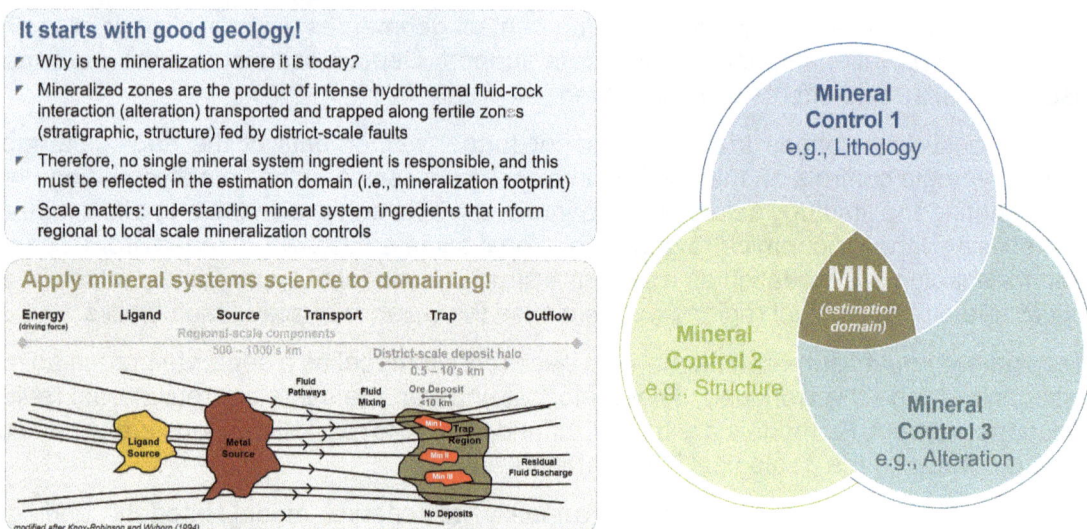

FIG 2 – A simple exercise to link back LSAM observations of mineral controlling features to a minerals system approach. Both complement the notion that estimation domains are the consequence of multiple geologic occurrences and thus, no single modelled geologic domain is truly representative of a mineralisation domain in epigenetic carlin-style deposits.

Mineralisation trend planes are not only critical to support the local anisotropy of implicit estimation domain models, but also to inform the estimate using dynamic anisotropy or otherwise known as locally varying anisotropy. The azimuth of the plane is flagged on each block it intersects with various set-ups to ultimately control its influence on the estimate. It is a pre-requisite to model all mineralisation trends as planes before starting the estimation domain process as it usually informs the user that there are many *known-unknowns* to address.

Estimation domains

A critical output of the geological modelling process are estimation domains of mineralised zones. Estimation domaining is based on internal guidelines that outline minimum standard practices of the construction, maintenance and version control of a three-dimensional estimation domain model using the Seequent Leapfrog Geo software package.

Estimation domains for use in grade estimation are geologically driven implicit models that require significant explicit control points to ensure the resultant geometry accurately honours the local mineralisation controls. The minimum standard to achieve this is to utilise implicit grade shells in Geo, though, some deposits have evolved to using an intrusion or vein geological model method to dominantly have explicit control.

Each deposit determines grade thresholds for estimation domains through visual analysis of the grade distribution in three-dimensions supported by exploratory data analysis. Analysis includes reviewing inflections in cumulative distribution frequency plots to help focus on specific grade ranges against local geological controls. A low-grade volume is constructed to isolate the mineralisation footprint that is inclusive of the main mineralised zones. Most deposits also utilise an internal high-grade mineralisation domain to constrain discrete mineralisation controls, such as breccias and complex structural intersections. Exceptions occur within local sub-domains of a deposit that may use single or additional thresholds to better honour the local geological controls.

Structural trends are constructed to represent the orientation of mineralisation controls, significantly influencing the final geometry of the domain. The construction uses a combination of existing lithostructural modelling products with additional structural inputs that align to the interpreted plunges of the local grade distribution. A series of estimation domain sensitivities are subsequently created to determine the appropriate strength and ranges for each structural trend, including the trend type.

In addition to structural trends, explicit points are used, particularly in areas of low data density, to help control the domain geometry to accurately honour the geologic controls and avoid model artifacts. All explicit functions are critically reviewed to not only minimise their usage but to reduce subjectivity while honouring the local interpretation. Parameter selections are typically standardised depending on the adopted Geo method and sensitivities are regularly explored.

Estimation domain logic and final products undergo an extensive Critical Checkpoint review process (Clark, Fiddes and Roux, 2025) to not only ensure compliance to Standards, but to verify that the project owner can adequately justify the volume against the interpreted mineralisation controls. Validation uses a suite of statistical analyses performed on raw and normal score transformed composite data, and an evaluation of summary statistics (including mean, variance, standard deviation, coefficient of variation, ranges and quantiles). This is balanced with visual measures such as histograms and probability plots. Contact analysis plots are used along with visual validation of the three-dimensional domains against the geologic understanding to guide the treatment of grades across domain boundaries.

Open pit deposits

All open pit deposits utilise an implicit grade shell approach using an Indicator interpolant method in Geo. An added complexity to these deposits is accurately representing the spectrum of material types including oxide leach, single, and double refractory ores. Leach usually has an economic cut-off grade close to the detection limit of a gold assay, so a very low-grade, global grade shell is constructed to not only constrain the mineralisation footprint, but to best represent the leach population without incorporating below detection-limit waste.

Low-grade leach domains typically represent oblate, stratabound mineralisation controls that is reflected in the lithostructural model. In the context of these deposits, favourable stratigraphic contacts are utilised creating a simplified structural trend interpretation constructed for the low-grade domain dominantly representing leach material types. Across the Carlin Trend, the chosen grade threshold to represent this domain is 0.1 ppm, as any lower approaches legacy detection limit issues, and higher typically exceeds leach cut-off grades. An alternative perspective is viewing this domain as the mineralisation footprint.

Once the grade thresholds are determined for each geological sub-domain, a series of indicator grade shell sensitivities are constructed for visual and statistical review before final selection. This is conducted for the remaining domain volumes, which ranges between two to three nested grade shells. Contiguous zones of moderate-grade that is typically refractory material is also common reflected by a medium-grade volume. These thresholds usually range from 1 ppm to 2 ppm while utilising locally attune structural trends to control the domain geometry.

All deposits maintain a high-grade volume utilising a grade threshold typically between 3 ppm and 7 ppm reflecting prolate mineralisation concentrated within the core of complex structural intersections. An example is at South Arturo, where structural trend planes were interpreted to follow the intersection lineation of two phases of recumbent folds that result in steep NW-trending grade plunges reflecting the mineralisation geometry well yet did not correlate to any single modelled lithology or fault.

Underground deposits

Underground deposits represent the most structurally complex and discrete zones of discontinuous high-grade mineralisation meaning domains are far more sensitive to local inflections in mineralisation controls and orientations. Additionally, they use a smaller selective mining unit further sensitising the importance of local accuracy. To control this reality requires a significant increase in explicit control points, as well as fine-tuning the local accuracy of structural trend interpretations. Thus, all underground deposits, including Goldstrike, Greater Leeville, Exodus and South Arturo, manage independent domain workflows for each geological sub-domain.

No global, low-grade domain of 0.1 ppm like open pits is used as there is no economic ore, oxide or not, that is mined underground at or near those grades. Each deposit has two domains, with one representing the relative mineralisation footprint that is typically around 3 ppm to 6 ppm, and the other, a high-grade domain nested within, typically using a grade threshold of between 7 ppm and 10 ppm. Each of these deposits used significant explicit control points on the high-grade domain ultimately shifting the expected population away from a strict grade shell approach.

Limitations and future improvements

The challenges with implicit grade shells (both indicators and RBFs in Geo) are plentiful. It requires a disciplined user to ensure the ultimate quality of the product is useable and geologically representative. Key limitations are firstly that they are at the mercy of data density no matter how much you adjust the parameters, meaning estimates tend to lean conservative and do not capture any growth upside in models unless explicitly commanded. And secondly, is a hyper-sensitivity to structural trends, also irrespective of the parameter selections, or explicit points, meaning if the trend is not locally nuanced to the interpretation, it will disaggregate into off-trend volumes.

The most material limitation to grade shells is that they are ultimately a proxy for a mineralisation controlling geometry not yet realised or geologically understood. As resource practitioners, we validate estimation domains by satisfying the assumption of stationarity. The issue is that the worst, most blobby and overly nested grade shells will satisfy a very low coefficient of variation sniff test. To avoid this, we introduced a secondary concept of satisfying 'geologic stationarity', that is, does it make sense geometrically in three-dimensions where two adjacent sub-populations aren't trying to globally connect using implicit 'handlebars.' And is there natural variance within the domain of samples that belong within the modelled sub-population, an example being a 12 m interval of lower-grade reflecting a truck-scale clast of limestone within a collapse-dissolution breccia host yet is part of that high-grade breccia sub-population.

The benefits of implicit modelling outweigh those of a traditional deterministic wireframing approach arguably for most epigenetic deposits but in this case, specifically for carlin-style deposits. Multiple sensitivities or hypotheses can be dynamically updated within minutes to less than an hour depending on the project size. Understanding the fundamental concepts is relatively simple for most experienced geologists and the software is intuitive making it easy to train teams fast. A grade shell approach is our minimum standard for these reasons; while accepting their limitations, they offer a relatively fail-safe, repeatable method that is easily auditable.

More recently, we transitioned towards geologically driven, dominantly explicitly controlled implicitly modelled domains at two of our flagship tier one projects (Figure 3). The initial workflow is similar in that the user determines appropriate grade thresholds that are evaluated onto an interval selection table flagged onto the drill holes. Geologists refine the domain intervals with metre-scale precision utilising all data sets in the drill hole correlation tool to ensure local accuracy to mineralisation controls. The resultant categorical variable is modelled as an intrusion geological model workflow that also utilises interpreted structural trends to honour local anisotropies, as well as explicit control points.

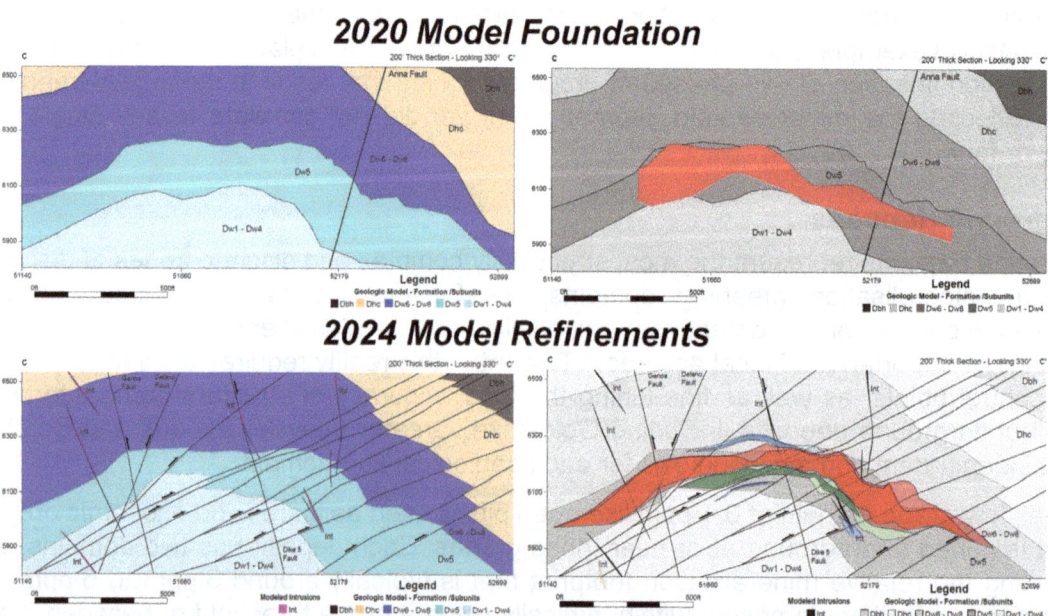

FIG 3 – Example of an explicit, geologically-driven estimation domain refinement and its evolution since a 2020 foundational model.

Estimation workflows

Two primary workflows were historically adopted across various NGM sites yet utilised in different enough ways to make the review process complicated. The first was a modified multiple-indicator-kriging workflow that used a localised ordinary kriging model to basically produce a recoverable resource model for each potential mineable panel. Developed over 25-years ago, this workflow remained largely static, nested within a dozen scripts that were challenging to the most seasoned resource practitioners reflecting the common 'black box' syndrome. Time and effort focused on unravelling complicated scripts to update lines of code, all while the underlying geologic model was too coarse a scale and fundamentally did not reflect the mineralisation controls leading to a somewhat flawed estimate, nonetheless.

The second common workflow was an inverse distance weighted to the power of three that utilised locally varying search to honour mineralisation controls and build growth upside into the models. The fatal flaw to these estimates was the intentional notion for the estimated distribution to match those of the raw assays producing locally inaccurate models. While these were mostly not used for grade control models during mining, these were the first models that we addressed as a critical priority.

The adopted new workflow is one of simplicity, heavily weighted to geologically-driven estimation domains and mineralisation trends, so practitioners focus on representing the mineralisation geometry and internal grade distribution. The absolute minimum baseline standard is globally nested implicit grade shells using various structural trends to help control their anisotropy, along with various parameter selections. Where appropriate, estimation domains are split to natural geological breaks where there is a clear change in the anisotropy/geometry to help reduce variogram noise. Though the preferred standard are strongly explicitly controlled implicit models where the estimation domains are manipulated by a geologist to represent the expected geometry.

Ordinary kriging is the estimator of choice, though, various non-linear estimation techniques compliment the standard workflow approach to solve a specific problem or explore continuous improvements (Samson *et al*, 2025). The key component through this process of change is encouraging lots of estimation sensitivities, for example, with sample selection, where multiple sensitivities were initially completed to help guide a decision when in doubt or spark discussion with the QPs.

An automated validation notebook was created in python. Validation of a block model should tell a story but to help frame the narrative, a standard set of plots and basic tabulations are required, and to help tell story. Figure 4 is an example of one standard plot, which is basically measuring how 'true' the estimate is while scaling it in normal score space factoring in the change of support correction. Generally unbiased models that appear to predict close to the 'truth' will form a 45° line on that plot. Conversely, where there is significant deviation, this usually implies an obvious error during the estimation set-up.

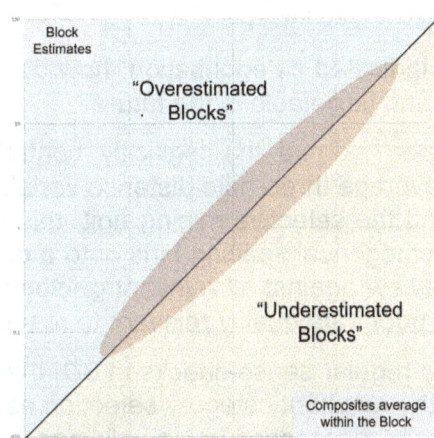

FIG 4 – Example of a standard validation plot that must be done but is not restricted to just this one graph.

Resource categorisation

Drill hole spacing studies support resource categorisation decisions as they quantify the value of information by relating revenue with relative confidence (eg Clarke *et al*, 2019). A reliable method is a conditional simulation of varying drill grids that result in a series of realisations to calculate a relative uncertainty for each block above chosen cut-off grade across varying drill hole spacings. A conditional simulation is only as good as the underlying geologic inputs and must be carefully considered. Moreover, the trade-off is that they're time consuming and require specialised resources to build the workflow correctly.

A Single Block Kriging study is a simple approach that computes an estimation variance of an attribute in a large production block given a set of samples to deduce a relative confidence (Verly, Postolski and Parker, 2014). This tool was exclusively used to define resource categorisation metrics across NGM. The biggest limitation is that results are highly sensitive to chosen quarterly or annual production volumes. The output is a series of drill hole spacings against confidence intervals. Moreover, it does not factor in other criteria such as estimation domain geometry, geologic certainty, data quality and assumes stationarity. The SBK workflow is designed to aid in resource categorisation decisions as it is fast and reproducible, but must be balanced by all other necessary criteria, and not abused by using the direct output as a literal definition.

The practical reality to resource categorisation is that local areas may have criteria that significantly influences the decision to class as Measured versus Indicated versus Inferred. Each of which may result in material impacts to Resources and Reserves disclosure, as well as life-of-mine planning. Examples of local criteria include consideration of geologic certainty of geometry and grade continuity, spatial clusters of modern data quality versus legacy areas of reverse circulation or conventional drilling. Such criteria cannot be represented in an automated workflow, especially not exclusively using data spacing, yet is typically recognised but ignored during final ResCat designation. Simplicity takes precedence.

Best practices across the mining industry maintain three-dimensional volumes that represent each categorisation. A non-exhaustive list of considerations include:

- Experience of drill hole spacing studies and reconciliation of similar deposits across carlin-style deposits to constrain the footprint to a reasonable data density. This will always serve as a core foundation.

- Orebody knowledge. Confidence in the geologic controls, grade continuity and physical geometry to support mine planning.

- Adequate data quality and three-dimensional data density to support the application of modifying factors for an economic evaluation.

- Within a 90 per cent confidence interval, will the estimate reconcile within 15 per cent over an annual production volume?

- If all the above support an Indicated categorisation, how do we convert this assessment into criteria that is reproducible and targetable in the future?

A simple workflow was developed akin to the explicitly controlled implicit estimation domain methodology. By estimating the average three-hole distance variable into an appropriate block size that factors both data support and the selective mining unit, this can be the basis for a resource categorisation volume. Flagging categorical spacing bins onto a drill hole interval selection table to begin the explicit manipulation review against all relevant geologic or metallurgical considerations where any query filters to restrict poor data quality relevant to resource categorisation is applied.

In appropriate section views, with regular sense-checks in 3D, the interval selections are adjusted accordingly. Implicit models use the resultant category selection as the base lithology to set contact relationships. Volumes are manually reassessed using polylines in section views and 3D to ensure geologic integrity. Once a final volume is generated, it is exported to a final block model for distribution (Figure 5).

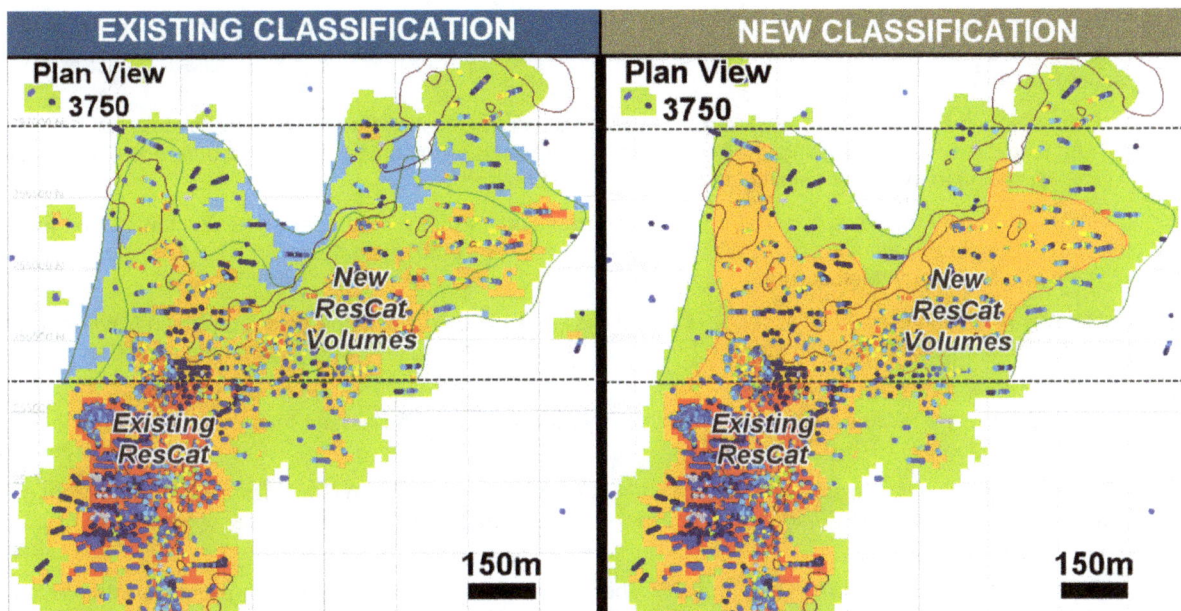

FIG 5 – Example of a spotted dog resource categorisation to considering the orebody knowledge through the explicit creation of a resource categorisation (ResCat) volume.

CONCLUSIONS

Tier one districts often suffer from tier one afflictions that are an oxymoron to the general perception that it automatically implies best practices. The reality is that poor practice becomes common practices that quickly propagates as a training manual for the next generation of geologists, who before long, is teaching a further diluted version of said practices. The onus is on all technical professionals to ensure we maintain geologic integrity through the highest quality three-dimensional models that underpin optimised mining plans. In mature operations, we create value through orebody knowledge to characterise quality as revenue-based material routing may surprise usual cut-off grade assumptions, as one example.

Nevada Gold Mines is the culmination of three mid-tier company cultures that were arguably immiscible. Through geologically-driven asset management underpinned by the Occam's Razor LSAM approach, we were able to drive significant cultural changes towards a geo-centric focus in everything we do with all stakeholders through the mining value chain. All while unifying a cohesive geologic understanding of carlin-style gold deposits that we are actively incorporating into growth targeting workshops to continue supporting our reserve replacement success.

The biggest changes within geoscience at NGM has been the resource estimation processes. While seemingly impossible, over 22 projects were overhauled within a two-year period, aligned to a new standard and underwent rigorous internal and external peer review and audits. Not all geologically modelled products are relevant to the estimate or mineralisation controls to that matter, but more significantly, it highlighted orebody knowledge gaps where trends were unexplainable. This focused fit-for-purpose interpretations that opened significant growth upside built into the estimates. It is a testament to a sustainable workflow that is dominantly weighted to geologically-driven estimation domains and mineralising trends that underpin a basic yet reliable ordinary krige estimate.

ACKNOWLEDGEMENTS

Hundreds of geoscientists have contributed to the understanding of the geologic evolution of carlin-type epigenetic gold deposits; we humbly stand on their shoulders. Special acknowledgements to Mark Bristow, Simon Bottoms, Grigore Simon, Tricia Evans, and Rod Quick for their supportive leadership and vision in driving a geo-centric culture across Nevada and appreciating the value of high-quality, three-dimensional models.

REFERENCES

Barrick Gold Corporation, 2020–2023. Annual reports. Available from: <www.barrick.com>

Bradley, M A and Eck, N, 2015. The Goldrush Discovery, Cortez District, Nevada – The Stratigraphic Story, in *Proceedings of the New Concepts and Discoveries, 2015 Geological Society of Nevada Symposium* (eds: W M Pennell and L J Garside), pp 435–452.

Clark, J M, Fiddes, C and Roux, M, 2025. Peer accountability and communication through a critical checkpoint framework – driving a culture of critical thinking, in *Proceedings of the Mineral Resource Estimation Conference 2025*, pp 35–44 (The Australasian Institute of Mining and Metallurgy: Melbourne).

Clarke, D, Wright, M and Minniakmetov, I, 2019. A practical assessment of drill hole spacing at Olympic Dam – juggling cost vs revenue, in *Proceedings of the Mining Geology Conference 2019*, pp 123–134 (The Australasian Institute of Mining and Metallurgy: Melbourne).

Cline, J S, Hofstra, A H, Muntean, J L, Tosdal, R M and Hickey, K A, 2005. Carlin-type gold deposits in Nevada: Critical geologic characteristics and viable Models, *Economic Geology 100th Anniversary Volume*, pp 451–484.

Cook, H E and Corboy, J J, 2004. Great Basin Paleozoic carbonate platform: Facies, facies transitions, depositional models, platform architecture, sequence stratigraphy and predictive mineral host models, US Geological Survey Open-File Report 2004–1078, 129 p.

Cowan, E J, 2014. X-ray Plunge Projection — Understanding Structural Geology from Grade Data, in *AusIMM Monograph 30: Mineral Resource and Ore Reserve Estimation — The AusIMM Guide to Good Practice*, second edition, pp 207–220.

Le Cornu, C, Clark, J M, Snider, L and Pari, J, 2025. The importance of geologically driven estimation domaining on resource estimation – an example from the Turquoise Ridge Gold Mine within the Getchell Trend in Nevada, in *Proceedings of the Mineral Resource Estimation Conference 2025*, pp 133–146 (The Australasian Institute of Mining and Metallurgy: Melbourne).

Muntean, J L, Cline, J S, Simon, A and Longo, A A, 2011. Origin of Carlin-type gold deposits, *Nature Geoscience*, 4(2):122–127.

Rhys, D, Valli, F, Burgess, R, Heitt, D, Griesel, G and Hart, K, 2015. Controls of fault and fold geometry on the distribution of gold mineralization on the Carlin trend, in *New concepts and discoveries, Geological Society of Nevada 2015 Symposium Proceedings* (eds: W M Pennell and L J Garside), pp 1245–1301.

Robert, F, Brommecker, R, Bourne, B T, Dobak, P J, McEwan, C J, Rowe, R R and Zhou, X, 2007. Models and Exploration Methods for Major Gold Deposit Types, in Proceedings of Exploration 07: Fifth Decennial International Conference on Mineral Exploration,

Samson, M, Benthen, G, Clark, J M, Le Cornu, C and Conn, D, 2025. Preg-robbing gold ores – an example of a non-linear estimation workflow when dealing with non-additive geometallurgical variables, in *Proceedings of the Mineral Resource Estimation Conference 2025*, pp 227–242 (The Australasian Institute of Mining and Metallurgy: Melbourne).

Verly, G, Postolski, T and Parker, H M, 2014. Assessing Uncertainty with Drill hole Spacing Studies – Applications to Mineral Resources, in *Proceedings of the Orebody Modelling and Strategic Mine Planning Symposium 2014*, pp 109–118 (The Australasian Institute of Mining and Metallurgy: Melbourne).

You're so vein – cognitive bias in the resource estimation process

D A Sims[1]

1. Director, Dale Sims Consulting P/L, Bolwarra NSW 2320.
   Email: dalesimsconsulting@gmail.com

ABSTRACT

Resource estimation is a human endeavour, complete with our innate foibles and frailty. As geoscientists we are trained in the 'scientific method' to some extent, yet as our personal experience and professionalism develop, we adopt different perspectives and practices which guide our approaches to estimation and its related geoscientific problems. Sound and valid judgement is at the core of competence, yet how do we recognise our own abilities or shortcomings, as well as assess the validity of our decisions?

Cognitive biases are mental errors and reasoning faults caused by our simplified information processing strategies. The geological profession relies on interpretation of mineralisation domains as part of the estimation process, and the opportunities for cognitive bias are abundant in domain development. This paper focuses on the interpretation of high-grade domains and their continuity, and case studies are presented from vein or lode dominant gold and base metal deposits.

Opportunities for mitigating the potential impact of cognitive bias on projects are discussed, although fundamental issues relating to how our industry functions stands in the way of significant progress. A desired outcome of this submission is to build wider awareness of the factors influencing cognitive bias in our estimates and broaden the consideration of their potential impacts.

INTRODUCTION

Decision-making is core to resource estimation, and under JORC Code (2012) public reporting the supervising Competent Person takes responsibility for the estimate and hence its constituent processes and decisions. Resource estimation is a human endeavour that is subject to, and limited by, our range of experiences and capabilities. The process may be affected by flaws, errors, biases, fixations and other human short-comings during our search for the 'truth' – the reliable estimation of expected metal content in the ground. Identifying the effects of those shortcomings will always be opaque, as the original metal content is not perfectly known, even after mining and reconciliation are completed. Recognition or discussion of cognitive bias in our sector is rare to non-existent, yet it commonly impacts our decision-making processes key to estimation.

In reviewing risk management in feasibility studies, McCarthy (2014) observed that projects are typically more sensitive to head grade variations than to capital costs, operating costs or metallurgical recoveries within their respective range of variation. Geology, resource and reserve estimation issues can be a major risk to project outcomes in transitioning from feasibility to production. McCarthy observed that 'if the geologic interpretations are wrong, the entire model will be wrong' and this potential outcome is irrespective of the interpolation methodology involved.

Domaining of lithology, alteration, physical character and, most usually, mineralisation tenor are common geologic constraints incorporated into resource estimations and hence block models. Domaining is best undertaken by the people who are closest to the project and the data. These are site or project geologists who have logged the core, mapped the mining exposures, managed the mining processes to deliver ore and reconciled existing resource models with production. Yet for many new projects, increasingly under cover and sight unseen, domain interpretations are based dominantly on drill core data. There is a view among some estimation practitioners that geologists are so poorly skilled at reliable domain interpretation that unconstrained geostatistical methods are preferrable in nearly all environments.

Love, Boland and Anderson (2016) presented an analysis of psychological factors affecting mineral project studies in their paper titled the 'Influence of cognitive biases on project evaluation' presented at the 2016 AusIMM Project Evaluation Conference. The paper focused on the question of not *what* we think, but *how* we think it. Leveraging a significant body of psychological research, the paper identified several cognitive processes or scenarios which develop biased and *optimistic* outcomes in project studies. Such biases can be conscious or unconscious, steeped in our own personal experience or circumstance, and influenced by the management structure which we work within.

The fundamental premise of Love, Boland and Anderson's 2016 paper is that cognitive biases exist in minerals project studies, mine developments and corporate acquisitions, and they can be significant contributors to stakeholder value destruction in our sector. By improving awareness of these issues in minerals industry practitioners and applying a few simple mitigating process steps, they proposed that the minerals sector could substantially reduce its propensity for catastrophic project failure and stakeholder loss.

Mineral estimation can have a wide range of impacts that are not always reflected in a company's share price. In a previous contribution to the MREC series, a case study was presented where poor decisions and flawed estimation processes applied by experienced and respected practitioners led to significant financial loss (Sims, 2023). The loss was not material for the company which owned the asset, but it was very material for the site and its employees.

Our success is ultimately measured through project viability in an environment where there are many moving parts, many beyond our control. In resource estimation we approximate, yet bias and noise also govern our results. This paper presents a summary of estimation cognitive biases as outlined by Love, Boland and Anderson (2016), then presents a series of case studies of vein/lode hosted deposits which illustrate aspects of such bias. Finally, it reflects on why we as a professional group are susceptible to cognitive bias and outlines opportunities to recognise adverse conditions before they occur.

COGNITIVE BIAS

Cognitive biases are mental errors and reasoning faults caused by our simplified information processing strategies. The study of cognitive bias has been undertaken over the last 50 years by research psychologists who have produced acclaimed books on the subjects of thinking and judgement written for the general public (Kahneman, Sibony and Sunstein, 2021).

The cognitive biases and potential mitigations discussed in the context of project evaluations by Love, Boland and Anderson (2016), and seen as relevant to resource estimation, include the following.

Planning Fallacy – the selection of the best case scenario as the base case. This is generally associated with applying an 'inside' view to decision-making due to lack of exterior or 'outside' frames of reference or perspectives. It delivers an optimistic bias. Potential mitigation of this bias may be achieved by the application of *Reference Class Forecasting* whereby similar external examples are compiled to develop a baseline prediction which can then be modified for local conditions evident in the project geology. This is the development of a base-case model through broader input and experience, modified with local tweaking. Commonly the favoured outcomes are already so well entrenched that any modifications may only be minor.

Anchoring Effects – presupposing the outcome then developing a conviction and associated actions to make it happen. This approach can be driven by dominant individuals in management roles via direct or indirect duress through declaration of a desired outcome. It induces practitioners to be careful with releasing potential modelling outcomes to their management, yet their actions can be driven by their desire to please these same people.

Confirmation Bias – seeking supporting evidence which fits our existing idea and model while ignoring contradictive data or ideas.

Availability Bias – adoption of easy-to-imagine outcomes, which may be our initial idea on scant information, then defending them even in the light of additional data or enquiry.

Catastrophes can result from the malfunction of complex systems. We are prone to develop unconscious assumptions that systems will work as planned so we can't easily imagine what can go wrong.

A mitigation strategy for this cognitive bias is the conduct of *premortems*. The process is to imagine that a major catastrophe has occurred, then seek written submissions from involved or arms-length practitioners as to how the event may have occurred and hence develop a range of potential contributing factors for further consideration. The key step is gaining a range of individuals' written submissions rather than through a meeting. Written submissions aim to avoid groupthink, conformity or dominance by a senior person or vocal minority – issues which can occur even with facilitated independent reviews.

Narrative Fallacy – the process of making up and believing stories about the past and the future, regardless of what we really know. This can lead to confident opinions that are ill-founded. Termed the *WYSIATI delusion (What You See Is All There Is),* it results in a failure to question data adequacy to support the assumptions or models being made from it.

THE ESTIMATION PROCESS AND GLOBAL TRENDS

Geoscientific resource estimation is an extension of our science, and a task often aspired to in a mining geologists' career path. Collectively, we develop data, analyse it, manage it and create one (or occasionally more) hypotheses based on the data, through applying the judgement of our experience. The hypothesis covers the spatial distribution of the material of interest as well as its quality – mineralised zone geometries and metal contents, ground conditions and mine-ability, treatability, density, adverse environmental conditions in waste rock etc. The models produced underpin analysis for assessment of financial benefit.

With specialisation in our sector, it is increasingly rare that the people undertaking the data collection are then creating the estimates. Positions are created in operational structures for core logging geologists, database geologists, QA/QC geologists, resource geologists – hence our estimation processes become increasingly segmented. Depending on an organisation's scale there may be a full suite of specialised professionals involved in-house in the process, or sections may be outsourced to external groups, largely experienced consultants.

Compilations of the estimation practices applied in the sector are rarely published, yet public reporting under international codes allows access to useful data on the topic. Opaxe and RSC Mining and Mineral Exploration (2019) produced a summary of global mineral resource reporting which was released as a special report to coincide with the AusIMM's International Mining Geologists' Conference. The data include public reports of mineral resources issued in the first ten months of 2019. The report listed a total of nearly 800 resource statements issued, with around half issued under the JORC Code. Ten per cent of the reports were for maiden or initial estimates.

Around 40 per cent of the estimates were delivered by in-house Competent/Qualified Persons compared to ~60 per cent produced by external estimators (consultants). Initial resource statements were dominated by external estimators who accounted for 85 per cent of these releases. Estimation methodologies were dominated by Ordinary Kriging (OK) at around 70 per cent of the total estimates, followed by Inverse Distance Weighting (IDW) methods at ~25 per cent, and Indicator methods at ~5 per cent. Initial reports were dominated by precious metals estimates at 65 per cent of the releases with base metals and industrial minerals making up the remainder.

In an update Opaxe (L Tooley, pers comm, 15 Jan 2025) have provided a database extract for the 2024 calendar year. Global mineral resource reporting has nearly halved since 2019 to around 400 estimates, with JORC Code (2012) compliant reporting around 55 per cent of reports and NI43–101 around 45 per cent of reports. Estimation methods are still dominated by OK with it used in around 65 per cent of estimates, followed by IDW methods at 25 per cent. Around 10 per cent of the reports do not list an estimation methodology, with only one estimate undertaken by the Nearest Neighbour method. Gold estimates accounted for around 50 per cent of the releases with initial estimates accounting for 15 per cent of total reports.

The dominance of OK and precious metal estimates within global mineral resource estimation practice underscores the importance of domaining in the estimation process. Unconstrained OK for interpolating significantly mineralised volumes within a deposit should be a very rare occurrence hence incorporation of mineralisation domains in estimation must be almost universal.

Given domaining is a key step in developing reliable resource estimates – how might cognitive bias affect this process?

DOMAINING VEIN/LODE HOSTED DEPOSITS

Estimates of vein/lode hosted deposits will commonly include high-grade domains which sit within low-grade envelopes which in turn are enclosed by surrounding waste. Each domain is likely to be applied as either a hard or soft boundary.

Fundamentally all geoscientists working on these styles of deposit are aware that domain continuity increases as grade threshold decreases, and so higher grade domains should be less continuous than lower grade domains. Structural controls for the lodes or veins, often associated with faults or rheological contrasts from host lithology variations, may govern the overall array and local boundaries of the veins or lodes. Higher grade regions within these structures or zones can contain significant metal content and hence have a much higher economic value and represent the 'cream' in a deposit or project.

For example, at the Kencana low-sulfidation epithermal gold and silver deposit in Indonesia, 40 per cent of the contained metal value was within a zone occupying only 10 per cent of the mineralised vein – a 'bonanza zone' rich in electrum and assaying between 80–200 ppm Au (Coupland *et al*, 2009). Interpretation and estimation of the 'bonanza zone' in this deposit had a significant impact on the global contained metal. Coupland *et al*, reconcile the interpreted geometry of the zone in the resource model with the actual distribution from mine development, mapping and grade control sampling, noting as well as the impact of the zone on the overall mine reconciliation. In this example, the high-grade domains had remarkable continuity, imparted from the mineralisation being controlled along major dilatational faults.

While some low-sulfidation epithermal systems can have remarkable continuity of vein geometry over large distances (Figure 1), others occur as stockworks and/or arrays of linking structures or ladders veins between more continuous bounding structures (Figure 2). Similarly orogenic deposits can contain discrete and relatively continuous lodes or reefs, or broader, less well defined stockwork zones (Figure 3). Commonly the specific detail of the deposit is only evident in grade control drilling and mapping data.

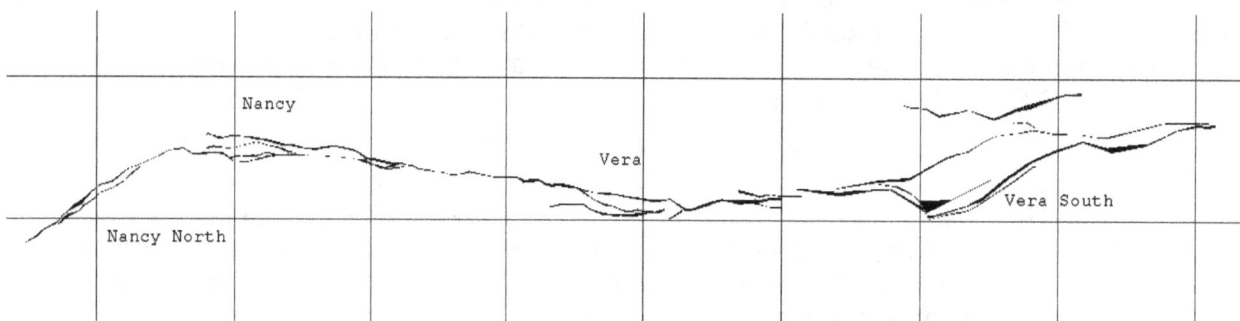

FIG 1 – A plan of the Pajingo Vera-Nancy low-sulfidation epithermal vein arrays in North Queensland, which extend over a 2 km strike (Sims, 2000). Grid space 250 m with drill spacing ~40–80 m within the main deposits.

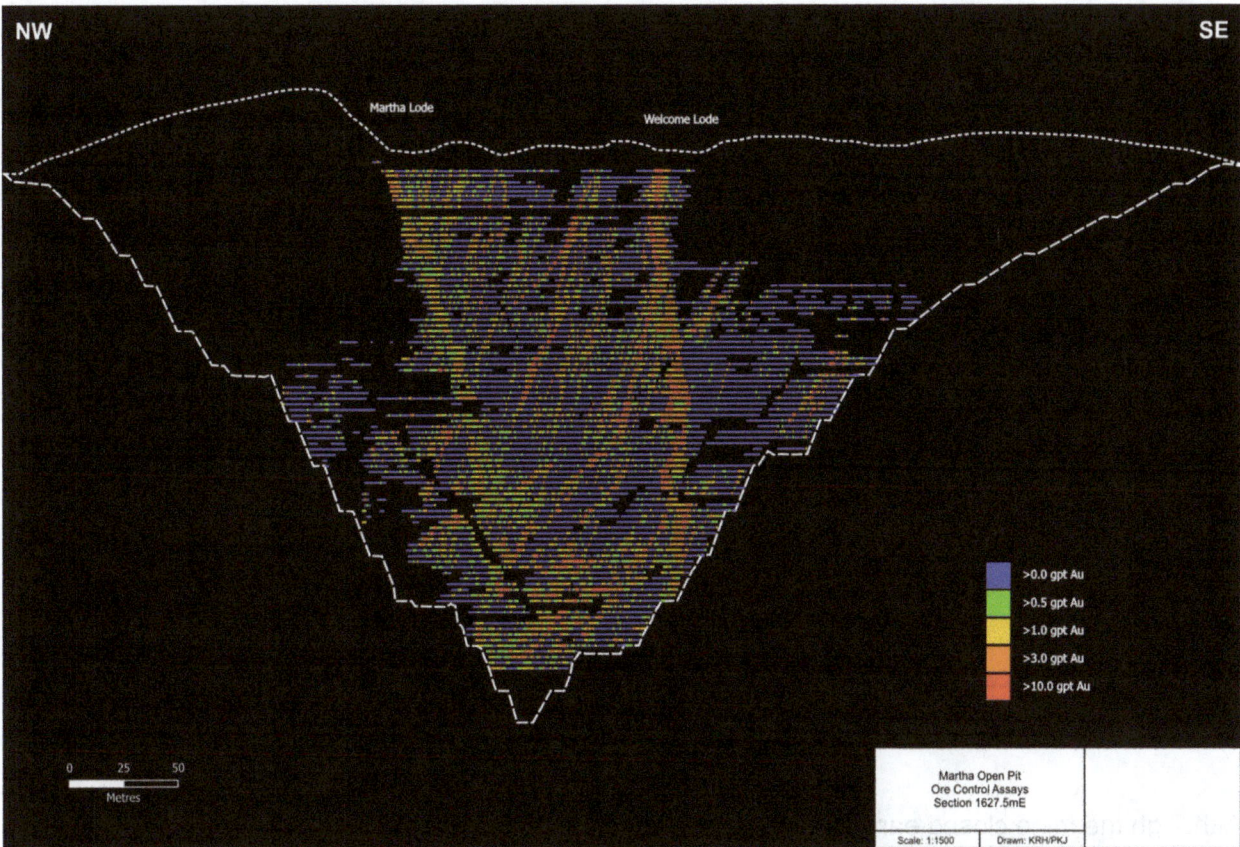

FIG 2 – Vein arrays between bounding structures – the Martha low-sulfidation epithermal Au Ag deposit at Waihi, New Zealand (L Torckler, pers comm).

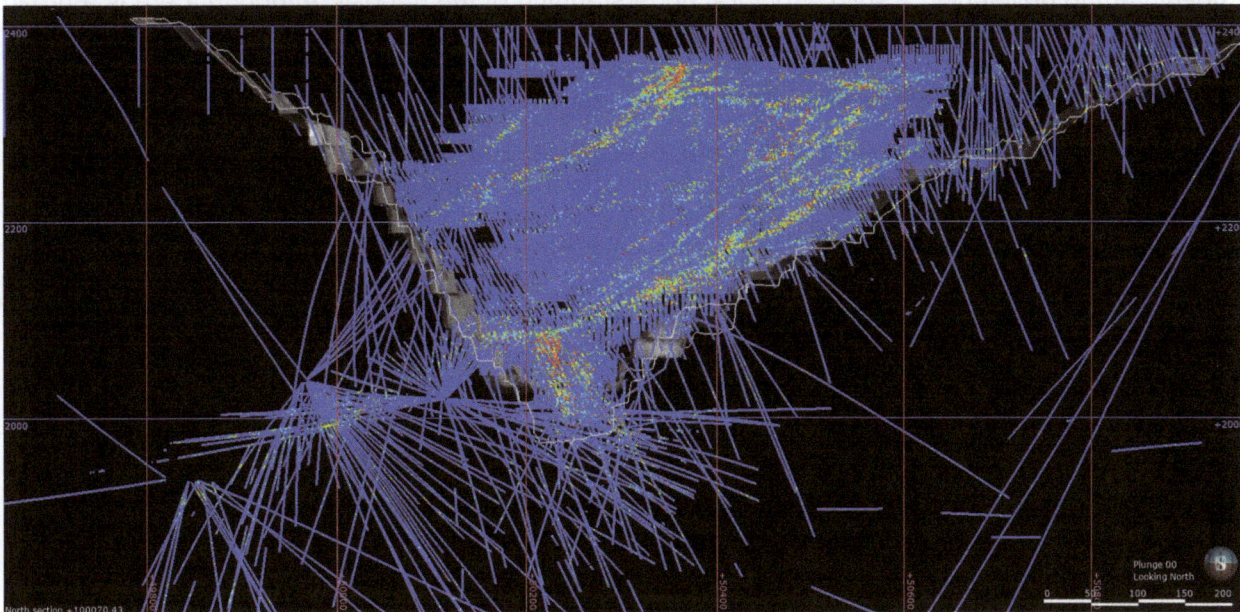

FIG 3 – Lode arrays within a Western Australia Archean orogenic gold deposit – low angle shear-hosted reefs and steeper less-defined stockwork zones.

Figure 4 shows two cross-sections through the Golden Cross low-sulfidation epithermal deposit in New Zealand, located approximately 8 km NNE of Waihi and hosted in the same geological terrain (Mauk and Purvis, 2006). The two sections are on northings just 25 m apart and are at the same scale. The cross-section on the left shows the interpretation of the mineralised system early in the project life following exploration in the 1980s (black veins are mineralised quartz veins), while the cross-section on the right shows the interpretation of the deposit (mineralised veins in red) from an internal company report written at the height of mining activity, when infill drilling, grade control and open cut and underground mine development and extraction had occurred up to the mid-1990s.

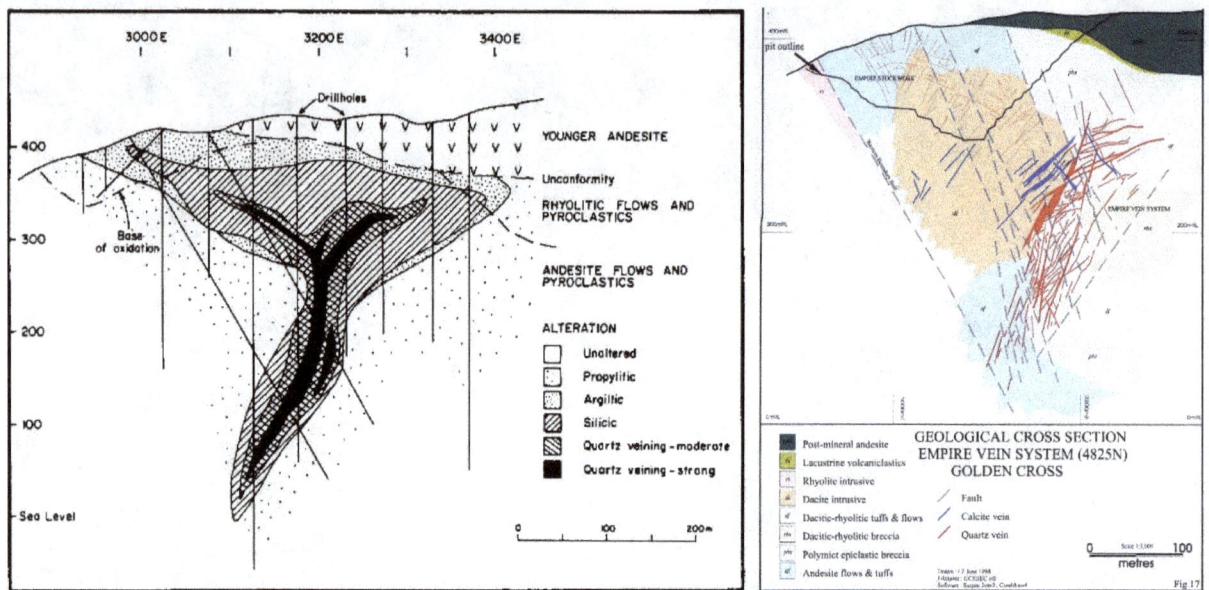

FIG 4 – Golden Cross deposit cross-sections – 1980s interpretation on the left (Hay, 1989), and 1990s interpretation on the right.

Although the mine closed early leaving significant metal in the ground due to tailings dam instability, the anecdotal evidence was that the mine plan was difficult to deliver. Given the dominance of vertically oriented diamond drill holes in the early data set (left), the interpretation of near vertical structures is problematic. Additionally, the vertical core holes were likely unoriented, a further impediment to developing a solid structural understanding of the deposit. Although there is a broad correlation of veining location, the nature of the structures and hence the continuity of the mineralisation is significantly different with increased data and understanding. If this deposit was discovered today, would we provide a significantly more reliable interpretation of the vein arrays?

The scale of observation, along with data orientation and spacing, are key determinants for interpretation outcomes as are the project stage, quality of exposure, and degree of experience in the mineralisation style. Although interpretations evolve with increasing data, exposure and understanding during project life, how can we ensure initial interpretations have the best opportunity for long-term success?

COGNITIVE BIAS IN THE INTERPRETATION OF HIGH-GRADE DOMAINS IN VEINS AND LODES – CASE STUDIES

Reflecting on cognitive bias in resource estimation and domain modelling through case studies requires firsthand knowledge of the examples. It is rare to have significant discussion of the domain creation process and rationale in public reports, and rarer still to see reconciliations of domain interpretations comparing initial interpretations with more advanced examples from increased data or from production experience.

We should expect that interpretations evolve with added data and exposure yet often cognitive bias has been a contributor to the initial interpretation which could have seen different outcomes from the original data with an alternative approach. Alternative interpretations are potentially dismissed as manifestations of 'differences of professional opinion' yet it needs to be science that governs the validity of our professional activity.

In general, the minerals industry is closed and secretive to the material details of domain interpretation in resource estimation for each deposit we publish, as we rarely share hard data. From experience reviewing projects for due diligence, investigation of the approach taken to interpretation and domaining only occurs with access to the informing data, the detailed documentary support, and the people involved.

These case studies have been in part anonymised yet are real-world occurrences.

Intrusion-related gold deposit

Hosted in granitic rocks this deposit had a phase of shallow open pit mining in the 1980s which did not end well. The project did not deliver the expected grade and the company went bankrupt. An expensive environmental cleanup was later funded by the state government. Stories of poor management surrounded the original operation and its failure.

Around a decade ago a junior explorer picked up the ground and commenced drilling along strike and around the flooded pits. Over a ~three year period they drilled 75 diamond core holes for around 17 km in HQ/NQ on 50 × 100 m to 100 × 100 m spacing and which were half-core sampled. No historical data was usable. No close-spaced or twinned holes were drilled.

There was a strong technical focus on the drilling program and data collection with around 23 000 structural measurements and 13 000 bulk density measurements collected as well as extensive four-acid digest multielement geochemistry. Staff turnover was relatively low, and the supervising geologist and their supporting team had significant experience on the project.

Mineralisation occurs in thin (2–50 mm), 1–2 m+ spaced, quartz-carbonate veins which contain visible gold and sulfides. Limited exposure of the deposit is available around exposed pit walls where the veins occur in wide-spaced and sheeted zones with vein continuity visible up to 10–20 m along strike.

An initial JORC Code (2012) compliant Inferred resource estimate was released to market totalling around 500 koz Au at ~2 g/t Au. Early drafts of the mineralisation domain interpretation were produced inhouse, while the estimate was interpolated by an external consulting group. The final model had the internally developed interpretation framework extended to the limits of the data by the consultants.

When the corporate management team, who did not have an exploration or mining background, interacted with potential industry investors on the project, they found that their estimate was poorly regarded and quickly dismissed. They sought external technical assistance to review the model and its processes, and later proved to be fully supportive of the revision process that followed.

The interpretation of the deposit for the initial resource consisted of an intersecting network of around 40 discrete vein package models/domains in sheeted arrays with around four dominant orientations (Figure 5). Individual vein packages/domains had modelled continuities of typically 200 to 600 m along strike and extending to around 250 m depth from surface. Each vein package was modelled on a ~0.5 ppm Au threshold and used as a hard boundary in the OK estimate. The interpretation was stated to be leveraging the considerable database of structural data. Although not stated in the public report, nugget values *within* these domains were 50 to 60 per cent.

FIG 5 – Interpreted vein array used in the initial mineral resource estimate.

Subsequent work to review the estimate and its interpretation included niche sampling to establish the gold was exclusively within the veins, second half core assaying to investigate sampling imprecision, assessment of structural data within 0.5 ppm Au assayed intervals to understand vein orientation diversity (Figure 6), and ultimately drilling of 40 reverse circulation (RC) percussion holes in an accessible area of the deposit on a 25 × 25 m scissored grid totalling 7500 m. Infill data (diamond and RC) had extensive QA/QC programs and did not support the mineralisation locations predicted by the existing interpretation, as well as confirming a high nugget persisted even with the improved RC sampling.

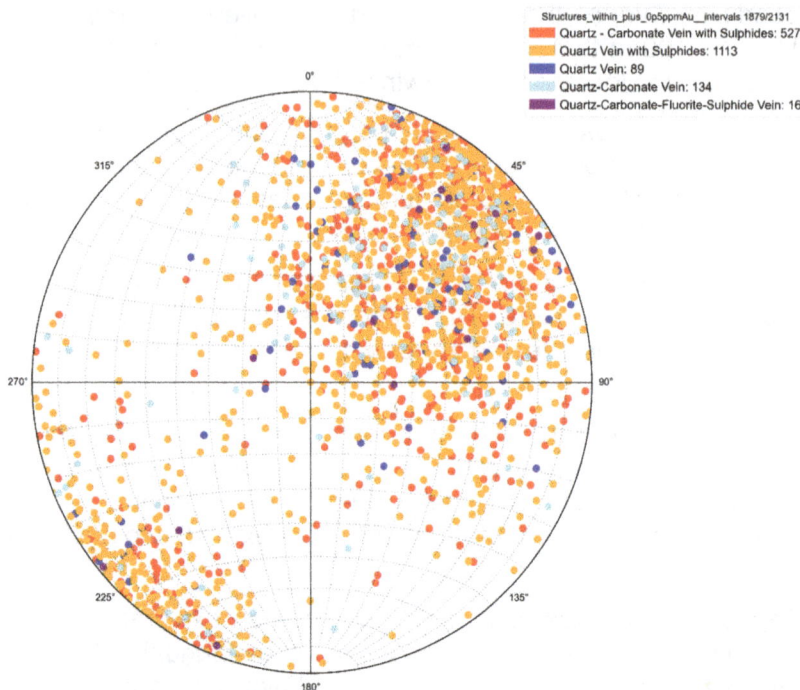

FIG 6 – Poles to quartz veins within plus 0.5 ppm Au intervals in the initial resource data set. Poles coloured by vein type.

The shortcomings of the sheeted vein array interpretation used in the initial estimate were not accepted by the supervising geologist during this process, who believed the continuity of the structures was significantly greater than evidenced by the data. The geologic team were considerably invested in the interpretation and showed their preference for maintaining it as the base case and so became anchored in their assumptions of the vein continuity. Eventually a campaign of drilling ~50 RC holes on a 5 m × 5 m pattern to simulate a grade control pattern clearly evidenced the lack of short range continuity. The deposit was recognised as a stockwork with a broadly NW trend and was re-estimated with Multiple Indicator Kriging. The contained metal increased slightly in the revised estimate and public statement following a scoping study, although the grade approximately halved. The project awaits development.

Base metal lode deposit

This deposit is a shallowly-dipping stratiform base metal deposit hosted in carbonate sediments. Reviewed as part of a due diligence assessment, which included a site visit, core inspection, data room access and meetings with the owner's staff and their estimation consultants. The resource estimate had been the subject of a prefeasibility study (PFS) undertaken by the owner with a significant proportion of the asset classified as Indicated resource under the JORC Code.

Drill spacing for Indicated resource was nominally 50 m × 50 m in the company's report yet closer to 80 × 80 m to 100 × 100 m in the data. The deposit held spectacular grades in some drilling intersections, but the core exhibited cross-cutting coarse sulfide textures in high-grade zones and general tectonic and sedimentological disruption of stratigraphy.

The company had undertaken further drilling following the PFS although this was to expand the footprint of the asset and not to infill any of the existing Indicated areas as they had a high level of confidence in that part of the resource. No close-spaced or twinned holes had been drilled, although a few holes were scissored and hence closed the pattern in a few locations to 10–20 m.

Figure 7 shows a cross-section through the higher grade upper part of the deposit. The lode domain interpretations undertaken by site geologists and developed with their consultants are shown as green outlines, and the resultant OK model for the section shown below. The 'BM_%' is the assay grade of the dominant base metal with the same legend in both the data and the model.

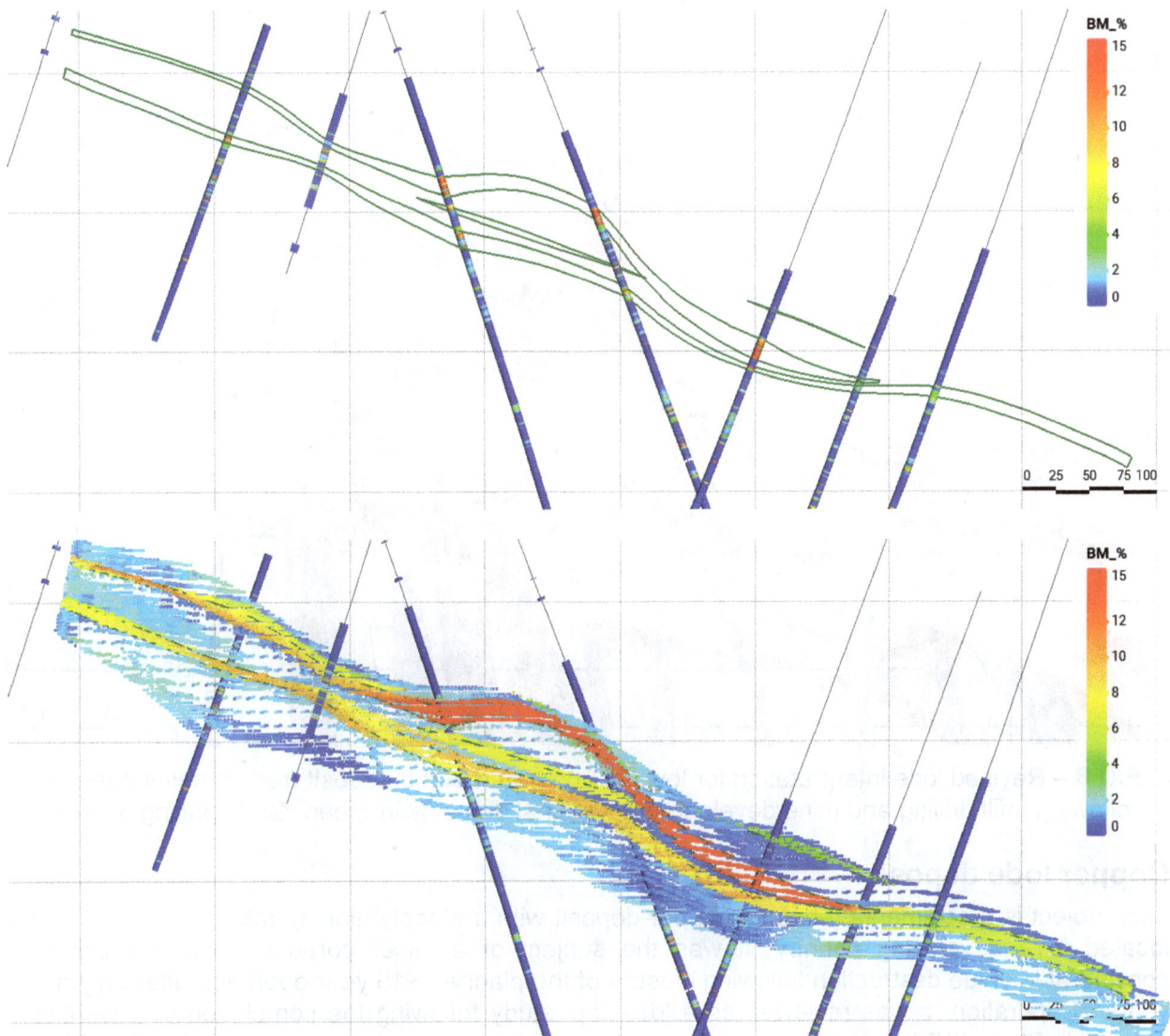

FIG 7 – Interpreted cross-section showing mineralisation domains (above) and associated block model (below). Hard boundary domains lead to lengthy high-grade sections of the resource model. Grid spacing 100 m.

With only rare opportunity to test short-range variability from the drilling data, a review of the interpretation was undertaken with core inspections during the site visit and access to the set of core photographs in follow up work. It was noted that the specific geology/stratigraphy and mineralisation appearance varied within and around the high-grade intervals between adjacent holes, and so the geological basis for the high-grade domain interpretation was queried with the supervising geologist.

The difference was relegated to 'professional opinion', yet the supporting data questioned the continuity between the high-grade intercepts as applied in the domain models. The application of these domains as hard boundaries then resulted in significant metal being estimated into the model, and the confidence in the interpretation used to support the Indicated classification over relatively wide drill spacings. Independent indicator-based domain modelling and estimation undertaken for the due diligence suggested around 30 per cent of the total high-grade metal had been overestimated due to the interpreted continuity.

The project proceeded to financing, construction and commenced production, but during ramp up it went into administration. The upper sections of the deposit have proved more complicated than anticipated in the resource model and infill grade control drilling is being undertaken to 24 × 24 m and in places 12 × 12 m.

The operation is experiencing 'ongoing modelling variances'. An updated interpretation displaying lode interpretation variances is shown in Figure 8.

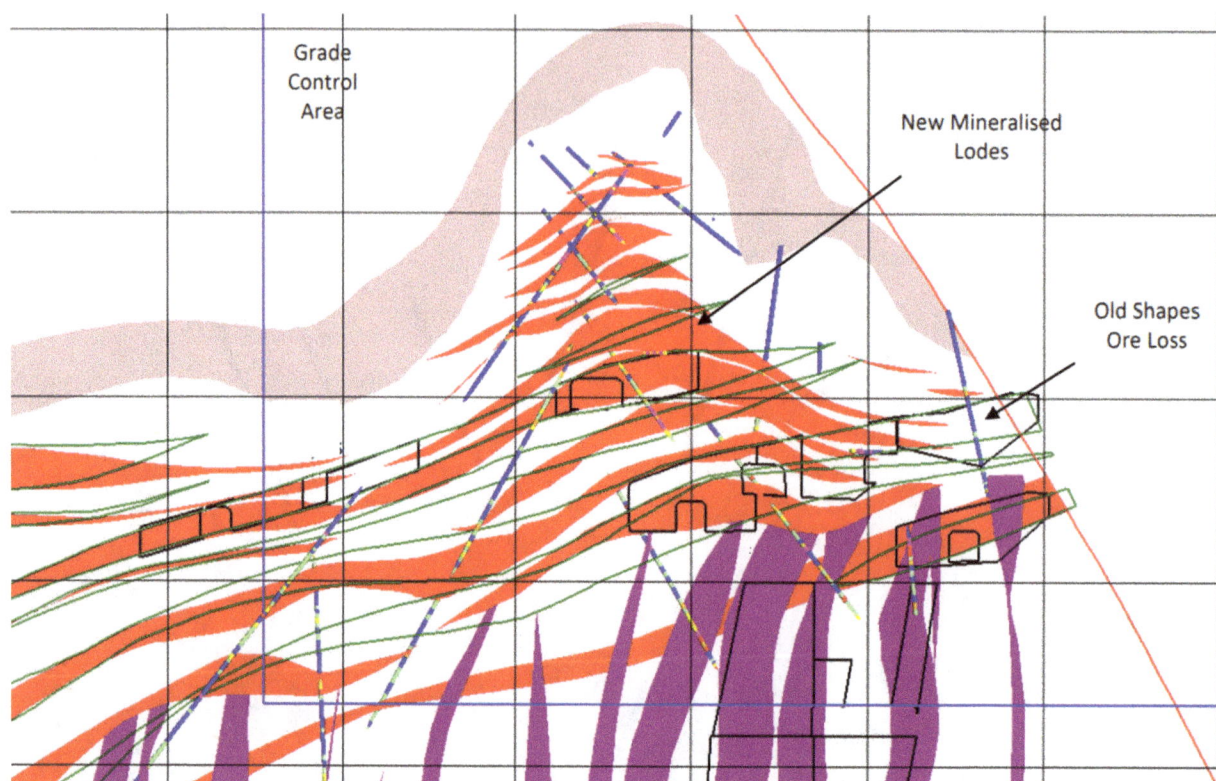

FIG 8 – Revised lode interpretation for low dip upper part of the deposit (red domains outlines) following infill drilling and mine development. Original domains in green. Grid spacing is 30 m.

Copper lode deposit

This project is a sedimentary-hosted copper deposit with a steeply dipping tabular lode geometry, located in a developing country. It was the subject of a major corporate failure which saw considerable value destruction following closure of the planned ~15 year operation after only a few years of operation. It was reviewed as a 'desktop' study following the completion of a Bankable Feasibility Study (BFS).

Located in the developing world and using national professional staff to undertake the bulk of the data collection from drill core, there was a level of concern by the owners regarding the accuracy of visual logging of copper species. The project was predicated on mining and treating sulfide copper mineralisation and the assumption was that weathering was relatively shallow on the project's deposits with expectations of oxidation to a maximum depth of around 30 m. Sequential copper analysis data had not been widely collected hence copper mineral species were not tested in routine analysis.

Oxide copper mineralisation had been noted deeper in some of the drill logs but that was presumed to be related to local structural features developing deeper spot oxidation. Oxide material was to be stockpiled given the mill was designed to treat sulfide ore only.

The consultants undertaking the estimation developed an approach to use the ratio of estimated Cu to S as an oxidation/mineralogy threshold, which was then used in mine planning to classify material for treatment.

The BFS included the consultant's 150 page estimation report for the major deposit in the project as an appendix with the very last pages of the report including swath plots for validation of the S estimate (Figure 9). The plots show a clear bias where the average block estimates (red) are significantly higher than the average informing samples (blue) both along the strike of the deposit as well as in the shallow levels of the deposit above ~900 RL. Given the approach taken to determine depth of oxidation from the S estimate, the bias in the estimate was anticipated to have impacts on the material type supply in the mine plan.

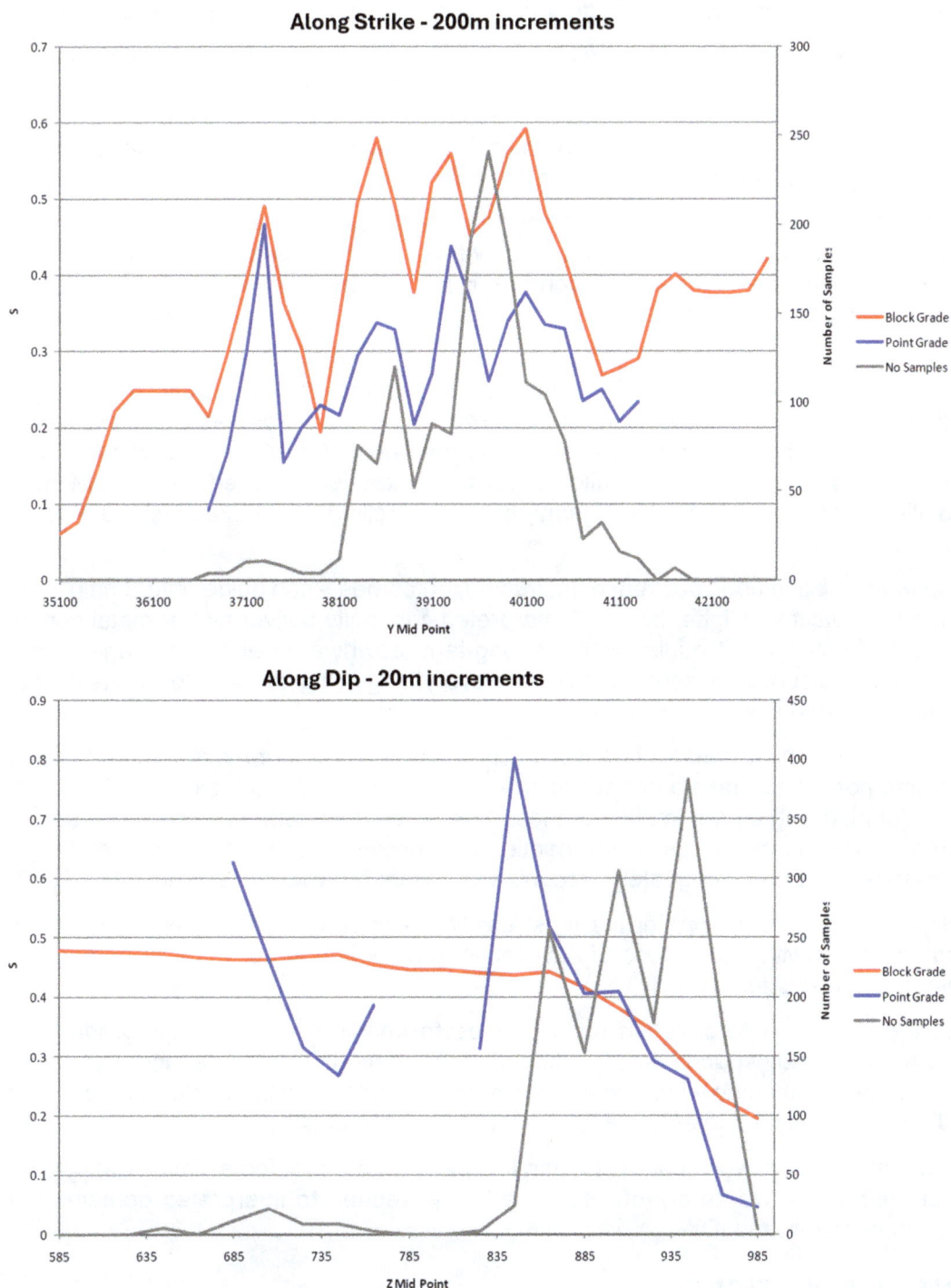

FIG 9 – Validation swath plots of sulfur estimate along strike (above) and by RL (lower).

Enquiry into the potential cause of the S model bias indicated that early shallow drilling had not been assayed for sulfur, so with a paucity of shallow low-value data to inform the model the OK estimate had pushed higher S values towards the surface 'oversmoothing' the model.

Both the owners and the consultants were informed of the issue, and it was suggested they assess the potential impacts on the mine plan. Both these groups roundly rejected there was any material impact to the BFS/project from the issue. The project proceeded to production yet closed within three years with deeper-than-anticipated copper oxidation levels being major contributing factor.

DISCUSSION

At this stage you may be asking yourself how do these case studies illustrate cognitive bias? Aren't these just examples of how we work in general in our industry? Are they really cognitive bias?

We commonly link high-grade intercepts together to form continuous high-grade domains that often anastomose through the geology. We generally construct them based on the judgement and experience of the modeller/interpreter and sometimes with production experience to support our assumptions. We believe in high-grade continuity and believe data on face value, because after all, look at the effort we put into QA/QC to make our data reliable.

We also commonly form strong opinions of our own technical work and its underlying judgements, and we tend to reject querying or criticism. Largely there is little opportunity available within our tight timelines to consider alternatives, and perhaps it is rare that someone asks the questions. We generally operate as a closed system – either at a project or site level, or more broadly at a corporate level.

Planning fallacy – where we select the best case scenario for the base case – is a cognitive bias applied during high-grade domain interpretation. *Reference Class Forecasting* is largely unavailable to us given the lack of public transparency around the detail of data, interpretation, and estimation inputs and outputs. And although we build our personal experience-based portfolios of interpretation and modelling examples as part of our judgement, reconciling those examples back to actuality is largely undervalued.

Anchoring effect – fixing upon desired, or instructed, outcomes – can guide our domain interpretation to reinforce the outcome. Higher levels of interpreted continuity deliver higher metal content in hard or soft bounded estimates. Coupled with the long-term downward trend in average grade of metal mined, project development success often needs everything going its way. Corporate duress through expectations on estimation outcomes can be real.

But largely it is the inherit nature of geoscientists which leads to our most common cognitive bias. We exist, and persist, to make order out of chaos – that is how we solve our earth science related problems. Coupled with a near total ignorance of geoscience in both the community and with some of our professional peers, we carry the mantle for geoscience truth and understanding, delivered with confidence. So do we adequately recognise or accept our own geoscientific fallibility?

Availability bias – where we opt for the most easily-imagined outcomes – is our *modus operandi*. Our imagined order within the chaos of data, although steeped in our experience and confidence, can prevent us thinking too laterally.

Selecting the 'easiest' interpretation option for us to imagine – linking high-grade vein or lode intercepts across widely spaced data – reflects the fundamental issue we face. But what else can we do? How can we imagine what domain geometries should be without informing drill data? And if we could, how can we then interpolate values into those domains?

The Opaxe data shows that globally less than five per cent of resource estimates apply a non-linear estimation method, so we as a profession are firmly wedded to interpreted domains in resource estimation through OK and IDW.

POTENTIAL SOLUTIONS

The fundamental problem is of our industries own making – we do not share, and we rarely learn collectively from our individual mistakes. Ours is an internally focused activity and any suggestion of applying independent or external peer review as an industry-wide practice is roundly decried as a 'money making exercise' spawned by said peer reviewers and their ilk. This is a disappointing position given external peer review is potentially one of the major opportunities for alternative perspective and input. Such review work needs to be at a detailed level which includes full data access and discussion with the people responsible for the decisions made during the estimation process. Overall, we need to be more open and realise if we make project-by-project improvements the whole industry will benefit.

There is significant potential benefit in using a *premortem* review as an internal/external exercise to deliver valuable perspectives on what could go wrong, but it requires a recognition and acceptance at senior levels of the organisation that a 'catastrophic failure' has potential to occur for them to even consider the process. Often the 'blind-faith' held either globally by an organisation, or within a specific project, overshadows consideration of the potential for failure, as the examples above illustrate.

Coupland *et al* (2009) document the comparison between a deterministic interpretation of the high-grade 'bonanza zone' mineralisation at Kencana with a simulation based approach. Using an assessment of both feasibility study scale resource drilling data as well as later infill and grade control information, the study aimed to better understand local variability over production-scale timelines. The paper supported the validity of using a simulation approach to considering uncertainty for mine planning expectations, and for resource classification assessment for the continuity of high-grade domains. It is a rare, published example.

Testing the short-range continuity of high-grade domains at the global estimation stage of a project is valuable. It can be undertaken through the deliberate application of closed spaced drill holes often in a specific valuable area, or through a wider-spaced program of twin-holes. With a trend towards deeper and totally blind deposits fully reliant on drilling data, such programs should be considered routine with the objective to adequately assess continuity of significantly valuable areas of the resource.

Improving the confidence of estimation practitioners in the quality of data provided along the estimation process chain requires a rethinking of data collection methodologies. Qualitative geological data collection contains significant potential for imprecision in output data where code-based logging systems are used to describe the rocks and their properties through manual logging. The current trends in collecting scanner-based or high-resolution image data on drill core have not so far translated into widely applied improvements in workflows or data quality in the industry. Although it is anticipated this will improve with time, the industry is largely committed to long established and generally flawed logging processes to its detriment.

Advanced computing methodologies are beginning to allow consideration of alternative model outcomes from data sets. The application of vein network simulation modelling via the 'DMX protocol' to simulate potential vein or lode arrays (Figure 10) was recently presented in the context of application in the gold industry (Munro and Van Dijk, 2024a, 2024b). Creating multiple interpretative options through data-driven analysis and simulation should allow alternative perspectives to influence our thinking processes. That is a future challenge – perhaps as we progress through and beyond the current OK/IDW era.

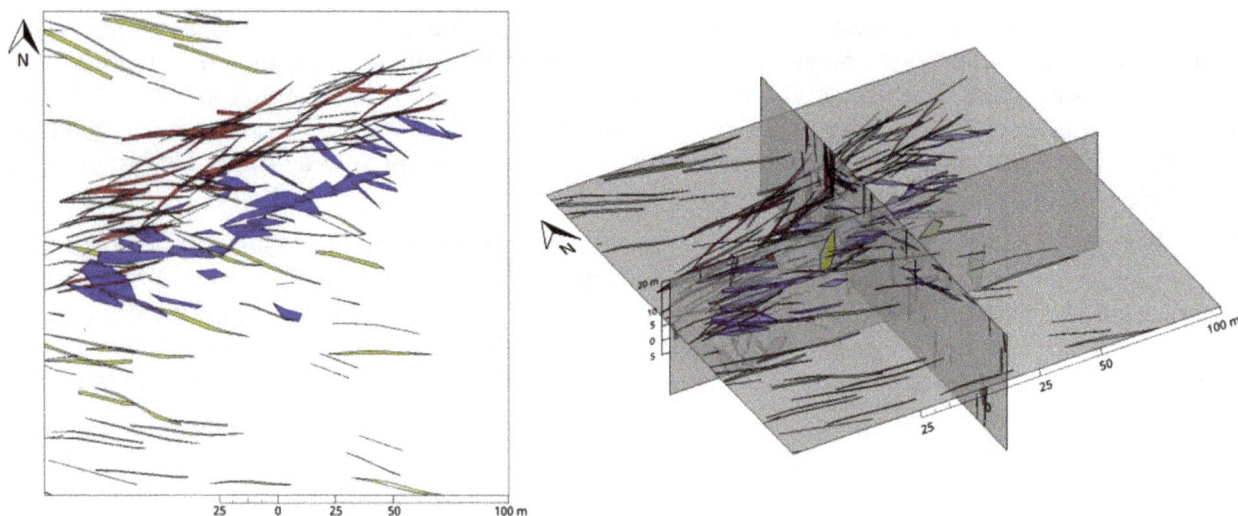

FIG 10 – Vein network simulation created using the 'DMX protocol' (from Munro and Van Dijk, 2024a, 2024b).

CONCLUSIONS

Cognitive biases affect how we think and the decisions we make, and their overall impact is a reduction in the consideration of viable alternative interpretations and domaining approaches, which then influence estimation outcomes.

If geoscientists who remain working in the profession for extended periods are by nature or need optimistic, then consideration of a fuller range of potential outcomes in an estimate may be overshadowed by positivity or overconfidence. Resource estimation is not a mechanical process.

Domains are ubiquitously applied at a global scale in estimation, and so the opportunity for bias to be applied in the process is widespread. The interpretation of high-grade domains in vein and lode style deposits is particularly exposed to bias.

There is considerable overlap in the influences which lead to the potential occurrence of bias in resource modelling and domain interpretation. The boundaries between them may not be apparent. Moreover, understanding the exact conditions, biases and influences at play in any specific technical project is difficult, even if you are at the core of the activity and in the thick of the action. None-the-less it is expected that readers may well recognise one or more of the biases discussed in this paper from experience, and hence consider these issues in their future work, or in the consideration of work of others.

ACKNOWLEDGEMENTS

Thanks to a number of long-term colleagues who undertook review of this paper prior to submission, and for suggestions from the conference reviewers. Kathryn Gall at Seequent-Bentley is thanked for providing access to their software for data review and image creation. Lauren Tooley at Opaxe.com is thanked for provision of both their 2019 report, and access to their 2024 database summary. I acknowledge that I recognise these biases have been present in my career both in my work and the way I think about geology and geoscience. We are only human after all.

REFERENCES

Coupland, T, Sims, D, Singh, V, Benton, R, Wardiman, D and Carr, T, 2009. Understanding geological variability and quantifying resource risk at the Kencana Underground Gold Mine, Indonesia, in *Proceedings of the Seventh International Mining Geology Conference*, pp 169–186 (The Australasian Institute of Mining and Metallurgy: Melbourne).

Hay, K R, 1989. Exploration case history of the Golden Cross Project Waihi, New Zealand, in *Mineral Deposits of New Zealand, Monograph 13* (ed: D Kear), pp 67–72 (The Australasian Institute of Mining and Metallurgy: Melbourne).

JORC, 2012. Australasian Code for Reporting of Exploration Results, Mineral Resources and Ore Reserves (The JORC Code) [online]. Available from: <http://www.jorc.org> (The Joint Ore Reserves Committee of The Australasian Institute of Mining and Metallurgy, Australian Institute of Geoscientists and Minerals Council of Australia).

Kahneman, D, Sibony, O and Sunstein, C, 2021. *Noise – a flaw in human judgement* (Harper Collins: London).

Love, S, Boland, M and Anderson, M, 2016. Influence of cognitive bias on project evaluation, in *Proceedings of the Project Evaluation Conference*, pp 226–235 (The Australasian Institute of Mining and Metallurgy: Melbourne).

Mauk, J L and Purvis, A, 2006. The Golden Cross epithermal Au-Ag deposit, Hauraki Goldfield, in *Geology and Exploration of New Zealand Mineral Deposits, Monograph 25* (ed: T Christie and R Brathwaite) pp 151–156 (The Australasian Institute of Mining and Metallurgy: Melbourne).

McCarthy, P L, 2014. Managing risk in feasibility studies, in *Mineral Resource and Ore Reserve Estimation; The AusIMM Guide to Good Practice*, pp 13–18 (The Australasian Institute of Mining and Metallurgy: Melbourne).

Munro, M and Van Dijk, J, 2024a. Discrete vein modelling – implications of a novel geology driven resource estimation methodology for vein hosted mineral deposits, in *Proceedings of the International Future Mining Conference 2024*, pp 281–285 (The Australasian Institute of Mining and Metallurgy: Melbourne).

Munro, M and Van Dijk, J, 2024b. Discrete vein modelling, in *Gold '24, Extended Abstracts AIG Bulletin*, 75:243–249 (Australian Institute of Geoscientists: Sydney).

Opaxe and RSC Mining and Mineral Exploration, 2019. Perth Mining Geology Conference special report: A Selection of Resource and Exploration Analytical Data, RSC Mining and Mineral Exploration. Available from: <https://www.rscmme.com/spotlight-report-2019-mining-geology-conference> [Accessed: 24 Jan 2025].

Sims, D A, 2000. Controls on high grade gold distribution at Vera Nancy gold mine, in *Proceedings of the 4th International Mining Geology Conference*, pp 5–36 (The Australasian Institute of Mining and Metallurgy: Melbourne).

Sims, D A, 2023. An estimation error, in *Proceedings of the Mineral Resource Estimation Conference*, pp 245–249 (The Australasian Institute of Mining and Metallurgy: Melbourne).

Data science, input data, machine learning and AI

Dare to transform – leveraging new orebody knowledge for next-generation resource estimation

J Coombes[1], P Hodkiewicz[2] and M Mattingley-Scott[3]

1. Managing Director, SageAbility, Perth WA 6000. Email: jacqui.coombes@sageability.org
2. Managing Director, Samarah Solutions, Perth WA 6000. Email: phodkiewicz@bigpond.com
3. Managing Director, Quantum Brilliance, Heidelberg 69115, Germany.
   Email: mark.mattingley-scott@quantum-brilliance.com

ABSTRACT

The minerals industry is experiencing a data revolution, driven by advancements in geoscience technologies that enable the collection of high-resolution, multivariate data sets. However, existing resource estimation workflows, including geological modelling, domaining, and classification, face challenges in fully harnessing the potential of these new data streams.

This paper explores the inherent constraints of current workflows in managing the scale and complexity of modern data sets and presents a framework for transformation. By incorporating advanced data processing techniques, multivariate geostatistics, and machine learning, resource geologists can enhance orebody models, improve classification decisions, and unlock greater value from collected data.

We also examine the evolving role of emerging technologies, such as artificial intelligence and quantum computing, and their potential to reshape resource estimation processes. Recognising the significant strides made by existing methodologies, this paper aims to build upon their foundation to enable more effective utilisation of modern geoscience data. Our call to action emphasises the importance of cross-disciplinary collaboration, workforce development, and investment in research to position the industry for a data-driven future.

The paper calls for the establishment of a Committee for the Future of Mineral Resource Estimation to drive cross-disciplinary collaboration, standardise practices, and guide competency development. By evolving workflows and leveraging emerging technologies, the resource estimation community can unlock significant value, enabling the minerals industry to harness deep geoscience intelligence and achieve precision mining outcomes.

INTRODUCTION

The minerals industry is undergoing a significant transformation, driven by unprecedented advancements in high-resolution geoscience and extraction technologies. Innovations in drill sample scanning, downhole sensing, and multivariate data acquisition are providing geologists with unparalleled insights into orebody characteristics, whilst technologies such as ore-sorting are reshaping how projects are optimised and valued. These advancements offer the potential to revolutionise the minerals industry with improved understanding of orebodies and more targeted minerals extraction and processing.

However, despite a dramatic influx of high-quality data, resource estimation workflows have generally not kept pace with the scale and complexity of modern data sets. Current practices, though robust and proven, often struggle to integrate large, multidimensional data sets effectively, creating bottlenecks that limit the industry's ability to fully leverage new orebody knowledge. These limitations impact not only the accuracy of resource models but also the downstream operations that depend on them, including mine design, processing optimisation, and feed strategy development.

At the same time, operational demands are evolving. The push for precision mining, in-line processing, modular systems, and real-time decision-making is placing greater emphasis on high-resolution resource models and geometallurgical insights.

Additionally, investment risks are heightened by uncertainties in resource-to-reserve conversions, making the quality of resource estimation critical to the industry's financial and operational success.

This paper argues that addressing these challenges and meeting emerging operational needs requires a fundamental evolution of resource estimation workflows. By leveraging cross-disciplinary innovation, advanced analytics, and cutting-edge technologies such as artificial intelligence and machine learning (AI/ML) and quantum computing, the resource estimation community has an unparalleled opportunity to reset and optimise its methodologies. Aligning workflows with the requirements of modern mining operations and the capabilities of emerging technologies can unlock significant value, enhance decision-making, and position the minerals industry as a global leader in data-driven innovation.

EMERGING TECHNOLOGIES IN GEOSCIENCES

Advancements in geoscience technologies are reshaping how the industry captures, interprets, and uses data, enabling a deeper understanding of orebodies and their variability. These technologies go beyond traditional methods, providing detailed, multivariate data sets that support more accurate geological modelling, resource estimation, and operational planning.

By capturing high-resolution information on mineralogy, chemistry, texture, and structure, emerging tools allow geologists to uncover insights that were previously hidden, transforming decision-making across the mining life cycle.

Modern data acquisition systems are at the forefront of this transformation. Drill sample scanning technologies, such as hyperspectral and XRF core imaging, now deliver continuous mineralogical and geochemical data with unprecedented precision. Similarly, downhole sensing and imaging tools collect high-resolution textural and petrophysical rock properties that enable real-time insights into subsurface geology. Automated core logging systems powered by machine learning ensure consistency and reduce the variability introduced by manual interpretation. High-resolution imaging techniques, including X-ray computed tomography and laser-induced breakdown spectroscopy (LIBS), provide micron-scale data that reveals subtle textural and mineralogical variations critical to understanding rock mass characteristics of the entire orebody.

These technologies enable geoscientists to uncover new insights. With closely-spaced data, geologists can better define geological controls on mineralisation, identifying subtle structural or geochemical trends that govern ore distribution. Multivariate data sets also reveal relationships between grade, texture, and mineralogy, allowing for more nuanced interpretations of ore variability. Such insights not only improve the reliability of resource estimates but also enhance confidence in downstream operations, from blending strategies to processing performance.

The integration of geological and geometallurgical data is becoming increasingly important. Emerging tools now link geoscience data with metallurgical performance metrics, enabling predictive models of recovery, grindability, and leachability. For example, mineralogical data from core scans can predict how ore will behave during comminution or flotation, supporting better process design. Real-time feedback loops between processing and geology further enhance this integration, allowing resource models to be updated dynamically as new operational data becomes available.

Exploration and early-stage modelling also benefit significantly from these technologies. High-resolution geochemical and geophysical data sets allow for more precise target definition, reducing exploration risks. Semi-automated (or 'assisted') 3D modelling tools powered by machine learning accelerate the creation of early-stage resource models, providing robust frameworks for project evaluation and de-risking.

For instance, scanning and imaging spectroscopy technologies generate continuous and high-resolution mineral-texture and mineral-chemistry maps of drill cores. Similarly, borehole televiewers and gamma-ray spectrometers deliver high-resolution structural and mineral-assemblage data in near real time, improving logging and interpretation workflows. Advanced geochemical sensors, such as portable XRF and LIBS systems, provide rapid, in-field assays that enable immediate decision-making, reducing delays associated with traditional laboratory analysis.

While these technologies offer immense potential, challenges remain. The costs of acquiring and processing such large data sets can be prohibitive, and the expertise required to integrate and interpret multivariate data is still developing within the industry. Collaborative efforts between geoscientists, data scientists, and technologists are critical to overcoming these barriers. Machine learning and artificial intelligence provide promising avenues to efficiently process and interpret the growing volume of geoscience data.

DRIVING OPERATIONAL EFFICIENCY AND FUTURE INDUSTRY NEEDS

Emerging technologies in minerals extraction are redefining how the industry approaches operational optimisation, focusing on feed and throughput improvements, waste reduction, and tailings management. These innovations aim to deliver greater value while addressing environmental, economic, and operational challenges.

The success of these transformative approaches hinges on detailed, high-resolution resource models that provide the critical data required to optimise these systems.

Opportunities in operational optimisation include:

- Feed and throughput improvements – Emerging technologies enable real-time adjustments to processing strategies, ensuring that ore blends are optimised to maximise recovery and maintain consistent throughput. Detailed resource models allow for better characterisation of ore variability, supporting dynamic processing that adapts to real-world conditions.

- Waste reduction and tailings management – Technologies focused on selective mining and ore upgrading minimise the extraction and processing of waste material, reducing tailings generation. By prioritising the quality of extracted material, these systems contribute to lower operational costs and a smaller environmental footprint.

Consider three examples that demonstrate the potential of emerging technologies:

1. In Line Recovery (ILR) – ILR incorporates selective mining and ore upgrading technologies that concentrate valuable material at or near the mining face. This reduces material movement and processing costs (Mining3, 2017).

2. In Mine Recovery (IMR) – IMR minimises rock movement by creating conditioned mineral blocks underground, combining chemical and/or biological leaching to extract valuable metals. This approach is particularly effective for remnant ore in sub-level or block caving operations. Hybrid approaches are also being explored, for example combining conventional processing for high-grade material with underground leaching for low-grade ore to maximise resource utilisation (Mining3, 2017).

3. *In situ* Recovery (ISR) – ISR eliminates rock movement entirely by processing ore in place, a technique well-established in uranium and potash mining. Advances in lixiviants, such as Curtin University's glycine leaching technology (Oraby, Eksteen and O'Connor, 2020), make ISR more environmentally and economically viable for a broader range of deposits. Glycine Technology is a non-toxic, reusable reagent operating in an alkaline environment, allowing for selective leaching of metals without significant gangue dissolution. Key benefits include lower energy requirements, minimal surface disruption, and reduced waste generation.

An essential step toward realising operational efficiencies is the integration of real-time or near-real-time data into resource models through robust measurement and material tracking systems, and by continuously updating the block model with production results, reconciled grade data, and actual throughput metrics, closing the feedback loop between planning and execution. Ultimately, this enables a more detailed and dynamic, data-driven approach to mining – one in which resource models evolve alongside real-world conditions, and every truckload of ore contributes to a more accurate, continuously improving geological and operational picture.

The success of these emerging technologies depends on the availability of detailed, content-rich resource models. High-resolution models provide the data required to:

- Accurately delineate ore and waste boundaries to support selective mining practices.

- Understand geometallurgical variability to optimise feed and throughput.

- Predict ore behaviour under leaching or processing conditions to enable precision or modular recovery strategies.

Without more detailed resource models, the potential of innovative extraction technologies cannot be fully realised. High-resolution resource estimation workflows are not just enablers of operational optimisation; they are prerequisites for the next generation of mining technologies.

MINERAL RESOURCE ESTIMATION

Mineral resource estimation serves as the critical conduit between geosciences and mine planning, forming the nexus of the mining value chain by translating geological data into actionable insights that define mine design, operational strategy, and ultimately, cash flow and value creation.

Established resource estimation workflows are reasonably robust and offer a structured approach to data collection, geological interpretation, resource estimation, and resource classification.

However, as the mining industry enters a new era of data abundance, these workflows are showing signs of strain. The primary bottleneck lies in their ability to integrate high-resolution, multivariate data sets into established practices.

Modern geoscience technologies now generate vast quantities of objective and detailed information ranging from hyperspectral core scans to downhole sensing data that capture a rich spectrum of orebody characteristics at multiple scales and densities. Much of this data remains underutilised, often stored without being effectively translated into practical-length drill hole intervals that support 3D geological modelling and resource estimation workflows.

Geological domaining, a cornerstone of resource estimation, exemplifies this challenge. Traditionally, the construction of 3D domains relies heavily on lower-dimensional data sets, usually just a few geochemical assays and subjective logging codes, making it difficult to incorporate the broader range of modern geoscience data. This reliance on subjective methods introduces variability and limits the resolution of resource models.

Similarly, resource classification practices often fail to reflect the granularity and variability captured by high-resolution data sets, leading to conservative or oversimplified outcomes that may not align either with collected information or with operational realities.

Another bottleneck arises in the translation of new insights into practical workflows. The integration of advanced geoscience technologies into resource estimation often requires significant changes to existing processes, which can be met with resistance due to the time, cost, and training involved. This disconnect between technological innovation and established practices slows the adoption of new methods and limits the industry's ability to fully leverage the potential of modern data sets.

While traditional workflows have been instrumental in shaping industry practices, they are increasingly challenged by the scale and complexity of modern data sets and the demands of emerging operational technologies.

Despite the challenges, the opportunities for improvement are clear. By addressing these bottlenecks, the industry can enhance the accuracy, resolution, and reliability of resource models, providing a stronger foundation for operational decision-making.

TRANSFORMING WORKFLOWS

By integrating practical changes into workflows, mining professionals can address bottlenecks, enhance efficiency, and improve the accuracy of resource models. This transformation bridges the gap between modern data acquisition technologies and actionable insights, ensuring resource estimation aligns with the demands of modern mining operations.

Practical steps for change

Step 1 – Preprocessing and data reduction

High-resolution data sets are invaluable but can be overwhelming. Preprocessing ensures they are manageable, actionable, and efficient:

- Data validation and cleaning – Standardise formats, address outliers, and remove inconsistencies to ensure accurate and reliable base data.

- Dimensionality reduction – Techniques like principal component analysis (PCA) focus on key variables, simplifying interpretation without losing critical insights.

- Spatial compositing – Aggregate raw data into meaningful drill hole intervals to align with the resolution needed for geological modelling and resource estimation.

These steps streamline workflows, preserve key insights, and make high-resolution data practical for resource estimation.

Step 2 – Geological interpretation and domaining

Geological interpretation and domaining create the structural and compositional framework for resource estimation. While advanced technologies and vast data sets can feel overwhelming, they offer geologists an unparalleled opportunity to radically enhance the quality and reliability of resource estimates. These include:

- Integrated approaches – Combining traditional expertise with advanced analytical methods such as clustering algorithms and machine learning enhances reproducibility and reduces subjectivity.

- Logging and interpreting drill holes in 3D – With the advanced spatial visualisation capabilities in most modern modelling software packages, geologists can consider drill holes and their high-resolution data as 'long, skinny outcrops' that can be used for 3D mapping and interpretation. This approach vastly improves the 3D understanding and modelling of complex controls on mineralisation and rock properties, compared to traditional 2D cross-sectional methods.

- Geometallurgical context – Domains are increasingly defined with geometallurgical attributes, such as hardness and recovery potential, linking geological features to operational performance. Examining a broad range of data and data sources improves the quality of the geological model.

- Defining domains – The use of multivariate analytics provides powerful tools for interpreting controls on mineralisation, enabling geologists to better understand and model the spatial and volumetric characteristics of orebodies. Typically, a resource geologist establishes 3D volumes (domains) that capture rock types, alteration zones, or mineralisation envelopes. However, as data becomes more multivariate (ie capturing grades, textures, mineralogical properties, geomechanical indices, metallurgical attributes etc), each variable may not align neatly with a single set of geological boundaries. Even in a single deposit, the boundary best suited for grade can differ from that for contaminants or metallurgical parameters, influencing both resource characterisation and mine planning. As data sets become more complex, variable-specific domains, which might entail separate wireframes or specialised clustering, become essential.

Step 3 – Incorporating multivariate geostatistics

It is not uncommon for resource geologists to rely on single-variable analyses. Modern data sets, however, capture multiple, interconnected orebody attributes that can enable a more nuanced understanding of these relationships, leading to better predictions and resource models.

Geostatistical techniques such as co-located kriging and soft indicator kriging have long been applied in the oil and gas sector to address similar challenges. Building on these and other experiences, recommendations to adapt current workflows include:

- Statistical analyses – Analyse relationships between variables using correlation techniques, including between transformed variables within and across domains, and examine these spatially.

- Variogram analysis – Examine multivariate spatial relationships with cross-variograms and soft indicator variograms and compare and contrast trends within and between domains.

- Multivariate kriging – Methods such as co-kriging, co-located kriging, universal kriging, and soft indicator kriging integrate multiple correlated variables and data types providing a more informed basis for decision-making.

- Geostatistical simulations – Similarly, multivariate simulations that deliberately incorporate point and spatial correlations across multiple attributes provide a richer result and opportunity to translate orebody characteristics into further optimisation processes.

Step 4 – Classification

Classification translates the outputs of resource estimation into confidence levels, ensuring resource categories align with data quality, geological understanding, and reporting standards. With the adoption of multivariate geostatistics, classification can leverage the interconnected nature of modern data sets to improve accuracy, transparency, and operational relevance. These include:

- Confidence and uncertainty variables – Contemporary approaches to classification incorporate data density, geological continuity, and modelling accuracy. There is opportunity to further enhance the classification process by leveraging multivariate outputs that capture more orebody complexities and rock mass characteristics.

- Quantitative and probabilistic methods – Advanced geostatistical techniques, such as conditional simulation and multivariate kriging, provide probabilistic frameworks for classification. These methods quantify uncertainty and variability across multiple attributes, offering a more objective and transparent foundation for resource categorisation.

- Operational relevance through multivariate insights – Incorporating geometallurgical outputs derived from multivariate analysis and modelling, such as hardness, recovery potential, and processing behaviour can help ensure classifications not only reflect geological confidence but also align with operational performance and strategic goals.

EMERGING TECHNOLOGIES

Looking to the future it is worth considering emerging technologies that can offer pathways for improved mineral resource estimation workflows.

This section examines the role of artificial intelligence (AI), machine learning (ML), quantum computing, and supercomputers as transformative tools for the minerals industry.

Artificial intelligence and machine learning

The integration of artificial intelligence (AI) and machine learning (ML) into resource estimation workflows is reshaping how the industry handles the vast and complex data sets generated by modern geoscience technologies.

These technologies go beyond traditional statistical methods, offering powerful tools to process multivariate data at scale. Their ability to identify patterns, uncover relationships, and predict outcomes has opened new opportunities to redefine workflows, enhance decision-making, and reduce reliance on manual and subjective processes.

One of the most promising applications of ML lies in geological modelling and domaining. High-resolution data sets, such as hyperspectral and geochemical imaging and scanning, capture intricate details about orebodies, but interpreting these data sets effectively is challenging. AI/ML approaches are powerful tools for 3D pattern recognition. However, the patterns must be geologically reasonable, which will require validation and modification by experienced geoscientists. As proven in other applications, such as AI-assisted radiology and medical diagnoses, and 'Centaur' (human + AI) teams in chess and Go, the combination of AI/ML and human intelligence will be the most powerful and reliable way forward.

ML algorithms, including clustering, decision trees, and neural networks, provide advanced methods to detect patterns and trends across multivariate data, offering a systematic and reproducible approach to domaining. Unlike manual interpretation, which often varies depending on individual expertise, these tools bring a level of consistency and precision that is difficult to achieve through traditional methods. For example, clustering algorithms can delineate geological domains by grouping data points with similar properties, while neural networks can uncover hidden relationships between structural and mineralogical features.

In predictive analytics and classification, supervised learning techniques have emerged as valuable tools for forecasting ore grades, mineralisation trends, and geometallurgical behaviour. These techniques rely on historical and training data sets to build predictive models that inform resource classification.

By incorporating probabilistic methods, ML enhances confidence in resource categories, ensuring they align with both geological understanding and operational needs. For instance, a supervised model could predict recovery rates based on ore composition, helping to optimise classification frameworks for processing efficiency. This shift toward data-driven classification reduces uncertainty and enhances transparency, building stakeholder confidence.

Another significant advantage of AI lies in its ability to facilitate real-time data integration. Modern mining operations generate a continuous stream of data from drilling rigs, processing plants, and monitoring systems. AI systems can dynamically ingest and analyse this data, updating resource models in real time. This capability ensures that models reflect current conditions, enabling geologists and engineers to make informed decisions quickly. For example, real-time updates to orebody models can optimise blending strategies or adjust processing parameters to maximise recovery. This dynamic integration bridges the gap between data collection and actionable insights, improving operational responsiveness and efficiency.

The potential of AI and ML to transform resource estimation workflows is immense, but their implementation requires careful consideration. The success of these technologies depends on the availability of high-quality training data, the selection of appropriate algorithms, and the expertise to interpret and validate results. While the promise of AI and ML is clear, their integration must be supported by workforce development and collaboration between geoscientists, data scientists, and mining engineers to ensure their application aligns with industry goals.

Quantum computing

Quantum computing is poised to redefine resource estimation by addressing computational challenges that are currently beyond the reach of classical systems (Gabor et al, 2020a). Unlike traditional computers, which process information in binary (0 or 1), quantum computers leverage qubits that exist in multiple states simultaneously through superposition. This capability allows quantum computers to explore vast combinations of variables in parallel, offering transformative potential for solving complex problems in multivariate geostatistics, simulation, and optimisation.

Quantum computing is advancing rapidly (Doherty, 2021). While traditional quantum computers rely on large-scale cryogenic systems, room-temperature quantum processors, offering a scalable and decentralised pathway for quantum adoption, are being pioneered to make quantum technology more accessible and practical.

Mineral resource estimation presents unique challenges that align well with the strengths of quantum computing:

- Large, multivariate data sets – Quantum processors can handle the computational demands of integrating multivariate data sets, providing richer and more reliable insights into orebody variability (Gabor et al, 2020b).

- Simulation and uncertainty modelling – Quantum-enhanced simulations can offer probabilistic resource models that capture complex interactions and uncertainty across spatial and geological parameters.

- Dynamic decision support – The real-time processing capabilities of quantum systems, especially as room-temperature processors become more accessible, can enable dynamic updates to resource models and operational plans.

For resource estimation, quantum computing holds promise in several key areas:

- Multivariate geostatistics – Quantum algorithms, such as quantum-enhanced co-kriging and variogram analysis, can process complex spatial relationships across multiple variables simultaneously. This capability allows geoscientists to integrate high-resolution data sets, such as grade, mineralogy, and recovery potential, into resource models with unprecedented precision.

- Simulation of orebody behaviour – Quantum systems excel at handling probabilistic problems, enabling more detailed geostatistical simulations that account for variability and uncertainty. These simulations offer richer insights into orebody behaviour, supporting risk assessments and decision-making for mine planning and optimisation.

- Optimisation of resource classification and domaining – Quantum computing's ability to evaluate numerous scenarios in parallel provides a powerful tool for refining resource classification and domaining workflows. This optimisation can improve the alignment of models with operational requirements, enhancing accuracy and efficiency.

- Quantum sensor data processing with quantum computers – Quantum systems are uniquely sensitive sensors of electromagnetism and gravity and provide immense refinements of data granularity and quality (Degen, Reinhard and Cappellaro, 2017; Stray *et al*, 2022). Such quantum sensors can be accompanied by ruggedised and miniaturised quantum computers for in-sensor data processing. This would, eg allow on-the-fly principal component analysis or singular value decomposition, both of which play an increasingly important role in geostatistics. Significant gains in speed and accuracy have been proven (Huang *et al*, 2022; Martyn *et al*, 2021) indicating that joint quantum sensor-computer could alleviate much of the burden of data preprocessing and data reduction.

A shift to quantum algorithms presents a unique opportunity for the resource estimation community to reimagine workflows and redefine geostatistical methodologies.

Supercomputing

As quantum computing continues to develop, advancements in supercomputing provide an immediate and transformative solution for addressing computational challenges in resource estimation. Technologies such as NVIDIA's cutting-edge GPU-driven systems, including the GeForce RTX 50 Series GPUs and the Project Digits personal AI supercomputer, offer unprecedented processing power that is increasingly accessible and scalable. These advancements enable the mining industry to handle massive geoscience data sets, perform complex analyses, and accelerate workflows that were previously constrained by computational limitations.

Supercomputing platforms play a crucial role in advancing machine learning integration for geological modelling, domaining, and resource classification. By processing vast data sets efficiently, these systems allow geoscientists to deploy sophisticated AI algorithms to identify patterns, improve predictive accuracy, and refine resource models. In addition, enhanced computational power supports the creation of detailed 3D models that incorporate geological, geochemical, and structural data, enabling real-time scenario analysis. This capability improves decision-making, fosters collaboration, and reduces the time required to evaluate resources. Furthermore, high-performance computing (HPC) platforms excel at running large-scale geostatistical simulations, which incorporate multivariate relationships and spatial variability, offering probabilistic assessments that enhance risk management and operational planning.

The recent announcement of tools like NVIDIA's Project Digits (a personal AI supercomputer capable of delivering up to one petaflop of performance) democratises supercomputing power. By making high-performance computing available to individual users and smaller organisations, these innovations empower resource geologists to process complex data sets and apply advanced analytics directly from their workstations.

Supercomputing, therefore, serves as an essential bridge technology, offering immediate opportunities to modernise workflows, integrate high-resolution data, and adopt AI-driven methodologies, while the mining industry prepares to embrace the transformative potential of quantum computing.

WORKFORCE DEVELOPMENT

The integration of advanced technologies such as AI, machine learning, quantum computing, and supercomputing into resource estimation workflows presents significant opportunities but also poses challenges related to workforce competency. As these technologies reshape the landscape of resource estimation, the skills and expertise required to effectively leverage them are evolving rapidly. The industry must address these changes through targeted training, cross-disciplinary collaboration, and the development of new frameworks for defining and assessing competency.

Evolving skillsets for a technological shift

The move toward high-resolution data sets and advanced computational tools demands new technical capabilities. Geoscientists and resource professionals must build proficiency in areas such as:

- Data science and analytics – Understanding multivariate statistics, machine learning algorithms, and data visualisation is critical for integrating high-resolution geoscience data into workflows.

- Computational proficiency – Familiarity with supercomputing platforms, programming languages (eg Python, R), and geostatistical software is becoming increasingly essential.

- Interdisciplinary thinking – Professionals must bridge the gap between geosciences, engineering, and computational fields to effectively collaborate on integrating emerging technologies (Coombes, 2025).

Competency frameworks and certification

Traditional competency frameworks, often focused on technical expertise and experience in geological modelling, domaining, and classification, must expand to include these emerging skills. Certification programs should evolve to:

- Incorporate technology training – Include modules on AI, ML, and quantum computing applications in resource estimation.

- Assess multivariate reasoning – Evaluate professionals' ability to interpret and apply high-dimensional data sets in geological contexts.

- Develop adaptive competency – Foster skills for working in dynamic environments where new technologies and workflows are continually emerging.

Upskilling through continuous learning

The rapid pace of technological advancements requires a commitment to lifelong learning. Companies and institutions should:

- Provide targeted training programs – Develop courses on advanced geostatistics, AI/ML driven modelling, and quantum applications tailored for geoscientists and engineers.

- Encourage cross-disciplinary collaboration – Promote exchanges between mining professionals, data scientists, and quantum computing experts to drive innovation and knowledge sharing.

- Support mentorship and peer learning – Establish networks for sharing expertise and experiences, fostering a culture of co-learning within the resource estimation community (Coombes, 2025).

Leadership and strategic vision

As workflows evolve, the definition of leadership within the resource estimation community must also adapt. Leaders will need to:

- Navigate technological change – Guide teams through the adoption of new tools and methodologies while maintaining alignment with strategic goals.

- Promote ethical and responsible use – Ensure that emerging technologies are applied transparently, consistently, and in accordance with reporting standards.

- Drive investment in innovation – Advocate for funding and resources to support the development and implementation of cutting-edge technologies.

A COMMITTEE FOR THE FUTURE OF MINERAL RESOURCE ESTIMATION

The minerals resource estimation community stands at a crossroads, where advancements in technology, data acquisition, and computational power offer unprecedented opportunities to transform resource estimation workflows. However, realising this potential requires a united effort to drive innovation, develop standards, and prepare the workforce for the future. To achieve this, we propose the establishment of a dedicated Committee for the Future of Mineral Resource Estimation.

Purpose and objectives

The proposed committee's mandate would be to guide the resource estimation community through the challenges and opportunities posed by emerging technologies, ensuring that practices evolve to meet the demands of a rapidly changing industry.

Key objectives should include:

- Setting a vision for innovation – Develop a forward-looking roadmap for integrating advanced technologies, such as AI, machine learning, quantum computing, and supercomputing, into resource estimation workflows.

- Standardising practices – Define best practices and guidelines for leveraging high-resolution data, multivariate geostatistics, and advanced modelling tools while maintaining alignment with international reporting standards.

- Building cross-disciplinary collaboration – Act as a platform to bring together geoscientists, data scientists, mining and processing engineers, technology providers, and regulatory bodies to address shared challenges and develop innovative solutions.

Core focus areas

The proposed committee should focus on critical areas that underpin the future of resource estimation:

- Workflow modernisation – Identify bottlenecks in current workflows and recommend technologies and methodologies to overcome them, including dynamic domaining, multivariate simulations, and AI-driven modelling.

- Technology integration – Evaluate emerging technologies for their applicability, scalability, and potential impact on resource estimation, fostering early adoption and adaptation within the industry.

- Competency development – Develop frameworks for training and certifying professionals, ensuring the workforce is equipped to leverage new tools and methodologies.

- Global collaboration – Facilitate dialogue between international stakeholders to harmonise standards and share insights, fostering a unified approach to innovation.

Structure and governance

The proposed committee should be an inclusive body representing diverse perspectives and expertise, with members drawn from:

- Industry leaders and practitioners in geosciences and resource estimation.

- Academic and research institutions advancing geostatistics and computational science.

- Technology vendors and innovators specialising in scanning and sensing systems, AI, ML, quantum computing, and supercomputing.

- Regulatory and reporting bodies to ensure alignment with global standards.

Regular working groups and conferences would enable the committee to address emerging challenges, publish findings, and refine its strategies.

Why now?

The rapid evolution of technology, combined with increasing demands for operational efficiency, sustainability, and transparency, makes this the ideal moment to act. Without coordinated efforts, the industry risks falling behind, unable to fully harness the transformative potential of modern tools and data sets. A dedicated committee could ensure that resource estimation remains at the cutting edge, capable of meeting the challenges of a data-driven future.

Call to action

We urge industry leaders, practitioners, researchers, and policymakers to support the formation of a Committee for the Future of Mineral Resource Estimation. By pooling resources, expertise, and vision, we can ensure that resource estimation workflows evolve to not only meet the demands of today but also anticipate the needs of tomorrow.

This initiative is a critical step toward positioning the resource estimation community as a leader in mining innovation, ensuring its contributions remain vital, accurate, and sustainable in an increasingly complex global landscape.

CONCLUSION

The mineral resource estimation community is on the brink of a transformative era, driven by advancements in technology, high-resolution data acquisition, and computational power. However, these opportunities bring with them the need for significant changes in workflows, methodologies, and professional competency.

This paper has explored how emerging technologies such as AI, machine learning, quantum computing, and supercomputing can address current bottlenecks and elevate resource estimation practices to meet the demands of modern mining.

Our key messages are:

- Modernise workflows – The integration of multivariate geostatistics, high-resolution data, and advanced computational tools is essential for improving accuracy, efficiency, and reliability in resource estimation. Transforming workflows to incorporate these technologies will enable better predictions, dynamic decision-making, and operational alignment.

- Empower professionals – Developing new competency frameworks, fostering interdisciplinary collaboration, and prioritising workforce upskilling are critical to equipping geoscientists and engineers to effectively leverage emerging technologies.

- Leverage innovation – Quantum computing and supercomputing offer unparalleled opportunities to address computational challenges, optimise workflows, and create richer, more actionable resource models. These technologies are not only enablers of efficiency but also catalysts for rethinking geostatistical algorithms and methodologies.

- Collaborative leadership – The establishment of a Committee for the Future of Mineral Resource Estimation provides a structured approach to drive innovation, standardise practices, and prepare the industry for the complexities of a data-driven future.

The resource estimation community must act decisively to embrace these advancements, foster collaboration across disciplines, and align with the evolving needs of the mining industry. By doing so, it can position itself as a leader in mining innovation, delivering smarter, more sustainable, and operationally relevant resource models.

We are at a pivotal nexus where we can fundamentally reimagine resource estimation. By aligning geological modelling, domaining, and classification with the capabilities of modern technologies, the community can set a new global standard for accuracy, transparency, and adaptability. Together, we can continue to ensure that resource estimation remains a cornerstone of the mining value chain, contributing to a more efficient, responsible, and data-driven industry.

ACKNOWLEDGEMENTS

The authors wish to acknowledge the contributions of the broader mining and geoscience community for their ongoing efforts in advancing resource estimation practices. Specifically, we extend our gratitude to Mining3 for their pioneering work in developing transformational mining technologies which have informed many of the concepts discussed in this paper.

Special thanks to Paul Hodkiewicz and Mark Mattingley-Scott for their expertise and perspectives on emerging technologies, particularly in the areas of innovative geoscience technologies and quantum computing, which have significantly enriched this work.

We are grateful for the technological advancements and insights provided by organisations such as Quantum Brilliance and Pawsey Supercomputing, whose innovations could shape the future of resource estimation workflows and geoscience data integration. Their dedication to pushing the boundaries of computational and analytical capabilities has inspired much of this discussion.

The views expressed in this paper are those of the authors and do not necessarily reflect the positions of the organisations mentioned. Any errors or omissions remain the responsibility of the authors.

REFERENCES

Coombes, J, 2025. Cracking the Competency Code: A Blueprint for Excellence in Minerals Reporting, SageAbility, Perth, Australia.

Degen, C L, Reinhard, F and Cappellaro, P, 2017. Quantum Sensing, *Rev Mod Phys*, 89(3):035002, American Physical Society. Available from: <https://journals.aps.org/rmp/abstract/10.1103/RevModPhys.89.035002>

Doherty, M, 2021. Quantum Accelerators: A New Trajectory of Quantum Computers, *Digitale Welt*, 5:74–79. https://doi.org/10.1007/s42354-021-0342-8

Gabor, T, Sünkel, L, Ritz, F, Phan, T, Belzner, L, Roch, C, Feld, S and Linnhoff-Popien, C, 2020a. The Holy Grail of Quantum Artificial Intelligence: Major Challenges in Accelerating the Machine Learning Pipeline, in Proceedings of the First International Workshop on Quantum Software Engineering, ICSEW'20.

Gabor, T, Zielinski, S, Roch, C, Feld, S and Linnhoff-Popien, C, 2020b. The UQ Platform: A Unified Approach to Quantum Annealing, in *Proceedings of the 5th International Conference on Computer and Communication Systems (ICCCS)*, pp 115–119. https://doi.org/10.1109/ICCCS49078.2020.9118547

Huang, H-Y, Broughton, M, Cotler, J, Chen, S, Li, J, Mohseni, M, Neve, H, Barbush, R, Kueng, R, Preskill, J and McClean, J R, 2022. Quantum advantage in learning from experiments, *Science*, 376(6598):1182–1186. https://doi.org/10.1126/science.abn7293

Martyn, J M, Rossi, Z M, Tan, A K and Chuang, I L, 2021. Grand Unification of Quantum Algorithms, *PRX Quantum*, 2(4):040203. https://doi.org/10.1103/PRXQuantum.2.040203

Mining3, 2017. In Place Mining: A transformational shift in metal extraction, Mining3.com. Available from: <https://www.mining3.com/place-mining-transformational-shift-metal-extraction/>

Oraby, E A, Eksteen, J J and O'Connor, G M, 2020. Gold leaching from oxide ores in alkaline glycine solutions in the presence of permanganate, *Hydrometallurgy*, vol 198.

Stray, B, Lamb, A, Kaushik, A, Vovrosh, J, Rodgers, A, Winch, J, Hayati, F, Boddice, D, Stabrawa, A, Niggebaum, A, Langlois, M, Lien, Y-L, Lellouch, S, Roshanmanesh, S, Ridley, K, de Villiers, G, Brown, G, Cross, T, Tuckwell, G, Faramarzi, A, Metje, N, Bongs, K and Holynski, M, 2022. Quantum sensing for gravity cartography, *Nature*, 602:590–594. https://doi.org/10.1038/s41586-021-04315-3

Use of machine learning to predict recovery of alumina involving multiple analytical methods at Worsley Alumina

J Dangin[1], P Soodi Shoar[2], T Wilson[3], J Harvey[4] and C Cavelius[5]

1. Data Scientist, DeepLime, Perth WA 6000. Email: johann.dangin@deeplime.io
2. Principal Resource Geology, South32, Perth WA 6000. Email: payam.soodishoar@south32.net
3. Data Scientist, DeepLime, Perth WA 6000. Email: tom.wilson@deeplime.io
4. Lead Resource Geology, South32, Perth WA 6000. Email: joshua.d.harvey@south32.net
5. Lead Developer, DeepLime, Pau France 64000. Email: claude.cavelius@deeplime.io

ABSTRACT

The bauxite deposit at South32's Worsley Alumina (Worsley) in Western Australia extends over ~4000 km^2 from Brookton in the north to Collie in the south (NB: South32 owns 86 per cent of Worsley Alumina, with 10 per cent held by Japan Alumina Associates (Australia) Pty Ltd and the remaining 4 per cent held by Sojitz Alumina Pty Ltd). The deposit has been explored and mined for over 40 years and contains more than 2.5 million analytical results. Over time, the analytical method to determine Available Alumina (AAl$_2$O$_3$) and Reactive Silica (RSiO$_2$) has been revised to better predict possible extraction of alumina through the refinery in response to the change in underlying mineralogy. Linear regressions between different methods are currently being used to convert all data into the latest analytical method (Worsley Laboratory Available Alumina – WLAA) to adjust for differences between the analytical methods. A significant amount of effort is currently needed to maintain this process as new data is added to the database, including evaluating sensitivities and impacts of updating the regressions.

An approach using Machine Learning (ML) has been developed to infer WLAA values for AAl$_2$O$_3$ and RSiO$_2$ where the measurements are based on different analytical methods. The automated workflow selects samples that have been analysed with multiple methods, to derive predictive models for each analyte in WLAA. These models are used to predict the missing WLAA results.

Additional workflows have been created to automate geological interpretation, using ML classified geological domains and implicitly modelled surfaces for each domain. This enables rapid assessment and validation of any changes resulting from updating the predicted WLAA values.

INTRODUCTION

Alumina refining relies on the Bayer Process, an industrial method to separate Alumina from bauxite ore (Habashi, 1994). This process was first developed in 1887 by Karl Josef Bayer though it has variations depending on differences in the source and quality of the ore (Hind, Bhargava and Grocott, 1999). The Bayer Process operates as a continuous cycle that involves four main stages: digestion, clarification, precipitation and calcination, it includes two key steps: 1) The pressure leaching of bauxite with NaOH (caustic soda) solution to yield sodium aluminate, and 2) The precipitation of pure aluminium hydroxide from this solution by seeding with Al(OH)$_3$ crystals.

Two critical measurements are used to assess the Alumina recovery potential of bauxite ore: Available Alumina (AAl$_2$O$_3$), which is a measurement of the Extractable Alumina and Reactive Silica (RSiO$_2$), which is the silica present in predominantly kaolinite that reacts in the extraction process. Accurate measurement of AAl$_2$O$_3$ and RSiO$_2$ is essential, as they influence nearly all aspects of bauxite mining operations. Mine planning, mineral processing, metallurgy, refinery operations, infrastructure, and the overall economics of the mine all depend on a thorough understanding of AAl$_2$O$_3$ and RSiO$_2$ within the orebody.

The Darling Range in South-west Western Australia hosts significant bauxite deposits and has been a key global source of bauxite ore for over 60 years (Hickman *et al*, 1992). South32 operates the Worsley mine on the eastern flank of the Darling Range, covering some 4000 km^2. Bauxite mining has been underway at Worsley since the 1980s (South32, 2024), with drilling data dating back to the 1950s. A vast database of historical drilling and mining data is therefore available. Analytical methods for determining AAl$_2$O$_3$ and RSiO$_2$ have evolved over the years to best represent the digestion process at the refinery, and to optimise mining and blending processes to maximise recovery. Reconciling historical data with more recent assay and analytical methods can be challenging but is critical for obtaining accurate estimates of ore grades and tonnages to optimise the recovery of AAl_2O_3 and $RSiO_2$. To date, simple regressions have been used to adjust between the analytical methods, however the correlation between the methods can be complex, and significant time and effort is spent generating statistical relationships between the methods to reduce bias. Developing a more robust method to help correlate between the different analytical methods was needed in order to ensure accurate reporting and modelling now and into the future.

Modern machine learning (ML) algorithms offer a unique and powerful solution to the challenges presented by decades of drilling data at Worsley. The current Worsley database includes over 370 000 drill holes with more than 2.5 million samples. The database contains missing and heterogenous data from decades of mining operations. Such a large and disparate database can create a significant challenge for traditional modelling and geostatistical workflows. ML algorithms are easily scaled and highly parallelised, allowing large database inputs such as the one present at Worsley (eg Schnitzler, Ross and Gloaguen, 2019). The flexibility and scalability of ML algorithms make it an ideal tool to assist in solving the unique challenges presented by the data at Worsley.

WORSLEY CASE STUDY

To manage the large size of the Worsley deposit and supporting databases, 13 lateral domains have been defined (Figure 1). The current routines to update geological models within each of these domains are carried out individually and involve extensive localised evaluation of the available data to ensure that the resulting mineralised zones are optimised on an un-biased data set. The aim of the Worsley case study was to utilise Python scripting and supervised AI models (ML) to significantly reduce the turnaround time of the geological models by automating key time-consuming processes and to enable rapid updates and assessments of changes from the addition of new results.

Geological setting

The bauxite deposits of the Darling Range province in Western Australia, including the deposits at Worsley, formed through intense weathering of Archaean granite and greenstones of the Yilgarn Craton during the Tertiary period (Hickman *et al*, 1992). These lateritic bauxites are rich in aluminium bearing minerals, including gibbsite, and generally low in reactive silica. Geology varies across the deposits, with two main basement types (granite and greenstone), each with distinct weathering characteristics, and networks of dolerite dykes causing significant heterogeneities across the area. This variability is a significant factor in mine planning and refining, with variations in silica content in particular leading to additional processing steps and increased costs. Despite this, the deposits remain economically significant, supporting large-scale open pit mining and alumina production.

Evolution of analytical methods for alumina recovery

AAl_2O_3 represents the proportion of aluminium oxide in bauxite which can be economically extracted using the Bayer process. Other forms of alumina may be present but are not economically viable to be extracted, for instance due to being bound in clay minerals which do not dissolve readily in caustic soda (Metson, 2011). $RSiO_2$, present in bauxite as kaolinite or other clay minerals, presents significant challenges for the Bayer method, as it forms insoluble sodium aluminosilicates when reacting with caustic soda. This reaction ties up aluminium in the insoluble sodium minerals, leading to a reduction in refinery efficiency. Additional caustic is then required to maintain extraction, leading to increased costs, and the insoluble waste requires long-term storage, resulting in significant containment and environmental remediation expenditure (Wang *et al*, 2019).

American Bayer Extractable Alumina (ABEA) is the primary industry method for measuring AAl_2O_3 and $RSiO_2$. However, since 2000, the Worsley Refinery has adjusted the sample preparation and analysis protocol to better represent the alumina extraction process at the refinery. The 'Worsley Laboratory Available Alumina' (WLAA) is the current representation of the Worsley refinery process and has been used since 2012. The historical changes to the different analytical methods used have often resulted in small differences in the results. The cost of re-assaying historical samples is prohibitive, and therefore for every method change, adjustments are made to standardise the result to the latest analytical method to allow the data being used to optimise downstream processes. For the purposes of this study, industry standard analytical methods are kept un-redacted, whilst proprietary methods are coded A though I.

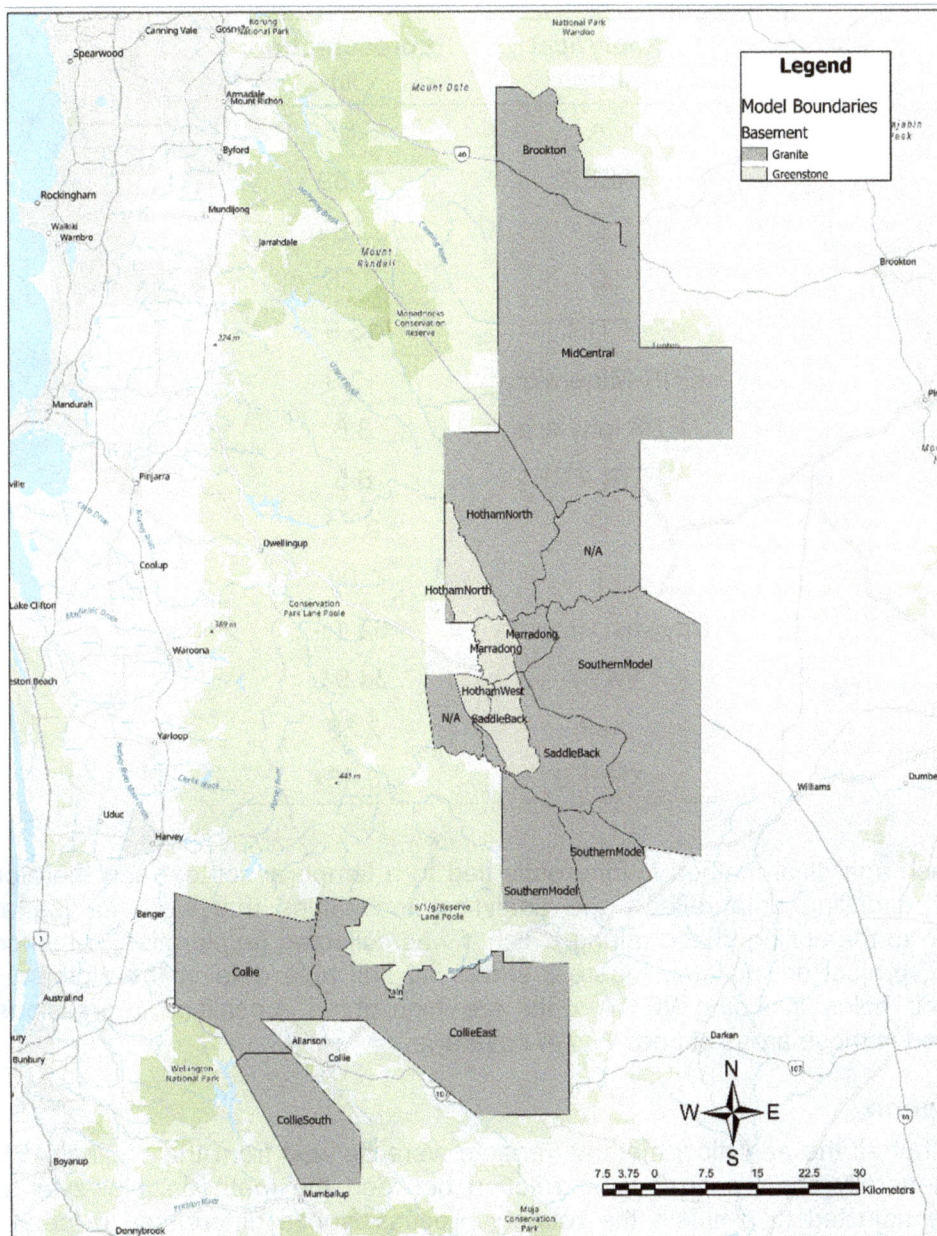

FIG 1 – Location of model areas for the Worsley deposit, Darling Range, Western Australia.

Data challenges

Data quantity and sparsity

The Worsley database contains over 2.7M records. A portion of the samples acquired have been analysed for key geochemical variables (Table 1). However, it is worth noting that not all samples have values for every chemical element, reflecting the evolution of necessities at the plant. The data also shows large groups of samples acquired through different analytical methods (A through I, WLAA, ABEA, XRF). In this data set approximately 31 per cent of samples have a value for AAl_2O_3 WLAA, whilst approximately ~20 per cent have a value for Method B $RSiO_2$. Whilst certain oxides like P_2O_5 have been analysed in a smaller subset of samples, these samples and this data still offers valuable insights into specific geochemical trends and provides constraint for the ML models to follow.

TABLE 1

Representation of analytical methods in the database. Industry methods shown un-redacted, proprietary methods coded A to I.

Analytical method	Representation (%)
A	6.58
ABEA	23.69
B	22.62
C	0.08
FTIR	13.9
FTIR-Mineralogy	13.9
FTIR-physical	3.8
FTIR-XRF	8.5
H	0.22
I	0.02
Physical Property	52.11
WLAA	34.94
XRF	2.38
XRFFS	0.43

Moreover, each analytical method is inherently tied to a temporal context and therefore a spatial dimension. The drilling data reflects the analytical techniques that were in use at the time, corresponding to the drilling phase during which it was collected (exploration, resource definition, grade control, project drilling etc). Figure 2 shows all drill hole data in the Hotham North area database. Drill holes including WLAA data are highlighted. Localised regressions could be misrepresented in those areas of poor WLAA coverage.

Data correlation

Considering that all the analytical method variants were derived from the original ABEA method, the results generally show a positive correlation between the methods. However, since these variants were adjusted to simulate the refinery process, subtle differences in performance are noted depending on the mineralogy. A significant challenge arises from the considerable overlap between the methods, which precludes the use of first-order transformations to align assay ranges effectively. Additionally, the management of outlier data remains a critical consideration in any regression or transformation process.

For both AAl_2O_3 and $RSiO_2$, most methods exhibited strong correlations. As shown in Figure 3, $RSiO_2\_WLAA\_ICP\_pc$ and $RSiO_2\_B\_pc$ demonstrate an overall good correlation, with an R^2 value of 0.961 for the full data set. However, a subset of points deviates significantly from this trend, particularly in the upper regions of the plot. These outliers, characterised by higher values on the y-axis ($RSiO_2\_B\_pc$) relative to their corresponding x-axis values ($RSiO_2\_WLAA\_ICP\_pc$), suggest some discrepancies between the two variables. Such deviations may result from differences in measurement techniques or sample preparation. The presence of these outliers complicates the prediction of WLAA values.

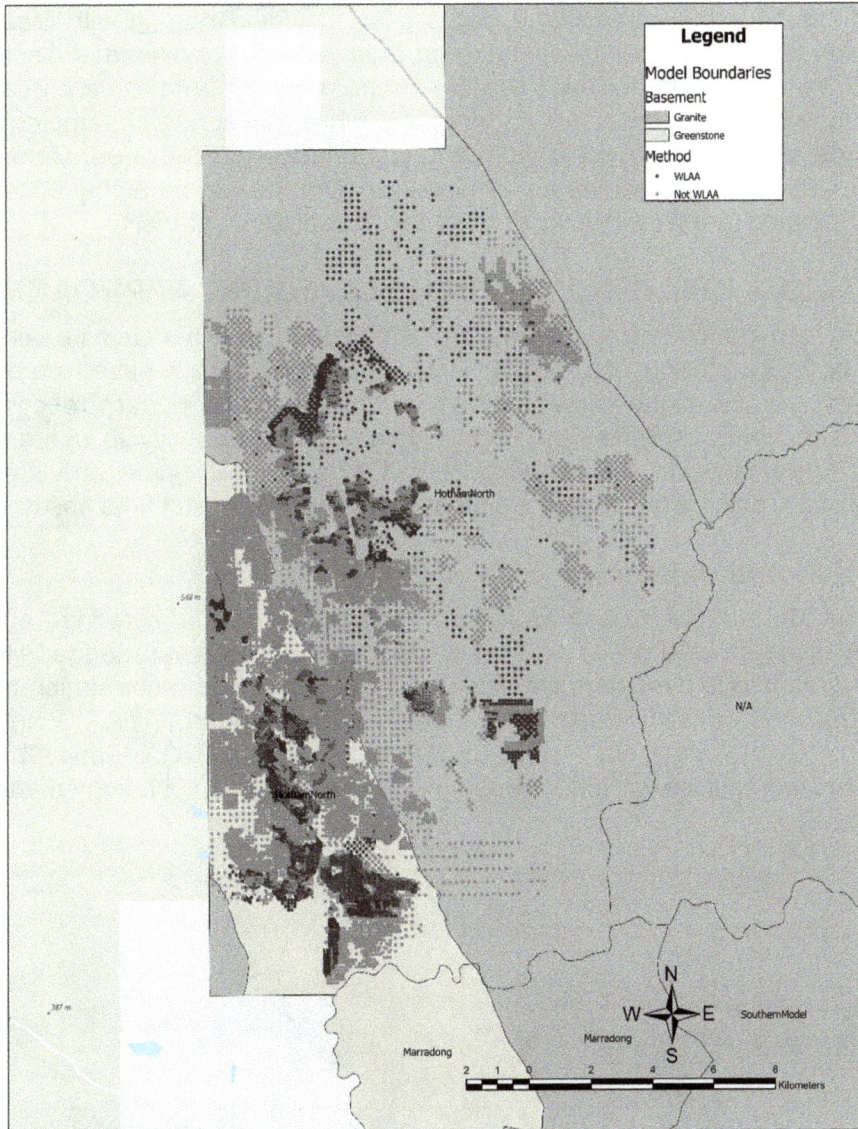

FIG 2 – All drill hole data used for training the ML models in Hotham North showing the spatial distribution of WLAA results.

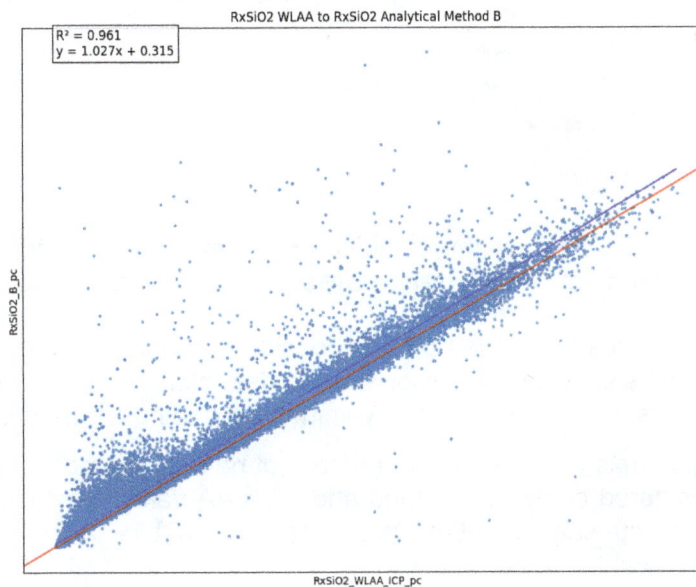

FIG 3 – The correspondence between analytical method B and WLAA values for $RSiO_2$ are globally on a linear regression but do show a number of outliers.

Other correlation methods have been tried in the past, including Kriging with Measuring Internal Drift (KMID), which is used for point estimation in data sets with systematic discrepancies. This method relies on the correlation between two data sets, requiring both to be present in the same area to estimate one based on the other. In our case, while thousands of samples were analysed using two methods, they were collected from a geographically limited area. Consequently, for a vast region, only one method's results are available, making co-kriging highly uncertain. Although co-kriging was attempted, the results did not meet the acceptance criteria.

PREDICTING WLAA USING THE MACHINE LEARNING APPROACH

Relationships between geochemical parameters are often unintuitive and non-linear (Jooshaki, Nad and Michaux, 2021; Zhang *et al*, 2024). The current regression-based prediction methods utilised at Worsley lack the higher-dimensionality and multi-variate pattern recognition capability that modern ML algorithms can offer. In practice, this means the ability to use a wide variety of input geochemical, geophysical and lithological variables to help guide and constrain the ML training and prediction, going well beyond a simple regression to predict final assay values.

Data preparation and selection

Before training the ML model for each analytical method, the data is screened and subjected to quality control. A model is created for each analytical method, for every logged lithology and for both granite and greenstone basement lithologies. Figure 4 shows the correlation between AAl_2O_3 values from a selection of analytical methods to the modern standard WLAA. Points are coloured by logged lithology (GR – Gravel, HC – Hard Cap, Friable Bauxite (BZ/BC), and CL – Clay). These figures indicate that regressions will differ depending on the basement lithology of the sample.

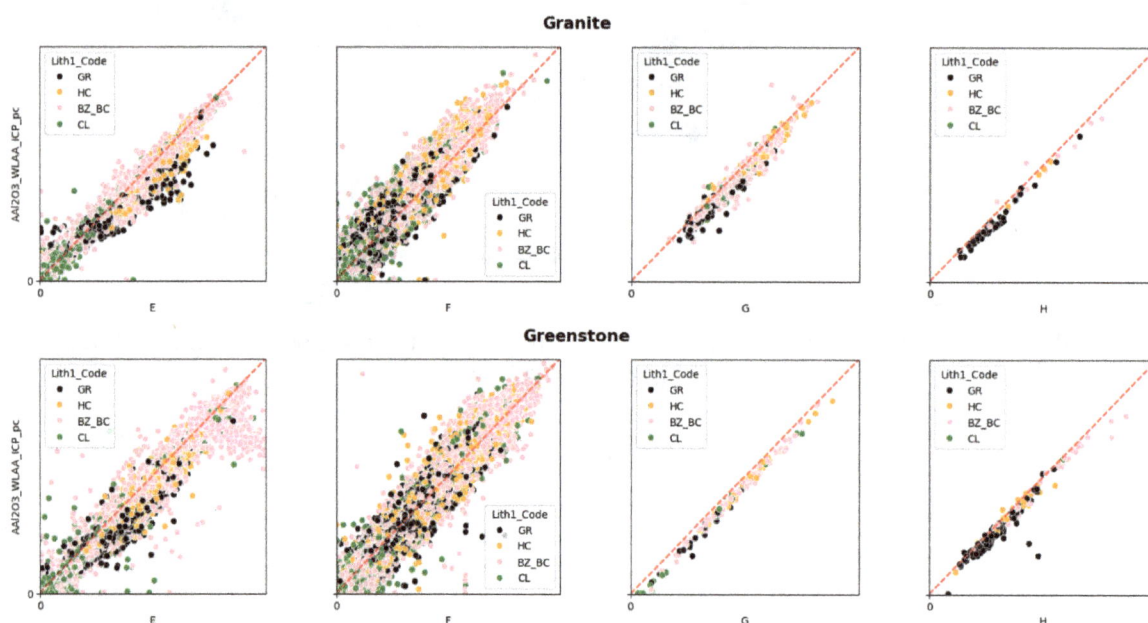

FIG 4 – Comparison of paired data in the combined Hotham North, Marradong and Saddleback areas across both basement types (Granite/Greenstone) and multiple analytical methods.

In selecting data, only samples that have both the considered analytical method and WLAA values are kept. Next, data points that fall below or on the detection limit are removed. An option is provided to remove outliers, which are defined by values exceeding a cut-off quantile.

Only machine learning models with a sufficient number of paired data points are retained (samples that have both the considered analytical method and a WLAA value), ensuring that the analysis is based on robust, reliable data sets for optimal performance.

Model training – XGBoost

We utilised the XGBoost regressor, implemented via the Python library *xgboost* (Chen and Guestrin, 2016), to model the relationship between legacy analytical methods and the WLAA values. The XGBoost Regressor algorithm was chosen for this project as it is one of the rare algorithms to natively deal with non-values. As any gradient boosted decision tree, it builds an ensemble of decision trees, with each tree aiming to minimise the errors made by the ensemble of previously built trees. This iterative process is guided by gradient descent principles, where the algorithm seeks to optimise a loss function, typically a Mean Squared Error (MSE), for regression tasks. Additionally, as decision-tree models are invariant to feature scales, and Gradient Boosting Decision Trees natively deal with null values, the effort on the feature engineering part was significantly reduced.

Feature selection and importance

The models are trained on all available geochemistry, including the proprietary analytical methods A through I, ABEA, WLAA, and XRF reads, along with mineralogy, geophysics (Magnetic Susceptibility), X and Y coordinates. Work has been conducted to test the influence of various features, particularly examining the impact of including versus excluding coordinates, as well as implementing different strategies for removing outliers within the training data.

Figures 5 and 6 show the features (variables) importance towards the model output based on their gain. The Gain measures the contribution of each feature to the models' prediction by calculating the improvement in the object function every time the feature is used to split a node. Features that lead to a larger reduction in the error are considered more important.

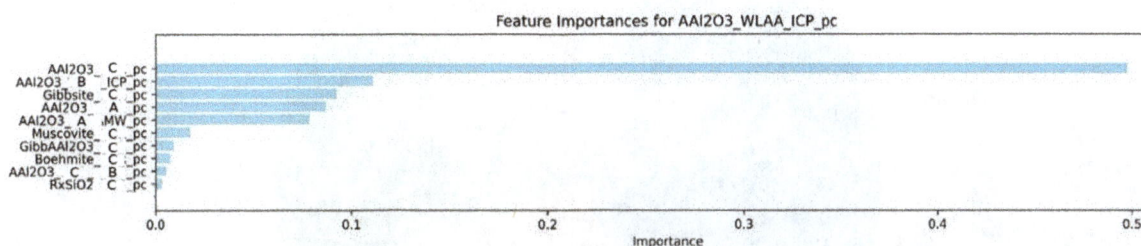

FIG 5 – Feature importance for AAl_2O_3 WLAA equivalent training in the bauxite zone within Granitic basement areas. Ten most important features shown.

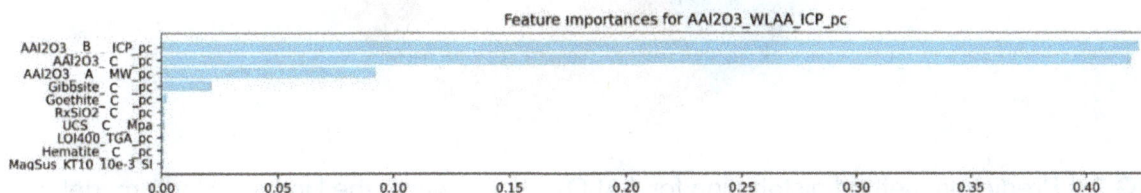

FIG 6 – Feature importance for AAl_2O_3 WLAA equivalent training in the bauxite zone within Greenstone basement areas. Ten most important features shown.

Figures 5 and 6 demonstrate that the prediction of AAl_2O_3 WLAA equivalence relies on different features for samples overlying Granite and Greenstone basement lithologies. This can be attributed to the fact that certain methods are more prevalent depending on the basement geology, and, to a lesser extent, some analytical methods may exhibit stronger correlations within a specific basement geology. AAl_2O_3 values from analytical method C impact more than 50 per cent of the predicted value of AAl_2O_3 WLAA values within the granitic basement zones. AAl_2O_3 values from Methods B and C impact more than 80 per cent of the predicted value of AAl_2O_3 WLAA within the greenstone basement areas.

Workflow for predicting WLAA values

While some analytical methods share enough samples in common with a WLAA value, other methods do not have enough 'paired' data. Yet, they have enough occurrences for them to be valuable in deriving a less robust WLAA value. In such a case, these analytical method values are used to infer WLAA values using a Quantile Transformation (QT). The distribution of the geochemistry is mapped to the one of the WLAA distribution. This transformation of values was achieved using NumPy's percentile and 1D-interpolation functions to compute quantiles and interpolate the transformed values, respectively (Harris *et al*, 2020).

In order of prevalence, the final WLAA value of a sample is given by:

- Its WLAA value if one exists, or
- Its Machine Learning predictive value if one exists, or
- Its Quantile Transformed value, or
- Its original value.

Each sample geochemical value for AAl_2O_3 and $RSiO_2$ is therefore given an index of confidence associated with the method that its value has been obtained. A breakdown of the distribution of prediction methods for the Hotham North model area is shown in Figure 7.

AAl2O3_WLAA Prediction Method - Hotham North Model Area

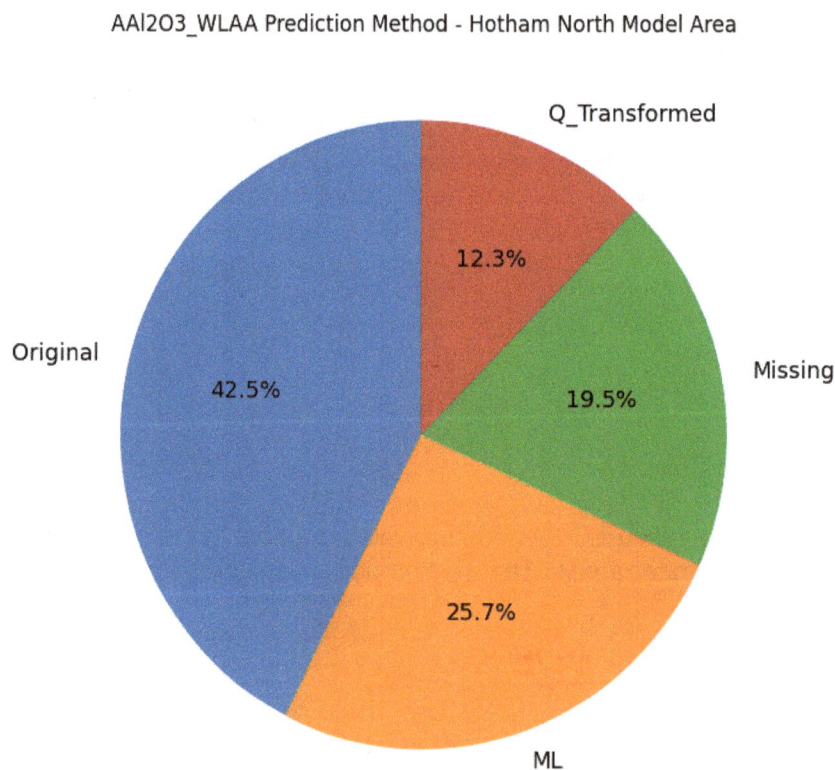

FIG 7 – Prediction method distribution for AAl_2O_3_WLAA within the Hotham North model area.

ML model validation

To validate the performance of each model, we use either a random split selection from the data, or data from a specific area. Random split selection ensures that the training data and test data set retain the same overall data distribution and provides an unbiased evaluation of the model's performance. Specific area selection tests the model's ability to generalise to specific geological contexts and gives a gauge on the robustness of the model. Lastly, the grade distributions of the predicted results were tested in select areas where infill drill samples were analysed using WLAA.

Validation of ML model against test data

WLAA predicted values are compared to a random selection of their true values which are hidden from the training test data set. The charts in Figure 8 show how predicted values (eval_) compare to the analytical method values. The test data area (Marradong) shows similar distributions between test and training data, with both cross plot and histograms giving close matches of their distributions.

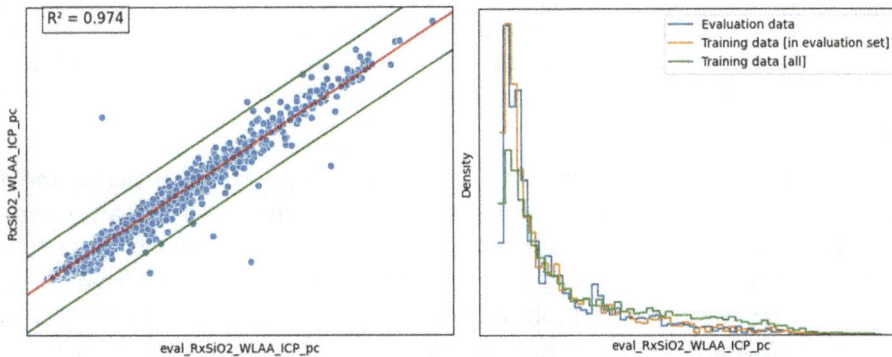

FIG 8 – The prediction of RSiO$_2$ WLAA-equivalent values is comparable to existing WLAA values on the test samples in Granitic bauxite zone. R^2 of 0.974.

The distribution of residuals shows that while most predicted values exhibit less than ±2 per cent deviation from their WLAA values, there are notable outliers with discrepancies exceeding 10 per cent. These larger deviations are consistent with the inter-method analytical variability observed for identical samples, as documented in section 3.2.

More generally, the Root Mean Squared Error (RMSE, Sahu, Srivastava and Saha, 2020), is used to quantify the performance of regression models. It measures the average magnitude of the errors between predicted and actual values, giving more weight to larger errors due to the squaring of differences. RMSE for the ML Models are shown in Figure 9. RMSE values for the predicted AAl$_2$O$_3$ and RSiO$_2$ across both Greenstone and Granite basements are indicative of a robust model. RMSE values for AAl$_2$O$_3$ of approximately 1.5 represent a ~3 per cent error given an AAl$_2$O$_3$ range of 0 to 60 per cent. For RSiO$_2$, the RMSE varies between ~0.5 to 1.8, an error between 1 to 4.5 per cent. RMSE appears consistently robust across both basement types and logged lithologies.

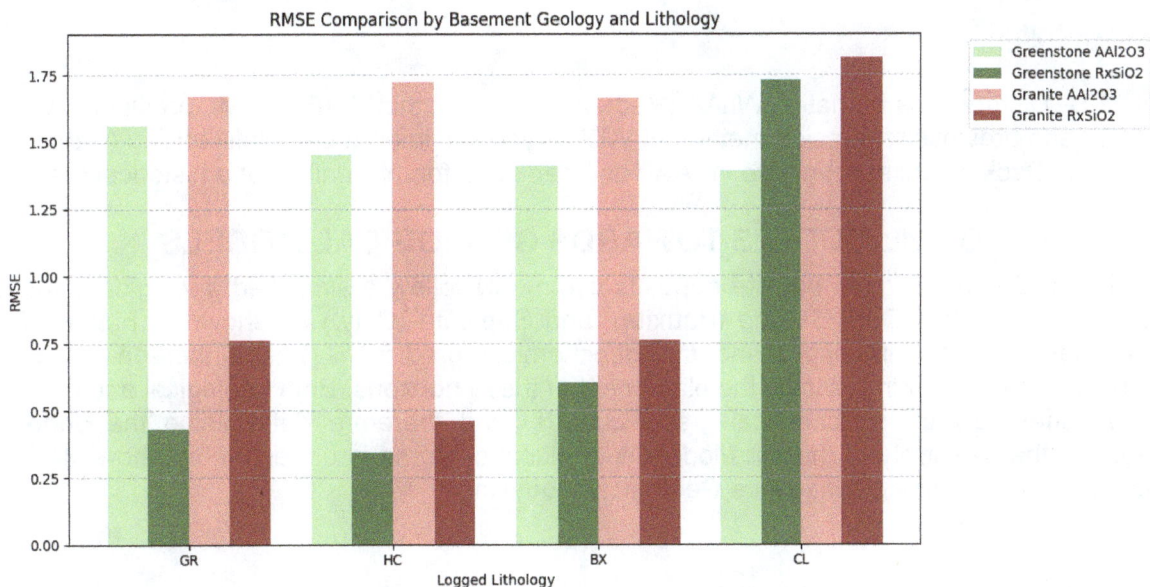

FIG 9 – RMSE of regression model per logged lithology and basement geology of sample (GR – Gravel, HC – Hard Cap, BX – Friable Bauxite, CL – Clay).

As demonstrated in section 3.2, a measurement error between WLAA and analytical method B is evident. The standard deviation of these intervals ranges from 1.62 to 1.81 for AAl$_2$O$_3$ and from 0.48 to 2.42 for RSiO$_2$ across the lithologies in the Marradong (Figure 1). The RMSE values fall well within the variability range of the analytical methods.

Using local area of infill drilling

To ensure the reliability of our analyses, it is crucial to locally validate the results of any transformations or regressions applied to the data. This validation process confirms that the statistical parameters and distribution of the transformed data closely align with those of the native WLAA data in the same region. Extensive investigations were conducted in ten areas within the Saddleback domain (Figure 1), utilising primary assays from analytical method B while analysing infill drill holes with the WLAA method.

The results from the infill drilling area are presented in Figure 10. The black line illustrates the native reference distribution from WLAA, whilst the blue line represents the primary data derived from analytical method B. Our objective is to transform the blue line to match the black line seamlessly. The graph clearly exhibits that the results generated by the AI method (red line) introduce significantly less variance in estimation than the current regression results (yellow line). This indicates the strength and effectiveness of our approach, paving the way for more accurate and reliable data analysis.

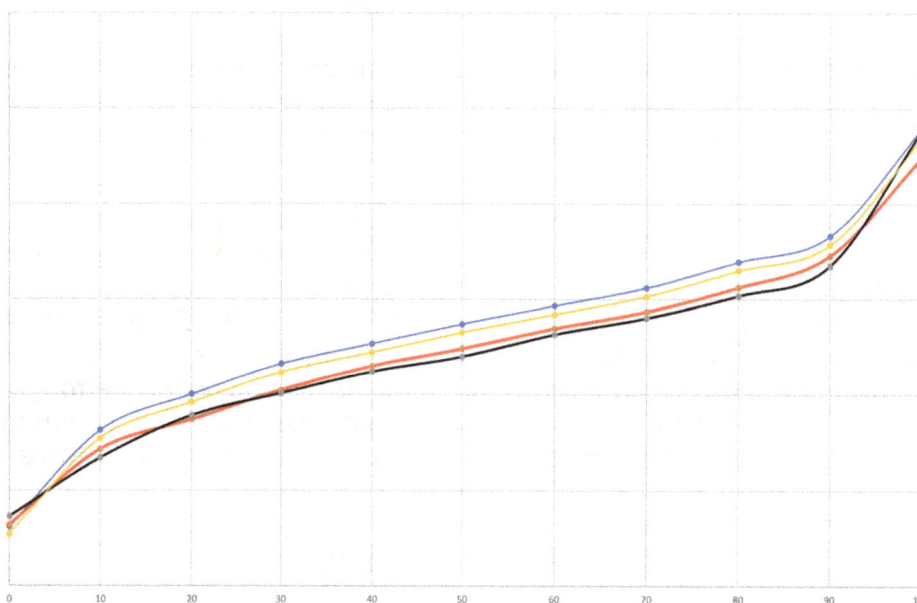

FIG 10 – Distributions for native WLAA (Red), analytical method B (Blue), ML predicted WLAA (black) and previous regression method of WLAA (yellow), in an area of infill drilling (within the SaddleBack model area) where WLAA has been used for infill drilling of a historical area.

SENSITIVITY OF ML MODELS TOWARDS GEOLOGICAL MODELS

Modelling of Darling Ranges bauxite deposits commonly follow a simplified geological sequence (Topsoil, Gravel, Hard Cap, B-Zone (Bauxite), and Saprolite (Clay)) as shown in Figure 11. ML Models described in this paper and the geochemical predictions based thereon will impact geological modelling by influencing the placement of these horizons. Both geological and economic bauxite zones rely on geochemical grade cut-offs, and therefore determining the underlying controls on the sensitivity of the ML Models is a critical outcome in understanding how ML based geochemical predictions will influence Geological Modelling.

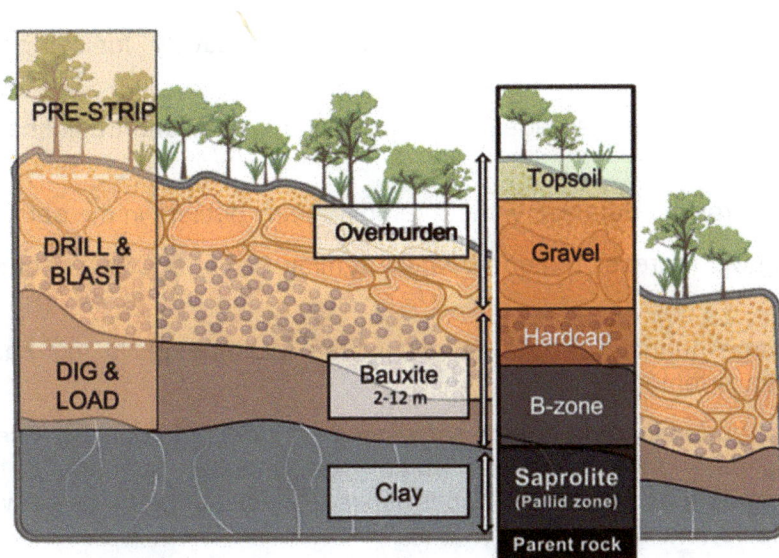

FIG 11 – Schematic of the geological horizons within the bauxite deposits of the Darling Range (Soltangheisi, George and Tibbett, 2023).

Automated workflow to geological models

An automated workflow is in place at Worsley to predict the lithology of each sample based on its geochemistry, geomechanical and geo-metallurgical attributes (Figure 12). While this is not discussed in-depth in this article, it is hereby mentioned to explain the impact of the WLAA predictive machine learning models on how geological volumes are estimated.

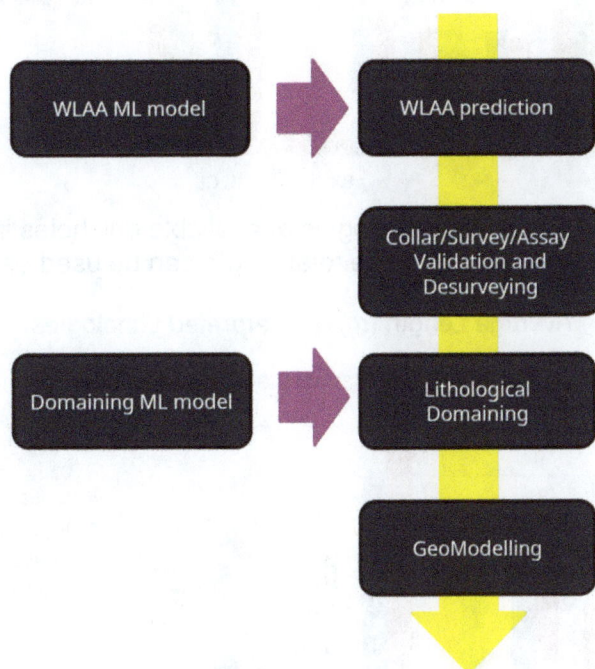

FIG 12 – Machine learning workflow for WLAA prediction, domaining geomodelling.

Impact of WLAA predictive ML model on geological model

To understand the main controls influencing the WLAA prediction ML models, the average and total length of each predicted lithology is compared on one of the model areas, Hotham North. Three scenarios were tested, each giving insight into model performance and generalisability under varying conditions. Each scenario was compared to the original scenario, and to a baseline model following the current interpretation procedures at Worsley. For this sensitivity study, we will use the 1) Total Length and 2) Average Length of interpreted lithologies in drill holes as a proxy to the 1) Ore volume and 2) Average domain thickness, respectively, for comparison with current methods.

The controls that are being checked against the original scenario and traditional geological model are:

- Using the entire training data set, without removing any outliers in the data.

- Only accepting ML regression models where a minimum of 5000 pairs of data can be found (up from 500).

- Excluding a large quantity of data as part of the ML training. In this case, data from the Marradong area is removed.

Scenario 1 allows an understanding of how sensitive the model is to outliers. Scenario 2 tests how the size of training data pairs influences model performance. Scenario 3 tests the model's robustness to missing data and its spatial dependence. Relative differences between the scenarios are shown in Figures 13 and 14.

Total Length (m) of Interpreted Lithologies

Comparison to Manual Geo Model Baseline

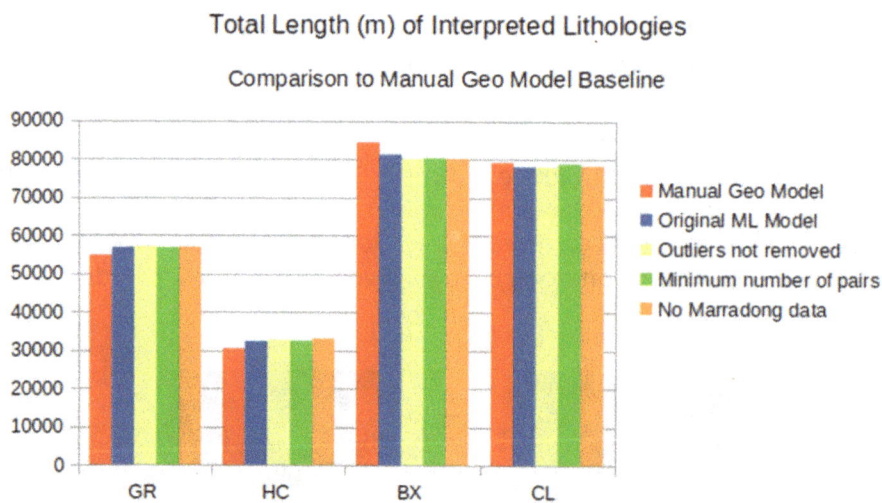

FIG 13 – Total length (m) of interpreted lithologies at available drill holes in Hotham North resulting from the four scenarios investigated. The total length can be used as a proxy for volume.

Average Length (m) of Interpreted Lithologies

Comparison to Manual Geo Model Baseline

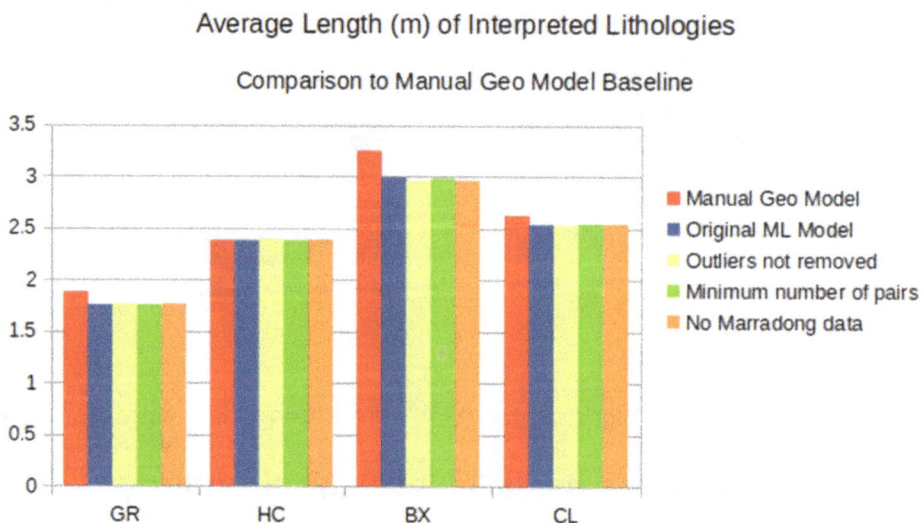

FIG 14 – Average length (m) of interpreted lithologies at available drill holes in Hotham North resulting from the four scenario investigated. The average length can be used as a proxy for predicted average domain thickness.

Figure 13 shows the total length of interpreted lithologies in each of the scenarios investigated. Compared to the manual geological model baseline, each ML scenario displays a close match with little variation. Figure 14 shows the average length of Interpreted lithologies in each scenario. Again, the ML scenarios closely fit the baseline geo model.

It can be established that the geological model, in its elementary form (Gravel, Hard Cap, Friable Bauxite (B-zone), Clay (Saprolite)), will not be overly sensitive to these controls for the creation of the Machine Learning Models.

Overall, the results demonstrate good correlation between models interpreted by the Geologist (reference model) and the ML derived models. A number of factors could be adjusted to improve the performance of the models. In particular, the ML classification of samples into lithological codes is a key factor in this analysis that is not investigated in this paper.

In summary, the result of the ML approach is achieved in a fraction of the time it takes for a manual geological model update using the current procedures at Worsley. This represents a significant opportunity for geologists to spend time on refining the models and optimising short-term mining activities.

Future work

Using ensemble methods, such as Gradient Boosting Decision Trees, provided an efficient way to evaluate whether a machine learning (ML) approach was suitable for the database at Worsley. These methods offered a reliable means to establish a solid baseline, which can be further refined through progressive model tuning in future work.

With a robust baseline now in place, interpreting feature importance alongside data correlation analysis can help narrow down feature selection and guide the choice of appropriate data imputation techniques. This step is especially important for exploring ML architectures that are incompatible with missing values, such as neural networks. Future work could focus on optimising hyperparameters to improve model performance, using techniques like grid search or Bayesian optimisation.

CONCLUSIONS

The application of Machine Learning into South32's Worsley Alumina database management and analytical workflows represents a significant advancement in addressing the challenges posed by decades of drilling data. By leveraging ML models, the prediction of WLAA-equivalent values for AAl_2O_3 and $RSiO_2$ from legacy analytical methods has been streamlined, offering a scalable, accurate, and automated solution. These predictive models reduce reliance on time-intensive regression analyses and enable more precise reconciliation of historical and modern data sets.

While the data presents significant challenges allowing for its sparsity and heterogeneity, the implementation of Python workflows to address this complex problem allows for quick analysis of the sensitivity of such ML models towards the final geological models. This framework also enables the future testing of other ML and AI algorithms, hyperparameter optimisation, advanced outlier handling, and the inclusion of any additional geochemical and geophysical features to further refine the models.

These workflows nest themselves within the automated process for geological modelling at Worsley and ensure a rapid and consistent update of any model in a matter of hours from the laboratory data to a block model. This approach has demonstrated adaptability to the complex and heterogeneous geology of the Worsley deposits, ensuring operational efficiency and informed decision-making at Worsley Alumina.

ACKNOWLEDGEMENTS

The authors would like to thank South32 and the Innovation team for supporting the work, and the Worsley geology team for their help reviewing the models.

REFERENCES

Chen, T and Guestrin, C, 2016. XGBoost: A scalable tree boosting system, *Proceedings of the 22nd ACM SIGKDD International Conference on Knowledge Discovery and Data Mining*, pp 785–794. Available from: <https://xgboost.readthedocs.io/>

Habashi, F, 1994. Bayer's process for alumina production--a historical perspective, *Cahiers d'Histoire de l'Aluminium*, 13.

Harris, C R, Millman, K J, van der Walt, S J, Gommers, R, Virtanen, P, Cournapeau, D, Wieser, E, Taylor, J, Berg, S, Smith, N J, Kern, R, Picus, M, Hoyer, S, van Kerkwijk, M H, Brett, M, Haldane, A, del Rio, J F, Wiebe, M, Peterson, P, Gerard-Marchant, P, Sheppard, K, Reddy, T, Weckesser, W, Abbasi, H, Gohlke, C, Oliphant, T E, 2020. Array programming with NumPy, *Nature*, 5857825:357–362. https://doi.org/10.1038/s41586-020-2649-2

Hickman, A H, Smurthwaite, A J, Brown, I M and Davy, R, 1992. Bauxite mineralisation in the Darling Range, Western Australia, Geological Survey of Western Australia, Report No 33:31:41.

Hind, A R, Bhargava, S K and Grocott, S C, 1999. The surface chemistry of Bayer process solids: a review, *Colloids and Surfaces A: Physicochemical and Engineering Aspects*, 146(1–3):359–374.

Jooshaki, M, Nad, A and Michaux, S, 2021. A systematic review on the application of machine learning in exploiting mineralogical data in mining and mineral industry, *Minerals*, 11(8):816.

Metson, J, 2011. Production of alumina, in *Fundamentals of aluminium metallurgy*, Woodhead Publishing, pp 23–48.

Sahu, S S, Srivastava, P K and Saha, S, 2020. Evaluation of machine learning models for multivariate regression, *Materials Today Proceedings*, 27(3):2549–2554.

Schnitzler, N, Ross, P S and Gloaguen, E, 2019. Using machine learning to estimate a key missing geochemical variable in mining exploration: Application of the Random Forest algorithm to multi-sensor core logging data, *Journal of Geochemical Exploration*, 205:106344.

Soltangheisi, A, George, S and Tibbett, M, 2023. Soil characteristics and fertility of the unique Jarrah forest of southwestern Australia, with particular consideration of plant nutrition and land rehabilitation, *Land*, 12:1236. Available from: <https://doi.org/10.3390/land12061236>

South32, 2024. Mine Development, South32 [online]. Available from: <https://www.south32.net/what-we-do/our-locations/australia/worsley-alumina/mine-development> [Accessed: 2 Jan 2025].

Wang, L, Sun, N, Tang, H and Sun, W, 2019. A review on comprehensive utilization of red mud and prospect analysis, *Minerals*, 9:362. Available from: <https://doi.org/10.3390/min9060362>

Zhang, S, Carranza, E J M, Fu, C, Zhang, W and Qin, X, 2024. Interpretable machine learning for geochemical anomaly delineation in the Yuanbo Nang District, Gansu Province, China, *Minerals*, 14(5):500.

Building on the shoulders of giants – geostatistics and geodata science

G Nwaila[1] and S Zhang[2]

1. Associate Professor, School of Geosciences, University of Witwatersrand, Braamfontein 2000 Johannesburg, South Africa. Email: glen.nwaila@wits.ac.za
2. Research Scientist, Geological Survey of Canada, Natural Resources Canada, Ottawa ON K1S 4L5, Canada. Email: steven.zhang@nrcan-rncan.gc.ca

ABSTRACT

Geostatistics is the theory and application of spatial statistics of regionalised variables. Its greatest success is its application to mineral resource and reservoir modelling, from orebody characterisation to mineral resource estimation. The geostatistical data modelling framework noticeably overlaps with the data science framework. The overlap presents an opportunity to advance both disciplines through hybridising their strengths. The guiding principle is to preserve the applicability and end-to-end explainability of geostatistics, while adopting the replicability, automatability and big data-suitability aspects of data science. This style of adaptation aims to ensure that:

- Geospatial data modelling will be practiced by a larger community, beyond its current confines.

- Geospatial modelling will be well-positioned to handle modern big data, particularly high-velocity and high-dimensional data.

- Geodata scientists and geostatisticians can achieve functional interoperability, providing a unifying framework and complementary methods to validate outcomes.

- There will be functional specialisation between data science and geostatistics.

This paper summarises various attempts to integrate geodata science and geostatistics through several examples:

- Domaining using machine learning (ML).

- A novel ML-based, geostatistics-inspired interpolation method (microblocking).

- Geo-manifold learning (GML).

Since the paper presents topics that are transdisciplinary, it adopts the broader and more general terminology in data science to describe various concepts, except where they are idiosyncratic to geostatistics. The examples illustrate that the key applications of geostatistics, such as domaining and interpolation are shared with geodata science and that the geodata science-based approaches are related to geostatistical formulations, but inherit the replicability, generality and automatability of data science methods. Lastly, the subdomain of research – GML is yet to fully emerge but holds tremendous potential for the modelling of non-Euclidean spatial data, by building on both geostatistical ideas and manifold learning.

INTRODUCTION

Geostatistics, like many other traditional statistical and scientific domains, has contributed substantially to the development of geodata science methods (Cressie,1990; Deutsch and Journel, 1998; Friedman, Hastie and Tibshirani, 2001; Abzalov, 2016; Erten, Yavuz and Deutsch, 2021). This is because:

- Geodata science is a general domain that studies the treatment of geodata, which contains all data that would be usable with geostatistical methods.

- Geostatistical development paved the way for many foundational algorithms and methods (eg Krige, 1951; Matheron, 1963, 1971; Deutsch and Journel, 1998; Rossi and Deutsch, 2013).

The overlap in algorithms between geostatistics and data science is undeniably increasing, as witnessed by a variety of algorithms becoming more general purpose, through increasing aspatial generalisations and mathematical and computational formalism (Chatterjee, Bandopadhyay and Rai, 2008; Chatterjee, Bandopadhyay and Machuca, 2010; Nwaila *et al*, 2024). For example, kriging, a quintessential class of interpolation methods in geostatistics, is the progenitor of a class of kernelisable data modelling methods in data science – Gaussian process (GP). In the geostatistical implementation, kriging makes use of a prior – which is the kernel, and a spatial similarity model – which can be in the form of a covariance function or a semi-variogram (Dutta *et al*, 2006; Samanta and Bandopadhyay, 2009). In GP, the notion of the kernel is generalised through kernelisation, while spatial similarity is replaced by a general notion of metric-driven similarity in the feature space (Matías *et al*, 2004). Closely related algorithms in data science include the support vector machine (SVM), which is another kernelisable algorithm with the property that instead of considering the global data set, it focuses on the marginal configurations. Another algorithm that could be considered to have a strong geostatistical resemblance are all the k-nearest neighbours (kNN) algorithms. For example, for the purpose of label propagation, kNN identifies the most similar neighbours to labelled samples, with feature similarity computed through a metric that could be Euclidean. It is clear that the route to generalisation from geostatistics is to:

- Consider generalised spatial geometry, away from Euclidean geometry or spatial features.

- Construct metrics of sample similarity that is more universal in response to the previous point.

- Consider arbitrary number of dimensions of the feature space (beyond 3) to ensure general suitability (Li *et al*, 2013; Jafrasteh and Fathianpour, 2017; Jafrasteh, Fathianpour and Suárez, 2018).

This route of specialisation to generalisation is the principal of development of the domain of data science – which considers the treatment of all data, including spatial data. Therefore, this is a foreseeable pathway of development.

In contrast, the reverse is seldom true – which is the retro-integration of geodata science concepts and methods, back into geostatistics in theory or at least in application. This direction of research would include:

- Adaptation of data science algorithms and more broadly, methods and approaches, back into geostatistics.

- Development of non-general, but highly useful methods and algorithms to suit geostatistical purposes.

- Inheritance of framework properties of data science into geostatistics, for example, the concept of highly replicable workflows and automated hyperparameter optimisation (Hastie, Tibshirani and Friedman, 2009; Witten *et al*, 2016).

In recent years, scholars have attempted to develop this pathway for several key applications of geostatistics: geodomain delineation; microblocking; and GML. Geodata is becoming more voluminous due to accumulation. However, the far bigger concern for the immediate future is the rise of geodata-as-a-stream, meaning that geodata generation and management will increasingly become more automated and higher velocity, in effect becoming big data proper through increased data velocity (Bourdeau *et al*, 2024; Ghorbani *et al*, 2022; Ghorbani *et al*, 2023b; Shimaponda-Nawa *et al*, 2023). Similar to the data velocity revolution in the biological and life sciences through the rise of the '-omics' fields, the trend towards higher velocity geodata is unlikely to reverse or stall. This is because of an increasing recognition of the future challenges of exploration and mining, for example, by moving deeper underground (Ghorbani *et al*, 2023a), alleviating supply bottlenecks and calming geopolitical tension (Ghorbani *et al*, 2024a; Zhang *et al*, 2023a), generating a broader diversity of geodata to suit analytics and geometallurgy (Bourdeau *et al*, 2024), satiating a general growth in material desire for energy and military needs (Zhang *et al*, 2023a; Ghorbani *et al*, 2024b), and remote operations (Shimaponda-Nawa *et al*, 2023). There is therefore an emergent need to evolve geostatistics to remain relevant in the era of big data-driven, remotely managed and potentially automated mining and exploration activities. Consequently, with the rise of bigger geodata, there must be a way forward to preserve geostatistical intentions and outcomes, while adopting the replicability, automatability and big data-suitability aspects of geodata science (and therefore also data science; eg Hosseini, Asghari and Emery, 2021).

Geodomaining is a class of methods in geostatistics that is concerned with the spatial partitioning of samples. In mineral resource estimation, geodomaining is used to separate contrasting types of rocks (eg ore or host rock). Formulating geodomaining as a proper geodata science task only requires the recognition that the objective is to assign each sample a label, in a manner that is subjected to operational constraints of spatially contiguity. There are studies that examine the use of data science methods to partition geodata for the purpose of geodomaining (eg Emery, Mery and Porcu, 2024; Fouedjio, Hill and Laukamp, 2018). Geospatial interpolation is a class of methods in geostatistics that invert sparse samples into dense representations (but also other related concepts, such as change-of-supports). Formulating spatial interpolation as a geodata science task requires the recognition that the objective is to construct interpolative models (Talebi, Mueller and Tolosana-Delgado, 2019; Talebi *et al,* 2020). Modelling data in generalised geometries is not a traditional subdomain of geostatistics but is under research (eg Hosseini *et al,* 2023; Jiang and Dimitrakopoulos, 2024). Essentially, many of the assumptions of the embedding geometry of the data, where such data is not Euclidean, are violated. This means that non-Euclidean data are generally more manifold-like, to the extent that their bounding surfaces resemble Riemannian manifolds. The clear implication is that Euclidean geostatistics are not useful. However, conducting geostatistics on manifolds can be thought of as a manifold learning task, whose formulation relies on the extension of manifold learning from data science to geodata (Boisvert, Manchuk and Deutsch, 2009; Caixeta and Costa, 2021). The combination of manifold learning and geostatistical goals, using geodata is GML proper.

METHODS

With the exception of GML, geodomain delineation and microblocking-based interpolation are detailed in Zhang *et al* (2023b) and Nwaila *et al* (2024), respectively. Readers are directed to those publications for full technical descriptions. Additionally, for GML, the algorithms were described fully in Caixeta and Costa (2021). Its integration with microblocking-based block modelling is novel as a showcase of the potential of GML and is described below. All workflows were implemented in Python with libraries such as scikit-learn, NumPy, pandas, GeoPandas, Matplotlib and GeostatsPy.

Point-wise geodomain and domain delineation

This study deploys the point-wise geodomaining method developed by Mohammadi *et al* (2022) (here on referred to as SDKMeans) and the domain delineation method developed by Zhang *et al* (2023b). The point-wise geodomaining algorithm is a distance-based adaptation of classic clustering methods. In particular, a distance matrix is used as spatial features for ML tasks, which is generated using all samples. The distance matrix, which has the dimensionality of the cardinality of the data set, is jointly reduced in dimensionality using multi-dimensional scaling (MDS). The resulting features in lower dimensions are then used in a parametric clustering task, using the k-means algorithm (Arthur and Vassilvitskii, 2007) (Figure 1). The method permits a balance of the importance of spatial continuity and sample attributes (eg geochemistry), which permits an explicit tuning of the spatial contiguity of cluster labels, making assigned domains more or less spatially contiguous. The major hyperparameters are associated with the MDS and k-means algorithms.

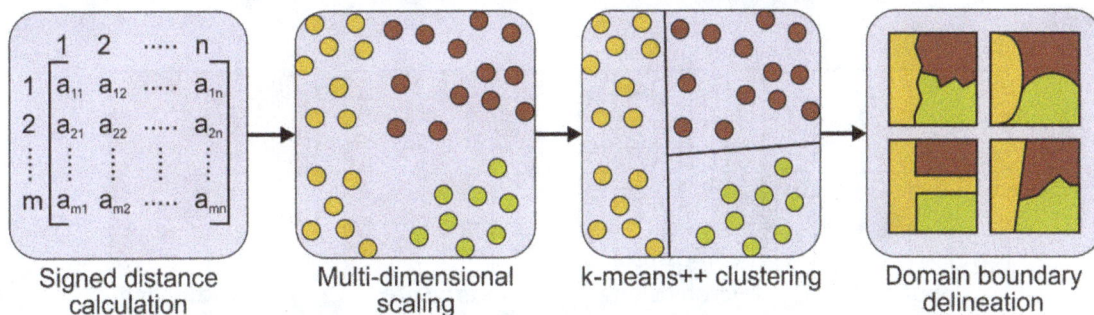

| Signed distance calculation | Multi-dimensional scaling | k-means++ clustering | Domain boundary delineation |

FIG 1 – Illustration of point-wise domaining.

Formulating geodomain delineation using geostatistical ideas and geodata science leverages the recognition that domain membership is a type of class label, which implies that geodomaining is formulate-able into a data science task. This task consists of two parts:

1. Assignment of class labels to all members.

2. Separation of classes in feature space using multi-class classification.

However, a ML-based formulation is necessary but insufficient, because there are additional spatial constraints to class distributions that are atypical of aspatial classification. There are two related requirements:

1. In order for the spatial separation to be meaningful, spatial mixing of class membership must be within tolerance of a downstream spatial task (eg extraction).

2. In order for the spatial class boundaries to be useful to downstream applications, there must be complexity constraints.

In Zhang *et al* (2023b), geodomain delineation was formulated as class boundary delineation using geodata science methods. In particular, given labelled samples, the task to spatially separate classes are formulated using a variety of ML algorithms. This formulation extends a parental geostatistical method for geodomain delineation using radial basis functions, to the data science framework. The implementation used typical ML-based workflows, experimenting with a variety of common algorithms and ultimately recommending SVM with the radial basis function as the kernel for geodomain delineation. The major hyperparameters are associated with kernel specifications for SVM.

The use of the data science framework in both point-wise geodomaining and domain delineation permits an automated solution with an explicitly controllable domain and boundary complexity. The benefits of the geodata science-based geodomaining method are:

- An adoption of a framework that is designed for replicability.

- Broader access to implementation platforms – the method uses standard algorithms, which are openly available in a variety of computing environments, such as Python, R and Matlab.

- A reduction in heuristic uncertainty induced by inter-practitioner subjectivity by simplifying hyperparameter tuning (Owusu and Dagdelen, 2024).

Microblocking

Spatial interpolation is possible using the geodata science framework. Although point-to-point interpolation is already a data science task, its point-to-block counterpart is more technically challenging and highly specific to geostatistics. No data science-based solutions were known prior to Nwaila *et al* (2024). Point-to-block interpolation requires a uniquely geostatistical concept known as change-of-support, which does not exist in data science or ML. All known algorithms capable of interpolation are static support only. Undoing the mathematical approximation of the punctuality of supports and using spatial super-sampling, a super-sampled grid (microblocks) can be constructed, such that each grid cell statistically resembles that of a punctual support. The prediction for all microblocks is performed using one or more ML models, using coordinates as features. Thereafter, down-sampling inferences across microblocks yields inferences for larger blocks (macroblocks) (Figure 2).

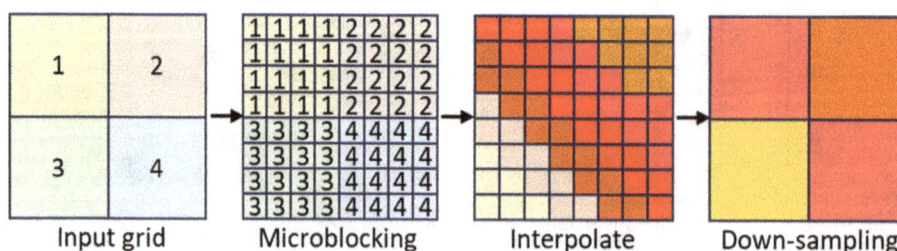

FIG 2 – Illustration of block modelling using microblocking.

The results were quantitatively and qualitatively compared with kriging results over several thousand trials that varied:

1. Sampling condition – biased random or regular.

2. Noise or the nugget effect.

3. Spatial anisotropy.

4. Sampling density.

It was found that the aggregate performance of the microblocks method is qualitatively and quantitatively comparable to kriging, in terms of the central, dispersion and extreme tendencies of the outcome (eg block mean and dynamic range). However, the microblocks method was far more computationally efficient with:

1. A minimum of just one meaningful hyperparameter (the number of neighbours for the kNN algorithm).

2. An algorithmic complexity of $O(n^1)$ compared to kriging's $O(n^3)$ (Zhong, Kealy and Duckham, 2016).

Therefore, the benefits of the microblocking method are:

1. Replicability, automation and metric-driven outcomes.

2. Implementation flexibility in terms of programming languages and software libraries.

3. A reduction in heuristic uncertainty via a reduction in inter-practitioner subjectivity.

4. A vast reduction in computational complexity while achieving a performance similar to that of kriging.

Geo-manifold learning

Within ML, there is a broad category of algorithms collectively called 'manifold learning'. Their entire premise is to project high dimensional data to lower dimensional spaces for the purpose of visualisation and modelling. Their objective is usually to maximally preserve some notion of distance in high dimensional space, to recreate a lower dimensional representation of the data, such that the distortion is minimised. These algorithms include, for example, isometric mapping (ISOMAP) (Tenenbaum, de Silva and Langford, 2000) or equivalently, metric MDS (Mardia, Kent and Bibby, 1979). Modelling non-Euclidean data, such as data from folded structures, faults or other similar geological objects is commonly required for geostatistical tasks such as mineral resource modelling and estimation. However, geostatistics like regular statistics, is formulated for Euclidean spaces. Notions of distance that are non-Euclidean, such as those occurring on generalised Riemannian manifolds, requires further adaptation of distance computations, both for grid distances and cell volumes. Locally varying anisotropy (LVA) and unfolding methods have been developed to address the challenges posed by non-Euclidean geological systems. LVA calculates geodesic distances and models local anisotropies to capture the spatial continuity of geological features, such as undulating orebodies and folded structures (Boisvert, Manchuk and Deutsch, 2009; Machuca-Mory, Rees and Leuangthong, 2015). This method has proven effective in certain scenarios, in resource estimation workflows that rely on geostatistics. However, the utility of LVA is limited by its reliance on extensive parameterisation and its framework reliance on geostatistics. These constraints reduce its adaptability to non-geostatistical workflows, making it unsuitable for broader applications in geodata science. Furthermore, it is unclear if any implementations of LVA in geostatistical workflows can address volumetric distortions that must occur within the grid or mesh due to the use of geodesic distances (eg such distances exist within and between grid cells, but are explicitly calculated only for cell-to-cell distances, not intra-cellular). Unfolding was introduced by Caixeta and Costa (2021), which is partly similar to LVA in that it also leveraged manifold learning for geostatistics-based mineral resource estimation. However, unlike LVA, because unfolding transforms the mesh instead of distance calculations, it is more general and can be implemented in the geodata science framework as a ML task. The functional procedures are:

- manifold extraction

- tangent surface generation

- de-skeletonisation

- mineral resource estimation.

Unfolding can be extended to demonstrate a fully integrated, end-to-end geodata science-based resource estimation workflow that incorporates:

- geodomaining
- microblocking-based interpolation
- unfolding as a GML method.

In particular, the integration of microblocking with unfolding uniquely and simultaneously solves both grid distance and cell volume distortions associated with non-Euclidean geometries.

During manifold extraction, an image-processing algorithm called skeletonisation is used to extract a skeleton from the grid. During tangent surface generation, a distance matrix (or graph) is constructed from the skeletal samples, which is then used as a constraint for metric MDS (or equivalently ISOMAP) to transform the skeleton to a low dimensional representation that maximally honours the original distance structure (Mardia, Kent and Bibby, 1979). Consequently, MDS produces an optimised, low-dimensional representation of the skeleton of the input model (Kruskal, 1964; Borg and Groenen, 2005). During de-skeletonisation, parts of the geometric model that are not already in the skeleton are attached back to the flattened skeleton. This process uses optimisation by minimising the mean squared-error (MSE) between the original distances (using a graph) and the transformed distances. Thereafter, iterative optimisation seeks to adjust coordinates that minimise the local distortion for all points using gradient descent. The process is illustrated in Figure 3.

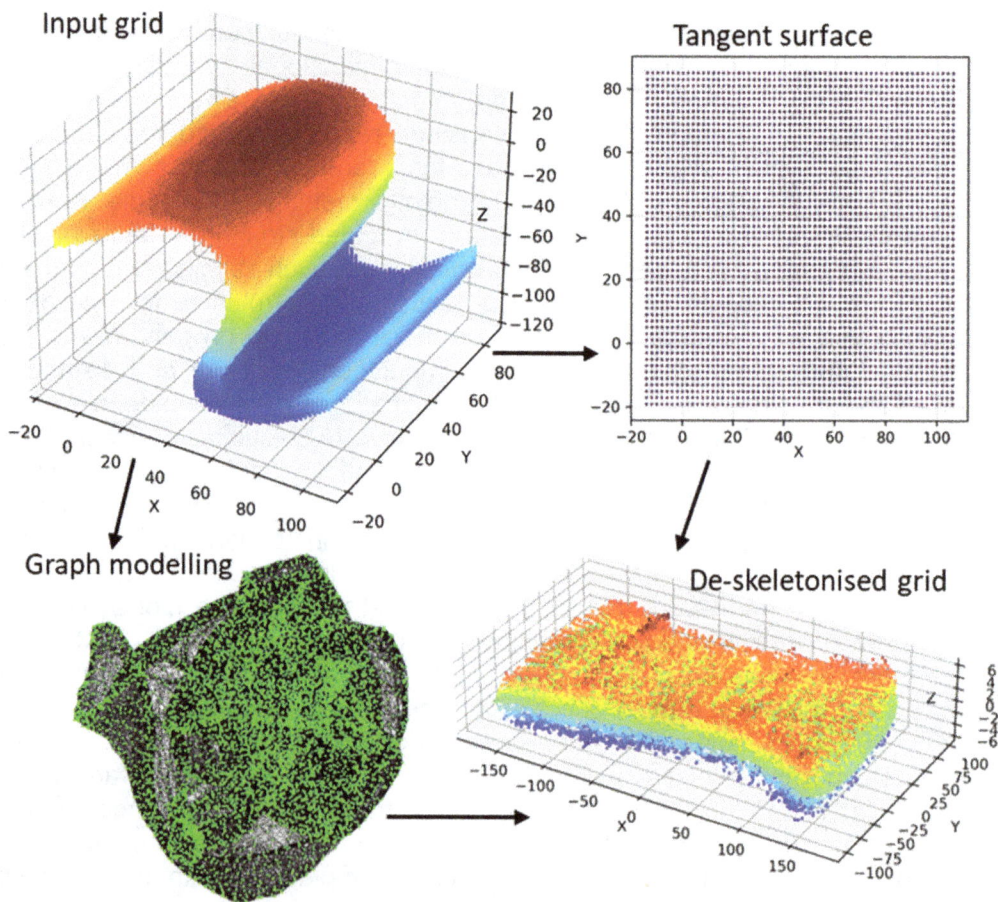

FIG 3 – Illustration of key components of geo-manifold learning.

RESULTS

Case Study 1 – A geodata science workflow for mineral resource estimation in Euclidean geometry

Overview

The first case study demonstrates a hybridisation of geodata science and geostatistics to create a scientifically rigorous and practical geodata science-based workflow for mineral resource evaluation and mine planning. The interpolation algorithm is ordinary kriging, making this first case study a mild integration. The method provides a reproducible geodata science workflow that integrates data preprocessing, ML-based geodomaining and geostatistical interpolation to estimate mineral resources using data from the Merensky Reef (detailed below). This type of integration is useful where objectivity and replicability is required in geodomaining for resource estimation.

The Merensky Reef is located in the northern portion of the Western Lobe of the Bushveld Complex (South Africa). The area is renowned for its platinum-group element (PGE) mineralisation. The data set comprises borehole samples that were used for mining operations. The workflow follows a typical data science framework, beginning with data preprocessing, including quality control to ensure the spatial accuracy of borehole coordinates and the consistency of geochemical data. Outlier detection and spatial validation checks were performed to isolate anomalies. Stratigraphic compositing was used to transform unevenly spaced samples into consistent intervals using a sliding window algorithm. This ensured comparability across samples and provided gridded data for further interpolation. Point-wise geodomaining was then performed using the method proposed in (Mohammadi *et al*, 2022). ML features were engineered through distance calculations, dimensionality reduction (via MDS). Clustering used the k-means algorithm combined with the silhouette score and Davies-Bouldin index. Hyperparameters were tuned iteratively through stochastic gradient descent to maximise cluster separability. Block Modelling used ordinary kriging, to create domain-specific mineral resource models. Variograms were fitted using automated least-squares methods, emphasizing the pre-range portion, following Zhang *et al* (2023b). Model performance was evaluated using kriging efficiency (KEFF) and slope of regression (SLOR) metrics (Deutsch, Szymanski and Deutsch, 2014).

Analysis

After compositing, PGE, Ni and Cu grades were determined within a 2 m ore zone of the Merensky Reef. The grades (Figure 4) exhibit significant variability, with a composite mean for PGE of 6.07 g/t and values ranging from 0.82 g/t to 13.39 g/t. The grades peak near the bottom of the ore zone, reflecting a concentration of mineralisation associated with the basal chromitite stringer and adjacent pyroxenitic units. Ni and Cu grades (Figure 4b and 4c) have composite means of 0.15 per cent and 0.03 per cent, respectively. Ni grades increase toward the base of the ore zone, while the opposite is true for Cu. Composited chemical analyses were then used for clustering tasks.

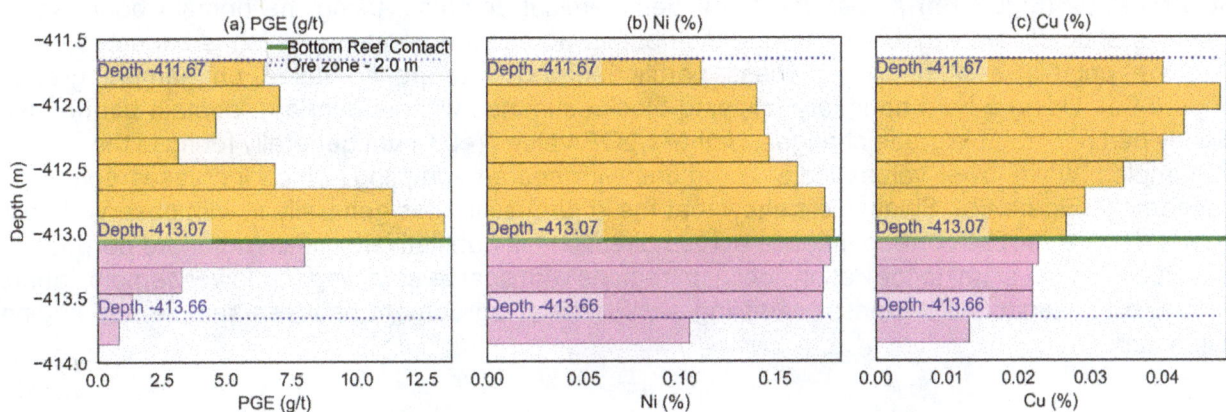

FIG 4 – Vertical stratigraphic distribution of (a) 4E PGE (g/t), (b) Ni (%) and (c) Cu (%) contents across the ore zone. Orange bars = mineralised zone above reef contact; violet bars = mineralised

zone at below reef contact; horizontal solid line = bottom reef contact; horizontal dotted lines delineate complete ore zone.

Using the product of the silhouette score and the Davies-Bouldin index, the optimal number of clusters is three (Figure 5). Unlike aspatial clustering, SDKMeans can create spatial continuity. The incorporation of spatial distances in SDKMeans permitted the assignment of class labels in a spatially contiguous manner (Figure 5). The three domains are geochemically distinct and spatially coherent. Once the samples were assigned domain labels, we performed geodomain delineation using the SVM algorithm (Figure 5).

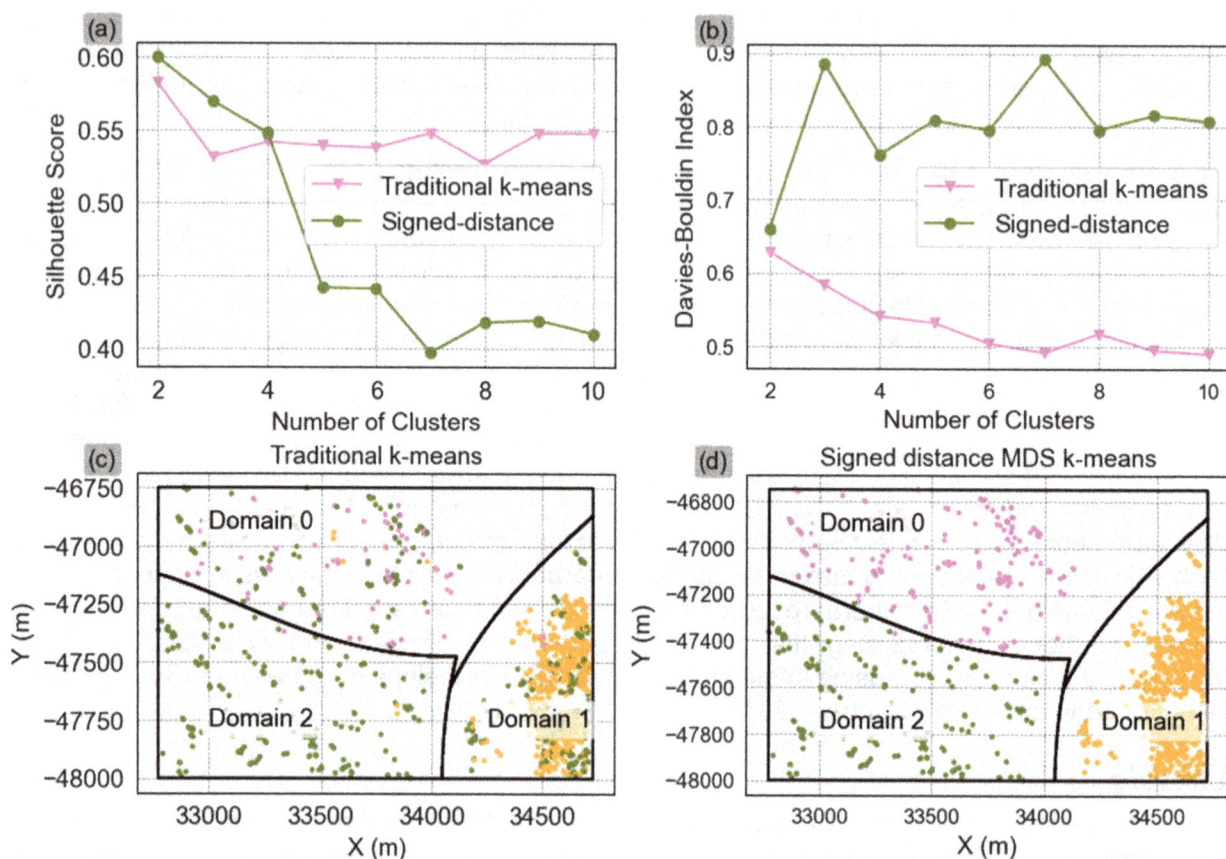

FIG 5 – (a) Silhouette scores for different numbers of clusters (k); (b) Davies-Bouldin Index for different numbers of clusters (k); (c) Spatial domain results using the traditional k-means clustering; and (d) SDKMeans results in the spatial domain.

The 4E PGE accumulation was estimated at the block level (Figure 6a), as well as the corresponding SLOR values (Figure 6b) across the three geochemical domains, using the domain boundaries. There is a significant variability in grade distribution across the domains, which is expected given that the point-wise domaining process partitioned the data partly based on sample grades (Figure 6a). Using a hard boundary (clipping block estimates without overlap), domain boundaries clearly partition notable grade changes. Higher SLOR values (>0.8) are generally found in the vicinity of samples, while lower values indicate regions with sparse sampling, which increases epistemic uncertainty (Figure 6b). Swath plots show that the kriging estimates generally fit with observed data trends (Figure 7). The correspondence across directions is qualitatively sufficient for the purpose of this study. However, minor deviations occur in regions with sparse sampling (eg lower sample counts in Figure 7). Regions with denser sampling show greater agreement between sample and kriging estimates.

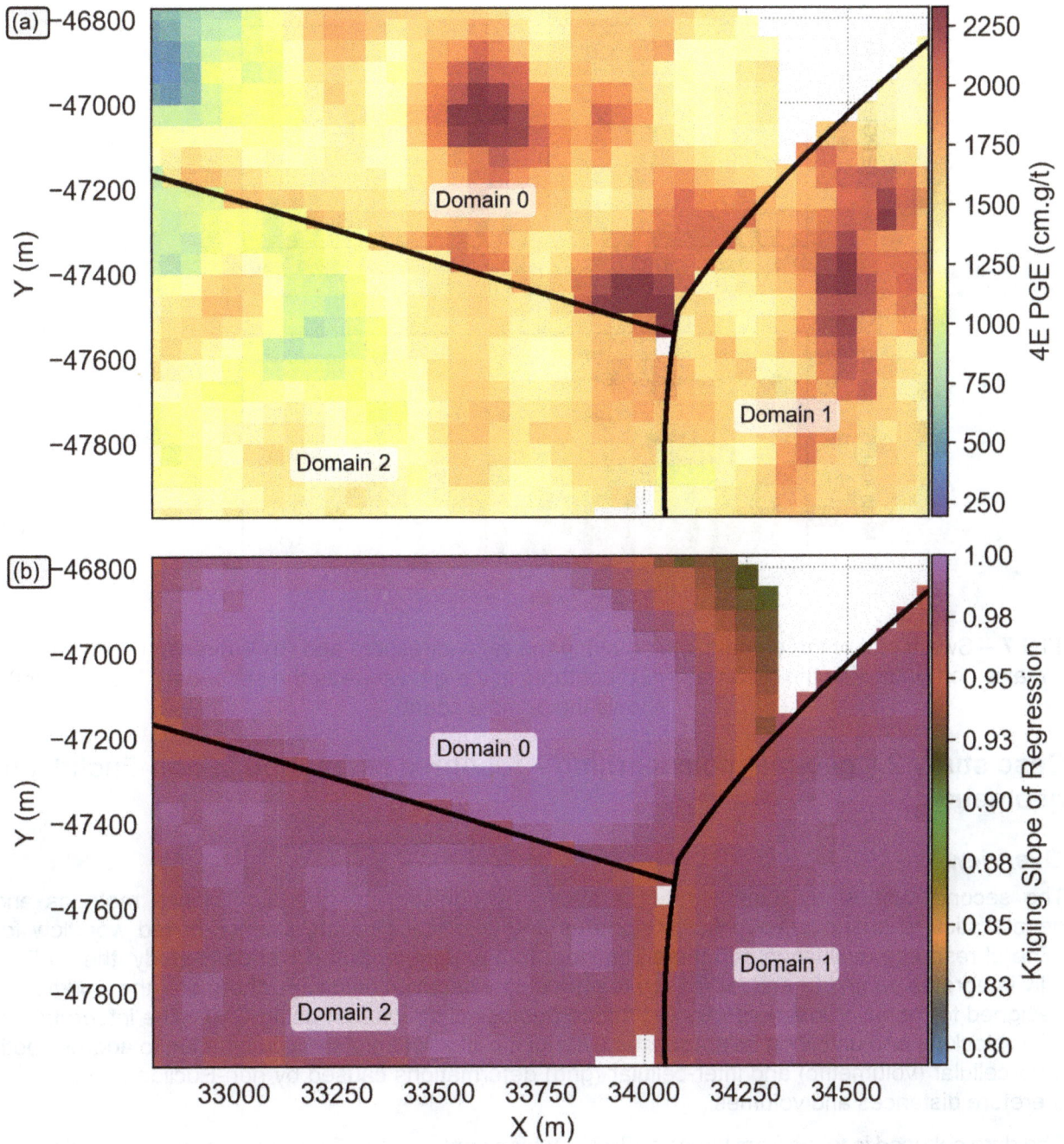

FIG 6 – (a) Mineral resource model of the Merensky Reef; and (b) the kriging slope of regression. The blocks are colour-mapped to grades. Domain boundaries are shown as lines. The scatter overlay represents the sample locations.

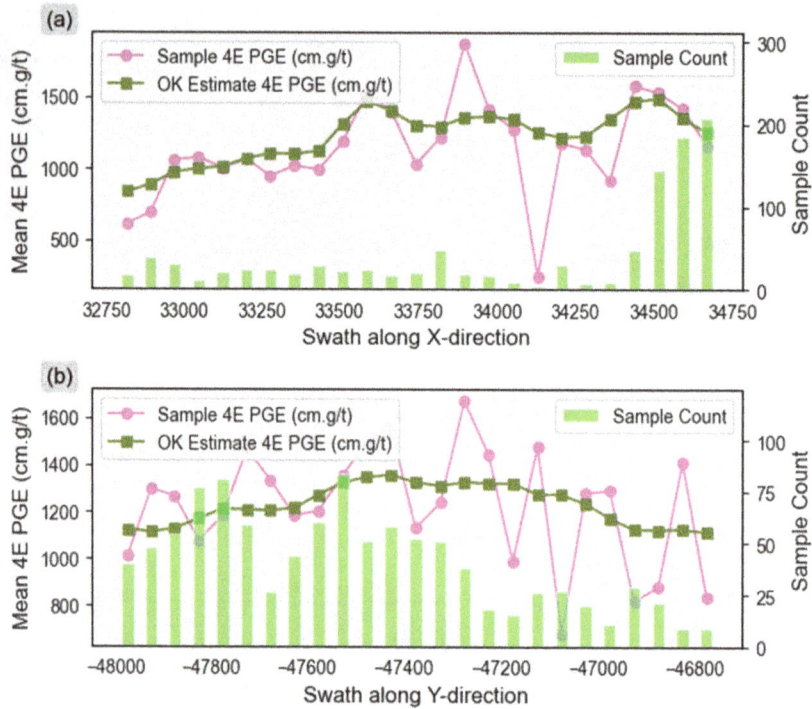

FIG 7 – Swath plot of the domain kriging along the (a) X-direction; and (b) Y-direction. The primary Y-axis (left) indicates the mean 4E PGE contents in cm·g/t, whereas the secondary Y-axis (right) shows the sample count.

Case study 2 – geo-manifold learning – resource estimation in non-Euclidean geometry

Overview

The second case study demonstrates a strong hybridisation of geodata science methods and geostatistics to create a scientifically rigorous and practical geodata science-based workflow for mineral resource evaluation and mine planning. In comparison to the first case study, the method fully integrates an end-to-end ML-based resource estimation workflow. Furthermore, the workflow is designed for non-Euclidean spaces (eg folded geological structures) using GML. The integration of microblocking and unfolding leverages the synergistic interaction of the combination to address both intra-cellular (volumetric) and inter-cellular (grid) deformations caused by non-Euclidean grids, and therefore distances and volumes.

The data set used is taken from Liu *et al* (2023), which captures a Li-Fe-Al-rich porcelain clay deposit. The deposit contains irregular rock veins or elongated nodules, whose general trend is north–south. The data set contains original borehole data, is fairly recent, and captures grades of Li, which is a critical material in modern batteries (eg Talens Peiró, Villalba Méndez and Ayres, 2013). The data is challenging to model in its native geometry, because the orebody contains a fold-like structure (Figure 8). Consequently, spatial continuity within the deposit is not Euclidean. The deposit is mainly composed of kaolinite, which formed through the weathering and/or hydrothermal alteration of alumino-silicate minerals (eg feldspar). In the vicinity of the deposit, only loose layers dating to the Quaternary Period are found, primarily in low-lying areas and valleys. Residual slope accretion and alluvial strata dominate the area. Known structural folds in the region are largely part of the Guyangzhai anticline (Southern China). Approximately 40 km of outcrop extend along the axis of the complex anticline. The orebody was geodomained based on a combination of field observations and numerical modelling (Figure 8).

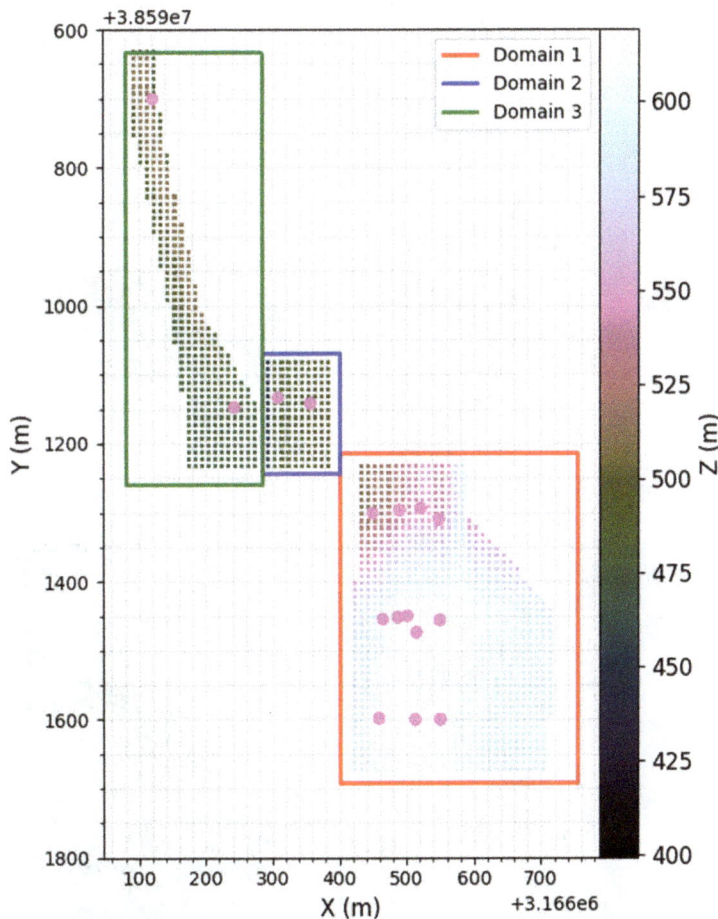

FIG 8 – Map of the orebody (porcelain clay) showing location of the boreholes (black circles), the elevation of the block model and domain configuration.

Analysis

There are a total of three block models to highlight the effects of correcting for:

1. Both grid and cell-volume distortion (microblocking before transformation, or pre-microblocking, Figure 9a).

2. Only grid corrections or volumetrically uncorrected (microblocking after transformation, or post-microblocking, Figure 9b).

3. Native-geometry estimation, assuming Euclidean geometry (no transformation, only microblocking, Figure 9c).

This is because gridding using pre-defined intervals can only occur in Euclidean geometry. On generalised manifolds, fixed-interval gridding cannot occur outside of local tangent spaces. Therefore, creating a grid to encapsulate the resource model over a non-Euclidean manifold means that if a portion of the manifold was flattened (eg to its tangent space), there would be both grid and cellular distortions. Grid distortions change cell-to-cell distances, while cellular distortions alter within-cell distances. Both types of distortions must be accounted for in downstream uses of such a model.

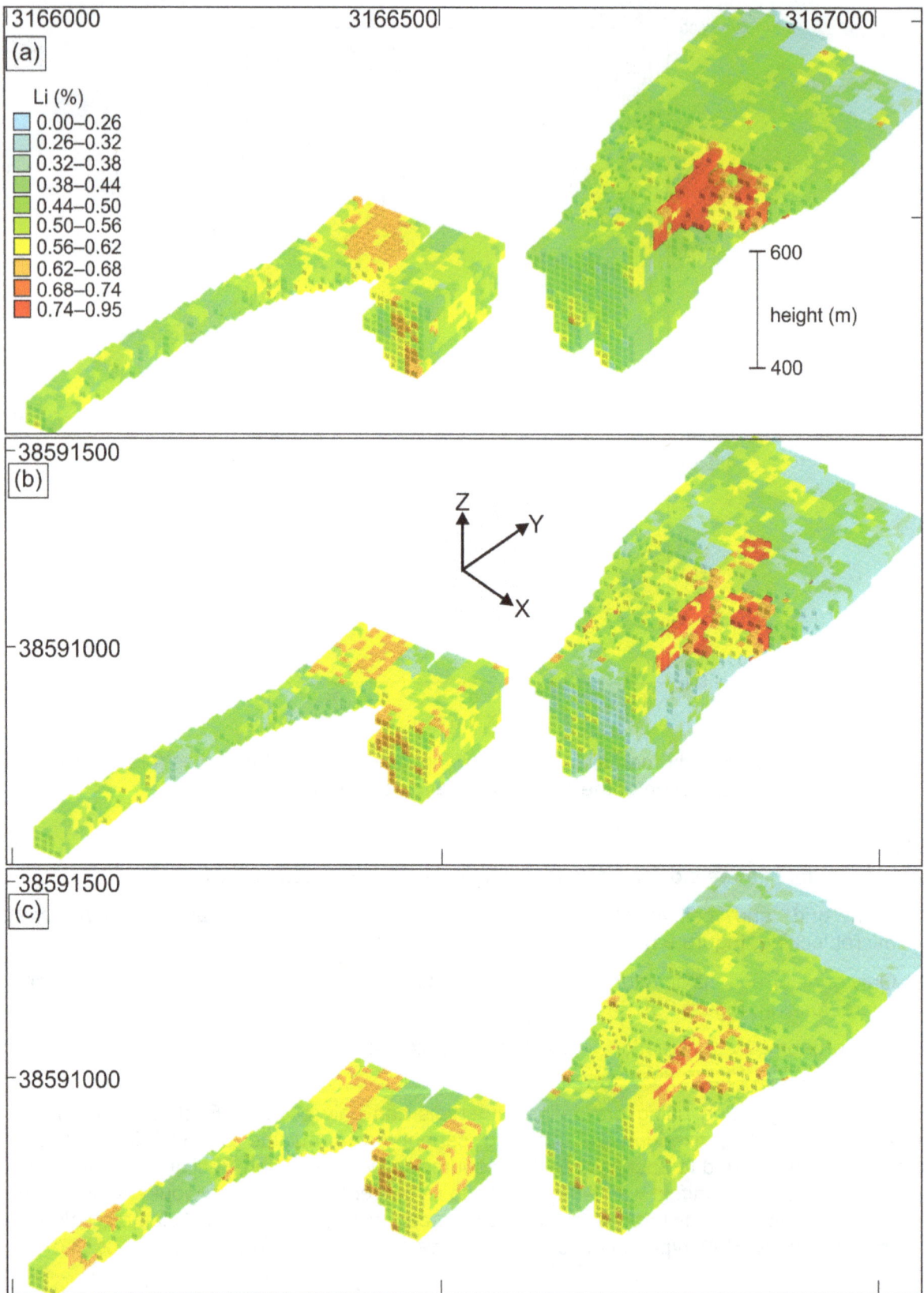

FIG 9 – Estimated blocks of the porcelain clay orebody using: a grid and volume-corrected unfolded model in (a); grid-corrected unfolded model in (b); and the original folded model (c).

Performing microblocking before applying the GML method of unfolding ensures that intra-cellular distance changes (volume changes inside each cell) are also modelled through the GML method, because each cell is modelled as a fine mesh instead of a point, which means that intra-cellular (or volumetric) distortions are modelled as microgrid deformations (cells treated as discretised continuous media, like in finite elements modelling). Therefore, estimation using microblocks displaced through GML would incorporate distance changes both within and between grid cells. Conversely, microblocking after GML only accounts for grid distortion but not intra-cellular changes. Models that were created using transformed coordinates exhibit a greater observable internal variation in the Li grade in domain 1, with lesser differences in domains 2 and 3 (Figure 9). Spatial continuity was visibly better in the pre-microblocked model. Model statistics suggest that the maximum grade is the most affected by the transformation (Table 1). This is consistent with a change in the volume of the grid cells hosting the highest block grades (see also Figure 9). The lowest standard deviation of block grades was in the pre-microblocked model, which can be expected if artificial sources of variability introduced by the geometry of the orebody was attenuated through volumetric correction.

TABLE 1

Statistics of the block models resulting from microblocking-based estimations using the volume-corrected (pre-microblocked), uncorrected (post-microblocked) and original folded models.

	Grid and volume-corrected	Only grid-corrected	Folded
Blocks	12894	12894	12894
Grade mean (%)	0.498	0.482	0.485
Grade standard deviation (%)	0.087	0.101	0.103
Minimum grade (%)	0.259	0.261	0.26
25th percentile (%)	0.445	0.414	0.428
50th percentile (%)	0.483	0.486	0.496
75th percentile (%)	0.538	0.555	0.558
Maximum grade (%)	0.919	0.938	0.844

DISCUSSION

Hybridising geodata science and geostatistical methods is mutually beneficial to both disciplines. The distinction between traditional geostatistics and geodata science is still substantial, since geostatistics only considers regionalised variables in geodata, whereas geodata science is general purpose, without a dedicated focus on spatial data (Erten, Yavuz and Deutsch, 2021; Abildin et al, 2022; Jiang and Dimitrakopoulos, 2024). The adaptation of data science methods into each domain is strictly required for its deployment (Hazzan and Mike, 2023). However, this type of adaptation is actually bi-directional, as the reverse can also occur, to enhance traditionally geostatistical tasks, methods or domains of practice. Some recent developments in the crossover of geostatistics and geodata science demonstrate numerous bipartite benefits. It is clear that geodomaining can be formulated into two separate geodata science tasks, whose solutions are two ML workflows (Fouedjio, Hill and Laukamp, 2018; Talebi et al, 2020). The replicability of ML workflows lies in its guiding framework, which makes use of objective, metric-driven performance evaluations to steer workflow design, much of which is automated (eg through out-of-sample testing). Model tuning only depends on data and hyperparameter choices, which makes replication simple. Similarly, microblocking reduces the number of hyperparameter choices compared to kriging for the purpose of both point-to-point and point-to-block interpolation (although it can be generalised to any type of cross-support use). Aside from the benefits already offered by the geodata (or data) science framework, the integration of microblocking and GML is a synergistic combination that does not yet exist in traditional geostatistics. This integration effectively addresses all modes of distortion that occur within and between grid cells due to distance transformations. These distortions are inevitable to permit distance computations (eg LVA), but whose intra-cellular (volumetric) effects remain unaddressed until now.

More fascinatingly, GML has great potential in overcoming the limitations of traditional Euclidean geostatistics in non-Euclidean structures. GML inherits the benefits of manifold learning, such as a diversity of algorithms, including ISOMAP, MDS and locally linear embedding (Peredo Andrade, 2022). The reason that a variety of such algorithms exists is because nonlinear dimensionality reduction does not generally admit unique solutions. Data properties (eg whether the data has a hole like a doughnut) control the feasibility of manifold learning algorithms, in pursuit of the least distorted low dimensional representation. Algorithmic experimentation is characteristic of data science, which is enabled by plug-and-play components, workflow automate-ability and metric-driven design. To the best of our knowledge, no method, including LVA (which also makes use of ISOMAP), is designed to remediate intra-cellular (volumetric) distortions (Boisvert, 2010; Lillah and Boisvert, 2015). Coupling microblocking with GML creates a synergistic method that can address both grid and cellular distortions. The results indicate a substantial difference in outcome, although further research is required to, including to extend GML to a variety of data geometries and algorithms.

As it pertains to geostatistics, the addition of geodata science concepts can enhance the scientific value of geostatistical outcomes, by leveraging the replicability of the data science framework and reducing algorithmic and workflow complexity, and manual effort. This type of adaptation can ensure the relevance of geostatistics in a world progressing to higher velocity geodata (Bourdeau *et al*, 2023) and more explicit differentiation of data roles (eg surveyors as data generators versus data modellers as data users, Ghorbani *et al*, 2023b). Such progress is necessary for the modernisation, expansion and security of future supply chains.

CONCLUSIONS

The overlap between geostatistics and geodata science is creating opportunities for the exchange of best practices, ideas and methods. This is a context that permits an unchecked proliferation of novelty, which is unproductively disruptive in the mineral industry, because the industry is highly conservative due to long timelines to profit, geopolitical tensions and extra-economic concerns. Routine novelty is a common theme in the time of exponential technologies and can easily create pervasive societal challenges in our ability to distinguish true innovation from merely ad hoc novelty. The better approach is to recognise that true innovation is robust, scientifically credible and withstands the test of time. Methods developed with robustness are those that feature basic scientific and statistical properties:

- replicability (eg using frameworks, baselines, open data and algorithms).

- a minimum amount of workflow and model complexity in relation to an objective.

- explainable workflow designs, beyond merely explainable models.

For geodata scientists, we can stand on the shoulders of giants to see beyond novelty. The giant that is data science provides us with a rigorous framework that facilitates scientific and statistical replicability. The giant that is geostatistics is time-tested through decades of applications, fully explainable methods, and canonical baselines for comparative analysis. Consequently, the definitive direction of progress is to explore ways to stand on the shoulders of both geodata science and geostatistics.

ACKNOWLEDGEMENTS

The authors acknowledge two anonymous reviewers for their constructive comments, which improved the readability and quality of the paper. We also acknowledge Julie E Bourdeau for her scientific and editorial contributions.

REFERENCES

Abildin, Y, Xu, C, Dowd, P and Adeli, A, 2022. A hybrid framework for modelling domains using quantitative covariates, *Appl Comput Geosci,* 16:100107. https://doi.org/10.1016/j.acags.2022.100107

Abzalov, M, 2016. *Applied mining geology,* 12 (Cham: Springer International Publishing).

Arthur, D and Vassilvitskii, S, 2007. k-means++: The advantages of careful seeding, in *Proceedings of the 18th Annual ACM-SIAM Symposium on Discrete Algorithms (SODA '07),* pp 1027–1035 (Society for Industrial and Applied Mathematics).

Boisvert, J, 2010. Geostatistics with locally varying anisotropy, PhD thesis (unpublished), University of Alberta, Canada. https://doi.org/10.7939/R31X5C

Boisvert, J, Manchuk, J and Deutsch, C, 2009. Kriging in the presence of locally varying anisotropy using non-Euclidean distances, *Math Geosci*, 41(6):585–601. http://doi.org/10.1007/s11004-009-9229-1

Borg, I and Groenen, P J F, 2005. Modern multidimensional scaling: theory and applications (2nd ed), (Springer). https://link.springer.com/book/10.1007/0-387-28981-X

Bourdeau, J E, Zhang, S E, Lawley, C J M, Parsa, M, Nwaila, G T and Ghorbani, Y, 2023. Predictive geochemical exploration: inferential generation of modern geochemical data, anomaly detection and application to northern Manitoba, *Nat Resour Res*, 32:2355–2386. https://doi.org/10.1007/s11053-023-10273-6

Bourdeau, J E, Zhang, S E, Nwaila, G T and Ghorbani, Y, 2024. Data generation for exploration geochemistry: past, present and future, *Appl Geochem*, 172:106124. https://doi.org/10.1016/j.apgeochem.2024.106124

Caixeta, R M and Costa, J F C L, 2021. A robust unfolding approach for 3-D domains, *Comput Geosci*, 155:104844. https://doi.org/10.1016/j.cageo.2021.104844

Chatterjee, S, Bandopadhyay, S and Machuca, D, 2010. Ore grade prediction using a genetic algorithm and clustering-based ensemble neural network model, *Math Geosci*, 42(3):309–326. https://doi.org/10.1007/s11004-010-9264-y

Chatterjee, S, Bandopadhyay, S and Rai, P, 2008. Genetic algorithm-based neural network learning parameter selection for ore grade evaluation of limestone deposit, *Min Technol*, 117(4):178–190. https://doi.org/10.1179/037178409X405732

Cressie, N, 1990. The origins of kriging, *Math Geol*, 22(3):239–252. https://doi.org/10.1007/BF00889887

Deutsch, C V and Journel, A G, 1998. *GSLIB: geostatistical software library and user's guide* (2nd ed), (New York: Oxford University Press).

Deutsch, J L, Szymanski, J and Deutsch, C V, 2014. Checks and measures of performance for kriging estimates, *J South Afr Inst Min Metall*, 114(3):223.

Dutta, S, Misra, D, Ganguli, R, Samanta, B and Bandopadhyay, S, 2006. A hybrid ensemble model of kriging and neural network for ore grade estimation, *Int J Min Reclamat Environ*, 20(1):33–45. https://doi.org/10.1080/13895260500322236

Emery, X, Mery, N and Porcu, E, 2024. Vector-valued Gaussian processes on non-Euclidean product spaces: constructive methods and fast simulations based on partial spectral inversion, *Stoch Environ Res Risk Assess*, 38:3411–3428. https://doi.org/10.1007/s00477-024-02755-7

Erten, G E, Yavuz, M and Deutsch, C V, 2021. Grade estimation by a machine learning model using coordinate rotations, *App Earth Sci*, 130(1):57–66. https://doi.org/10.1080/25726838.2021.1872822

Fouedjio, F, Hill, E J and Laukamp, C, 2018. Geostatistical clustering as an aid for mineralisation body domaining: case study at the Rocklea Dome channel iron mineralisation deposit, Western Australia, *App Earth Sci*, 127:15–29. https://doi.org/10.1080/03717453.2017.1415114

Friedman, J, Hastie, T and Tibshirani, R, 2001. The elements of statistical learning (vol. 1), (New York: Springer Series in Statistics).

Ghorbani, Y, Nwaila, G T, Zhang, S E, Bourdeau, J E, Cánovas, M, Arzua, J and Nikadat, N, 2023a. Moving towards deep underground mineral resources: Drivers, challenges and potential solutions, *Resour Policy*, 80:103222. https://doi.org/10.1016/j.resourpol.2022.103222

Ghorbani, Y, Zhang, S E, Bourdeau, J E, Chipangamate, N S, Rose, D H, Valodia, I and Nwaila, G T, 2024a. The strategic role of lithium in the green energy transition: Towards an OPEC-style framework for green energy-mineral exporting countries (GEMEC), *Resour Policy*, 90:104737. https://doi.org/10.1016/j.resourpol.2024.104737

Ghorbani, Y, Zhang, S E, Nwaila, G T and Bourdeau, J E, 2022. Framework components for data-centric dry laboratories in the minerals industry: a path to science-and-technology-led innovation, *Extr Ind Soc*, 10:101089. https://doi.org/10.1016/j.exis.2022.101089

Ghorbani, Y, Zhang, S E, Nwaila, G T, Bourdeau, J E and Rose, D H, 2024b. Embracing a diverse approach to a globally inclusive green energy transition: Moving beyond decarbonisation and recognising realistic carbon reduction strategies, *J Clean Prod*, 434:140414. https://doi.org/10.1016/j.jclepro.2023.140414

Ghorbani, Y, Zhang, S E, Nwaila, G T, Bourdeau, J E, Safari, M, Hoseinie, S H, Nwaila, P and Ruuska, J, 2023b. Dry laboratories–mapping the required instrumentation and infrastructure for online monitoring, analysis and characterisation in the mineral industry, *Miner Eng*, 191:107971. https://doi.org/10.1016/j.mineng.2022.107971

Hastie, T, Tibshirani, R and Friedman, J, 2009. *The elements of statistical learning: data mining, inference and prediction* (New York: Springer).

Hazzan, O and Mike, K, 2023. *Guide to Teaching Data Science: An Interdisciplinary Approach* (Springer Nature).

Hosseini, S T, Asghari, O, Benndorf, J and Emery, X, 2023. Real-time uncertain geological boundaries updating for improved block model quality control based on blast hole data: a case study for Golgohar iron ore mine in Southeastern Iran, *Math Geosci*, 55:541–562. https://doi.org/10.1007/s11004-022-10030-0

Hosseini, S T, Asghari, O and Emery, X, 2021. An enhanced direct sampling (DS) approach to model the geological domain with locally varying proportions: Application to Golgohar iron ore mine, Iran, *Ore Geol Rev*, 139:104452. https://doi.org/10.1016/j.oregeorev.2021.104452

Jafrasteh, B and Fathianpour, N, 2017. A hybrid simultaneous perturbation artificial bee colony and back-propagation algorithm for training a local linear radial basis neural network on ore grade estimation, *Neurocomputing*, 235:217–227. https://doi.org/10.1016/j.neucom.2017.01.016

Jafrasteh, B, Fathianpour, N and Suárez, A, 2018. Comparison of machine learning methods for copper ore grade estimation, *Comput Geosci*, 22(5):1371–1388. https://doi.org/10.1007/s10596-018-9758-0

Jiang, Y and Dimitrakopoulos, R, 2024. Simultaneous stochastic optimization of mining complexes with equipment uncertainty: application at an open-pit copper mining complex, *Min Technol*, 133:241–256. https://doi.org/10.1177/25726668241263408

Krige, D G, 1951. A statistical approach to some basic mine valuation problems on the Witwatersrand, *J South Afr Inst Min Metall*, 52(6):119–139.

Kruskal, J B, 1964. Multidimensional scaling by optimizing goodness of fit to a nonmetric hypothesis, *Psychometrika*, 29:1–27. https://doi.org/10.1007/BF02289565

Li, X L, Li, L H, Zhang, B L and Guo, Q J, 2013. Hybrid self-adaptive learning-based particle swarm optimization and support vector regression model for grade estimation, *Neurocomputing*, 118:179–190. https://doi.org/10.1016/j.neucom.2013.03.002

Lillah, M and Boisvert, J B, 2015. Inference of locally varying anisotropy fields from diverse data sources, *Comput Geosci*, 82:170–182. https://doi.org/10.1016/j.cageo.2015.05.015

Liu, Z N, Deng, Y Y, Tian, R, Liu, A H and Zhang, P W, 2023. A new method for estimating ore grade based on sample length weighting, *Sci Rep*, 13:6208. https://doi.org/10.1038/s41598-023-33509-0

Machuca-Mory, D F, Rees, H and Leuangthong, O, 2015. Grade modelling with local anisotropy angles: a practical point of view, in 37th Application of Computers and Operations Research in the Mineral Industry (APCOM, 2015).

Mardia, K V, Kent, J T and Bibby, J M, 1979. *Multivariate analysis* (Academic Press).

Matheron, G, 1963. Principles of geostatistics, *Econ Geol*, 58(8):1246–1266.

Matheron, G, 1971. *The theory of regionalized variables and its applications,* vol 5 (Paris: École Nationale Supérieure des Mines).

Matías, J M, Vaamonde, A, Taboada, J and González-Manteiga, W, 2004. Comparison of kriging and neural networks with application to the exploitation of a slate mine, *Math Geol* 36:463–486. https://doi.org/10.1023/B:MATG.0000029300.66381.dd

Mohammadi, H, Hosseini, S T, Asghari, O and Asadi Harouni, P, 2022. Ore body domaining by clustering of multiple-point data events; a case study from the Dalli porphyry copper-gold deposit, central Iran, *Ore Energy Resour Geol*, 10–13:100018. https://doi.org/10.1016/j.oreoa.2022.100018

Nwaila, G T, Zhang, S E, Bourdeau, J E, Frimmel, H E and Ghorbani, Y, 2024. Spatial interpolation using machine learning: from patterns and regularities to block models, *Nat Resour Res*, 33(1):129–161. https://doi.org/10.1007/s11053-023-10280-7

Owusu, S K A and Dagdelen, K, 2024. Impact of Competent Persons' judgements in Mineral Resources classification, *J South Afr Inst Min Metall*, 124(7):371–382. http://doi.org/10.17159/2411-9717/1538/2024

Peredo Andrade, O F, 2022. Large scale geostatistics with locally varying anisotropy, PhD Thesis (unpublished), Universitat Politècnica de Catalunya, Barcelona, Spain.

Rossi, M E and Deutsch, C V, 2013. *Mineral resource estimation* (New York: Springer Science and Business Media).

Samanta, B and Bandopadhyay, S, 2009. Construction of a radial basis function network using an evolutionary algorithm for grade estimation in a placer gold deposit, *Comput Geosci* 35(8):1592–1602. https://doi.org/10.1016/j.cageo.2009.01.006

Shimaponda-Nawa, M, Nwaila, G T, Zhang, S E and Bourdeau, J E, 2023. A framework for measuring the maturity of real-time information management systems (RTIMS) in the mining industry, *Extr Ind Soc*, 16:101368. https://doi.org/10.1016/j.exis.2023.101368

Talebi, H, Mueller, U and Tolosana-Delgado, R, 2019. Joint simulation of compositional and categorical data via direct sampling technique–application to improve mineral resource confidence, *Comput Geosci*, 122:87–102. https://doi.org/10.1016/j.cageo.2018.10.013

Talebi, H, Peeters, L J M, Mueller, U, Tolosana-Delgado, R and van den Boogaart, K G, 2020. Towards geostatistical learning for the geosciences: a case study in improving the spatial awareness of spectral clustering, *Math Geosci*, 52:1035–1048. https://doi.org/10.1007/s11004-020-09867-0

Talens Peiró, L, Villalba Méndez, G and Ayres, R U, 2013. Lithium: sources, production, uses and recovery outlook, *JOM*, 65:986–996. https://doi.org/10.1007/s11837-013-0666-4

Tenenbaum, J B, de Silva, V and Langford, J C, 2000. A global geometric framework for nonlinear dimensionality reduction, *Sci,* 290(5500):2319. https://doi.org/10.1126/science.290.5500.2319

Witten, I, Frank, E, Hall, M and Pal, C, 2016. Data mining: practical machine learning tools and techniques (4th ed), (Cambridge: Todd Green).

Zhang, S E, Bourdeau, J E, Nwaila, G T and Ghorbani, Y, 2023a. Emerging criticality: Unraveling shifting dynamics of the EU's critical raw materials and their implications on Canada and South Africa, *Resour Policy,* 86:104247. https://doi.org/10.1016/j.resourpol.2023.104247

Zhang, S E, Nwaila, G T, Bourdeau, J E, Ghorbani, Y and Carranza, E J M, 2023b. Machine learning-based delineation of geodomain boundaries: a proof-of-concept study using data from the Witwatersrand goldfields, *Nat Resour Res,* 32:879–900. https://doi.org/10.1007/s11053-023-10159-7

Zhong, X, Kealy, A and Duckham, M, 2016. Stream Kriging: Incremental and recursive ordinary Kriging over spatiotemporal data streams, *Comput Geosci,* 90:134–143. https://doi.org/10.1016/j.cageo.2016.03.004

A practical use of machine learning to aid gold recovery studies at Liberty Gold's Black Pine Oxide Gold Project in Southern Idaho

V Wilson[1], C Gomes[2] and A Barrios[3]

1. Principal Resource Geologist, SLR Consulting Australia Pty Ltd, Perth WA 6008.
   Email: vwilson@slrconsulting.com
2. Senior Geostatistician, SLR Consulting (Canada) Ltd, Edmonton AB T6B 3H9, Canada.
   Email: cgomes@slrconsulting.com
3. Senior Resource Geologist, Liberty Gold, Vancouver BC V6E 1B4, Canada.
   Email: abarrios@libertygold.ca

ABSTRACT

As Liberty Gold advances its Black Pine Oxide Gold project in southern Idaho towards feasibility, developing a robust and reliable geometallurgical domain model that accurately describes variability in predicted gold recovery is a critical part of project derisking. At Black Pine, 176 column tests across the deposit inform recovery equations partitioned by deposit area, formation, gold grade and metallurgical domain, defined by gold cyanide solubility or the ratio of drill assay gold cyanide leach to gold fire assay. Liberty Gold drilling at Black Pine has generated approximately 64 800 cyanide leach assay and fire assay pairs from over 1000 drill holes in support of this work. However, due a handling error, approximately 23 600 cyanide leach assays were contaminated, resulting in database gaps and reduced confidence in metallurgical domain modelling. While re-assaying was undertaken where possible, large data gaps in important resource areas remained. Working closely with Liberty Gold, SLR applied a machine learning (ML) approach to predict gold cyanide solubilities in drill holes with contaminated or missing data. Three ML algorithms (Neural Network, Random Forest and Gradient Boosting) were iteratively trained on select drill hole data parameters, primarily including: adjacent valid gold cyanide leach assays; distance functions, lithology, and logged carbon intensity. The ML models were tested on predicting the missing recovery data classes, and the best ML model was able to correctly classify the geometallurgical domains with a 91 per cent weighted accuracy. The resulting infilled database allowed for metallurgical domain modelling to proceed across the deposit without the need for costly and time-prohibitive re-drilling of data gaps.

INTRODUCTION

The Black Pine deposit is a sedimentary rock-hosted, Carlin-style gold deposit located in Idaho, approximately 29 km north-west of the town of Snowville, Utah. The deposit has seen intermittent mining; small scale operations producing gold and sometimes silver from 1915 to 1955 using open pit and underground methods, and a larger scale open pit operation by Pegasus Mining from 1991 to 1997, placing 31 Mt and 665 640 gold ounces on a series of cyanide heap-leach pads, recovering gold using carbon adsorption methods, and producing doré bars after solvent electrowinning for a total recovered gold of approximately 435 000 oz (Sawyer, Scott and Comba, 1997).

Liberty Gold acquired the Project in 2016 and has conducted extensive verification work, drilling and exploration, and technical studies, culminating in their filing of a National Instrument 43–101 – Standards of Disclosure for Mineral Projects ('NI 43–101') technical report entitled 'Technical Report and Pre-feasibility Study for the Black Pine Gold Project, Cassia and Oneida Counties, Idaho, USA', effective June 1, 2024, and dated November 21, 2024, on the Company's profile on SEDAR (M3, 2024).

GEOLOGY AND MINERALISATION

The Black Pine gold mineralisation can be best classified as sedimentary rock-hosted, Carlin-style mineralisation.

As presently understood, the Black Pine (BP) property geology is comprised of a lower structural plate that includes the Devonian Jefferson Formation and Mississippian Manning Canyon Shale (PMmc), a middle plate characterised by Pennsylvanian carbonate rocks of the Oquirrh Group, and an upper plate (PPos) predominantly consisting of Permian siltstones and sandstones of the Oquirrh

Group. The lithologic contact between the lower plate and middle plate is sheared and brecciated, and middle plate units are complexly structurally interleaved. Middle plate strata are considerably more deformed than strata in the upper and lower plates. Liberty Gold has differentiated the middle plate rocks into six discrete formations: Pola (limestone, siltstone), Polb (siltstone), Polc (limestone, dolostone, siltstone), Pold (dolostone, limestone, sandstone), and Pols (siltstone, limestone) (Figure 1). The lowest member of the middle plate is the PMmx, interpreted to be a large structural mélange zone incorporating middle and lower plate rocks. Irregular carbonaceous lenses (Poc) throughout the middle plate are recognised by their black shaley appearance, these lenses are logged and modelled in detail.

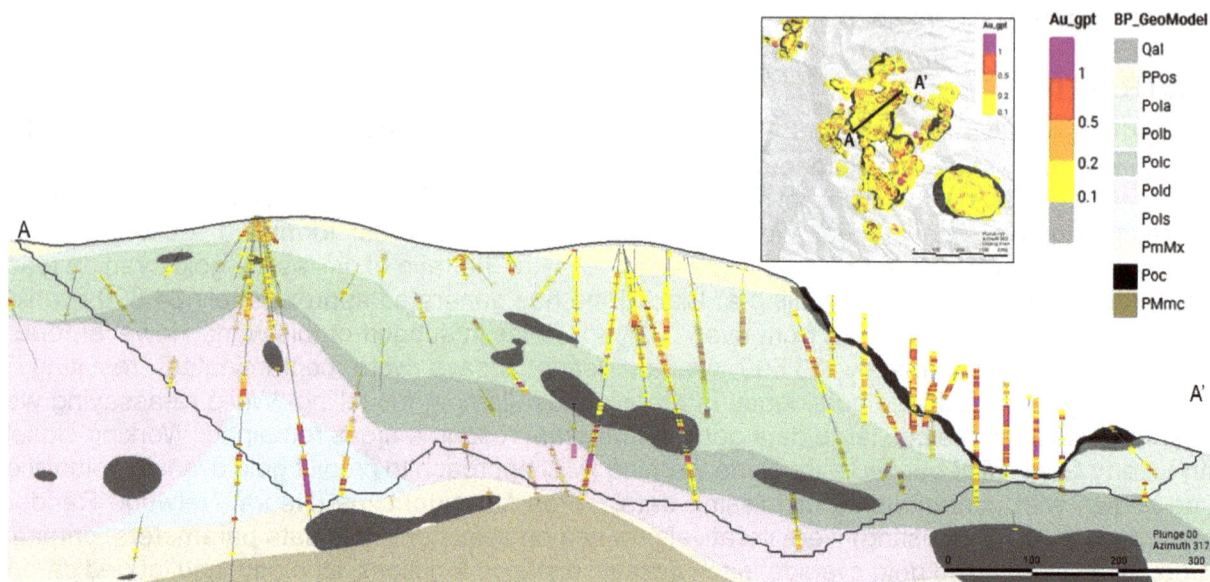

FIG 1 – Black Pine geological model.

The middle plate, which hosts the gold mineralisation of interest, has a structural thickness ranging from approximately 200 m to 500 m. At least two major deformational events are evident, manifested by Mesozoic thrust faults and tight to open folds, overprinted by Cenozoic, low- to high-angle normal faults. Gold is distributed throughout the middle structural plate, with higher-grade mineralisation occurring within favourable stratigraphic units, such as calcareous siltstones, as well as in and adjacent to breccia bodies and along variously orientated low- to high-angle brittle faults.

Three-dimensional modelling by Liberty Gold, utilising surface mapping and drill data, envisions relatively flat faults separating the lower and middle plates, with a structurally thickened middle plate centred on the outcropping area of mineralisation and diminishing in thickness to the north and south. The distribution of higher-grade gold mineralisation is controlled largely by favourable stratigraphy as well as a series of north- to north-west-striking listric normal faults that bound the east side of an overthickened zone of massive limestone and dolostone.

GEOMETALLURGICAL MODEL

Understanding gold recovery, particularly in a low-grade oxide deposits such as Black Pine, is critical. At Black Pine, eight historical and 176 Liberty Gold column tests across the deposit (Figure 2) inform recovery equations partitioned by deposit area, geologic formation, gold grade and four metallurgical domains (Met 1–4). In addition to these variables, final metallurgical equations are dependent on:

1. Head grade (higher head grades are typically found in strongly decalcified rock and on average have a higher recovery).

2. Presence of preg-borrowing clays (clayey zones that exhibit lower cyanide solubility in laboratory assay that do not show up in a column leach test).

3. Presence of smoky carbon (exhibits similarity's to the preg-borrowing clays).

4. Presence of carbon (Poc) (carbon is preg-robbing) (M3, 2024).

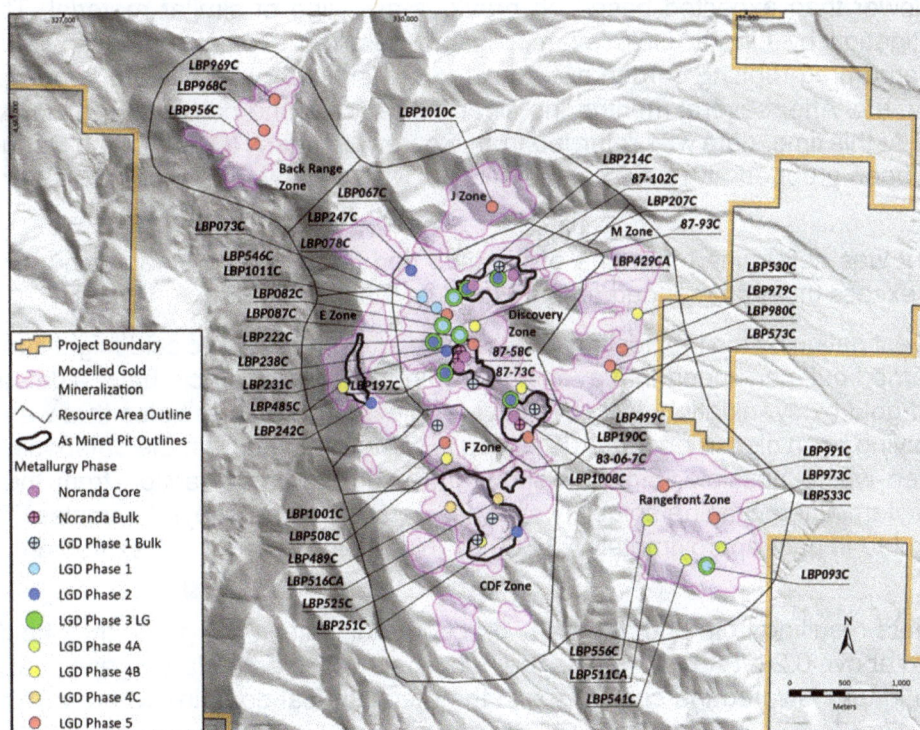

FIG 2 – Distribution of metallurgical testing at Black Pine.

Metallurgical domains are defined at Black Pine spatially and volumetrically using wireframes, and divide categorically based on gold cyanide solubility, or the ratio of drill assay gold cyanide leach to gold fire assay (%AuCN) as shown in Table 1. Met 1–3 are all considered oxidised material, Met 2 and Met 3 have a lower %AuCN response largely due to lower head grade, increased clays (preg-borrowing), or presence of silica rich typically dolomitic rocks where gold may be encapsulated.

Met 4, as well as all domains overprinted by modelled presence of carbon (Poc) are not considered leachable and are given a 0 per cent recovery in the resource model. Within the block model, metallurgical domains are represented as a code value (1 to 4).

TABLE 1

Black Pine project geometallurgical domains.

Metallurgical domain	Category	Cyanide solubility (%AuCN)
Met 1	Oxide	65 > %AuCN
Met 2	High transition	50 > % AuCN < 65
Met 3	Low transition	25 > % AuCN < 50
Met 4	Non-leach	% AuCN < 25

To support this work, Liberty Gold drilling at Black Pine has generated approximately 64 800 cyanide leach assay and fire assay pairs from over 1000 drill holes. During the first years that Liberty Gold was drilling at Black Pine, a cyanide leach test was only run on gold samples with a fire assay gold value greater than 0.2 g/t Au, but as the project grew, and the significance of the lower grade halo was realised, Liberty Gold expanded the cyanide leach test to include all samples returning gold values greater than 0.1 g/t Au.

CYANIDE SOLUBLE GOLD ASSAY INVESTIGATION

In 2021, it was noticed that results for cyanide soluble (AuCN) assaying of some pulps were considerably lower than expected based on previous assaying of similar material. This led to an internal investigation by Liberty Gold examining the possible causes of the anomalous results, including elevated preg-robbing carbon not detected in visual logging, elevated and possibly preg-borrowing clay, etc. After several months, no clear indication of the nature of the low AuCN results was identified. At this time, data was returned from a core twin of an RC hole in the Rangefront Zone. This hole had been prepped at a different lab and AuCN results were materially higher for the core hole.

By mid-2022, it was determined that one of the labs had been adding stearic acid (an anti-caking agent) well in excess of recommended amounts during the pulping step.

A follow-up investigation by the lab, testing whether any amount of stearic acid might suppress AuCN results was carried out, concluded that even small amounts of stearic acid have a deleterious effect on recovery of gold by cyanidation. With the coarse rejects from earlier holes that were prepped in this manner having been discarded, the AuCN results were removed from the database. Reassaying was undertaken where possible. Figure 3 shows an example cross-section from the Rangefront Zone of the contaminated cyanide solubility samples versus the re-assayed samples. This handling error was found to impact approximately 23 600 cyanide leach assays across 465 drill holes at the Project.

The missing data resulting from both the early procedure of limiting the grade threshold for cyanide leach testing to above 0.2 g/t Au, as well as the handling error, created large gaps spatially over the Project, which in turn lowered the confidence in the metallurgical domains where proximal to those gaps.

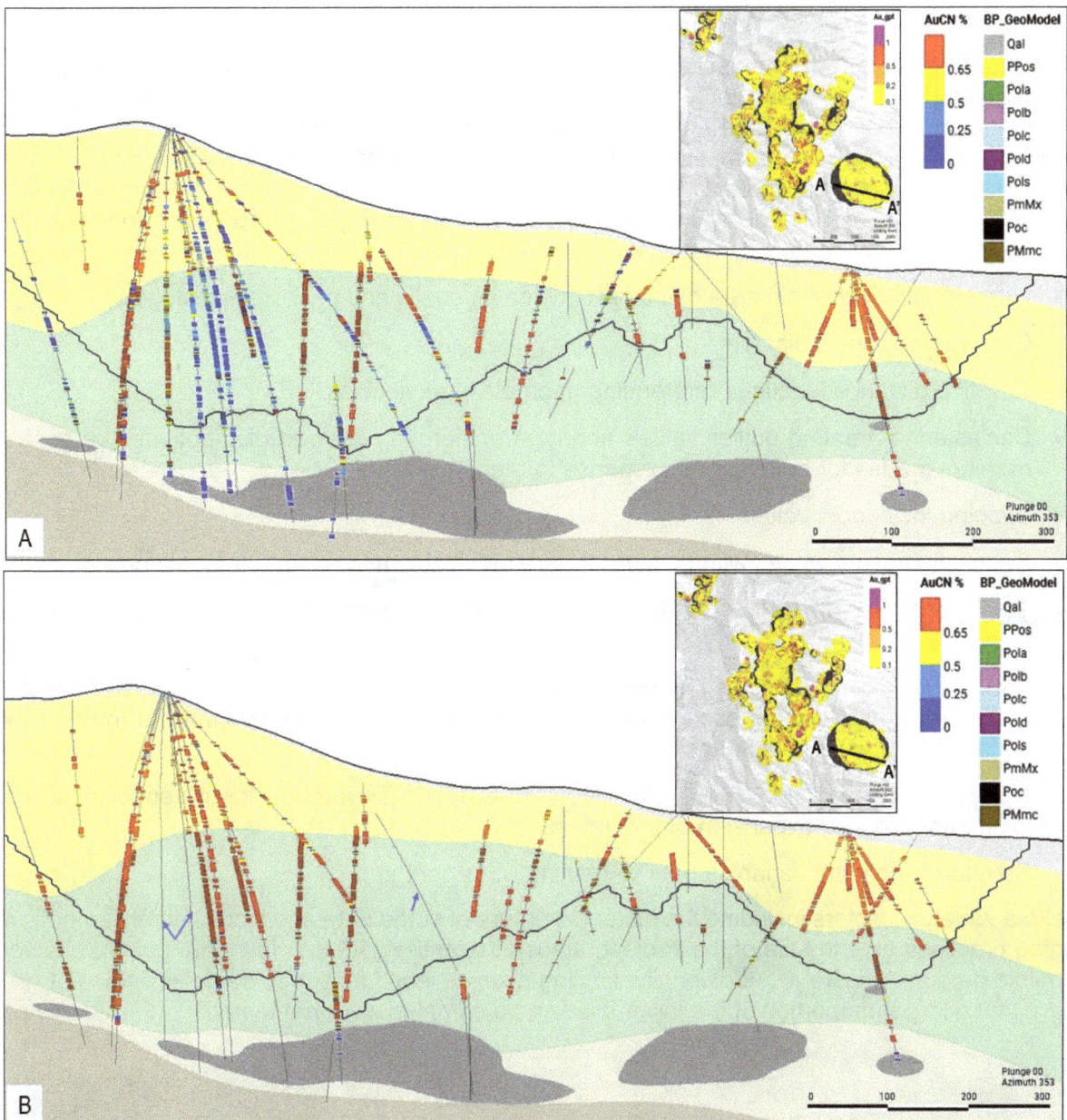

FIG 3 – Comparing: (a) contaminated AuCN results, to (b) re-assayed results in the Rangefront zone. Note: where samples could not be re-assayed the hole has no values (as shown by the blue arrows).

PREDICTING CYANIDE SOLUBLE GOLD ASSAY VALUES

SLR worked closely with Liberty Gold geologists and metallurgists to use machine learning to predict gold cyanide solubilities in drill holes with contaminated or missing data. Three ML algorithms, Neural Network (NN), Random Forest (RF) and eXtreme Gradient Boosting (XGB) (LeCun, Bengio and Hinton, 2015; Chen and Guestrin, 2016; Pedregosa *et al*, 2011) were iteratively trained on selected drill hole data available entries. Prior to ML modelling, multiple pre-processing steps were performed to suit the data to aid in supervised learning tasks. The ML modelling aimed at predicting both metallurgical domains through classification, and actual gold recovery values through regression.

The description of the ML modelling process is kept superficial, as the goal is not to detail the prediction strategy in-depth. Instead, this paper highlights a broader view of a successful case study combining ML prediction engines with geological and geostatistical tools to establish reliable geometallurgical domains, critical for advancing and assessing mineral projects.

Pre-processing and exploratory data analysis

Pre-processing is one of the most relevant steps in ML modelling since results depend heavily on having good predictors which display meaningful relationships with the target. The BP drilling database comprises a significant volume of variables available which derive from various source including chemical fire-assays, mass-spectrometry multi-elemental analysis and a diverse range of logging variables including lithology, alterations, colour, and mineralogical concentrations. Aiming at finding value related to the target (gold recovery), several pre-processing and exploratory steps have been conducted on the data set, including:

- Filtering samples within area of interest (inside pit shells and on *in situ* lithologies).

- Creating lithology indicators.

- Creating distance functions referencing geometallurgical class.

- Declustering, treating outlier values and log-transforming skewed distributions like gold and rhenium grades for better linear response to the target.

- Clipping lower gold values.

- Replacing absent values with 0 values where applicable and reliable (non-historic logging).

- Target-encoding of well-correlated categorical logged variables like carbon, domain, and colour.

- Variable reduction, including super-secondary variable aggregation (Babak and Deutsch, 2009), multi-dimensional scaling (MDS) and correlation matrixes to limit the inputs to key variables and avoid redundancy.

- Multiple Imputation (MI) (Silva and Deutsch, 2017) to fill missing input variables at a few locations, enabling expansion of prediction.

- Standardising input variables prior to training ML algorithms.

Due to a variety of factors including the historical nature of some data and inconsistent sampling and logging practices over the life of the Project, across the entire data set, few data points reflected a complete data set (Figure 4). Additionally, limiting cyanide leach testing to samples with >0.1 g/t Au meant that a larger proportion of the lower grade gold samples were not tested.

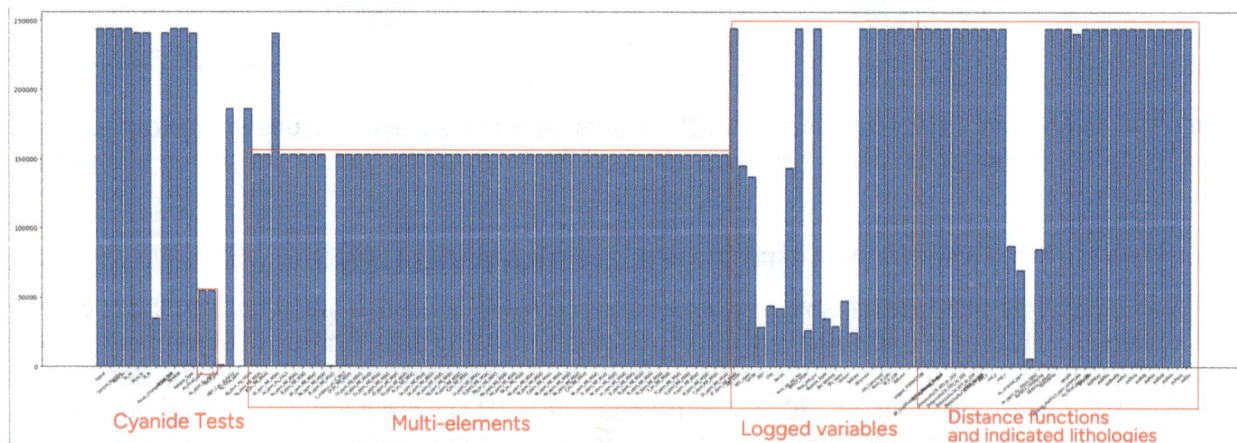

FIG 4 – Number of recorded instances of available variables within the data set.

Using a variety of visual, statistical, and geostatistical techniques, the data set was explored to help understand key relationships between the available variables and cyanide soluble gold values. Key relationships were identified with logged variables including alteration colour, presence of carbon, oxidation, lithology unit and subunit, as well as built distance function.

Selecting input variables

Variables were selected according to their correlation with the target (cyanide leaching recovery), their correlation with each other, effectiveness, reliability and availability. Intentionally, the set of final variables were diverse to control and limit the influence on the result of any single variable.

Final variables included analytical values nickel (Ni), sulfur (S), uranium (U), log transformed gold fire assay (logAu), and log transformed rhenium (logRe); logged variables carbon intensity, colour, and lithology; distance to categorised metallurgical domain values (1 to 4), and a created supersecondary variable (supersec), an aggregated variable created to maximise correlation with the primary variable and based on the combination of several lower correlation variables including selenium (Se), copper (Cu), phosphorus (P), molybdenum (Mo), and germanium (Ge).

Using logAu as a reference distribution, logRe, Ni, S, U, and the previously created super-secondary variable were jointly imputed. This process extended predictions to an additional 5000 sample locations where carbon data was available, but the imputed values were missing – accounting for slightly more than 5 per cent of the training set.

Multiple Imputation (MI) generated realisations for these missing values while preserving their multivariate relationships and variance as observed in the original data. The inclusion of imputed data improved ML model performance on the test set, suggesting its reliability.

When applied to small percentages of missing data and blended with a diverse set of predictors, MI remains a robust and safe solution for handling missing values in ML applications. The streamlined ML modelling workflow is summarised and illustrated in Figure 5.

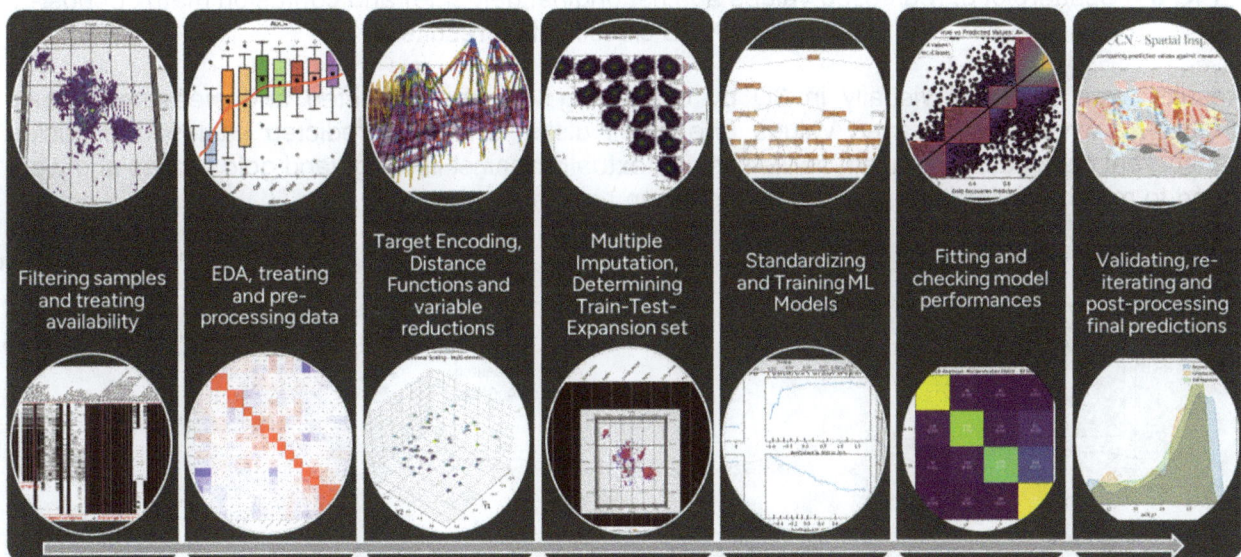

FIG 5 – Overview of ML model workflow.

Some categorical logged variables such as lithology, domain and alteration colour have been treated in a process called 'Target Encoding', where classes are replaced by the mean of the target variable, since categorical values cannot be read by most ML models. Predicted results have been checked through analysis of scatterplots (regressions), confusion matrixes (classification) and 3D dimensional checks (both).

Machine learning model results and performance

Classification and regression tasks were performed for all three ML methods (NN, RF and XGB). XGB reached the highest accuracy among all models and thus was selected as the final resultant data set. Actual and predicted values total approximately 82 000 samples. XGB performance plots are seen in Figure 6 of the train and test sets jointly. The regression results on the test set only reach an R^2 of 0.69, while the training set performs at 0.75, meaning the model does not overfit while also being capable of generalising to unseen data. A confusion matrix (Figure 6, right) of XGB post-processed regression classes shows high performance of the oxide (Met 1; 99 per cent accuracy) and non-leach (Met 4; 94 per cent accuracy) domains. Transitional domains were harder to predict; Met domain 2 and 3 reporting 74 per cent and 82 per cent accuracy, respectively. The XGradient Booster (XGB) method overall returned 94 per cent (classifier) weighted accuracy, and the XGB regression was post-processed to consistently match classifications, reaching also 94 per cent accuracy. RF and NN reported classification accuracy of 85 per cent and 86 per cent. The cloud of regressed values (Figure 6, left) is continuous without visual artefacts and achieves a high R^2 value of 0.81.

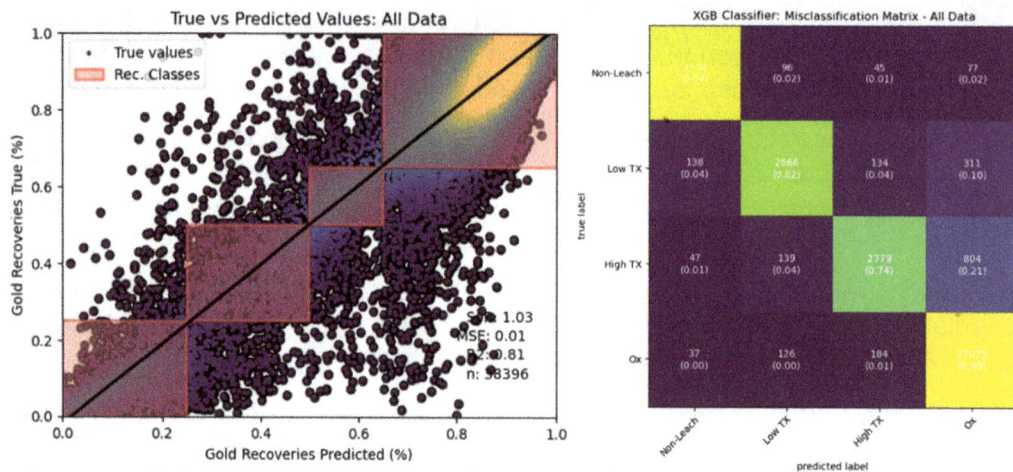

FIG 6 – Scatterplot of regressed values against original (left-side) and confusion matrix of post-processed XGB classification results.

Results were validated visually in 3D by comparing with the predicted and measured values (Figure 7), where the assayed values (the small disks) and the predicted values (the thicker transparent disks) are overlayed, as well as contextually against lithology and gold grades and were found to match very well, highlighted by the green arrows.

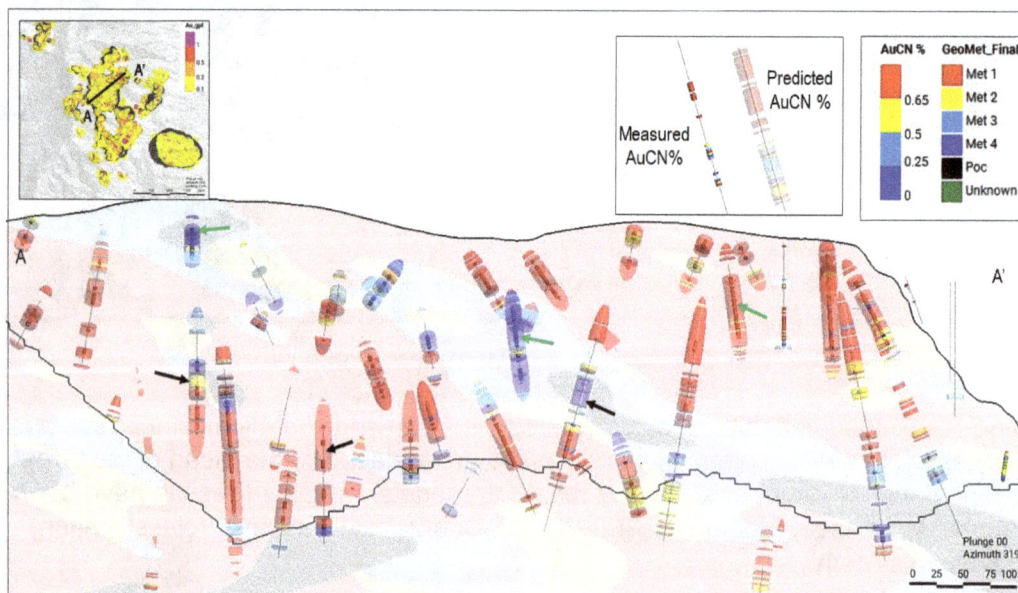

FIG 7 – Cross-section comparing predicted values (XGB_regression) against measured data.

Knowing where the carbonaceous or lower recovery zones are consequential in this type of deposit; the black arrows in Figure 7 highlight several areas where the predicted values expanded on the assayed values improving the GeoMet modelling.

The predicted metallurgical domain classes of the drill hole data were used as input to build three dimensional wireframes (Figure 8b), updating the previous model which did not access the predicted data set. A cross-section of the original and final model is presented in Figure 8.

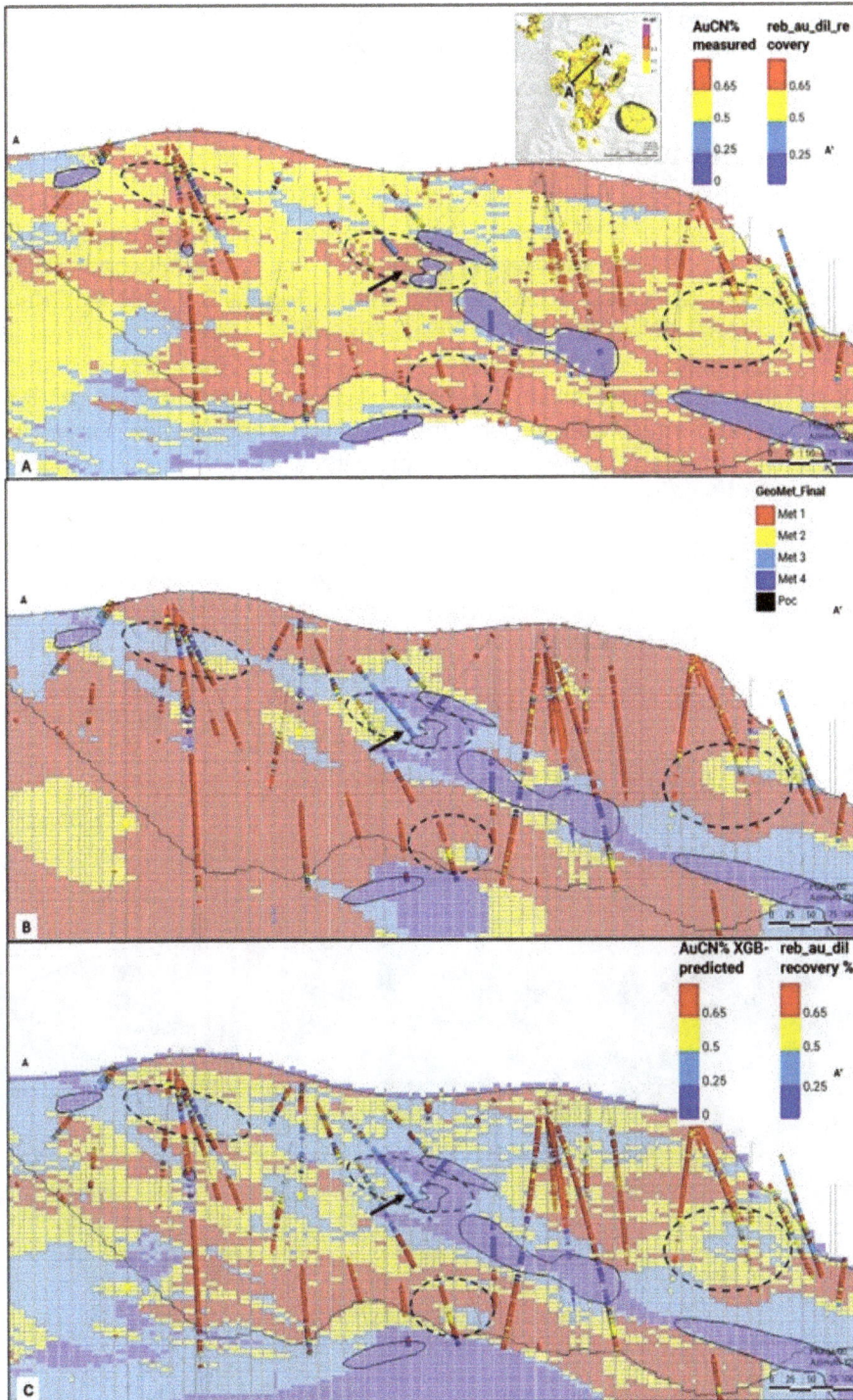

FIG 8 – (a) Cross-section comparing the original recovery model and original valid AuCN drill hole samples prior to any machine learning work; (b) the metallurgical domain model using the predicted AuCN values; and (c) the final recovery model, AuCN drill hole samples, and imputed sample data.

The recovery blocks in Figure 8a were defined using gold grade and geologic formation, and all units except Poc were assumed to be oxide. The recovery blocks in Figure 8c used the gold grade, geologic formation and the GeoMet domains built from the predicted XGB classifier shown in Figure 8b; representing an expansion on both the method by which recovery was determined, through consideration to degree of oxidation, and the data set available to generate the GeoMet domains, through machine learning. The dashed lines highlight areas that are notably improved between the early Figure 8a and resultant Figure 8c models. The confidence in building the GeoMet model Figure 8b that ultimately lead to the resultant model Figure 8c would not have been possible without the predicted values. We can see specific formations, such as the Polb unit, that is clay rich and generally preg-borrowing, stands out in the updated recovery domain model. The black arrow in the middle of the sections highlights a zone of imputed lower %AuCN values which correlates well with the clay and proximity to known carbon in the Polb.

Consideration to other techniques

Prior to this approach, SLR and Liberty Gold geologists explored several avenues to try to best represent recovery values across the Project. Using valid (original and re-assayed) %AuCN results, met domain modelling was first attempted using a standard wireframe approach. While in areas with generally consistent recovery values, such as Rangefront (Figure 3b), this approach worked well. However, in areas such as Talman, shown in Figure 9a, where valid recovery values show higher variability over short distances due to the complicating presence of both clay and smoky carbon, and large volumes of missing data, the domain extension distances beyond valid results were found to be subject to user bias and interpretation, and the resultant shapes (not shown) were thought to under-represent transition material upon comparison with assumed presence of clay and smoky carbon.

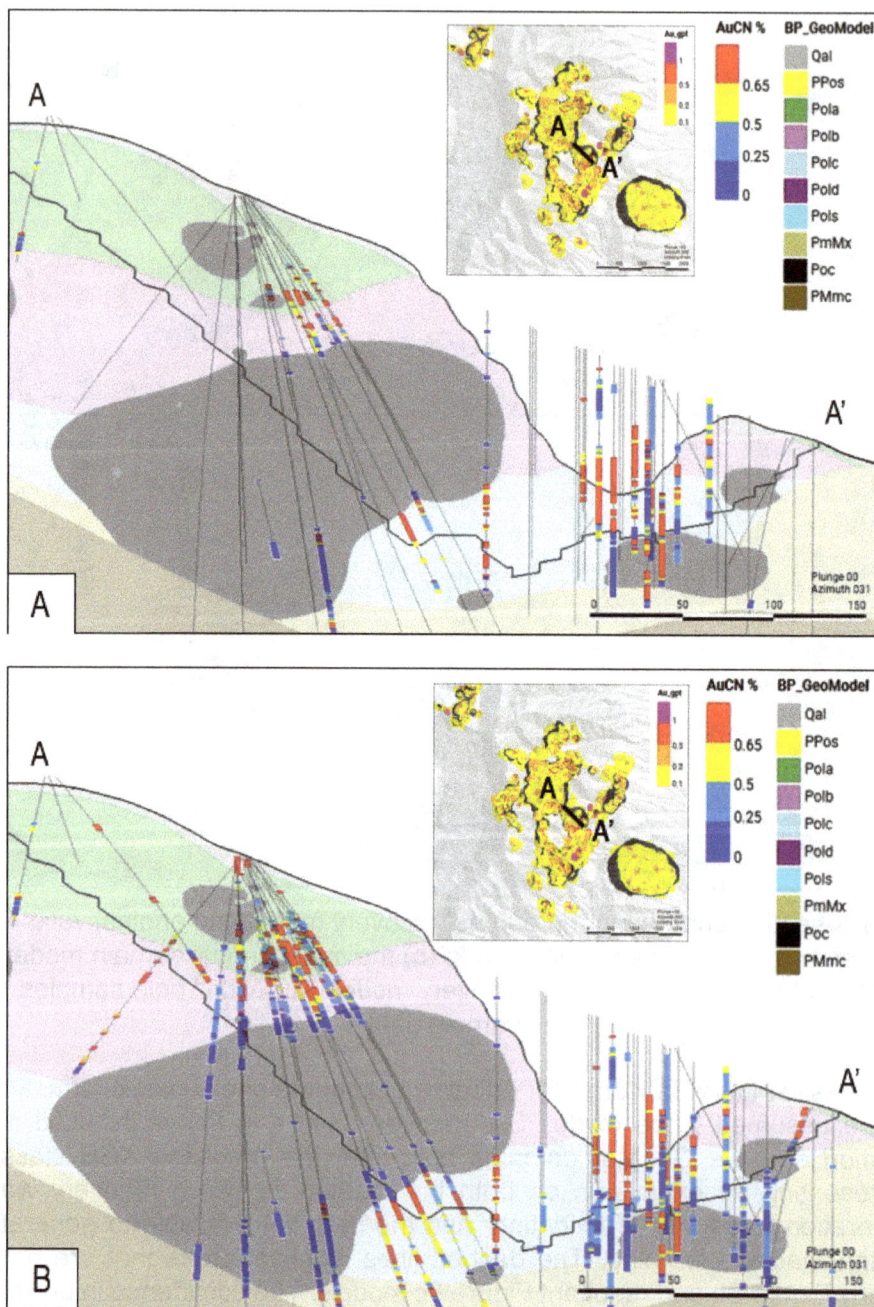

FIG 9 – Comparing: (a) valid AuCN results, to (b) imputed AuCN results in the Talman zone.

Other methods, such as co-kriging or indicator kriging, were initially considered but it was determined that these approaches would require a complex array of domains isolating data populations by lithology unit, mineralisation domain, presence of clay, and presence of smoky carbon to represent the varying relationship of gold with recovery. Considering that clay was not modelled discretely at the deposit, specific modelling criteria was still being developed for clay, and that the spatial distribution of data would lead some domains to have no or very sparse data density, these methods were not considered further.

CONCLUSIONS

Metallurgical test work at Black Pine recognised that various oxidised domains existed ranging from oxide to non-leachable material (Met 1–4). Cyanide solubility ratios were identified as a good proxy for these metallurgical domains, however gaps in the database due largely to a handling error at the prep lab reduced the confidence in modelling Met 1–4 domains. Machine learning was able to predict met domain values to 94 per cent weighted accuracy.

The largest impact of the study was the increased resolution and prediction of transitional categories over earlier models which did not benefit from the imputed data set. These domains proved more difficult to accurately predict, likely due to the smaller sample pool compared to oxide and non-leach sample availability.

Recommendations

While not included in the study, applying locally varying anisotropy (LVA) based on lithological contacts during the distance function calculation may enhance the accuracy of recovery predictions in areas of the deposit with slightly offset geological orientations.

A limitation of the categorical approach is the inability to adjust the recovery model if the category bins limits change. At the sample level, the categorical and numerical results match, and at the time of writing, Liberty was exploring estimating AuCN values using the full data suite, including predicted values. Employing a numerical approach at Black Pine would allow recovery values to be estimated or simulated for individual blocks, could improve resolution for the next phase of study, and would be more flexible to accommodate changes in other aspects of the recovery model and equations.

Additional work to understand the relationship between clay speciation, presence, and quantity may allow discrete modelling of interfering clays and additional visual and statistical validation of the recovery model.

ACKNOWLEDGEMENTS

The authors would like to thank Liberty Gold Corp for allowing publication of this study. In addition, this study was supported by Gary Simmons, Will Lepore, Sarah Conolly, James Catley, and Renan Lopes.

REFERENCES

Babak, O and Deutsch, C V, 2009. Collocated cokriging based on merged secondary attributes, *Mathematical Geosciences*, 41:921–926.

Chen, T and Guestrin, C, 2016. XGBoost: A scalable tree boosting system, in Proceedings of the 22nd ACM SIGKDD International Conference on Knowledge Discovery and Data Mining.

LeCun, Y, Bengio, Y and Hinton, G, 2015. Deep learning, *Nature*, 521(7553):436–444.

M3, 2024. Technical Report and Pre-feasibility Study for the Black Pine Gold Project, Cassia and Oneida Counties, Idaho, USA, effective June 1, 2024, dated November 21, 2024.

Pedregosa, F, Varoquaux, G, Gramfort, A, Michel, V, Thirion, B, Grisel, O, Blondel, M, Prettenhofer, P, Weiss, R, Dubourg, V, Vanderplas, J, Passos, A, Cournapeau, D, Brucher, M, Perrot, M and Duchesnay, E, 2011. Scikit-learn: Machine Learning in Python, *JMLR*, 12:2825–2830.

Sawyer, V, Scott, C C and Comba, P, 1997. Black Pine Mine Heap Leach Closure Pre-Closure Planning and First Year Closure Activities, internal report for Pegasus Mining Corp, 20 p.

Silva, D S and Deutsch, C V, 2017. Multiple imputation framework for data assignment in truncated pluri-Gaussian simulation, *Stochastic Environmental Research and Risk Assessment*, 31(9):2251–2263.

Domaining and
geological modelling

The importance of geologically driven estimation domaining on resource estimation – an example from the Turquoise Ridge Gold Mine within the Getchell Trend in Nevada

C Le Cornu[1], J M Clark[2], L Snider[3] and J Pari[4]

1. Manager, Resource Geology, Nevada Gold Mines, Elko NV 89801, USA.
   Email: christopher.lecornu@nevadagoldmines.com
2. Chief Geologist, Nevada Gold Mines, Elko NV 89801, USA.
   Email: jesse.clark@nevadagoldmines.com
3. Superintendent, Resource Geology, Nevada Gold Mines, Elko NV 89801, USA.
   Email: lsnider@nevadagoldmines.com
4. Senior Resource Geologist, Nevada Gold Mines, Elko NV 89801, USA.
   Email: jpari@nevadagoldmines.com

ABSTRACT

Geologic domains are not estimation domains. This is particularly true in epigenetic gold deposits as they fail the test of stationarity. The three-dimensional distribution of grade is too often ignored, and mineralisation trends and geometry is neglected leading to unexplained blow-outs in resource estimates. While contouring grade above selected cut-off grades is a common practice to avoid these issues, they introduce other artifacts, especially if the mineral controls understanding is poor.

Turquoise Ridge is a high-grade Carlin-type gold deposit located within the Getchell Trend in northern Nevada. The resource model used quasi-stationary domains requiring extensive manipulation of estimation parameters to overcompensate for the lack of orebody knowledge leading to inadvertently smearing high-grades inside a low-grade envelope using subpar mineralisation trends. Globally, it validated well, but on a monthly and quarterly production basis, it reconciled poorly, losing up to 30 per cent of ore headings as new levels opened up with grade control drilling.

The geological understanding was challenged leading to a significant evolution over the past two years, from a predominant steep dipping fault zone control to a low angle fold and thrust architecture. This triggered an overhaul to the estimation domaining strategy. Implementation of explicitly-controlled implicit modelling techniques were chosen to:

- honour the new mineral controls
- appropriately constrain high-grade mineralisation
- allow continuity along interpreted geological controls.

The key to a robust resource model are geologically driven estimation domains that are balanced with explicit controls appropriately scaled to the deposit and mining method, and passes the geostatistical assumption of stationarity. This paper focuses on the explicit-implicit modelling techniques implemented and the impact to resource estimation, briefly touching on the methods used to evolve the understanding of the controls on mineralisation. The challenges and limitations of different domaining strategies are also discussed.

INTRODUCTION

Geologic domains are not estimation domains. This is particularly true in epigenetic gold deposits as they fail the test of stationarity. The three-dimensional distribution of grade is too often ignored, and mineralisation trends and geometry is neglected, leading to unexplained blow-outs in resource estimates, or worse, missed opportunities that become sterilised. While contouring grade above selected cut-off grades is a common practice to avoid these issues, they introduce other artifacts, especially if the mineral controls understanding is poor.

Carlin-style deposits are an example of epigenetic, hydrothermal mineral systems hosted by complexly deformed sedimentary rocks. They form some of the largest concentrations of gold deposits in the world. The prolific Carlin-trend in Nevada is the premier example, which has poured over 100 Moz of gold since the discovery of the archetype Carlin deposit in 1961. Together with the

Carlin-style gold deposits of the Battle Mountain and Getchell trends, including the Turquoise Ridge deposit, Nevada Gold Mines operates the world's largest gold producing complex, with a total Mineral Resource base of greater than 80 Moz (Barrick Gold, 2023).

Turquoise Ridge is a high-grade, Carlin-style deposit located within the Getchell trend of northern Nevada. The geological controls on mineralisation has continuously evolved since the discovery of the Getchell deposits in the 1930s. Cassinerio et al (2011) document the most recent interpretation based on various underground observations that is represented in a single two-dimensional cross-section. While this section captured key characteristics of the Turquoise Ridge deposit, the authors admit that it did not adequately explain the 3D variability, of which profoundly impacts the structural interpretation and controls on mineralisation.

The two-dimensional section is one of thousands of examples that is symbolic to the traditional methods of geologic analysis that is transplanted to graduate geologists globally. While generally easier to teach, or convey a complicated message, they consistently fail to holistically describe the geologic relationships that influence the most fundamental components that a geologist should be focused on: the three-dimensional orebody geometry and distribution of grade.

This paper presents a new model of the Turquoise Ridge deposit to explain the complex three-dimensional structural controls on mineralisation, as well as documenting the adopted modelling process to best honour these relationships.

CARLIN-STYLE MINERAL SYSTEMS

Carlin-style mineral systems are a unique class of epigenetic deposits that are typically high-grade and high-tonnage. Deposits are hosted where feeder faults to gold bearing fluids interact with calcareous, carbonaceous silty mudstones. This typically forms a combination of fault/breccia hosted gold mineralisation, surrounded by more disseminated stratabound gold mineralisation (Figure 1). Sulfidation is the principal mechanism for gold precipitation, incorporating gold in the crystal lattice of hydrothermal arsenian pyrite rims. This forms both refractory and double refractory material types depending on the abundance and type of carbon associated with gold.

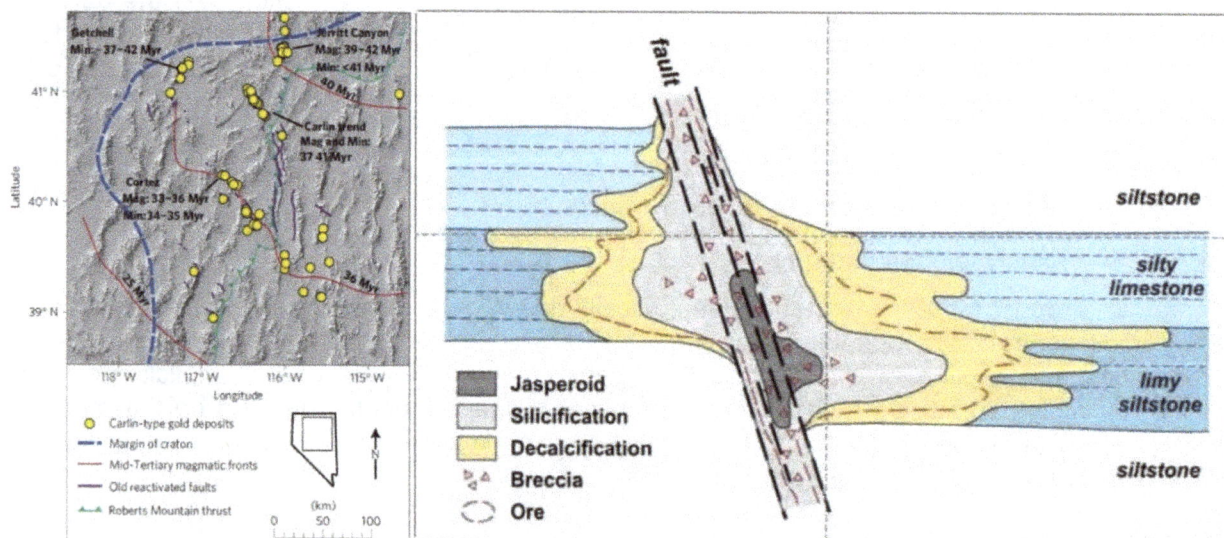

FIG 1 – Left: location of Carlin-type gold deposits and associated trends. Right: schematic simplified diagram of typical Carlin-type gold systems (Robert et al, 2007).

Sedimentary units hosting Carlin-style gold mineralisation in Nevada were deposited within the slope and carbonate platform environments of a passive margin during the Ordovician, Silurian and Devonian. These rocks have subsequently undergone a prolonged history of deformation, summarised by Rhys et al (2015):

- Phase I: thick-skinned contraction thrusted the deep marine sediments of the Roberts Mountain Allochthon eastwardly over and collapsing the passive margin shelf and platform deposition in the late Devonian.

- Phase II: thin-skinned contraction evidenced by low-angle thrust faults propagated east-verging inclined to recumbent folds that trend north–south (local north-west to west vergence also observed). This Jurassic event refolded Phase I folds.

- Phase III: far-field contraction in the hinterland of the Cretaceous Sevier Orogeny resulted in north-east to north-west-trending upright open folds refolding both Phase II and I folding.

- Phase IV: Eocene extension and magmatism introduced Carlin gold mineralisation.

This prolonged history of compression resulted in the thickening of preferential host rocks locally, creating ideal locations to form well-developed mineralisation.

GEOLOGY OF THE TURQUOISE RIDGE DEPOSIT

The Turquoise Ridge Underground deposit is located along the Getchell trend, which also includes the Twin Creeks deposit, as well as a series of other deposits along the Getchell fault zone, interpreted to be the primary control on gold mineralisation in the district. The Getchell fault zone dips at 30–50° to the east and is interpreted to have undergone a prolonged history of normal, reserve and strike-slip kinematics (Joralemon, 1951).

The Turquoise Ridge underground operation is the highest-grade operation in Nevada Gold mines, with 2023 reported measured and indicated resources of 15 Moz grading at 9.57 g/t (Barrick Gold, 2023). This makes the Turquoise Ridge deposit unique within the Nevada Gold mines portfolio, as well as Carlin systems in general.

The underlying stratigraphy at Turquoise Ridge is somewhat different to the Carlin systems along the Carlin and Battle Mountain trend. Figure 2 illustrates the tectonostratigraphic column, with the most significant gold mineralisation being hosted within the Cambrian-Ordovician Comus formation. This formation consists of a series of calcareous, carbonaceous silty mudstones (Lower Comus), overlain by a package of interbedded limestone-mudstone and basalt (Middle-Upper Comus).

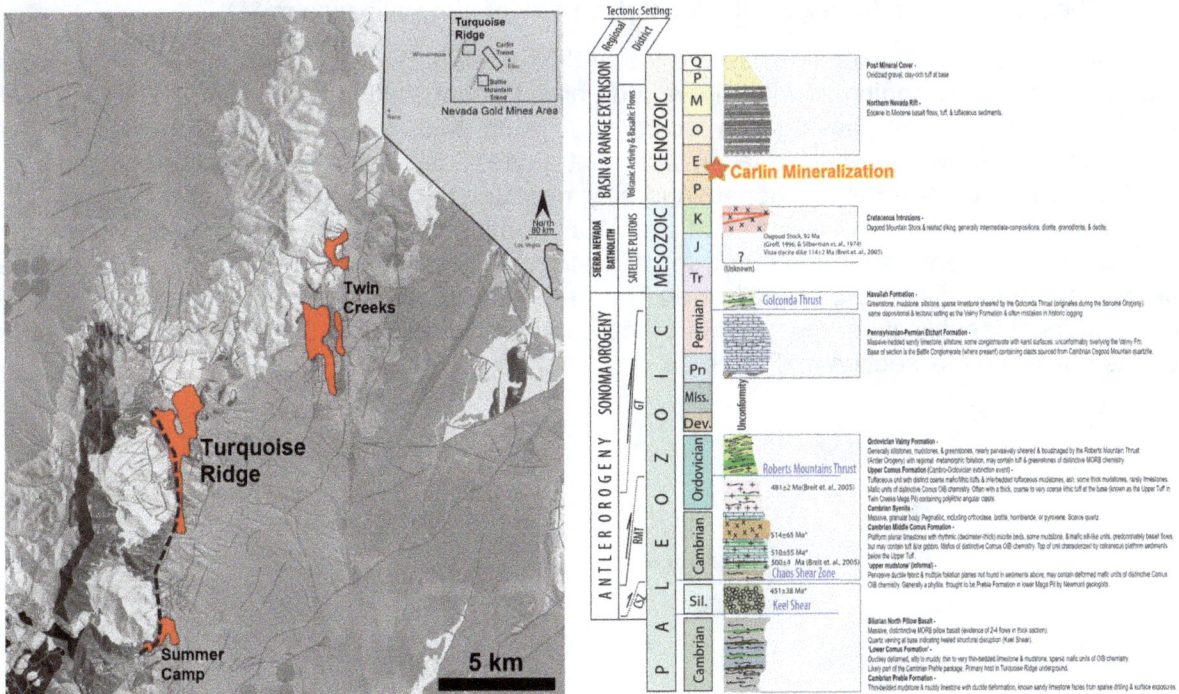

FIG 2 – Getchell district geological map showing location of gold deposits (left), and District stratigraphic column (right).

The Comus formation is overlain by the Valmy formation, primarily consisting of basalt, chert and siltstone. These units have undergone the same regional deformation events highlighted above, and folding is a dominant feature throughout the Getchell district. These rocks are intruded by the Cretaceous Osgood stock and related dacite dykes and sills that also utilised pre-existing structures, dominantly parallel to the Getchell Fault.

While the gold mineralisation at Turquoise Ridge is of typical Carlin-style, similar to those within the Carlin and Battle Mountain trends, the local controls is what sets it apart. Cassinerio *et al* (2011) observed the gold mineralisation to occur along complex structural intersections between:

- North-east to north-west-trending high angle faults and fractures.
- West-north-west trending Palaeozoic basin margin.
- Synforms and antiforms.
- Calcareous lithologies.

Stratabound mineralisation is restricted to the Lower Comus Formation, but has a sharp boundary with very high-grades that are exclusively controlled within these intersections.

Like most Carlin systems, decalcification, argillisation and silicification are all closely associated with gold mineralisation. Though, unlike most Carlin systems, is the strong presence of realgar and orpiment mineralisation, often spatially related to gold mineralisation. Cline and Hofstra (2000) interpreted two-stages of realgar and orpiment, one associated with early ore-stages, likely related to a switch from high fluid-rock interaction to cooling of the still-sulfur and gold-enrich fluid, precipitating orpiment and later realgar in void space. Most of the realgar and orpiment is, however, paragenetically later than gold, yet their significance is poorly documented or understood (Cline and Hofstra, and references therein).

Several hypotheses emerged over the last decade, most notable of which was the dominance of steeply dipping structures, striking NE-SW, dipping 70° to the NW, interpreted to be major fluid conduits. Gold mineralisation was interpreted to reflect this orientation, particularly where they intercept Getchell fault parallel structures and Cretaceous dacite dykes. Ponding of gold bearing fluids under dacite dykes was interpreted to be an ideal trap for mineralisation. North-east striking, steeply dipping faults lacked factual evidence, but helped to explain trends in gold mineralisation at the deposit scale. This led to a series of linear structural 'corridors' interpreted to explain gold grades, however, they did not help to resolve the local geometry of mineralisation in 3D.

Hypotheses were also developed to evolve the understanding of the stratigraphy within the Lower Comus formation. Re-logging and multi-element interpretation defined two populations within the Lower Comus, named the 'Basal-slope facies' (BSF) and 'Mid-slope facies' (MSF), names given based on their interpreted depositional environment along a carbonate slope platform. These units are composed of calcareous, carbonaceous silty mudstones, with the BSF containing a higher composition of carbonaceous muds, and the MSF dominantly siltstone with minor interbeds of limestone. Gold mineralisation appeared to be preferentially hosted within the Basal-slope facies, however, due to the difficulty of visually identifying the difference between these two sub-units and lack of historical identification, the geometry and continuity of the BSF unit was poorly defined, as evidenced by the cross-section on Figure 3a.

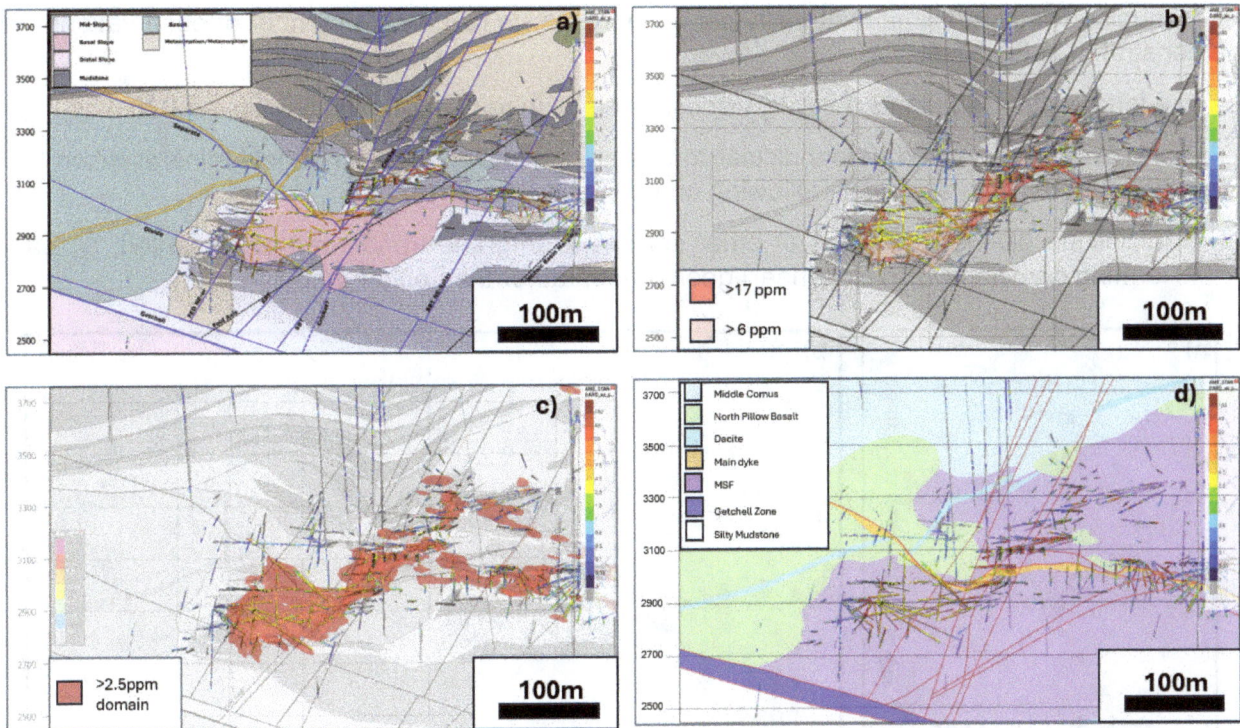

FIG 3 – (a) Detailed geological cross-section across the deposit, (b) Interpretation of controls on gold mineralisation, (c) Resource domains, (d) Cross-section from 3D geological model.

Folding has been documented within the Getchell district for some time, most notably (and visually obviously) in the Twin Creeks Mega pit, where it has a dominant control on gold mineralisation (Stenger *et al*, 1998). Despite the large amount of structural data collected at Turquoise Ridge, a fold interpretation was not applied to the Turquoise Ridge underground geologic model.

The interpretations above were stitched together over several years through basically one-dimensional observations, that is, single point-scale observations made during core logging and underground mapping. When compiled in 2D sections, ruler-straight lines connected these observations creating several hypotheses explaining the mineralisation trends. The culmination of which were exposed as deeply inadequate when an attempt to construct three-dimensional estimation domains.

The resultant grade shell domains used the interpreted controls as structural trends ultimately producing poorly constrained, discontinuous 'blobs' that were at the mercy of the information effect. Grade shell cut-offs were chosen in a geostatistical attempt to create stationary domains, using histograms and log probability plots to differentiate different grade populations. As a result, a significant effort was made to unravel the local 3D controls on mineralisation and apply this knowledge to resource domains and the estimation.

Figure 3b illustrates a detailed cross-section of the interpreted controls and geometries of mineralisation as of 2022. Figure 3d illustrates a cross-section from the 2022 geological model, developed in leapfrog along the same section.

GEOLOGIC CONTROLS ON MINERALISATION

Gold trends analysis

Mineralisation geometry and its grade distribution are the foundations to explaining mineral controls and is a three-dimensional problem to solve. One method to do this is discussed by Cowan (2022), which considers assay data as structural data. By resolving potential trends in assay data, hypotheses can be developed to explain those trends. The maximum intensity projection tool in Leapfrog Geo is particularly useful in resolving complex 3D distributions of mineralisation.

At Turquoise Ridge, maximum intensity projection analysis revealed a number of observations, as demonstrated in Figure 4:

- a long section highlights a strong deposit scale plunge of approximately $25° \to 025°$

- a cross-section sliced perpendicular to this plunge shows an area of gold mineralisation that appears to pinch and swell, following an undulating pattern

- in oblique 3D view, a series of other discrete high-grade trends can be observed, typically forming dominantly prolate (linear) geometries, rather than oblate (tabular).

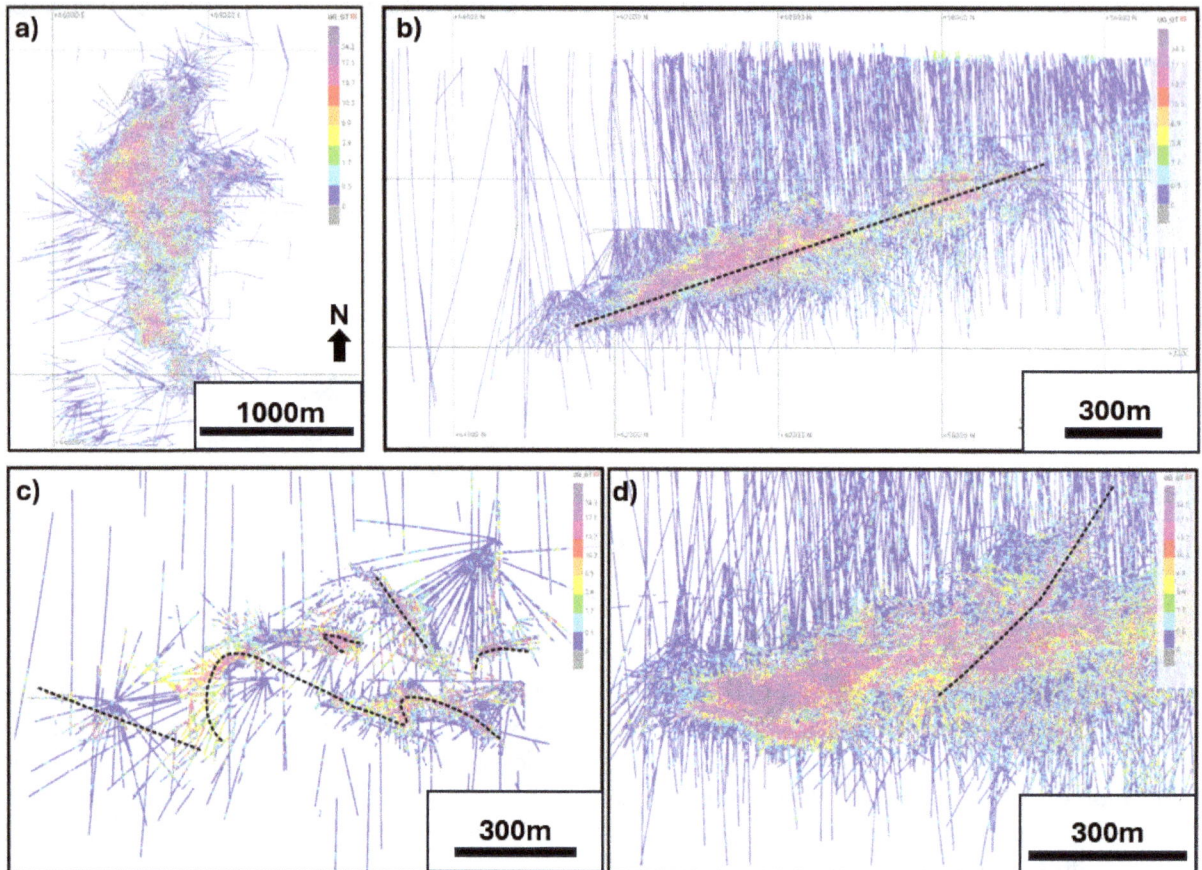

FIG 4 – Trends in gold mineralisation: (a) Plan view; (b) Long section looking east; (c) Oblique cross-section looking NNE; (d) Oblique 3D view of mineralisation.

The combination of strongly plunging controls and prolate geometries allude to complex 3D structural controls, which may be a result of folds and/or intersection lineations (Cowan, 2014). Furthermore, local mineralisation geometries do not follow the interpreted 045° striking interpreted corridors.

Structural analysis

An abundance of structural data has been collected from underground mapping historically. Strong decalcification and alteration of the host lithologies make acquiring oriented core in Carlin systems problematic, however, acoustic televiewer (ATV) is widely adopted to collect structural data in its absence (Hartman, 1998).

Bedding and fault data collected from five core holes using ATV is displayed and interpreted in Figure 5a. Throughout the core in these holes, bedding appears to be tightly folded, as observed in Figure 6, which also shows folding observed underground. Two main orientations of bedding can be observed in this data set, including an east dipping orientation and a north-west dipping orientation, interpreted to globally represent the two limbs of the folds. The intersection point of these two limbs suggests a fold plunge of $24° \to 026°$, coinciding with the same plunge of the deposit scale mineralisation.

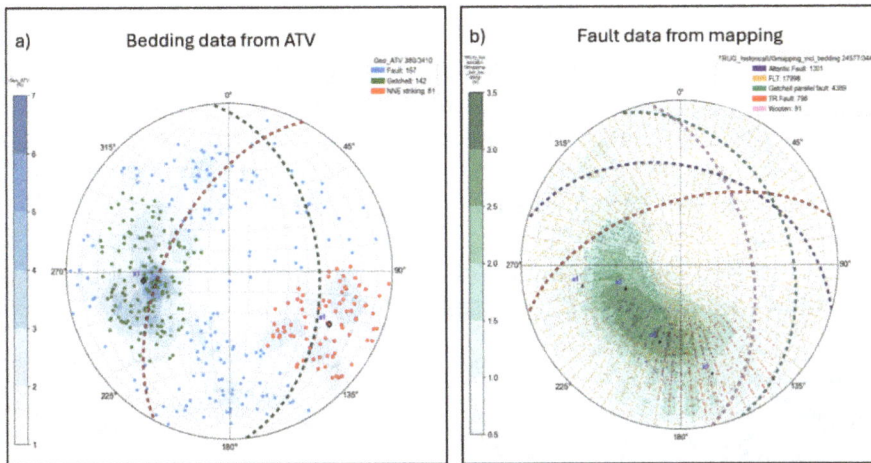

FIG 5 – Stereonets showing: (a) bedding data from drill hole ATV data, (b) fault data from underground mapping.

FIG 6 – (a) underground photo of folding looking north; (b) folding observed in core.

Fault data collected from underground mapping can be observed on Figure 6b. While all fault data seems chaotic, when viewed locally, four main trends associated with mineralisation become apparent:

1. Getchell parallel (157/30ENE).

2. TR fault parallel (072/58NNW).

3. Atlantic fault parallel (115/40NNE).

4. Wooten parallel (168/55E).

With Getchell parallel being the globally dominant trend, and the rest being constrained to local areas. Areas where these faults intercept, such as Atlantic-TR, form linear high-grade shoots of mineralisation.

Lithology

To resolve folding possibly related to mineralisation, lithology needs to be refined. The most obvious lithologic contact present at Turquoise Ridge is the contact between a basaltic unit, known as the 'North Pillow Basalt', which generally overlies the Lower Comus formation. This contact is well defined by drilling, resulting in a high certainty in its geometry, which has an undulatory nature to it, Figure 7. These undulations can be modelled to a reasonable accuracy for over 300 m strike length and display a global plunge of 24 -> 025. Figure 7, consistent with the deposit scale plunge of mineralisation and interpreted fold plunge. Well-developed mineralisation is typically observed within these undulations, which are interpreted to represent fold hinges.

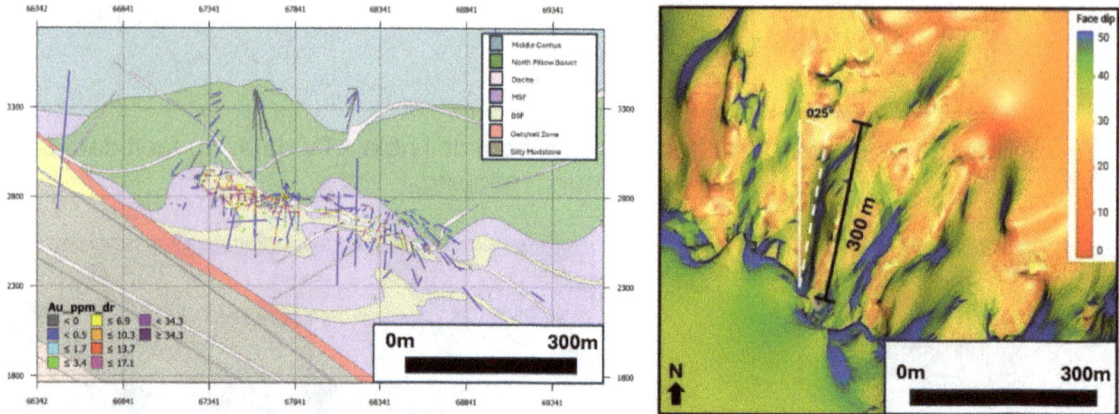

FIG 7 – Cross-section showing undulations in lower north pillow basalt contact (left), dip angles along the base of the north pillow basalt surface, displaying trends of undulations (right).

As previously mentioned, the Lower Comus can be divided into two subunits using four-acid ICP multi-element interpretation and visual logging. Several immobile elemental plots can be used to distinguish these units, most notably V-Cr, Sc-Zr and Sc-Al (Figure 8). It was observed from this work that the BSF appears to thicken in some areas and thin or disappear in others (Figure 7). Thickening is typically observed in areas close to the undulations with the North Pillow Basalt, interpreted to be fold hinges.

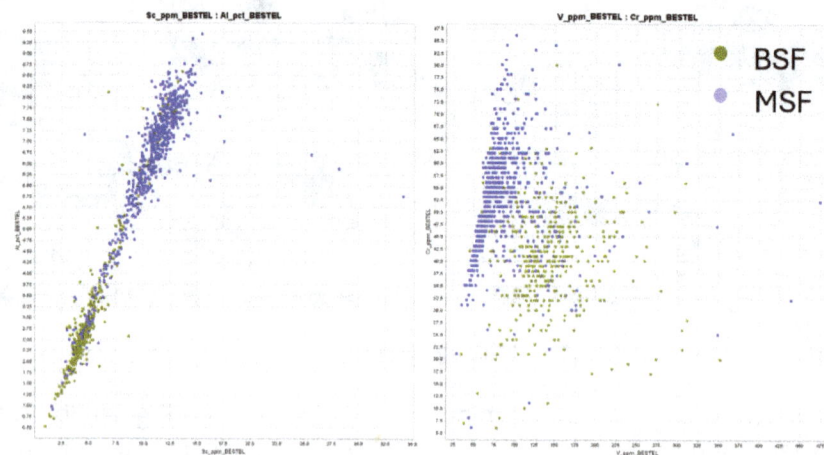

FIG 8 – Sc-Al and V-Cr plots showing different sub-unit populations.

The resultant interpretation therefore relies heavily on understanding the underlying fold and thrust architecture, where folding and associated thrusting has thickened preferential host rocks within hinges. Subsequent deformation events have formed other discrete structures that act as localised fluid conduits. This interpretation has been built into an updated 3D geological model for Turquoise Ridge Underground, represented by the cross-section in Figure 7.

DOMAINING STRATEGY

Methodology

A shift in the interpretation of mineral controls from dominantly steeply dipping structures to a lower angle fold and thrust system has significant impacts on the gold resource domains. Areas where steep structural corridors were interpreted to directly control mineralisation are now interpreted as west limbs of west verging fold hinges, and therefore need to be represented as such in the domains and estimation. Resource domains were built with implicit-explicit techniques in a series of steps.

The first step was to create a structural trend model, which honours local geometries of mineralisation according to the new geological interpretation, to help push anisotropy of the resultant domains. It was determined that due to the prolate geometry of the mineralisation, the final domains would have to be built using leapfrogs RBF numeric model and intrusion tools, rather than vein systems which are more appropriate for tabular geometries. Isotropic grade shells were initially created with a series of grade cut-offs which, along with the maximum intensity project tool, helped to visualise gold trends. A series of surfaces were created using structural disks placed through trends in mineralisation, primarily honouring the fold and thrust architecture that controls mineralisation. Appropriate strengths and ranges were applied to the surfaces in a structural trend model, and the model was applied to the grade shells to push anisotropy. Meshes were added and modified in an iterative process to ensure all shells were behaving appropriately according to the interpretation of their controls.

During this process, it became apparent that contacts between high-grade (+10 ppm) and lower grade (3–10 ppm) appear sharp in some areas, with higher grade being much more discrete. Histograms were plotted using the above grade shells in localised areas in an attempt to observe approximant grade cut-off thresholds. Figure 9 illustrates local histograms from 'FED' and 'MBD' mining areas within a grade shell cut-off of 3 ppm. Higher grade 'bimodal' distributions with a mean of around 20 ppm can be observed in these histograms, suggesting that a 3 ppm cut-off for domains in high-grade areas may not be locally stationary, consistent with the visual observations of discrete high-grade zones located within broader low-medium grade packages. This in turn poses another problem with the grade shell domaining method, as RBF numeric models struggle to push continuity of discrete styles of mineralisation, often resulting in isolated, discontinuous shells even with the strongest anisotropy applied. Figure 10a illustrates a global 10 ppm grade shell with an anisotropic trend applied, showing lack of continuity along known mineral controls.

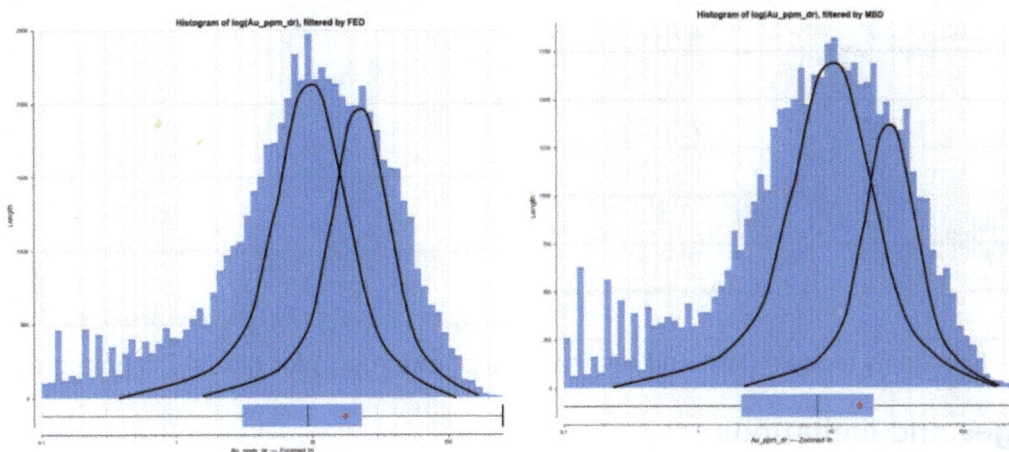

FIG 9 – Histogram of the log for two local mining areas; FED (left), MBD (right).

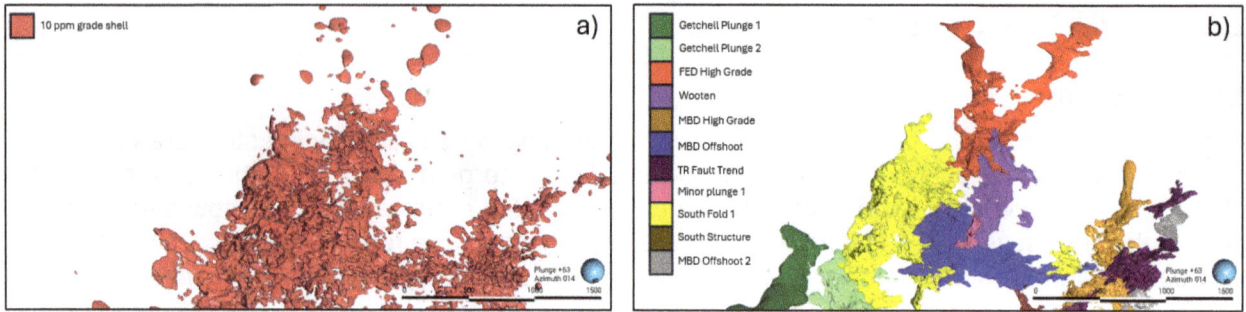

FIG 10 – 3D view of domain volumes looking down to the NNE: (a) 10 ppm grade shell; (b) Geologically driven 'intrusion' generated domains split by geological controls.

As well as a lack of statistical stationarity, global grade shell domains include multiple geological controls and orientations and therefore don't represent geological stationarity. This is particularly problematic for variogram modelling, where multiple contradicting orientations included in the same domain result in isotropic variogram models with short ranges, Figure 11. To solve these problems, intrusion models were generated in leapfrog based on interval selection of assay data which were manually selected based on the interpreted geological control. This was initially based on a cut-off grade of 10 ppm, however, also included lower grades (5–10 ppm) which were part of the same geological control, helping to push continuity of the domains where appropriate. The structural trend model was applied to push anisotropy, and explicit lines were used in areas of lesser data density to help extrapolate continuity along interpreted mineral controls. The result of this is illustrated in Figure 10b, with each domain coloured by its interpreted geological control. These domain splits resulted in domains showing normal distributions, whilst appropriately constraining high-grade and pushing continuity along geologic controls.

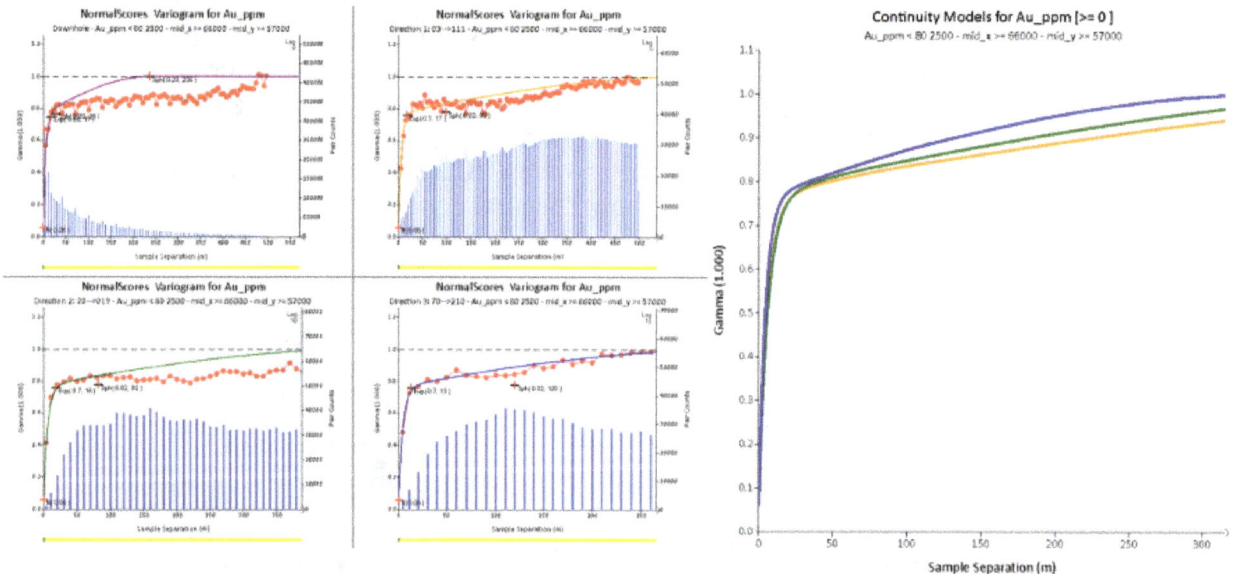

FIG 11 – Continuity models (left) and resultant variogram (right) for global 2.5 ppm grade shell.

Challenges and limitations

There are a number of challenges and limitations faced during the implementation of this domaining method. First, the intrusion tool in Leapfrog uses the same RBF algorithm as RBF numeric models, which struggles to honour discrete mineralisation geometries as highlighted above. Therefore, a significant amount of explicit linework is required in areas of lesser data density in order to extrapolate domains effectively. In contrast, vein system tools would do a better job with this, however, they run the risk of pushing too much continuity in a tabular rather than linear geometry, as well as being harder to manipulate within tightly folded systems.

Secondly, by domaining the high-grade population separately from the medium to low-grade population, variance is removed from the data set which presents challenges when variogram modelling. This, combined with splitting domains further by geological controls and therefore reducing the data available to model, produces an artificially high nugget (due to lack of variance) within continuity models. However, the benefits of being able to provide more robust, geologically valid directionality to the variograms as a result of splitting by orientations and controls of mineralisation, as well as the ability to geologically constrain and control high-grade through the domaining process were thought to outweigh these limitations in this case.

ESTIMATION IMPACT

The above changes to the resource domains have resulted in significant impacts to the resource model locally. Figure 12 demonstrates a typical scenario in a mined-out area of the block model. The old block model (Figure 12a), using only a low-grade (2.5 ppm) shell, resulted in more tonnage over a cut-off of 7 ppm (approximant economic cut-off). As more data is acquired through infill drilling and channel sampling, the volume of >7 ppm material systematically drops by 20–30 per cent in some cases. This mine is dominantly mined via cut-and-fill due to the shallow dipping nature of the orebody and poor rock mass quality. From a production standpoint, cut-and-fill panels were therefore systematically being removed from the mine plan on the edges of the orebody as new data and visual ore calls reduced the extent of ore-grade mineralisation. Figure 12b shows the model with updated domains in the same plan view section, along with newly acquired infill drilling and channel sampling data, demonstrating the reduced extent of the >7 ppm material, honouring the extent of 7 ppm in the channel samples much more accurately.

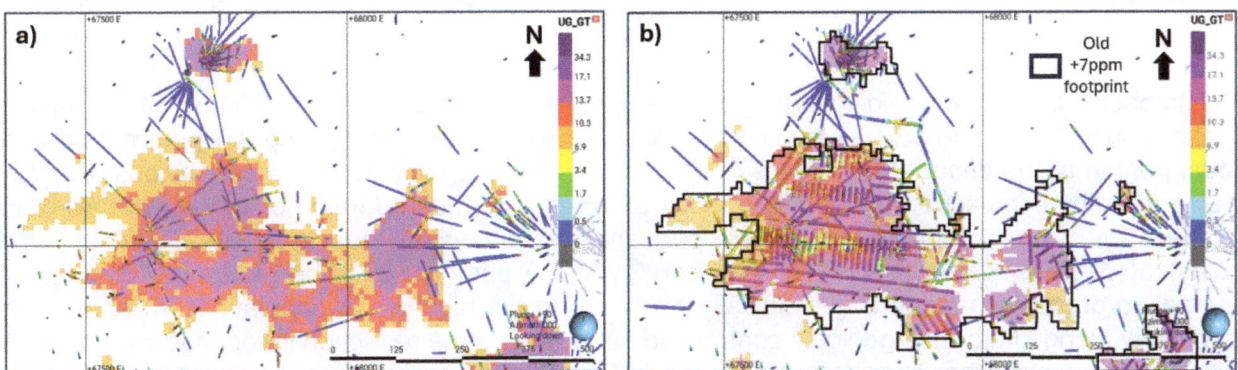

FIG 12 – 2D plan view of block model: (a) using only a 2.5 ppm grade shell; (b) using a high-grade domain (10 ppm) nested within a low-grade (2.5 ppm) domain.

The application of these interpretations and domains can also be observed in the mine plans around the remaining resource. Figure 13 illustrates an example of this, where a better understanding of the controls on high-grade mineralisation has allowed for the addition of high-grade domains, resulting in a block model that represents infill drilling and channel sampling at a much higher resolution.

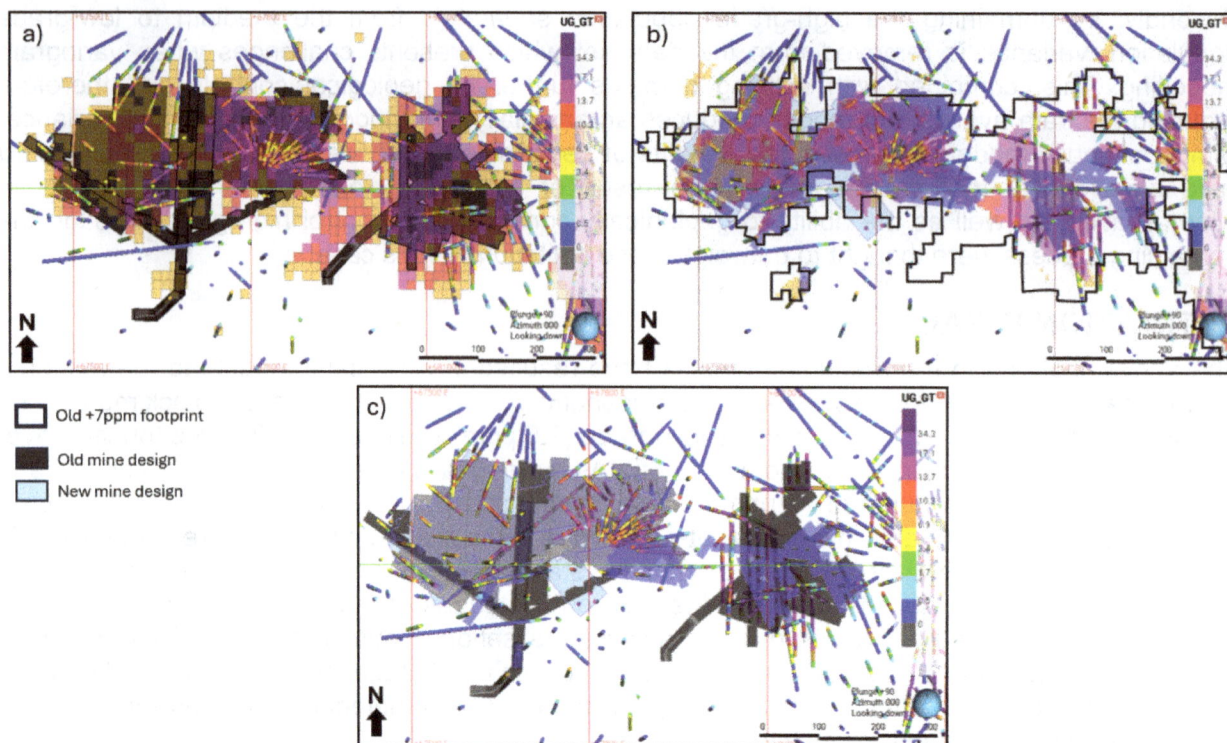

FIG 13 – 2D plan view of block model showing planned panels on old model (a); versus new model (b).

The impact of this new domaining strategy can also be observed numerically on Figure 14, displaying global ounce and grade profiles at differing sample count sensitivities, using a 7 ppm cut-off in a local portion of the deposit. This figure illustrates a number of impacts, the most obvious of which being the global impact on grade and ounces as a result of these domain changes, with average grades increasing by approximately 25–30 per cent, and ounces decreasing by 15–20 per cent as a result of decreased tonnes (-35–40 per cent) over a 7 ppm cut-off. Therefore, the benefits of increased grade as a result of geologically constraining the high-grade population, whilst pushing continuity along interpreted geologic controls do not outweigh the reduction in tonnage over the cut-off. This suggests either overestimation in the old model as a result of unconstrained high-grades smearing into lower grade material, or underestimation in the new model. Visual validation of the block model versus geological controls and existing data suggests the former is more dominant, Figure 15, however, there may be some conservatism in the new model in less common, local areas due to this domaining technique.

Figure 14 also illustrates the sensitivity of the two models when changing the minimum and maximum number of samples used in the estimation. When comparing an estimate using 14 samples versus 30 samples, there's only a -0.5 per cent difference in total ounces and -0.5 per cent difference in grade with the new domaining strategy, compared to a -5.5 per cent difference in total ounces and -4 per cent difference in grade with the old strategy. This makes sense, as the new domaining strategy reduces the variance of the grade in the estimation domains, and given the interpretation that this is a high-grade, discrete, structurally controlled gold deposit, it highlights the importance of reflecting the geologic understanding in the domaining.

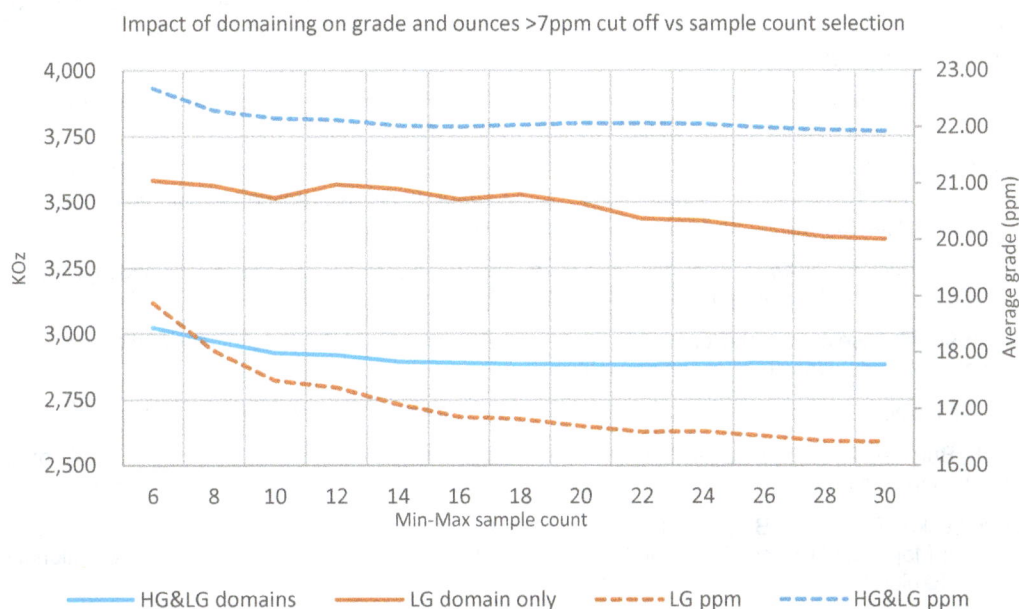

Impact of domaining on grade and ounces >7ppm cut off vs sample count selection

FIG 14 – Estimation sensitivities when comparing the old model versus new model at different sample selections.

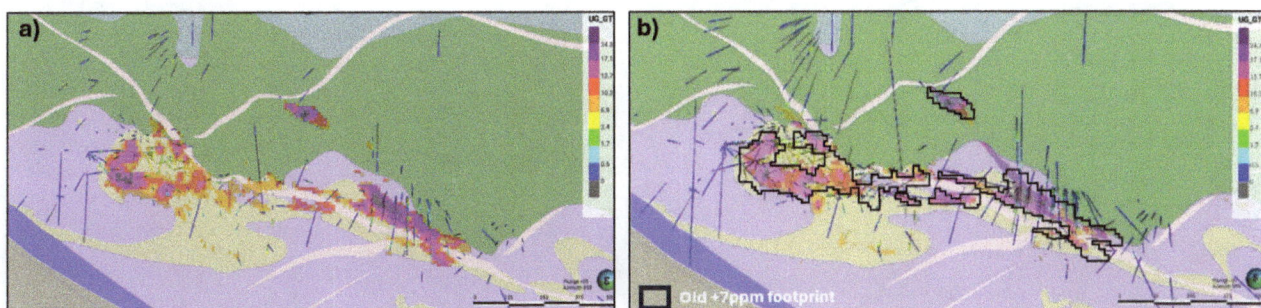

FIG 15 – New geological cross-section with: (a) old block model; versus (b) new block model, looking north.

CONCLUSION

Resource domaining forms a critical part of the resource estimation process. If the underlying controls on mineralisation are not well understood, it's impossible to build reliable domains and estimations that reconcile well on the scale needed for underground mining. In complex deposits with multiple controls, it's of upmost importance that the geological interpretation holds together in 3D. Turquoise Ridge is a complex gold deposit, influenced by the interactions between folding of preferential stratigraphy and intersections with faults of multiple orientations. Understanding the distribution of gold mineralisation in 3D, along with using structural data, geochemical data and underground mapping data has helped to develop a solid geological interpretation which explains the local controls of mineralisation. Resource domains can be built using a combination of explicit and implicit methods to represent both geostatistical stationarity and geological stationarity, which constrains grade populations appropriately whilst honouring geological continuity. Whilst the techniques highlighted above work well for Turquoise Ridge, other tools may be more appropriate for different deposit types. In any case, a high degree of explicit control driven by a solid understanding of the controls on mineralisation is required.

REFERENCES

Barrick Gold, 2023. Barrick Annual Report 2023, pp 150–157. Available from: <www.barrick.com/English/investors/quarterly-reports>

Cassinerio, M D, Muntean, J L, Steininger, R and Pennell, B, 2011. Patterns of lithology, structure, alteration and trace elements around high-grade ore zones at the Turquoise Ridge gold deposit, Getchell district, Nevada, *Great Basin evolution and metallogeny*, pp 949–977.

Cline, J S and Hofstra, A A, 2000. Ore-fluid evolution at the Getchell Carlin-type gold deposit, Nevada, USA, *European Journal Of Mineralogy-Stuttgart*, 12(1):195–212.

Cowan, E J, 2014. X-ray Plunge Projection – Understanding Structural Geology from Grade Data, *Monograph 30: Mineral Resource and Ore Reserve Estimation – The AusIMM Guide to Good Practice*, second edition, pp 207–220 (The Australasian Institute of Mining and Metallurgy: Melbourne).

Cowan, E J, 2022. Structural analysis from drill hole assay data using locally varying anisotropies (LVA)–a catalyst for structural geological enlightenment, *Structural geology and resources, Kalgoorlie*, pp 12–19.

Hartman, S O, 1998. The use of acoustic log data in structural interpretation at the Turquoise Ridge deposit, Getchell Mine, north-central Nevada, University of Nevada, Reno.

Joralemon, P, 1951. The occurrence of gold at the Getchell mine, Nevada, *Economic Geology*, 46(3):267–310.

Rhys, D, Valli, F, Burgess, R, Heitt, D, Greisel, G, Hart, K, Pennell W M and Garside, L J, 2015. Controls of fault and fold geometry on the distribution of gold mineralisation on the Carlin trend, *New concepts and discoveries*, 1:333–389.

Robert, F, Brommecker, R, Bourne, B T, Dobak, P J, McEwan, C J, Rowe, R R and Zhou, X, 2007. Models and Exploration Methods for Major Gold Deposit Types, in Proceedings of Exploration 07: Fifth Decennial International Conference on Mineral Exploration.

Stenger, D P, Kesler, S E, Peltonen, D R and Tapper, C J, 1998. Deposition of gold in Carlin-type deposits; the role of sulfidation and decarbonation at Twin Creeks, Nevada, *Economic geology*, 93(2):201–215.

Advances in geologic and resource modelling methods for locally-varying controls on orogenic gold mineralisation

J A McDivitt[1], D Greene[2], E Hart[3], B Wilson[4] and M Hadavand[5]

1. Senior Resource Geologist, Evolution Mining, Kalgoorlie WA 6430.
   Email: jordan.mcdivitt@evolutionmining.com
2. Specialist – Resource Management, Newmont Corporation, Perth WA 6008.
   Email: dave.greene@newmont.com
3. Exploration Manager, Newmont Corporation, Perth WA 6008. Email: erin.hart@newmont.com
4. Project Geostatistician, GeologicAI, Edmonton AB T5K0G9, Canada.
   Email: bwilson@geologicai.com
5. Senior Geostatistician, GeologicAI, Edmonton AB T5K0G9, Canada.
   Email: mhadavand@geologicai.com

ABSTRACT

Orogenic gold deposits can display complex, non-stationary ore geometries reflecting factors such as host rock heterogeneity, progressive deformation, and multiple mineralisation events. Recent advances in modelling and estimation methods help to represent such local controls on mineralisation at the Palaeoproterozoic (circa 1788 Ma) Oberon gold deposit (Tanami, Northern Territory). Currently in the pre-feasibility study stage, the deposit exhibits a complex geometry that manifests as an elongate, doubly-plunging dome of metasedimentary and intrusive rocks.

Lithostratigraphic domains are defined by interpreting conventional drill core logging data supplemented by natural gamma and multi-element geochemical data. Unsupervised machine learning algorithms, including K-means and density-based clustering, refine and validate these domains. Mineralised domains are characterised by planar envelopes of stockwork-type veins or disseminated sulfides, aligning with limb or axial-planar geometries. Gaussian mixture models of Au leverage As and S data to help define latent Au thresholds for the mineralised domains. Secondary indicators, such as vein frequency and alteration intensity, are encoded based on indicator thresholds and aggregated to a secondary indicator score to support the definition of the mineralised domains. Local, high-grade ore shoots reflect the hinge of the dome, smaller nested folds, and lithologic contacts. Such trends define anisotropic variogram models within the principal planes of mineralisation. Python-based resource estimation workflows utilise locally varying anisotropy for experimental variography and local capping methods for calibrated Ordinary Kriging.

The results of this case study highlight the benefit of supplementing conventional geological approaches to domain definition with high-quality, high-dimensional data sets, which provide the basis for domain validation and refinement using unsupervised machine learning algorithms. The application of domaining strategies, variogram methods, and capping techniques tailored to honour local plunge controls and local search neighbourhoods enhance the strength of the Ordinary Kriging estimator in its ability as a non-stationary estimation algorithm.

INTRODUCTION

Orogenic gold deposits contain ~40 per cent of global gold resources within deformed and metamorphosed orogenic belts developed across the Earth's geologic history (Goldfarb, 2024). The deposits manifest as structurally complex, narrow-vein, high-grade (~1–10 g/t), Au-only ore systems often amenable to selective mining methods (Goldfarb *et al*, 2005). Orogenic gold mineralisation is commonly present as quartz-carbonate veins associated with free gold and sulfides (± Au), with high-grade shoots developed in linear structural zones (eg fault jogs, fold hinges; Dubé and Gosselin, 2007). These ore shoots represent local controls on mineralisation that reflect factors such as the syn-mineral deformation regime, anisotropy in the host rock package, and the degree of overprinting deformation (Robert and Poulsen, 2001). High-grade ore shoots often represent a transition from a planar mineralised domain to a linear mineralised domain (Figure 1).

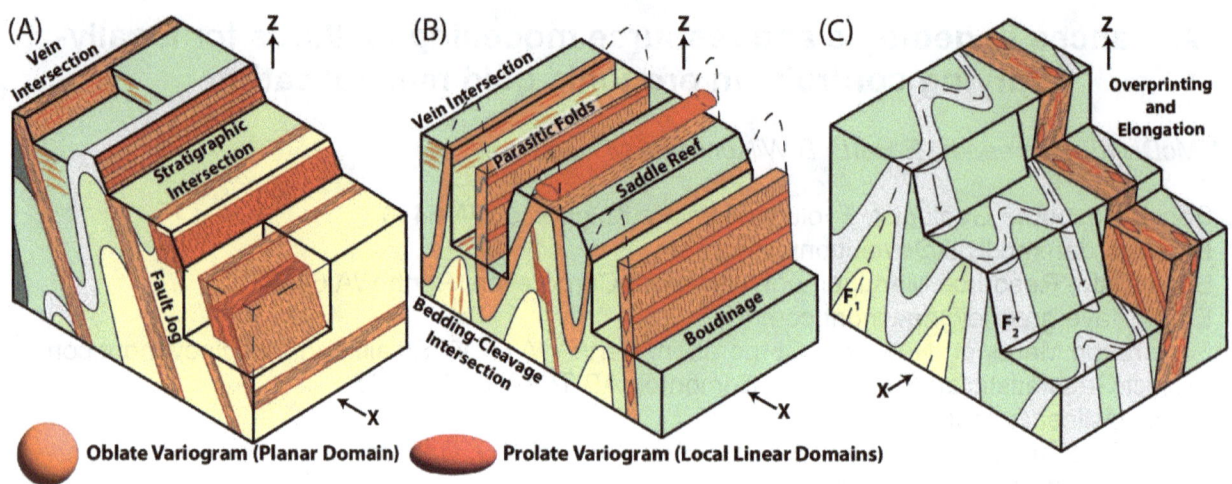

FIG 1 – Examples of different structural styles of orogenic gold systems and different mechanisms of ore shoot formation. **(a)** Shear-zone dominated orogenic gold systems where ore shoots may be defined at structural or stratigraphic intersections and along jogs or bends in fault zones. Examples include the Sigma-Lamaque deposit and other deposits of the Val d'Or region (Robert and Brown, 1986; Robert, 1990; Robert and Poulsen, 2001) and the Revenge deposit in the Kambalda region (Nguyen *et al*, 1995). **(b)** Fold-dominated orogenic gold system showing ore shoots developed as saddle reefs, in zones of boudinage, and in areas of parasitic folding or at structural intersections. Examples include deposits of the Bendigo and Ballarat goldfields (Cox *et al*, 1991), those of the Meguma Terrane (Sangster and Smith, 2007), and deposits of the Tanami Region. **(c)** Overprinted orogenic gold systems represent early systems overprinted by later ductile deformation. Ore shoots may develop as mineralisation is folded, boudinaged, and rotated towards a preferred direction of elongation. Examples include the Renabie deposit of the Wawa Subprovince in the Superior Craton (McDivitt *et al*, 2017).

Despite the variety in the structural styles of orogenic gold systems (Figure 1), a common expression of mineralisation is local, high-grade linear ore shoots nested within larger planar structures (eg faults, contacts). The transition from planar to linear structural domains highlights a change in grade continuity and the theoretical variogram, with planar domains represented by an oblate variogram and a prolate variogram applicable to the linear domains (Figure 1). The delineation of overarching planar controls on mineralisation can help to constrain the presence of local ore shoots by defining planar boundary constraints on a mineralised domain. This may be straightforward in lode-dominated systems, where clear hanging wall and footwall contacts are evident. For deposits characterised by stockwork mineralisation or vein arrays, the overarching planar controls may be cryptic due to the absence of clear bounding structures and gradual transitions away from the mineralised zones.

The Oberon gold deposit, an open pit and underground mineral resource, is located 28 km NNE of the world-class Dead Bullock Soak (DBS) deposit in the Palaeoproterozoic Granites-Tanami Orogen (Northern Territory, Australia; Figure 2a). Like DBS, the Oberon deposit represents a fold-dominated orogenic gold system hosted in metasedimentary rocks, where mineralisation is associated with narrow quartz veins related to the regional orogenic Stafford Event (D_2; ca 1810–1790 Ma; Maidment *et al*, 2020; Petrella *et al*, 2020; Crawford, 2025). While DBS is characterised by regularly oriented '70/70' quartz-albite vein arrays with free-milling gold along S_2 cleavage domains (Miller, 2010; Petrella *et al*, 2020), mineralised veins at Oberon define stockwork zones with gold occurring predominantly as micron-scale inclusions in arsenopyrite and pyrite within quartz ± K-feldspar ± sulfide veins (Meria, 2011; McDivitt, 2024a; Crawford, 2025). Stratabound mineralisation represents a second, less-significant, sulfide-rich, vein-poor mineralisation style at both Oberon and DBS (eg Orac Beds, Auron Beds, Schist Hills Iron Member) that developed contemporaneously with the vein-hosted mineralisation in Fe-rich stratigraphic units (Pendergast, 2011; Petrella *et al*, 2020).

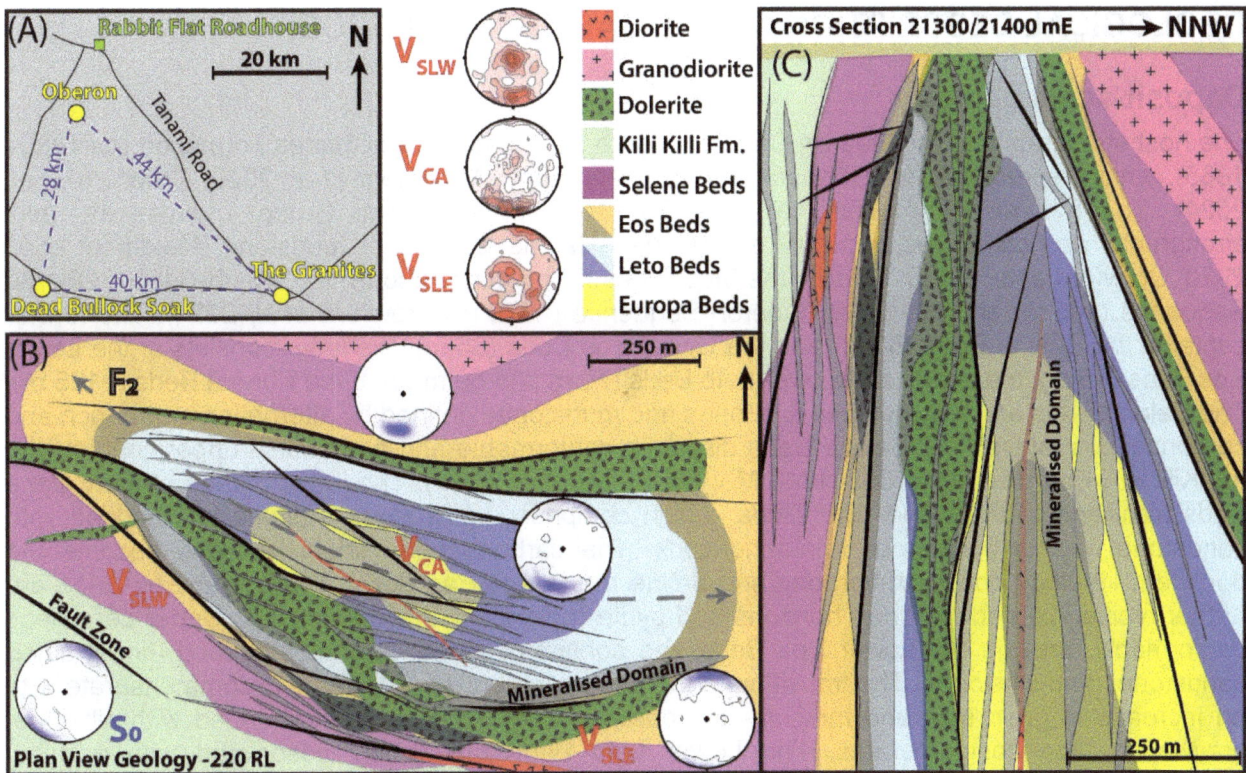

FIG 2 – Location and local geology of the Oberon deposit. **(a)** Location of the Oberon deposit with respect to the Dead Bullock Soak and Granites mines in the Tanami Region. **(b)** Plan section displaying the geology, brittle fault network, and interpreted mineralised domains of the Oberon deposit. The blue-contoured stereonets indicate bedding measurements. The red-contoured stereonets show vein measurements. Different domains of the deposit are represented by CA (Central Axis), SLW (Southern Limb West), and SLE (Southern Limb East). **(c)** Cross-section displaying the geology, brittle fault network, and interpreted mineralised domains of the Oberon deposit.

During recent resource definition campaigns, geological modelling of the Oberon deposit faced challenges due to uncertain boundaries and geometries associated with mineralised domains defined by stockwork zones. These challenges led to significant efforts in exploring various methodologies and techniques for a more robust definition of mineralised domains and representing local controls on mineralisation. In this manuscript, we present the outcomes of this case study and highlight the implications that the results have for advances in domain definition and resource estimation techniques for locally-varying controls on mineralisation in deposits with non-stationary ore geometries. Considering the common themes concerning controls on mineralisation in orogenic gold deposits, the applicability of these or similar methods is likely widespread. The implementation of such techniques has the potential to enhance the veracity of geology and resource models with respect to bedrock geology.

The content of this manuscript is organised into the following themes:

- A description of the geology and mineralisation of the case study deposit (Oberon).
- Advances in geological modelling methods for domain definition:
  - lithostratigraphic domains
  - mineralised domains.
- Advances in resource modelling methods for local controls on mineralisation.

Together, these different themes establish the necessary geological context for the study area while offering a comprehensive overview of advances in geological modelling and resource estimation methods applied to the deposit. The key outcomes of the study are summarised in the final section of the manuscript.

GEOLOGICAL DESCRIPTION OF THE CASE STUDY DEPOSIT

Deposit-scale geology

The host stratigraphy to the Oberon Deposit (Figure 2b, 2c) is the Mt Charles Formation (ca 1885–1865), which represents lower Tanami Group stratigraphy (Maidment *et al*, 2020; Crawford *et al*, 2024). The Killi Killi Formation (ca 1865–1840 Ma; upper Tanami Group) overlies the host stratigraphy and comprises immature sandstone, siltstone, and minor mudstone (Maidment *et al*, 2020; Crawford *et al*, 2024). Locally, the Mt Charles Formation is subdivided into the Upper Oberon Sequence (145 m) and the Lower Oberon Sequence (>270 m; Crawford *et al*, 2024). The former contains the Selene Beds (80 m) and the Upper Eos Beds (65 m); the latter consists of the Lower Eos Beds (15 m), the Upper and Lower Leto Beds (70 m and 60 m), and the Europa Beds (>125 m). The Selene Beds are zebra-striped siltstones and mudstones defined by alternating quartz-rich and carbonaceous layers. The Selene Beds display a conformable and gradational upper contact with the Killi Killi Formation (ca 1865–1840 Ma; Maidment *et al*, 2020). The lower portion of the Selene Beds transition to the Upper Eos Beds, which comprise argillaceous siltstone interbedded with sandstone and carbonaceous shale horizons. The carbonaceous shale horizons become the dominant lithology down stratigraphy and define the Lower Eos Beds. The Lower Eos Beds are further typified by nodular chert horizons and pyrite (potentially diagenetic). The Upper Leto Beds conformably underlie the Lower Eos Beds and consist of planar- to cross-bedded, fine-grained sandstones and siltstones. The transition to the underlying Lower Leto Beds is commensurate with an increase in grain size and medium- to coarse-grained sandstone interbedded with siltstone, greywacke, and conglomerates. The Lower Leto Beds are locally associated with a basal conglomerate, representing an erosional, unconformable contact with the underlying Europa Beds. The Europa Beds are a monotonous package of chloritic siltstone and represent the lowermost stratigraphic unit recorded at Oberon.

The sedimentary rock package at Oberon is host to a gabbroic dyke and sill network—referred to as the Nemesis Dolerite—which strikes along the extent of the deposit (Figure 2b, 2c). The dyke and sill network is most pronounced along the southern limb, where it may reach ~100 m in thickness. A Mg-rich gabbroic sill (up to ~50 m thick) named the Iris Dolerite is hosted in the southern limb within the Upper Eos Beds. The gabbroic intrusions are cross-cut by gold mineralisation and deformed in association with the folded host rock stratigraphy. Pre-ore granodiorite intrusions (ca. 1815 Ma; Crawford, 2025) occur further outboard in the Selene Beds and Killi Killi Formation (Figure 2b, 2c). Small-scale diorite dykes and sills are commonly developed in zones of strong gold mineralisation, such as the southern limb and in the core of the deposit. These diorite dykes are also cross-cut by ore-stage veins (McDivitt, 2024a); however, they are not constrained by geochronology.

The geometry of the Oberon deposit dominantly reflects F_2 folding associated with the Stafford Event (Figure 2b, 2c). Broadly north–south bulk shortening and subvertical elongation during regional compression formed the tight to isoclinal, slightly south-vergent, doubly-plunging Oberon Dome. The dome is elongate, with a steeply north-dipping axial plane that varies south-east to east-striking across the deposit defining an arcuate geometry. The plunge of the dome varies from a steep westerly plunge in the west, to a shallow easterly plunge in the core of the deposit, with a steep easterly plunge in the east. The Stafford Event overprints the earlier D_1 Tanami Event, which was associated with regional, greenschist facies metamorphism of the host rock package (Crispe, Vandenberg and Scrimgeour, 2007; Crawford, 2025).

Deposit-scale controls on mineralisation

Lithostratigraphic controls

The transition between the upper and lower Tanami Group stratigraphy is marked by an abrupt change in sedimentary provenance and lithogeochemistry (Lambeck *et al*, 2008; Maidment *et al*, 2020; Crawford *et al*, 2024). At the district scale, the stratigraphic boundary between the upper and lower Tanami Group acts as a first-order control on the development of major gold deposits (Lambeck *et al*, 2008). In the Oberon area, the geochemical transition from the upper to lower Tanami Group is associated with the boundary between the Upper Oberon Sequence and the Lower Oberon Sequence. Mineralisation is preferentially hosted in the Lower Oberon Sequence, which is more mafic in composition (Figure 3a). This relationship emphasises the boundary between the upper and lower Tanami Group as a deposit-scale, lithostratigraphic control on mineralisation at Oberon. The Nemesis Dolerite defines an additional deposit-scale, lithostratigraphic control on mineralisation. The bulk of mineralisation in the deposit is hosted close to the thickest portion of the Nemesis Dolerite system (Figure 3b).

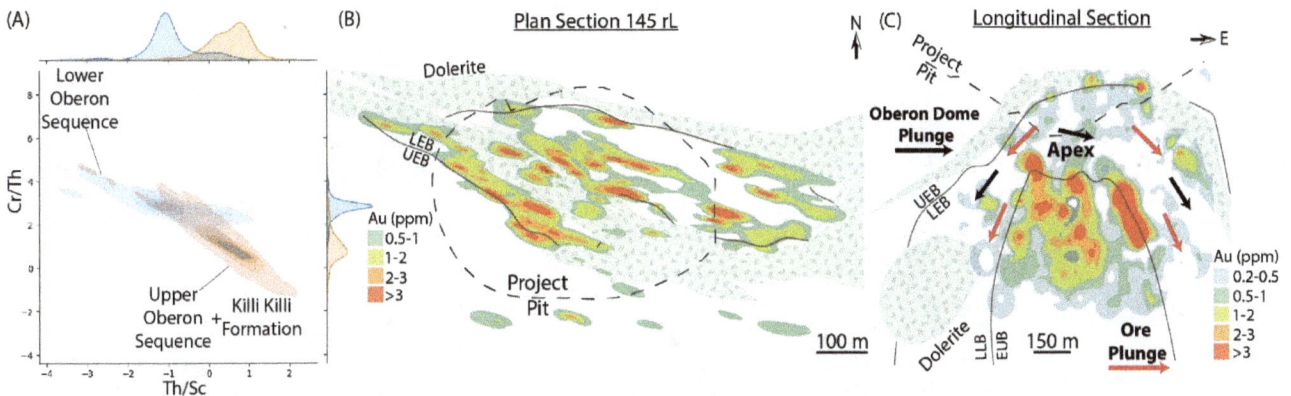

FIG 3 – Deposit-scale controls on mineralisation. **(a)** Bivariate and univariate kernel density estimates (KDEs) for Cr/Th and Th/Sc highlight geochemical differences between the lower and upper stratigraphy. **(b)** Plan section of the Oberon deposit displaying a grade shell model with respect to the Nemesis Dolerite and the project pit. The grade shell highlights preferential mineralisation along the thickest portion of the Nemesis Dolerite, where the dolerite strikes to the south-east. **(c)** Longitudinal section of the Oberon deposit displaying a grade shell model in the context of the Nemesis Dolerite and the apex of the Oberon Dome. The grade shell highlights the preferential location of mineralisation in the apex of the dome, with ore plunges to the west and east that parallel the dome's geometry.

Structural controls

Deformation related to the regional D_2 Stafford event is a first-order control on gold mineralisation. The Oberon Dome is a deposit-scale, F_2 fold that is the main controlling structure of mineralisation, which is preferentially concentrated in the apex of the doubling-plunging fold (Figure 3b). Across the deposit the Nemesis Dolerite defines a sigmoidal geometry by a change in preferred strike from ESE-, to SE, to E, from west to east across the deposit. This sigmoidal geometry may record a component of sinistral transpression during D_2 deformation that acted as an additional control on gold mineralisation (McDivitt, 2024a).

Local controls on mineralisation

Lithostratigraphic controls

Within the project pit, the Lower Leto Beds, Upper Leto Beds, Lower Eos Beds, and Upper Eos Beds are the highest-grade units and account for ~22 per cent of the pit volume (Figure 4a). In the underground project, the Europa Beds represent the highest proportion of mineable shape optimisers (MSOs; ~32 vol%). The Lower Eos Beds, Upper Leto Beds, and Lower Leto Beds account for the next highest proportions of stratigraphic units in the underground MSOs (~23 vol%, ~17 vol%, and ~12 vol%, respectively).

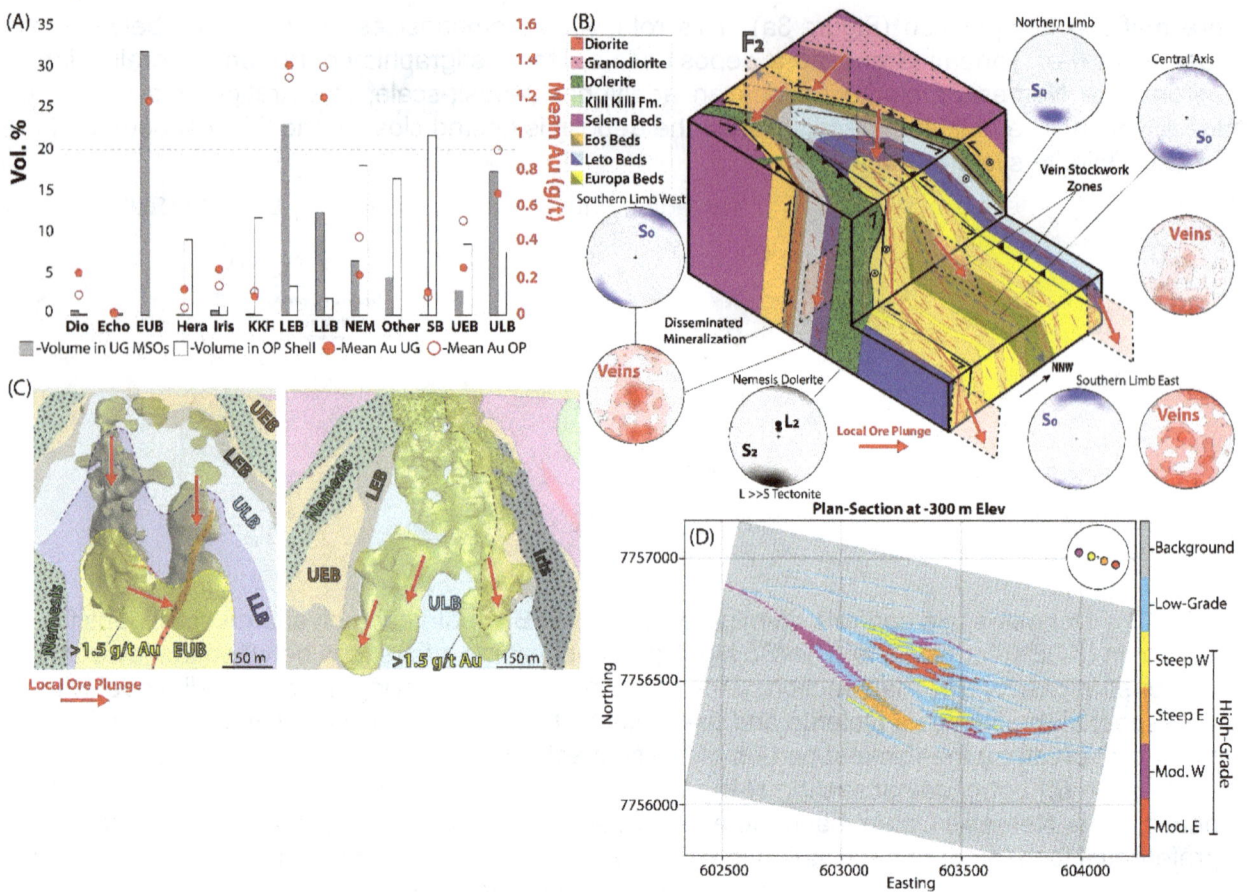

FIG 4 – **(a)** Volumetric proportions of different stratigraphic units for the open pit shell and MSOs are shown in the context of declustered mean Au grades. **(b)** Schematic 3D diagram of the Oberon deposit illustrating the presence of different mineralisation styles (ie disseminated and stockwork) and variable ore plunges across the deposit that reflect different local controls on mineralisation. **(c)** Plane-of-vein sections displaying ≥1.5 g/t grade shells within planar mineralised domains. The left section is from the core of the deposit and highlights sub-vertical plunges related to nested anticlines in the upper portions of the deposit. A shallow E plunge at the contact between the Lower Leto Beds (LLB) and the Europa Beds (EUB) mirrors the gross plunge of the Oberon Dome and becomes the dominant plunge orientation at depth. The right section is from the western southern limb and illustrates W-plunging ore shoots that parallel the gross plunge of the Oberon Dome. Steeply E-plunging shoots are observed where the Lower Eos Beds (LEB) are folded in the footwall of the Iris Dolerite. **(d)** Plan section of the Oberon block model displaying mineralised and background domains. The planar, parent mineralised domains represented by oblate variograms are shown in blue. Local, high-grade domains with a plunge component are shown in yellow, orange, red, and magenta. The variable plunge of the high-grade domains across the deposit reflects different local controls on ore shoot geometry.

Within these preferentially mineralised units, different mineralisation styles are expressed in accordance with local rheological, geochemical, and structural controls (Figure 4b). Disseminated mineralisation favours incompetent units where grain-scale dilational zones accommodated diffuse fluid migration (eg Cox, 2005). Disseminated mineralisation occurs dominantly as arsenopyrite mineralisation in the Lower Eos Beds. Stockwork-type vein mineralisation manifests as a dominant ESE-striking, steeply N-dipping vein set and subordinate, subhorizontal and NNE- to ENE striking sets (Figure 2b; McDivitt, 2024a; Crawford, 2025). Stockwork-type mineralisation is preferentially hosted in and near stratigraphic units that have facilitated fluid overpressuring and brittle failure (eg Cox, 2005). Intense zones of stockwork mineralisation are developed at the interface of the Lower Leto Beds and the Europa Beds, where the apex of the Oberon Dome and the composition of the stratigraphic units facilitated hydrothermal fluid build-up, fluid overpressuring, and brittle failure. Notwithstanding small, dolerite-hosted, high-grade lodes, the dolerite and diorite intrusions are not preferred host rocks and are typically poorly mineralised. Mineralisation is often concentrated at the contacts of the dolerite and diorite intrusions in adjacent metasedimentary rock (McDivitt, 2024a; Crawford *et al*, 2024). The preferred expression of different mineralisation types in specific lithostratigraphic units and the disparate gold endowment of different lithostratigraphic units across the deposit highlight the importance of local lithostratigraphic controls on gold mineralisation.

Structural controls

The variable architecture of the Oberon Dome resulted in non-stationary ore geometries that reflect local controls on mineralisation (Figure 4b). Along the northern and southern limbs of the dome, planar mineralised domains parallel the limbs of the fold and stratigraphic contacts. In the core of the dome, along the hinge of the fold, planar mineralised domains are interpreted to strike ESE and dip sub-vertically, with a geometry similar to the axial plane of the fold and the dominant ore-stage vein orientation. Within the planar mineralised domains, local ore plunges show changes commensurate with the local geometry of the dome (Figure 4b). In the western portion of the dome westerly ore plunges are common, with easterly ore plunges displayed in the eastern portion of the dome. Smaller folds nested within the dome exert additional controls on ore plunge, with sub-vertical plunge geometries observed. Such nested folds may reflect parasitic folding during the formation of the Oberon Dome; alternatively, they may reflect vestigial structures of an earlier D_1 architecture.

Specific examples

Specific examples of local ore plunges within planar mineralised domains are shown in Figure 4c as plane-of-vein sections. In the core of the dome, along an ESE-striking, sub-vertical mineralised domain, ore plunge varies from sub-vertical in the higher levels of the deposit to shallowly-east plunging at depth. The sub-vertical plunges reflect controls by smaller, nested anticlines developed in the dome's core. At depth, the shallow east plunge of the Oberon Dome at the interface between the Lower Leto Beds and the Europa Beds in the dome's apex correlates to an east-plunging ore shoot (Figure 4c).

In the western portion of the deposit along the southern limb, mineralisation occurs as a planar domain subparallel to the geometry of the limb. Within the mineralised domain, local ore shoots vary from steeply west plunging to steeply east plunging (Figure 4c). The former geometry is defined in the western portion of the domain, where the Oberon Dome plunges to the west. The steep easterly plunge is defined where the Lower Eos Beds display small-scale folding along the footwall of the Iris Dolerite.

The examples above highlight the interplay of lithostratigraphic and structural controls on the geometry of local ore shoots. Figure 4d shows a summary section that displays the plunge geometry of linear, high-grade shoots across the deposit relative to planar, low-grade parent domains.

ADVANCES IN GEOLOGICAL MODELLING METHODS FOR DOMAIN DEFINITION

Lithostratigraphic domains

Machine learning methods applied to multi-element data

The refinement and validation of lithostratigraphic domains benefited from applying unsupervised machine learning methods to multi-element (ME) data collected at a frequency of 1 m per every 5 m of drill core (McDivitt, 2024b). The ME data results from a 4-acid digestion with a combination of ICP-AES and ICP-MS for analysis from ALS laboratory yielding 60 elements. The benefits of the systematic collection of such data are highlighted in Halley (2020). The ME data set was limited to fresh rock to avoid weathering effects, and initially, all lithostratigraphic domains in the fresh rock were evaluated together.

Data preprocessing and machine-learning algorithms were completed using Python, with Scikit-learn and SciPy as the main libraries. Preprocessing of the data set included despiking to mitigate the impacts of detection limits and a quantile transformation to fit the variables to a normal distribution. Principal component analysis (PCA) was used as a dimensional reduction technique. Univariate loading on the first and second principal components was generally dominated by elements related to lithogeochemistry (eg rare earth elements) as opposed to those related to mineralisation and hydrothermal alteration (eg Au, As, Ag, Te).

Initial agglomerative hierarchical clustering (Müllner, 2011) was applied using the Ward variance minimisation algorithm (Ward, 1963). Dendrograms were used to evaluate the hierarchical clustering results to help understand the number of clusters inherent to the data set. This was done in conjunction with the Elbow Method (Thorndike, 1953) and Silhouette Analysis (Rousseeuw, 1987) to help determine the optimal number of clusters for K-means clustering. For K-means clustering (cf Jain, 2010), the K-means ++ algorithm (Arthur and Vassilvitskii, 2007) was used to initialise the centroids of the clusters. The methods above yielded equivocal results for determining the optimal number of clusters within the combined fresh rock data set. Therefore, k=2, k=4, and k=7 clusters were all tested, and the results were compared to known lithostratigraphic domains (Figure 5a–5c). A decision tree classifier was trained on the normalised data using the K-means cluster labels to understand univariate influences on the clusters. Understanding the univariate influences on the clusters provided additional geochemical context to the cluster results (Figure 5c).

The initial cluster results from the combined fresh rock data set provided a good first-pass check on the validity of generalised lithostratigraphic domains, including the Upper and Lower Oberon Sequence and granodiorite, diorite, and dolerite intrusions (Figure 5a, 5b). Increasing the number of clusters facilitated the validation of more detailed lithostratigraphic subdivisions (Figure 5c). Whereas certain lithostratigraphic domains showed strong to moderate cohesion across the different clusters (eg granodiorite; Figure 5c), other units showed significant spread across multiple clusters. Given the heterogeneous nature of some metasedimentary units—defined by visually distinct layers (eg zebra striping in the Selene Beds) or zones of mineralisation and alteration—the intraformational, across-cluster spread was not unexpected.

Potential differentiation of the Europa Beds was a key focus given the unit's importance to underground gold endowment (Figure 4a). The combined fresh rock data set was reduced to only the Europa Beds and subjected to the same workflow described above. The results for optimal cluster determination were less equivocal, and k=4 was selected for K-means clustering. A spatial review of the 3D distribution of the clusters showed that Cluster 1 (C1) and Cluster 4 (C4) are closely intermingled in the upper horizon of the Europa Beds (Figure 5d). Cluster 2 (C2) is generally restricted to an undefined lower horizon (Figure 5d). Cluster 3 (C3) is associated with the highest gold grades and is interpreted to reflect the geochemical signature of mineralisation. A review of the clusters on a Hallberg (1984) plot using Ti and Zr concentrations was done to provide lithogeochemical context using immobile trace elements (Figure 5d). Whereas C1, C3, and C4 are generally andesitic on this basis, C2 is associated with a more mafic composition (Figure 5d). In concert with the more mafic composition, the 3D spatial cohesion of C2 as a lower horizon in the Europa Beds highlights a lithogeochemically distinct horizon of the Europa Beds, now referred to as the Lower Europa Beds. This example emphasises how the use of clustering in association with 3D spatial analysis in a lithogeochemical context can aid in defining new lithogeochemical subdomains of key mineralised units.

Differentiation of intrusive rock units initially applied the same workflow described above, but only samples of intrusive rock (ie granodiorite, dolerite, diorite) were used. As part of the workflow, the clusters were inspected in the context of principal components (Figure 5e, 5f), and t-SNE (van der Maaten and Hinton, 2008) was used as a visualisation tool (Figure 5g). Within the intrusive rock data set, the presence of irregular clusters highlighted the suboptimal implementation of K-means clustering (Figure 5e). To better align the clusters with the data structure, alternative, density-based clustering methods were tested. Density-based spatial clustering of applications with noise (DBSCAN; Ester et al, 1996) and Hierarchical density-based spatial clustering of applications with noise (HDBSCAN; McInnes, Healy and Astels, 2017) were both applied to the data, with hyperparameters such as epsilon, minimum samples, and minimum cluster size tested heuristically, and the results reviewed in the context of PCA and t-SNE. Subsequent to this testing phase, HDBSCAN was chosen as the preferred algorithm due to its ability to define clusters of variable density. The implementation of HDBSCAN resulted in five clusters reasonably aligned with the irregular structure of the data, in addition to noise (Figure 5f, 5g). A review of the clusters in a lithogeochemical context (Figure 5h, 5i) showed reasonable alignment of the clusters with different rock compositions (C1=granodiorite, C2/C3=diorite, C4/C5=dolerite). The results showed that C2 and C3 are distinct diorite types (high-P, low-P; Figure 5i), with C2 diorite preferentially located in the core of the Oberon Dome and C3 along the southern limb (Figure 5j). Cluster C4 was identified as a high-Mg dolerite (the Iris Dolerite; Figure 5i) located in the hanging wall of the Nemesis Dolerite (Cluster C5; Figure 5j). A review of the C1 cluster in 3D showed strong spatial alignment with the known granodiorite bodies (Figure 5j).

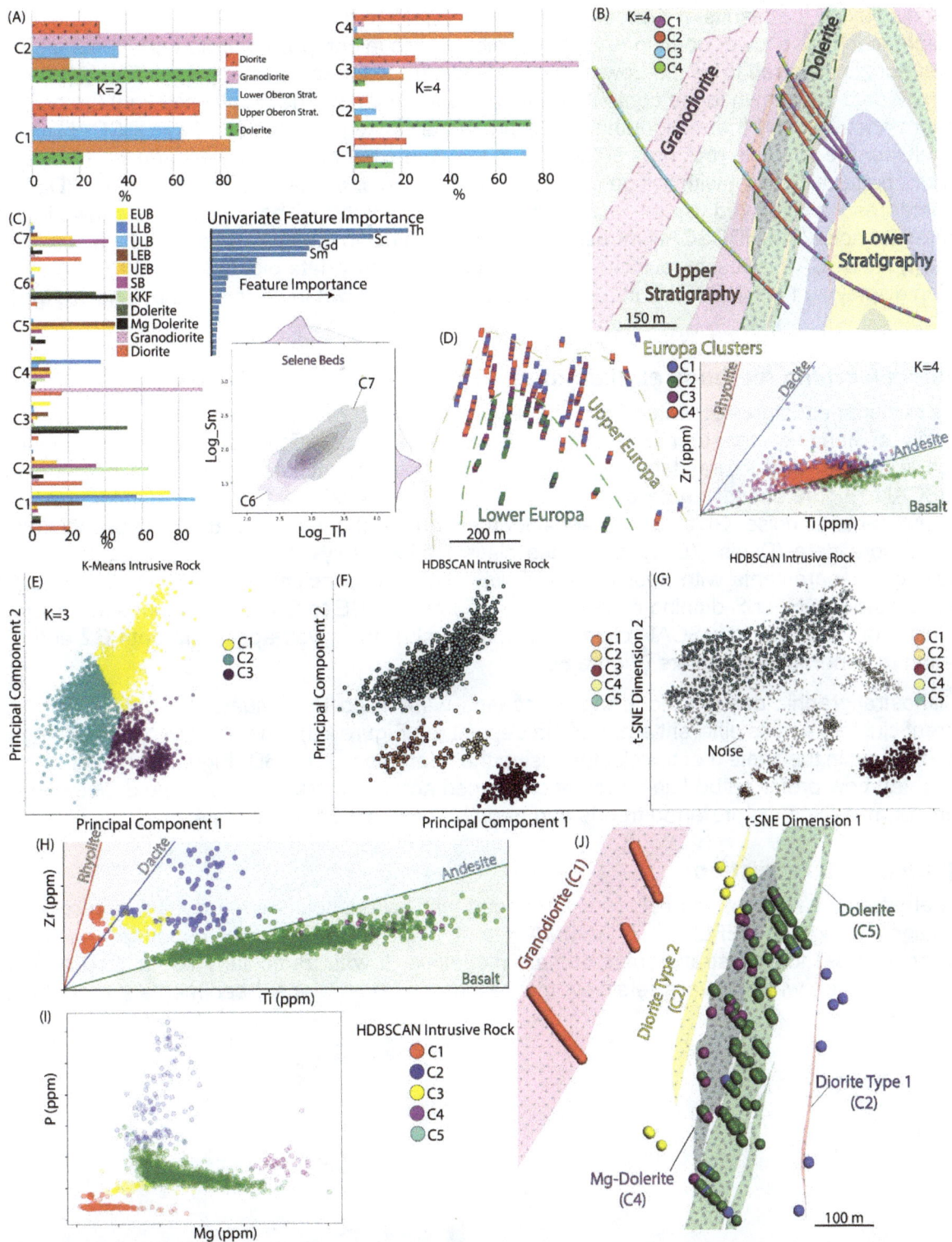

FIG 5 – Machine learning methods applied to lithostratigraphic domain modelling. **(a)** Results of K-means clustering using k=2 and k=4 clusters in the combined fresh rock data set. The cluster results are shown in the context of the generalised lithostratigraphic domains. **(b)** A cross-section displaying K-means clustering results using k=4 clusters in the combined fresh rock data set. The results are shown in the context of the different lithostratigraphic domains. **(c)** Results of K-means clustering for k=7 clusters in the combined fresh rock data set. The cluster results are shown in the context of lithostratigraphic domains. The bar plot illustrates an example of the importance of the univariate feature as determined by a decision tree classifier applied to the cluster labels. The importance of the univariate features is useful for providing lithogeochemical context to the results. For example, the bivariate and univariate KDEs for the Selene Beds highlight different geochemical components in that unit that have affinities for different clusters. **(d)** Long section of the Europa Beds displaying the results of K-means clustering applied to the Europa Bed data set. The results highlight a cohesive lower domain (C2) that is more mafic in composition and represents a distinct lower horizon referred to as the Lower Europa Beds. **(e)** K-means cluster results for the intrusive rock data set shown in the context of principal components. The results highlight the poor performance of the K-mean algorithm in the presence of irregular clusters. **(f)** HDBSCAN cluster results for the intrusive rock data set shown in the context of principal components. The results highlight better alignment with the data structure and irregular cluster morphologies. **(g)** HDBSCAN cluster results displayed using t-SNE as a visualisation method. The clusters are shown in the context of noise. **(h)** HDBSCAN intrusive rock clusters shown on a Hallberg (1984) classification plot. **(i)** HDBSCAN intrusive rock clusters are shown on a bivariate plot of Mg and P. **(j)** A cross-section displaying the HDBSCAN intrusive rock clusters is shown in the context of lithostratigraphic domain volumes.

Machine learning methods applied to structural data

Poles to planar structures collected from oriented core are shown in Figure 6a. The data set consists primarily of bedding measurements, with vein measurements being the second most abundant structure type. The structural data was subjected to a two-stage clustering workflow beginning with HDBSCAN (Figure 6b) to create clusters based on data density while classifying the remaining measurements as noise. Given the potential for measurement uncertainty and error associated with oriented core data (Davis, 2014), the noise classification likely helped to clean the data set by removing measurements with poor reproducibility. The two principal clusters (C2, C4; Figure 6b) represent steeply N- or S-dipping planes that strike from the SE to ENE. To enhance the geometric resolution of the initial HDBSCAN clusters, K-means clustering was applied to both C2 and C4 in isolation to generate subclusters (Figure 6c).

For lithostratigraphic domain modelling, heatmaps were used to evaluate the proportions of the different clusters across different areas of the deposit (eg Figure 6d). The orientation of bedding data was reviewed in the context of the clusters using the heatmaps and in 3D (Figure 6e). This approach highlighted how preferred bedding orientations varied across different areas of the deposit and was useful for modelling the preferred trends and geometries of stratigraphic units.

Natural gamma radiation

Natural gamma radiation was collected in concert with gyroscopic survey data during recent drill campaigns using the REFLEX EZ GAMMA™ tool. The natural gamma response was used as another data set for lithostratigraphic domain modelling. It was particularly useful for highlighting areas of dolerite, which have a relatively low response, from adjacent sedimentary rock domains (Figure 6f).

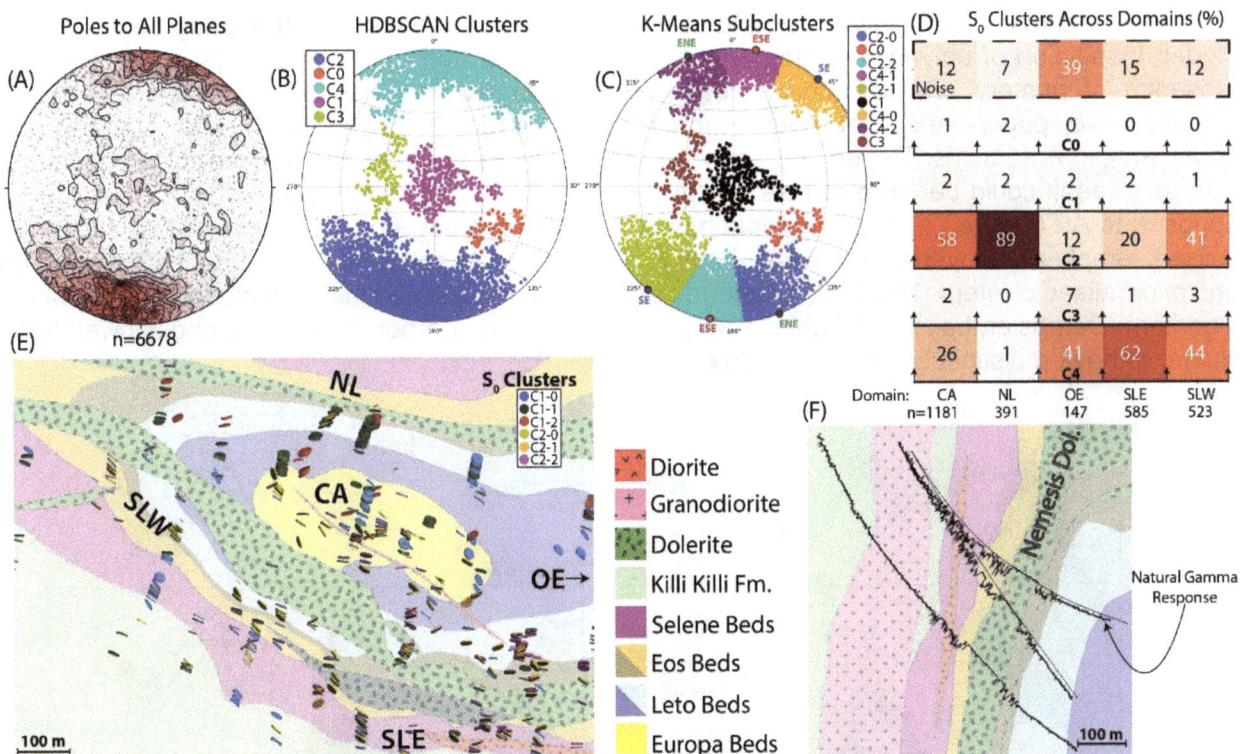

FIG 6 – (a) Poles to all planar structures in the oriented core data set contoured on a stereonet. **(b)** Results of HDBSCAN clustering applied to the structural data. **(c)** Second-stage K-means clustering results provide further geometric resolution of the two main clusters defined by HDBSCAN. **(d)** Heatmap of bedding data cluster proportions and noise across different domains of the deposit. CA = Central Axis, NL = Northern Limb, OE = Oberon East, SLE = Southern Limb East, SLW = Southern Limb West. **(e)** Plan section displaying bedding data colour-coded according to its cluster assignment. Noise is not shown. Such visual examination helps to highlight which orientations are dominant in different domains of the deposit. CA = Central Axis, NL = Northern Limb, OE = Oberon East, SLE = Southern Limb East, SLW = Southern Limb West. **(f)** Cross-section displaying natural gamma radiation response with respect to lithostratigraphic domains. The image highlights the diminished gamma response across the Nemesis Dolerite.

Mineralised domains

Defining mineralised domains from sparse data in early-stage projects can be challenging due to the information effect and anticipated changes to domaining strategies and domain geometries as additional data become available. Certainty in the definition of mineralised domains can be further compromised by a lack of clear geological boundaries to constrain the boundaries of the mineralised domains. The Oberon deposit is both an early-stage project and a deposit with a strong component of stockwork-type vein mineralisation. Therefore, it presents an opportunity to test different strategies for defining mineralised domains in the context of both constraints. Deterministic mineralised domains were validated and informed by Gaussian Mixture Models (GMM) and Secondary Indicator Scores. While neither method singularly defined the mineralised domains, the methods served as independent checks on the deterministic solution and can be used to improve ongoing domain definition.

Gaussian mixture models and latent Au thresholds

Gaussian mixture models (GMMs) have multiple applications in geostatistics, including fitting probability density functions, multiple imputation, and clustering (Gomes, Boisvert and Deutsch, 2022). A GMM clustering approach was applied to Au, As, and S assay data to help define latent Au thresholds to separate mineralised and background domains.

In concert with Au assay data, As and S data were routinely collected and overlapped with 80 per cent and 74 per cent of the Au assay data, respectively. The raw assay intervals were composited to 3 m using lithostratigraphic domain boundaries as split codes. The combined Au-As-

S data set was restricted to fresh rock and further subdivided based on the lithostratigraphic domain model. Inspection of univariate histograms from log-transformed data consistently highlighted the presence of primary and secondary peaks (Figure 7a), emphasising the potential for mixed, Gaussian-like populations that could represent low-grade (ie background) and higher-grade (ie mineralised) domains. Depending on which element was evaluated, evidence for a secondary histogram peak could be more or less subtle. The latter scenario often occurred with As and S as opposed to Au (Figure 7a). For this reason, and to take advantage of covariance, GMM clustering utilised samples with overlapping Au, As, and S data instead of solely Au data to define background and mineralised clusters. The GMM clustering approach implemented the Expectation-Maximisation algorithm (Dempster, Laird and Rubin, 1977), assumed a two-component system, and allowed each cluster to have a distinct covariance matrix.

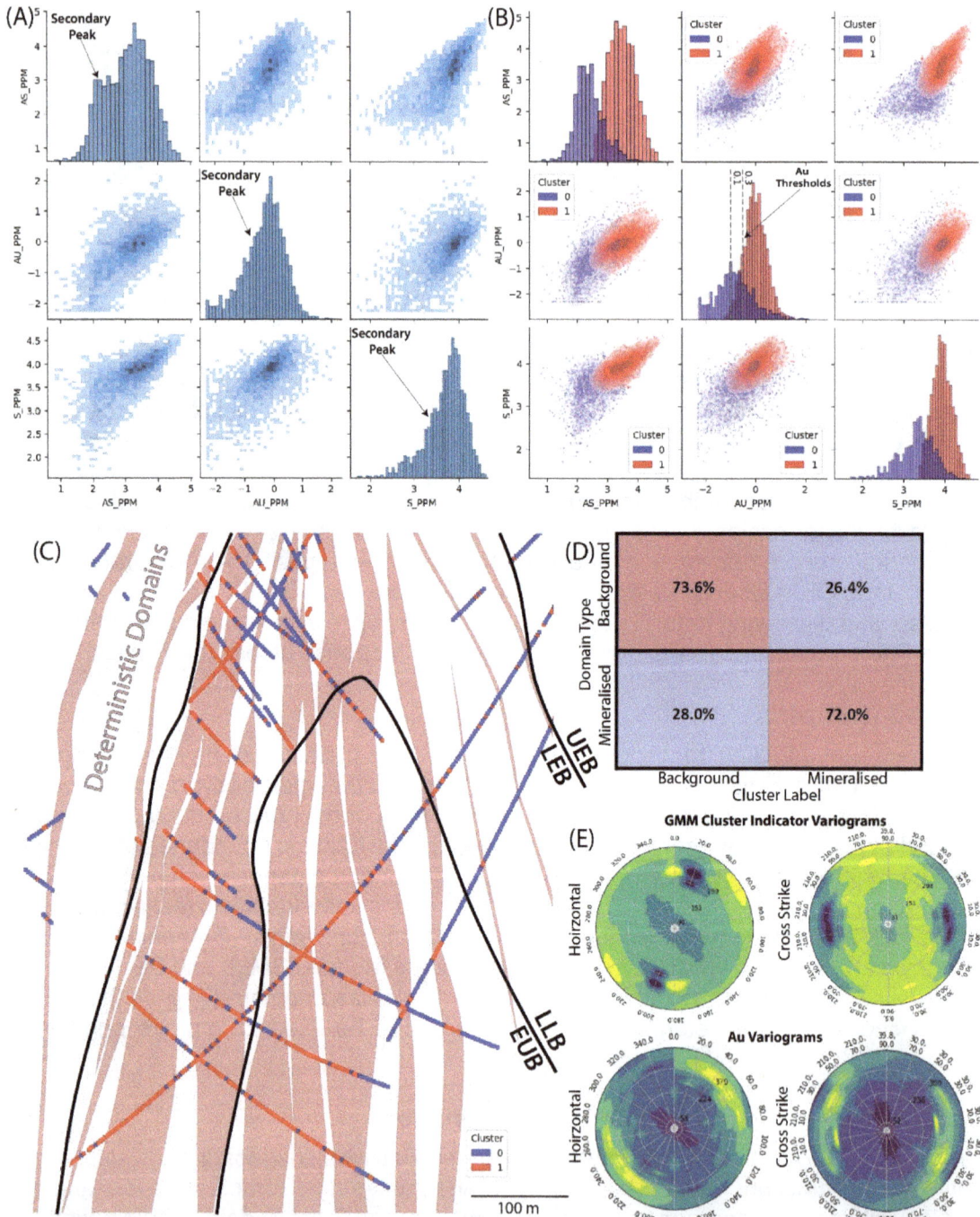

FIG 7 – Gaussian mixture model (GMM) clustering: **(a)** Matrix displaying univariate and bivariate histograms of Au, As, and S assay data in the Europa Beds. Evidence for mixed populations is highlighted by the secondary peak most clearly displayed by the As data. **(b)** The results of GMM clustering applied to the Europa Beds Au, As, and S data set. The red cluster (1) is interpreted as a mineralised population, and the blue cluster (0) is interpreted as a background population. **(c)** Cross-section displaying interpreted, planar, deterministic mineralised domains in the context of GMM cluster results. **(d)** A heatmap displaying the proportions of the different cluster types (ie mineralised versus background) in the context of the deterministic domains. The results highlight ~70 per cent concordance between the deterministic domain interpretation and the GMM clusters for mineralised and background domains. **(e)** Horizontal and cross-strike radar plots displaying the result of indicator variography for the GMM clusters (top) and experimental Au variography (bottom).

The results of GMM clustering in the Europa Beds are visible in Figure 7b, where Au thresholds of 0.1 ppm Au and 0.3 ppm Au are shown in the context of the clusters. The GMM cluster results for all lithostratigraphic domains are shown in the cross-section of Figure 7c with respect to the deterministic mineralised domains. The GMM clusters aligned reasonably with the interpreted mineralised and background domains. Within the mineralised domains, 72 per cent of the composites were assigned to the higher-grade, 'mineralised' cluster (Figure 7d). Within the background domain, 74 per cent of the composites were assigned to the lower-grade, 'background' cluster (Figure 7d).

The clusters were also used in indicator variography to understand their spatial variance with respect to structural features and Au assay data. Cluster indicator variography results for the Europa Beds are shown in the context of experimental Au variography in Figure 7e. The directions indicated for maximum continuity are similar for both the cluster labels and Au data with a SE- to ESE-strike and vertical- to steep-N dip (Figure 7e). These geometries are also similar to the dominant orientations of bedding and mineralised veins in the Europa Beds (McDivitt, 2024a).

Secondary indicator score

Exploratory data analysis was applied to numeric and categorical drill core data sets to evaluate secondary indicators of gold mineralisation. The ME data were included in the evaluation but not used in the final calculations of the secondary indicator score due to the discontinuous nature of the data set. The results of the EDA highlight that As and S assay data (Figure 8a)—as well as select alteration types (K-feldspar, silica, sericite; Figure 8b) and vein frequency—showed elevated expressions in zones of gold mineralisation. However, the importance of the different secondary indicators varied across the lithostratigraphic domains. For example, in the Europa Beds, sulfur, pyrite, and vein frequency were elevated in association with gold mineralisation, but these indicators did not apply to the Lower Eos Beds (Figure 8a). The disseminated nature of mineralisation in the Lower Eos Beds and potentially diagenetic pyrite precluded a quantifiable association among gold mineralisation, vein frequency, and sulfur in that unit. Calculations of indicator variables from numeric (As, S, vein frequency) and categorical variables (alteration types) for the different lithostratigraphic domains are detailed below.

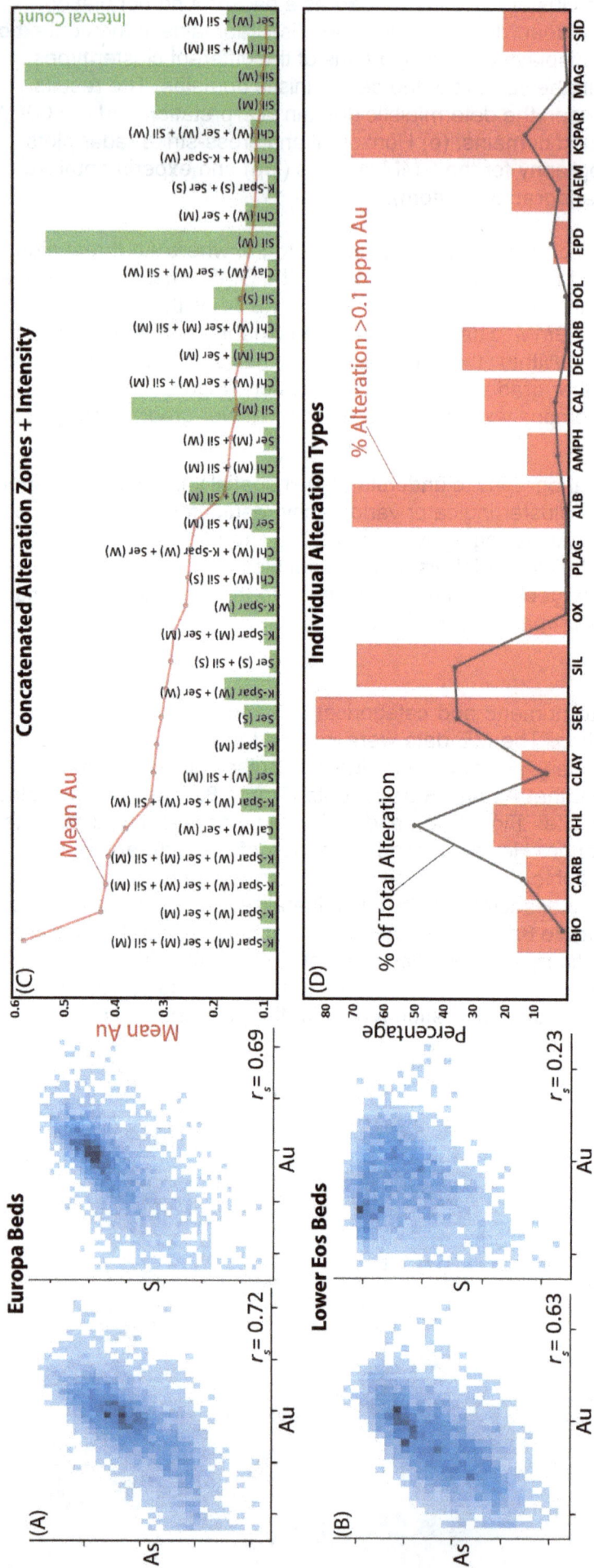

FIG 8 – Secondary indicators of Au mineralisation. (**a**) Bivariate histograms of Au-As and Au-S for the Europa Beds. Strong Spearman correlations (r_s) are defined between Au, As, and S. (**b**) Bivariate histograms of Au-As and Au-S for the Lower Eos Beds. While a strong Spearman correlation (r_s) is shown for Au and As, the correlation for Au and S is much weaker. (**c**) Bar plot (green) displaying the interval count of distinct overlapping alteration types (eg K-spar, silica, sericite) with mean Au >=0.1 ppm and the alteration intensities. The alteration types are sorted according to their mean Au grade (red line). The plot highlights that the highest grades occur where moderate K-spar, silica, and sericite alteration overlap. (**d**) Bar plot (red) displaying the percentage of different alteration types present where Au >=0.1 ppm. For example, sericite (SER) occurs in 80 per cent of alteration zones >=0.1 ppm Au. The black line shows the percentage of total alteration that the alteration type occurs in. For the case of sericite, that alteration type occurs in ~40 per cent of all alteration zones of the deposit.

Indicator variables $I_k(\boldsymbol{u}; Z_k)$ for each continuous secondary variable $Z_k(\boldsymbol{u})$ were calculated as follows:

$$I_k(\boldsymbol{u}; Z_k) = \begin{cases} 1, & if\ Z_k(\boldsymbol{u}) \geq t_{d,k} \\ 0, & otherwise \end{cases}, \boldsymbol{u} \in d, k = 1, \dots, K \tag{1}$$

Where K is the number of numeric secondary variables, d is the lithostratigraphic domain, and $t_{d,k}$ is the indicator threshold associated with the domain.

Indicator variables $I_k(\boldsymbol{u}; Z_k)$ for each categorical secondary variable $Z_k(\boldsymbol{u})$ are determined by:

$$I_k(\boldsymbol{u}; Z_k) = \begin{cases} 1, & if\ Z_k(\boldsymbol{u}) = c_{d,k} \\ 0, & otherwise \end{cases}, \boldsymbol{u} \in d, k = 1, \dots, K \tag{2}$$

Where K is the number of categorical secondary variables, d is the lithostratigraphic domain, and $c_{d,k}$ is the indicator category associated with the domain.

The resulting indicator variables are then assigned into $g = 1, \dots, G$ groups aligned with the input data types. Each group will contain J number of indicator variables. In our example, G=3 groups will be used: (Group 1) geochemical data (ie As, S; $J = 2$), (Group 2) vein frequency data ($J = 2$), and (Group 3) alteration data (ie K-feldspar, silica, sericite ($J = 3$). These three groups were used to generate three corresponding composite scores:

$$S_g(\boldsymbol{u}) = \frac{1}{J} \sum_{n=1}^{J} I_k(\boldsymbol{u}; Z_k), \boldsymbol{u} \in d, I_k(\boldsymbol{u}; Z_k) \in g \tag{3}$$

Yielding $S_1, \dots, S_G$ mineralised scores. These individual scores are then aggregated into a final secondary indicator score:

$$\hat{S}(\boldsymbol{u}) = \frac{1}{G} \sum_{g=1}^{G} S_g(\boldsymbol{u}), \boldsymbol{u} \in d \tag{4}$$

The secondary indicator variables, composite scores, and secondary indicator score were used in addition to the Au assay data to help evaluate the presence of mineralised domains. These data were evaluated on both a sample-by-sample basis, along the drill traces, and via numeric modelling using indicator shells (Figure 9).

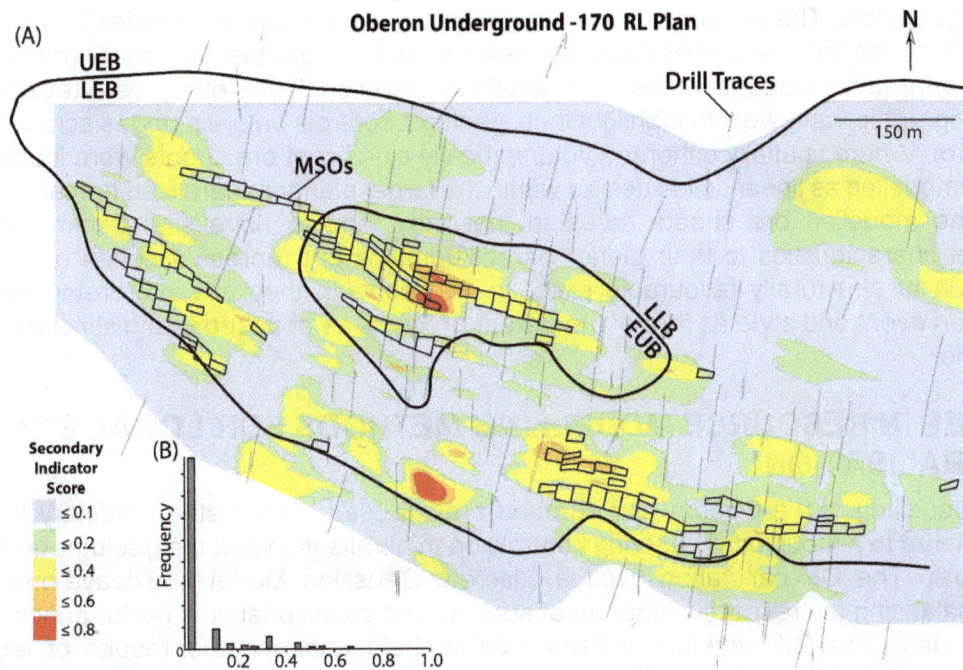

FIG 9 – Secondary indicator score. **(a)** Plan section showing a numeric model of the secondary indicator score in the context of key stratigraphic contacts (UEB/LEB, LLB/EUB), drill traces, and MSOs. **(b)** Histogram displaying the distribution of the secondary indicator score.

Machine learning methods applied to structural data

The same two-stage clustering workflow (HDBSCAN/K-means) and related analysis detailed above were used to help constrain the geometries of mineralised domains. In addition to the bedding data, the orientations of clustered vein measurement data were reviewed across the deposit. This approach helped to support the geometries of mineralised domains in different areas of the deposit.

Definition of mineralised domains and subdomains

Mineralised domains were interpreted subsequent to the definition of a lithostratigraphic framework and a model of brittle faults for the deposit (eg Figure 2b, 2c). These products provided the groundwork for understanding the key geologic controls on mineralisation. The results of univariate statistical analysis and GMM clustering provided support for which Au threshold(s) may be suitable for separating mineralised domains from background. Using these results, 0.3 ppm Au and 0.1 ppm Au thresholds (Figure 7b) were evaluated hierarchically to evaluate the presence of a mineralised zone. The secondary indicator variables, composite variables, and secondary indicator score were also used to assess the presence and continuity of mineralised domains. In conjunction with the geometries of different stratigraphic horizons and intrusive units, the oriented core data and the results of the two-stage clustering approach helped to constrain the geometries of the mineralised domains. Variography applied to Au assay data and cluster labels (eg Figure 7e) provided additional geometric constraints for the mineralised domains. This holistic approach allowed for mineralised domains to be defined across the deposit as planar volumes (Figure 2b, 2c). These planar, mineralised domains preferentially parallel stratigraphic contacts along the northern and southern limbs of the Oberon Dome. Within the dome's core, planar mineralised domains are interpreted to have ESE-striking, sub-vertically-dipping geometries similar to the axial plane of the dome and the most prominent vein orientation in the deposit. Where the mineralised domains parallel stratigraphic contacts along the limbs of the Oberon Dome, there is less uncertainty associated with the boundaries of the domains due to the stratigraphic control. Conversely, the mineralised domains in the dome's core cross-cut stratigraphy in the hinge of the fold. Because these domains comprise stockwork mineralisation, as opposed to discrete lode structures, and are interpreted to cross-cut stratigraphy, there is a higher level of uncertainty associated with the boundary definition of these domains.

Establishing planar mineralised domains across the deposit allowed for the evaluation of local, linear ore plunges within the planar domains. Statistical analysis of Au within the mineralised domains revealed a potential secondary grade population at ≥1.5 g/t Au evidenced by a subtle flexure in log-scale probability plots. The planar mineralised domains were used as boundaries for numeric grade shell modelling, with the grade shell threshold selected to highlight the secondary grade population within the planar mineralised domains. The results of grade shell modelling were reviewed against the lithostratigraphic framework to highlight local geologic controls on ore plunges across the deposit (eg Figure 4c). Where spatially coherent, volumetrically significant ore shoots were identified, these zones were modelled as linear solids nested within the larger planar mineralised domains. A geologic review of the modelled ore shoots revealed that they tend to have similar mineralogical and geochemical characteristics to their planar parent domains but represent zones of more intense mineralisation at structurally favourable sites. In that context, they are interpreted as the same mineralisation event and style as that of the parent domains as opposed to a distinct mineralisation event or style.

ADVANCES IN RESOURCE MODELLING METHODS FOR LOCAL CONTROLS ON MINERALISATION

Resource modelling of the Oberon underground project used a calibrated, Ordinary Kriging (OK) workflow tailored to honour locally-varying controls on mineralisation and the geologic understanding of the deposit. The OK calibration used a Discrete Gaussian Model and leave-one-out cross-validation, balancing the respective change of support and conditional bias performance to optimise the kriging plan. The OK workflow utilised LVA to help enhance the results of experimental variography, model variograms, and local search neighbourhoods. A local capping strategy helped to mitigate non-stationary Au grades by evaluating the impact of high-grade samples in the context of local sample populations. The estimation approach employed the local, linear ore shoots to define nested high-grade domains that utilise prolate variograms aligned with the plunge direction of the shoot. These nested high-grade domains were modelled using soft contacts with their host parent domains, which used oblate variograms to represent grade continuity along the planar structures. These methods helped advance the Oberon deposit's resource modelling workflow and translate locally-varying controls on mineralisation into the final resource model. Resource Modelling Solutions Platform (RMSP) was the primary software package used for the resource modelling workflow.

Variogram methods

Supervised clustering

Planar mineralised domains and lithologic boundaries were used to define LVA controls. The LVA aligned local searches and variogram models to interpreted planar controls on mineralisation, which were either lithostratigraphic boundaries or the axial plane of the Oberon Dome. Supervised clustering of the LVA (Figure 10a) was used to group different portions of the estimation domains based on similar LVA orientations. Experimental variography was completed separately within each of the clusters. This allowed for areas of similar geometry to be evaluated individually to assess different model variograms across the domain. This approach to variography was most commonly used for background domains, which are characterised by irregular morphologies that can vary significantly in their geometries.

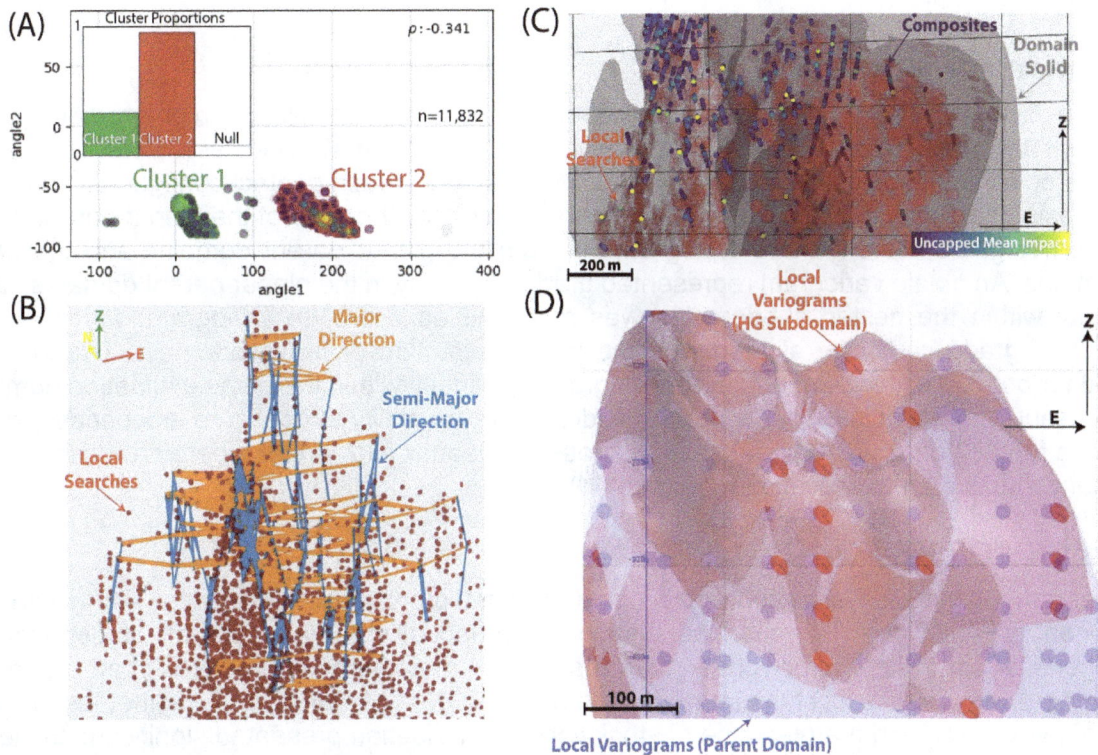

FIG 10 – **(a)** Supervised clustering of estimation grid nodes using LVA angles to group areas of similar orientation within a domain for experimental variography. The proportions of the clusters (Cluster 1, Cluster 2) are shown on the inset. **(b)** Plane-of-vein section for a planar estimation domain illustrating LVA-guided experimental variography. Local searches are shown as the small red ellipses, with the principal search directions aligned using the LVA control. **(c)** Plane-of-vein section showing the mineralised Lower Eos Beds with local searches defined by LVA. The composites of the domain area are coloured according to their uncapped mean impact, which is a parameter used to implement the local capping strategy applied to the underground estimation workflow. **(d)** Example of the estimation approach applied in the Europa Beds to capture an E-plunging shoot of high-grade material within a lower grade planar structure. The planar structure is similar in geometry to the axial plane of the Oberon Dome, and has grade continuity represented by an oblate variogram model. The E-plunging shoot is nested in the planar domain with a soft boundary configuration and uses a prolate variogram model for grade continuity. Note that local searches in this figure are scaled to visualise anisotropy, but are not representative of the true search ranges.

LVA-guided experimental variography

The LVA was also used directly in experimental variography by creating local search neighbourhoods at grid nodes to search for sample pairs at lag distances along principal directions oriented by the LVA (Figure 10b). This approach has the benefit of using all composite data in an estimation domain without distortion of the sample configuration that occurs with flattening workflows. This method was used primarily for variography within planar mineralised domains that showed geometric variance in accordance with the geometry of the Oberon Dome.

Local capping

A local capping strategy was used to help mitigate grade non-stationarity within estimation domains and better honour local grade populations. The local capping utilised the LVA to generate local search neighbourhoods. Using samples within the local search neighbourhoods, the impacts of individual composites on the search neighbourhoods were evaluated by their maximum mean impact. If an individual composite has an effect on the mean of the local sample population greater than the set threshold, the composite is capped on the basis of the threshold (Figure 10c). Compared to global capping methods, this approach avoids penalising outliers that are locally supported within high-grade trends, focusing instead on outliers in marginal grade areas of the non-stationary domains.

Estimation approach

The OK estimation approach was set-up to account for the presence of linear, high-grade ore shoots nested within larger planar domains of mineralisation (eg Figure 1). The nested ore shoots were represented as subdomains within the planar parent domains. A soft contact configuration was used on the basis of a geological review. An LVA control defined by the solid of the planar parent domain was used to guide searches and variograms in both the planar parent domains and the nested subdomains. An oblate variogram represented grade continuity in the planar parent domains. Grade continuity within the nested subdomains was modelled as a prolate variogram, with the major direction of grade continuity aligned with the plunge direction of the subdomain. The approach allowed for both planar and linear controls on grade continuity within the same estimation domain to be represented in the final resource model. Additionally, the ability to configure independent plunge directions for the various high-grade shoots across the deposit allowed for a better representation of the local controls on mineralisation (*cf* Figure 4d).

SUMMARY OF KEY OUTCOMES

The Oberon deposit is a Palaeoproterozoic, fold-dominated orogenic gold system with a significant component of stockwork-type vein mineralisation. The deposit's domal geometry, lithostratigraphic heterogeneity, and history of polyphase deformation and metamorphism resulted in an ore-forming environment with various local controls on mineralisation. Non-stationary ore geometries, the early stage of the project, and the presence of stockwork mineralisation presented significant challenges in defining domains suitable for resource estimation.

Although the Oberon deposit is a gold-only system, the systematic collection of ostensibly redundant data across the deposit (ie multielement data and the Au-As-S assay suite) proved considerably beneficial for domain definition by enabling the application of unsupervised machine learning algorithms. While these algorithms are helpful for domain validation and refinement, developing a robust lithostratigraphic framework via conventional geological methods is a key precursor. Such a framework will provide a platform for 3D, spatial analysis of the results and provide the necessary context for the outcomes of unsupervised learning, particularly in cases of equivocal results during hyperparameter selection. The systematic acquisition of high-dimensional, high-quality geochemical data sets and the application of unsupervised learning algorithms to such data will help to validate, refine, and improve lithostratigraphic frameworks, but such an approach is not a substitute for fundamental geological knowledge and conventional modelling workflows.

While the methods implemented here have resulted in geological and resource models of the Oberon deposit that align with our geologic understanding, the advantages and disadvantages of such methods may not be fully clear until they are tested in a production environment. Nevertheless, applying these methods has proven helpful in resource definition and exploration drilling by providing accurate representations of local ore plunges and the geometries of mineralised zones.

Lithostratigraphic domain modelling

The study's best outcomes associated with unsupervised learning were when the data was subdomained by the existing lithostratigraphic framework (eg Figure 5d, 5j). This approach resulted in less uncertainty around hyperparameter selection and proved helpful in defining new domains that had not been constrained based on geological characteristics. The definition of the Lower Europa Beds (Figure 5d) is an example where unsupervised learning highlighted a clear geochemical subdivision in the stratigraphy that was not obvious based on visual rock characteristics. Assessing the unsupervised learning outcomes in a lithogeochemical and 3D context provides important geological insight when interpreting the results.

The K-means clustering algorithm and uncertainty in the number of clusters to select emphasised challenges associated with hyperparameter selection (Figure 5a–5c). Without underlying geological knowledge, unsupervised learning can provide ambiguous results in terms of domain definition. With the underlying geological knowledge, the results validate and refine the lithostratigraphic domains. Limitations of the K-means algorithm were shown by suboptimal performance in the presence of non-spherical clusters, which were evaluated using PCA and t-SNE as visualisation tools (Figure 5e–5g). Density-based spatial clustering methods (DBSCAN, HDBSCAN) were utilised to improve the alignment of clusters with the data structure (Figure 5f, 5g). The density-based spatial clustering methods have the added benefit of noise definition, interpretable as mixed populations sampled across geological contacts (likely case for multi-element data) or erroneous data (not uncommon in oriented core measurements). Applying HDBSCAN to structural data sets, including oriented core data sets, is likely an under-utilised technique to cluster structural databased on orientation and remove erroneous measurements. This approach then allows for the quantitative evaluation of structural clusters across different deposit areas to highlight principal geometries and trends to guide domain modelling.

Mineralised domain modelling

In the absence of clear geological boundaries for mineralised domains, deterministic models often evaluate grade thresholds to differentiate mineralisation from background domains. These thresholds may be determined by inspecting histograms or log probability plots under implicit assumptions of mixed Gaussian populations. Gaussian mixture models provide an unsupervised learning technique for clustering data into mineralised and background populations. This technique helps define latent grade thresholds, assigns cluster labels to samples, and facilitates the validation and refinement of deterministic domains (Figure 7).

Secondary variables, such as geochemical indicators and alteration, are also often considered in the definition of mineralised domains. Exploratory data analysis identifies secondary variables of interest and indicator thresholds (eg Figure 8). Within the lithostratigraphic framework secondary indicator variables can be grouped and aggregated into composite scores and a total secondary indicator score (Eqn 1 – Eqn 4). This approach shows some similarities to indicator kriging algorithms and provides scores as outcomes from aggregated indicator variables that are used as a decision-making tool for domain definition. This workflow allows for the inclusion of continuous and categorical variables with flexibility in grouping and dimensional reduction while evaluating the likelihood that a sample belongs to mineralised domain based on secondary indicators of mineralisation.

In orogenic gold systems, structural controls on mineralisation can be viewed within a framework where linear, high-grade ore shoots, represented by prolate variogram models, are nested within higher-order planar mineralised domains or structures, represented by oblate variogram models (Figure 1). Translating this framework into a methodology for modelling mineralised domains requires first modelling the overarching planar controls on mineralisation, then evaluating the presence of linear high-grade ore shoots within the planar domains. When successful this hierarchical approach highlights local high-grade ore shoots nested within larger domains that may not be obvious in the absence of planar boundary constraints.

Resource modelling methods

The concept of stationarity is central to resource modelling (Dias and Deutsch, 2022). The popularity of Ordinary Kriging stems from its quasi-stationary estimation capabilities (Rossi and Deutsch, 2014; Dias and Deutsch, 2022), which highlights the need for flexible estimation techniques that can overcome non-stationary domains. The use of local constraints such as LVA in experimental variography (Figure 10b), local capping (Figure 10c), and nested, linear subdomains (Figure 10d) helps to facilitate the estimation of non-stationarity domains. Improved representation of local controls on mineralisation within both geology and resource models will refine the models and enhance the ability of the Ordinary Kriging estimator within non-stationary domains.

ACKNOWLEDGEMENTS

The authors acknowledge the Warlpiri people, Traditional Custodians of the land on which the exploration activities were undertaken, and pay respects to their Elders, past, present and emerging. Both Newmont Corporation and GeologicAI are thanked for their permission to publish the results. The Tanami Near Mine Exploration team is commended for their geological expertise and strong dedication to the development of the Oberon orebody, which has made this study possible. Colin Carey, Andrew Crawford, David Maidment, Adrian Diaz Petit, Sam Vine, and Ryan Barnett are thanked for their comments on an early draft of the manuscript.

REFERENCES

Arthur, D and Vassilvitskii, S, 2007. k-means++: The advantages of careful seeding, in *Proceedings of the 18th Annual ACM-SIAM Symposium on Discrete Algorithms (SODA '07)*, pp 1027–1035 (Society for Industrial and Applied Mathematics: Louisiana).

Cox, S F, 2005. Coupling between deformation, fluid pressures and fluid flow in ore-producing hydrothermal systems at depth in the crust, in *Economic Geology One Hundredth Anniversary Volume* (eds: J W Hedenquist, J F H Thompson, R J Goldfarb and J P Richards), pp 39–75 (Society of Economic Geologists).

Cox, S F, Wall, V J, Etheridge, M A and Potter, T F, 1991. Deformational and metamorphic processes in the formation of mesothermal vein-hosted gold deposits — examples from the Lachlan Fold Belt in central Victoria, Australia, *Ore Geology Reviews*, 6(5):391–423. https://doi.org/10.1016/0169-1368(91)90038-9

Crawford, A F, 2025. The Critical Controls on Mineralisation and the Stratigraphic Position of the Oberon Orogenic Gold Deposit (Northern Territory, Australia), PhD thesis, University of Western Australia.

Crawford, A F, Maidment, D W, Thebaud, N, Masurel, Q and Evans, N J, 2024. A revised stratigraphic model for the ~1910–1835 Ma Tanami Group, the Northern Territory, Australia: Implications for exploration targeting, *Precambrian Research*, 411:107510. https://doi.org/10.1016/j.precamres.2024.107510

Crispe, A J, Vandenberg, L C and Scrimgeour, I R, 2007. Geological framework of the Archean and Paleoproterozoic Tanami Region, Northern Territory, *Miner Deposita,* 42:3–26. https://doi.org/10.1007/s00126-006-0107-1

Davis, B K, 2014. Use and abuse of oriented drill core, *Monograph 30 – Mineral Resource and Ore Reserve Estimation - The AusIMM Guide to Good Practice*, pp 121–134 (The Australasian Institute of Mining and Metallurgy: Melbourne).

Dempster, A P, Laird, N M and Rubin, D B, 1977. Maximum likelihood from incomplete data via the EM algorithm, Journal of the Royal Statistical Society, Series B (Methodological), 39(1):1–38.

Dias, P M and Deutsch, C V, 2022. The Decision of Stationarity, in *Geostatistics Lessons* (ed: J L Deutsch). Available from: <http://www.geostatisticslessons.com/lessons/stationarity> [Accessed: March 2025].

Dubé, B and Gosselin, P, 2007. Greenstone-hosted quartz-carbonate vein deposits, in *Mineral Deposits of Canada: A Synthesis of Major Deposit-Types, District Metallogeny, the Evolution of Geological Provinces and Exploration Methods* (ed: W D Goodfellow), Geological Association of Canada, Mineral Deposits Division, Special Publication, 5:49–73.

Ester, M, Kriegel, H P, Sander, J and Xu, X, 1996. A density-based algorithm for discovering clusters in large spatial databases with noise, in *Proceedings of the 2nd International Conference on Knowledge Discovery and Data Mining (KDD-96)*, pp 226–231.

Goldfarb, R J, 2024. Gold deposits—Where, when and why?, Proceedings of the Gold24 International Symposium, pp 127–132.

Goldfarb, R J, Baker, T, Dubé, B, Groves, D, Hart, C and Gosselin, P, 2005. Distribution, character and genesis of gold deposits in metamorphic terranes, in *Economic Geology One Hundredth Anniversary Volume* (eds: J W Hedenquist, J F H Thompson, R J Goldfarb and J P Richards), pp 407–450 (Society of Economic Geologists).

Gomes, C G, Boisvert, J and Deutsch, C V, 2022. Gaussian Mixture Models, in *Geostatistics Lessons* (ed: J L Deutsch). Available from: <http://www.geostatisticslessons.com/lessons/gmm> [Accessed: January 2024].

Hallberg, J A, 1984. A geochemical aid to igneous rock type identification in deeply weathered terrain, *Journal of Geochemical Exploration*, 20(1):1–8. https://doi.org/10.1016/0375-6742(84)90085-2

Halley, S, 2020. Mapping magmatic and hydrothermal processes from routine exploration geochemical analyses, *Economic Geology*, 115(3):489–503. https://doi.org/10.5382/econgeo.4722

Jain, A K, 2010. Data clustering: 50 years beyond K-means, *Pattern Recognition Letters*, 31(8):651–666. https://doi.org/10.1016/j.patrec.2009.09.011

Lambeck, A, Huston, D, Maidment, D and Southgate, P, 2008. Sedimentary geochemistry, geochronology and sequence stratigraphy as tools to typecast stratigraphic units and constrain basin evolution in the gold mineralised Palaeoproterozoic Tanami Region, Northern Australia, *Precambrian Research*, 166(1–4):185–203. https://doi.org/10.1016/j.precamres.2007.10.012

Maidment, D W, Wingate, M T D, Claoué-Long, J C, Bodorkos, S, Huston, D, Whelan, J A, Bagas, L, Lambeck, A and Lu, Y, 2020. Geochronology of metasedimentary and granitic rocks in the Granites–Tanami Orogen: 1885–1790 Ma geodynamic evolution, Geological Survey of Western Australia, Report 196, 50 p.

McDivitt, J A, 2024a. A Structural Review of the Oberon Au Deposit and Evidence for Sinistral Transpression in a Restraining Bend, Internal report, Newmont Corporation, Tanami [Unpublished].

McDivitt, J A, 2024b. Unsupervised Learning for Refined Geological Domains at Oberon, Internal report, Newmont Corporation, Tanami [Unpublished].

McDivitt, J A, Lafrance, B, Kontak, D J and Robichaud, L, 2017. The structural evolution of the Missanabie-Renabie gold district: Pre-orogenic veins in an orogenic gold setting and their influence on the formation of hybrid deposits, *Economic Geology*, 112(8):1959–1975. https://doi.org/10.5382/econgeo.2017.4536

McInnes, L, Healy, J and Astels, S, 2017. HDBSCAN: Hierarchical density-based clustering, *The Journal of Open Source Software*, 2(11). https://doi.org/10.21105/joss.00205

Meria, D, 2011. Characterisation of gold mineralisation, Oberon prospect, Tanami region, NT, BSc thesis, The University of Adelaide, South Australia.

Miller, J, 2010. Structural study of the Callie Deposit, Tanami region, NT, Internal report, Newmont Corporation, Tanami [Unpublished].

Müllner, D, 2011. Modern hierarchical, agglomerative clustering algorithms, arXiv preprint, https://doi.org/10.48550/arXiv.1109.2378

Nguyen, P T, Harris, L B, Powell, C M A and Cox, S F, 1998. Fault-valve behaviour in optimally oriented shear zones: An example at the Revenge gold mine, Kambalda, Western Australia, *Journal of Structural Geology*, 20(12):1625–1640. https://doi.org/10.1016/S0191-8141(98)00054-6

Pendergast, J, 2011. Tanami operations, Callie mine: discovery of the Auron orebody, a +2Moz discovery in a mature goldfield, NewGenGold, Perth, p 9.

Petrella, L, Thébaud, N, LaFlamme, C, Miller, J, McFarlane, C, Occhipinti, S, Turner, S and Perazzo, S, 2020. Contemporaneous formation of vein-hosted and stratabound gold mineralisation at the world-class Dead Bullock Soak mining camp, Australia, *Mineralium Deposita*, 55(5):845–862. https://doi.org/10.1007/s00126-019-00902-7

Robert, F and Brown, A C, 1986. Archean gold-bearing quartz veins at the Sigma mine, Abitibi greenstone belt, Quebec, Part, I, Geologic relations and formation of the vein system, *Economic Geology*, 81:578–592.

Robert, F and Poulsen, K H, 2001. Vein formation and deformation in greenstone gold deposits, *Structural Controls on Ore Genesis*, pp 111–155 (Society of Economic Geologists). https://doi.org/10.5382/Rev.14.05

Robert, F, 1990. Structural setting and control of gold-quartz veins in the Val d'Or area, southeastern Abitibi Subprovince, in (eds: S E Ho, F Robert and D I Groves) Gold and base metal mineralisation in the Abitibi Subprovince, Canada, with emphasis on the Quebec segment, University of Western Australia, Geology Key Centre and University Extension, Publication 24, pp 164–209.

Rossi, M E and Deutsch, C V, 2014. *Mineral Resource Estimation* (Springer Dordrecht). https://doi.org/10.1007/978-1-4020-5717-5

Rousseeuw, P J, 1987. Silhouettes: A graphical aid to the interpretation and validation of cluster analysis, *Journal of Computational and Applied Mathematics*, 20:53–65. https://doi.org/10.1016/0377-0427(87)90125-7

Sangster, A L and Smith, P K, 2007. Metallogenic summary of the Meguma gold deposits, Nova Scotia, in *Mineral Deposits of Canada: A Synthesis of Major Deposit-Types, District Metallogeny, the Evolution of Geological Provinces and Exploration Methods* (ed: W D Goodfellow), Geological Association of Canada, Mineral Deposits Division, Special Publication No 5, pp 723–732.

Thorndike, R L, 1953. Who belongs in the family?, *Psychometrika*, 18(4):267–276. https://doi.org/10.1007/BF02289263

van der Maaten, L J P and Hinton, G E, 2008. Visualizing high-dimensional data using t-SNE, *Journal of Machine Learning Research*, 9:2579–2605.

Ward, J H, 1963. Hierarchical grouping to optimize an objective function, *Journal of the American Statistical Association*, 58:236–244.

Estimation

DIY – implementing customised Python workflows for flexible and effective mineral resource estimation

L Bertossi[1] and D Carvalho[2]

1. Senior Resource Geologist, Glencore, Brisbane Qld 4123.
   Email: laerciobertossi@glencore.com.au
2. Principal Geologist, GeologicAI, Brisbane Qld 4123. Email: dcarvalho@geologicai.com

ABSTRACT

Traditional and conventional commercial mineral resource estimation (MRE) software rarely offer the flexibility needed to create modern customised workflows and address specific requirements of unique deposits and operations. These software packages are typically designed with a set of predefined tools and functionalities (sometimes with scripting capabilities) but often lack the required adaptability when dealing with more complex deposits, unconventional settings, different data sources and types. When scripting is available, most of these packages offer specific programming languages, making it challenging for users to learn and integrate different platforms and tools.

This paper advocates for a 'Do It Yourself' (DIY) approach to generate MRE workflows using Python, a general-purpose programming language that has become one of the most popular worldwide. Python workflows are transparent, easy to learn, self-documenting, capable of processing large data sets and integrating many machine learning routines. Such workflows improve reproducibility, validation and revision of results, independent of the commercial software used for modelling.

Python's open-source nature allows for greater flexibility, integration and control over the entire MRE workflow. By integrating various free-to-use libraries and capable commercial geostatistical/resource modelling modules, it is possible to create dynamic and interactive routines that incorporate all key steps: data validation, exploratory data analysis, variography, block modelling, and grade estimation. In a scripted environment, many iterations are possible, and validation becomes more visual, featuring a variety of customised plots, statistics, grade-thickness relationships, 3D views, grade-tonnage curves and swath plots in a streamlined fashion.

This flexibility is highly effective for large-scale MRE projects, as it automates repetitive tasks and reduces time spent on manual processes. This allows practitioners to focus on critical geological interpretation, refining models, and testing estimation scenarios, leading to more accurate and realistic MRE.

This work explores insights into the advantages of Python over other programming languages, practical examples, and a selection of open-source and commercial Python-based tools for MRE and machine learning, aiming to inspire geologists and modellers to innovate and approach challenges creatively.

INTRODUCTION

Despite the widespread use of conventional commercial software for MRE, many tools lack the flexibility and customisation required for complex geological modelling. This rigidity often results in time-consuming workflows and limited adaptability, leaving geologists and resource modellers without the tools needed to streamline processes or explore innovative approaches.

Advances in programming, automation and IT infrastructure, have opened new opportunities to address these challenges. Workflows such as those presented by Minniakhmetov and Clarke (2023), Ward (2023), and Wilson (2019) serve as impressive examples of comprehensive MRE systems that drive both innovation and efficiency for multiple operations and projects. These highly technological workflows incorporate features such as automation, cloud computing, grade-control estimation, categorical and continuous simulation, and more. A key commonality across these examples, in addition to the required cultural shift and technical expertise, is the adoption of Python as a central tool.

This paper aims to showcase the role of Python-based workflows in modernising MRE practices. It begins by addressing the challenges associated with conventional commercial software, followed by a discussion on how pioneers in geostatistics and MRE leveraged scripting and programming languages to drive innovation. The paper then highlights the advantages of Python over other programming languages, demonstrates practical applications in resource modelling, and provides an overview of Python-based tools available for geological modelling, geostatistics, and machine learning.

THE 'CHALLENGE' OF MANY SOFTWARE OPTIONS

The mining industry has relied on specialised software for geological modelling, resource estimation, mine planning, and scheduling since the early days of computing. These specialised and commercial software are indispensable and responsible for supporting the evolution of the industry. The tools offered in such platforms allows geologists and engineers to better understand the geology, characterise deposits, and use this information to develop strategies to extract resources from the ground.

Throughout their career, geologists and mining engineers frequently navigate multiple software platforms to explore and validate data, create and process workflows, visualise results, and generate reports. This practice is common due to several factors, including the historical adoption of different tools across organisations, cost considerations, and the unique strengths and capabilities that each commercial software offers compared to its competitors. To address this, many platforms aim to provide a comprehensive suite of tools that cover all aspects of resource modelling, from data exploration to final reporting. However, certain platforms still excel in specific areas such as geological modelling, exploratory data analysis (EDA), geostatistics, and resource estimation. As a result, professionals often rely on a combination of software to leverage these specialised features, which can pose challenges in terms of data integration, workflow consistency, and training requirements.

Additionally, when present, many of these software platforms employ proprietary scripting syntaxes (or languages) that differ significantly between providers and are tailored for specific tasks within their individual ecosystems. This creates a substantial barrier for the user, limiting portability and the reusability of skills or workflows across platforms, a challenge we can refer to as 'syntax lock-in'. Those unique tools and proprietary scripting languages require users to learn software-specific syntaxes that are usually not transferable between platforms. This leads to several inefficiencies: significant time lost adapting workflows across projects, limited flexibility in transferring scripts or tools between systems, and reliance on external manual interventions, which ultimately reduces overall efficiency.

Learning and mastering traditional software and proprietary scripting languages demands significant time and effort from users, representing a considerable investment for individuals and businesses. When users transition to a different company or software, much of their software-specific expertise may become obsolete, requiring substantial retraining and potentially months or even years of practice to achieve the same level of proficiency as with their previous platform. This is a complex challenge inherent to the nature of software development and providers, as software ecosystems often evolve independently with unique standards and practices. It is not unique to the mining industry and users must adapt to these changes to continue evolving and advancing in their careers.

LEARNING FROM LEGENDS – THE POWER OF SCRIPTING IN MRE

In recent years, the mining software industry has increasingly shifted towards developing more user-friendly platforms, focusing on building better interfaces and simplifying workflows. These efforts are of utmost importance for democratising technology, shifting the industry towards more advanced techniques, and integrating modern thinking into geological modelling and resource estimation.

However, these advancements can come at a cost when user-friendliness is prioritised over flexibility, customisation, and automation. This is what we can refer as 'IPhonisation' of resource modelling. This 'IPhonisation' implies that software tools are being simplified to attract a broader audience but may sacrifice the flexibility, transparency, and deeper functionality that experts often need. It also implies that updates are becoming increasingly incremental, making them less disruptive and potentially limiting users from reaching their full potential or explore more.

As previously mentioned, the concept of democratising technology is undoubtedly important for driving the evolution of the industry as a whole. However, it is crucial that these advancements also accommodate customisation and flexibility, ensuring that experts and inquisitive geologists have the tools to delve deeper and explore beyond standard workflows.

Simplified workflows and pre-packaged tools, while convenient, often reduce the ability to tailor processes to specific geological challenges or unique deposit characteristics. This trade-off can limit geologists' control over key decisions, such as fine-tuning estimation parameters, implementing automated workflows for more complex scenarios, creating thorough validation routines or venturing into technologies not yet established in our field. In a way, the lack of openness not only constrains our individual ability to innovate but also hinder accelerated progress of the industry.

The roots of traditional and modern geostatistics within MRE frameworks are deeply tied to the use of programming languages, scripted environments, and commonly open-source tools. When reviewing Goovaerts' (2010) list of geostatistical software, several interesting insights emerge. The list highlights notable contributions from pioneers who developed geostatistics and software tailored to specific purposes, such as Clayton Deutsch and Andre Journel (1992, GSLIB), Isobel Clark (Geostokos Toolkit), Pierre Goovaerts (2010, AUTO-IK), Jaime Gomez-Hernandez and R Mohan Srivastava (GCOSIM3D/ISIM3D), and Edward Isaaks (SAGE2001), among others.

Many widely used commercial software still relies on algorithms developed by such visionaries. Additionally, many of the individuals mentioned are authors of foundational books in the field and were key innovators of their time. A more recent example is Michael Pyrcz, who extensively leverages Python for his innovative work and teaching in spatial data analytics, geostatistics, data science, and machine learning.

This evolution highlights that innovation in geostatistics requires critical thinking and is greatly facilitated by the flexibility and adaptability of programming languages and scripted environments. It also raises a critical question: if the pioneers of geostatistics and resource modelling regarded scripting and programming as indispensable, should we accept the limitations of conventional commercial software without exploring the potential of programming?

WHY PYTHON?

It is certainly unfeasible for all practitioners to develop their own software and comprehensive toolsets. However, leveraging a programming language within resource modelling is essential for driving innovation, enabling automation, ensuring auditability and reproducibility, avoid manual errors and achieving a deeper understanding of data and results. Given its popularity, continuous growth and available libraries for data science, statistics and plotting, the authors recommend the use of Python programming language. Additionally, Python is often considered easier than other programming languages due to its simple, readable syntax and a vast ecosystem of libraries that simplify complex tasks, making it more accessible to beginners and efficient for experts.

According to the TIOBE Index, Python is currently the preferred programming language among users (TIOBE, 2025). In 2024, it was awarded the title of 'Language of the Year' due to its remarkable 9.3 per cent increase in popularity, far exceeding other languages such as Java (+2.3 per cent), JavaScript (+1.4 per cent), and Go (+1.2 per cent). Recognised as the default choice in numerous fields, Python's prominence continues to grow, with the potential to achieve the highest ranking ever recorded in the index (Figure 1).

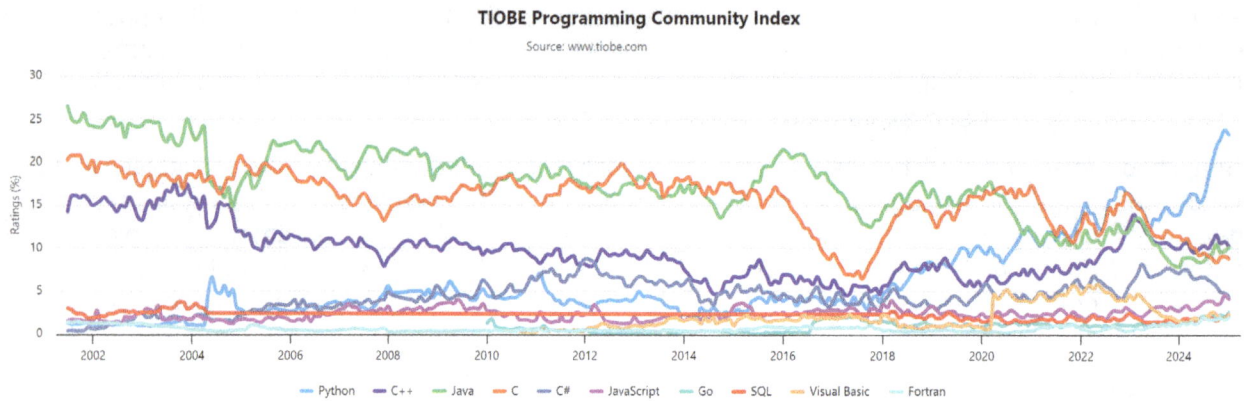

FIG 1 – TIOBE index ratings from 2002 to 2025 showing the increase of python ratings and usage within the programming community.

As of January 2025, the Python Package Index (PyPI) hosts over 600 000 libraries, highlighting the extensive range of tools and libraries available to Python developers. This vast ecosystem supports applications across diverse domains, including scientific computing, data visualisation, machine learning, and web development, contributing to Python's versatility and widespread adoption. The continuous growth of PyPI reflects the active engagement of the Python community in developing and sharing resources that address a wide array of computational challenges.

One of Python's key advantages over conventional commercial software is its seamless integration with Git and platforms like GitHub, which facilitates robust version and document control. This integration enables geologists and resource modellers to track and manage changes in their workflows systematically, providing a clear history of EDA, statistics, variograms, estimation parameters, and differences in volumes and tonnages. By maintaining a well-documented and versioned record of these critical components, users can ensure reproducibility, transparency, and accountability in their resource modelling processes.

One notable downside of Python is its slower execution speed compared to some compiled languages; however, this limitation is far outweighed by its flexibility, extensive library support, and ease of use. Nevertheless, for many tasks involving structured data (e.g.: drill holes and block models) manipulation, processing, and plotting, Python is significantly faster and more customisable than conventional commercial software available in the industry. This advantage arises from the fact that many widely used Python libraries for data science and machine learning are built on wrapped C and C++ code, enabling high-performance processing while maintaining Python's flexibility and ease of use. Additionally, numerous Python libraries and open-source tools are available for resource modelling tasks, including those written in Python and other programming languages (to be discussed in a subsequent section).

Some conventional commercial software platforms offer limited Python scripting integration within their ecosystems, while others are still in the process of implementing it. In most cases, this support is partial, restricting access to certain functionalities and often forcing users to revert to the graphical user interface (GUI) to complete specific tasks, thereby interrupting the workflow with unnecessary steps. Furthermore, many software vendors charge additional fees for Python scripting capabilities, making these features less accessible to users.

The next section will present examples and ideas from the authors, demonstrating how Python has enhanced resource modelling workflows, contributed better understanding of results and avoided manual rework. These examples range from simple calculations to more advanced tasks, such as block model or wireframe manipulation, showcasing the versatility and power of Python as a tool in this domain.

SOME PRACTICAL PYTHON APPLICATIONS IN RESOURCE MODELLING

The examples provided below showcase solutions developed in situations where conventional commercial software lacked the necessary tools, flexibility, or customisation to address specific cases and deposit characteristics. These examples highlight how Python's versatility enables users to overcome such limitations. It is important to note that these are merely illustrative samples; the potential applications are virtually limitless, thanks to the extensive range of resources, libraries, and community support available online.

Ore and waste misclassification

In their classic book, Isaaks and Srivastava (1989) presented theoretical plots comparing true values to predicted values as a method of evaluating the performance of estimation methods. As highlighted by Deraisme (2004), this type of analysis reveals the impact of misclassification on resource estimates and emphasises the critical role of accurate predictive models in supporting optimal decision-making. As mentioned, there is a disconnect between the advancements presented in scientific literature and the tools and functionalities offered by commercial software to their users. This scatter plot for conditioning checking, comparing composites versus blocks at selective mining unit (SMU) scale are mostly not present in conventional commercial software, even though it was described decades ago.

Any practitioner, from beginner to advanced, would benefit from routinely evaluating results in this manner for both grade-control and long-term models. Such evaluations can help identify potential issues in estimation parameters, data quality, outlier management, and other critical aspects of the modelling process.

The plot on Figure 2 shows the classic setting. All aspects of this plot were customised: the dashed 1:1 and the regression lines, the correlation coefficient, the labels for each quadrant, and the correct and incorrect classification proportions. The customised elements help the practitioner quickly understand and communicate results. The idea is similar to a cross-validation but comparing SMUs versus composites. The estimated block model and composites can be loaded as comma-separated values (CSV) files using the Pandas package. Once loaded, the composites within each SMU can be identified using the cKDTree class (from SciPy), which facilitates efficient spatial searches. The SMU index is then assigned to the corresponding group of composites and averaged using NumPy or Pandas. With the block model containing averaged composites, plotting and customisations can be performed using Matplotlib.

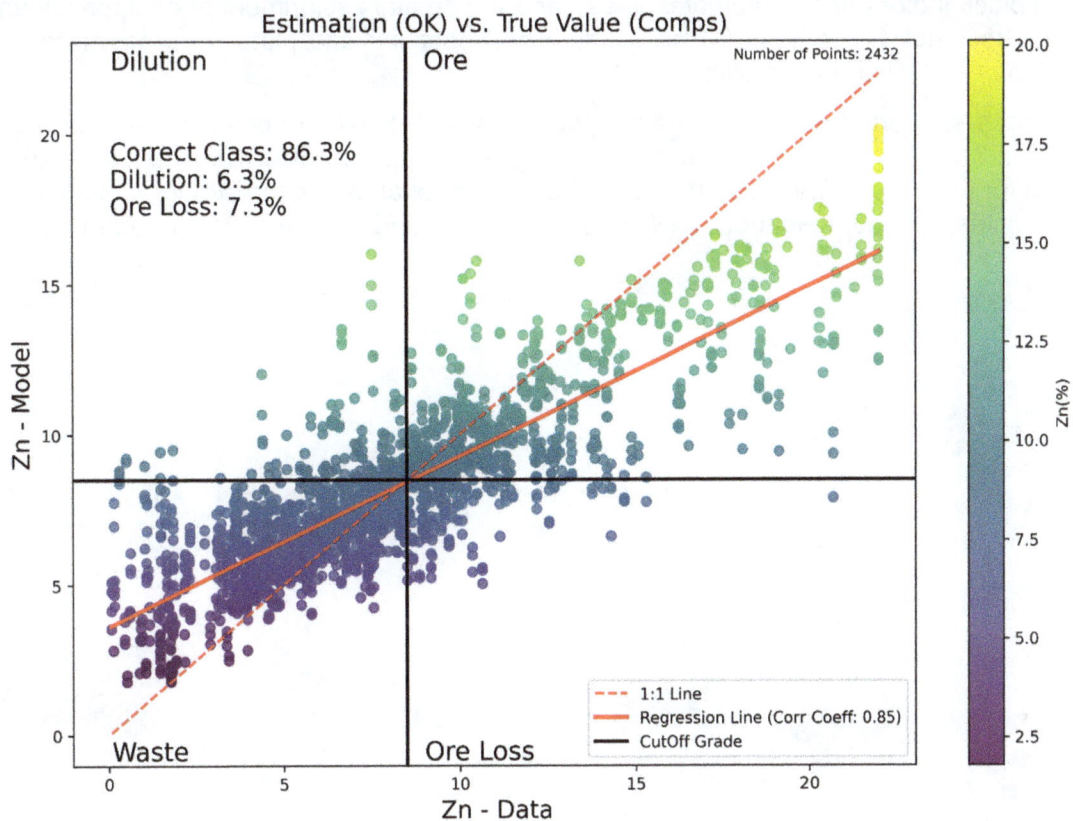

FIG 2 – Estimation versus 'True' values and the misclassification proportions.

Estimation validation in stratigraphic deposits – stratigraphic grade profiles

For many deposits, grade estimation requires local control over search parameters and variography to accurately reflect the geometry of mineralisation and the grade distribution. The concept of Locally Anisotropic Kriging (LAK) allows spatial anisotropy to vary locally, aligning more effectively with the geometry of the deposit, thereby improving local accuracy. Similarly, Locally Varying Anisotropy (LVA) addresses the complexities of non-linear geological structures, such as folds and faults, by enabling anisotropy parameters to vary locally (Boisvert, 2010). Most conventional commercial software includes tools that are derived from these techniques.

Later, Edward Isaaks introduced, as described by Cardwell and Cartwright (2016), practical solutions for structurally intricate deposits to capture the bedding-controlled nature of sediment-hosted deposits. This approach involves a layering code strategy to subdivide each bedding unit, restricting the use of samples for grade estimation within each sublayer, and assigning local dip and dip azimuth to define a search ellipsoid. This program was developed by Isaaks and used in some important deposits around the world.

Although this approach is typically applied to unique deposits with high-grade variability within each geological layer, its validation proves highly valuable by displaying the stratigraphic grade profile of the deposit. This stratigraphic grade profile can reveal mineralisation characteristics or pose critical questions to guide the estimation strategy, such as: Is the high-grade mineralisation concentrated in the hanging wall, footwall, or at the centre of the layer? Are there contacts between units where mineralisation is stronger? With the standard tools available in conventional commercial software, such validation is non-trivial and may not be feasible to implement. Traditional tools such as swath plots may average out and mask results. None of the conventional commercial software offering LAK functionality includes this validation feature.

To validate LAK estimates for a sediment-hosted massive sulfide deposit, the layering technique and stratigraphic grade profile analysis were implemented using Python scripts and the Resource Modelling Platform (RMSP) software. In this example, the estimation workflow was executed with local searches in RMSP. Block models and composites were processed using Pandas, while numerical calculations were performed with NumPy. The layering was achieved using these libraries (based on block indices and coordinates axes), and the spatial assignment of composites and sub-layers was accomplished with cKDTree. Customised sections and plots were generated using a combination of RMSP and Matplotlib.

This approach supports the validation and evaluation of estimation accuracy within a lithological layer. Figure 3 shows an example of the layering technique, in which original units are further subdivided for estimation or validation. Figure 4 shows an example of the stratigraphic grade profile with composites, nearest neighbour (NN) and ordinary kriging (OK) results by each unit and sub-layer.

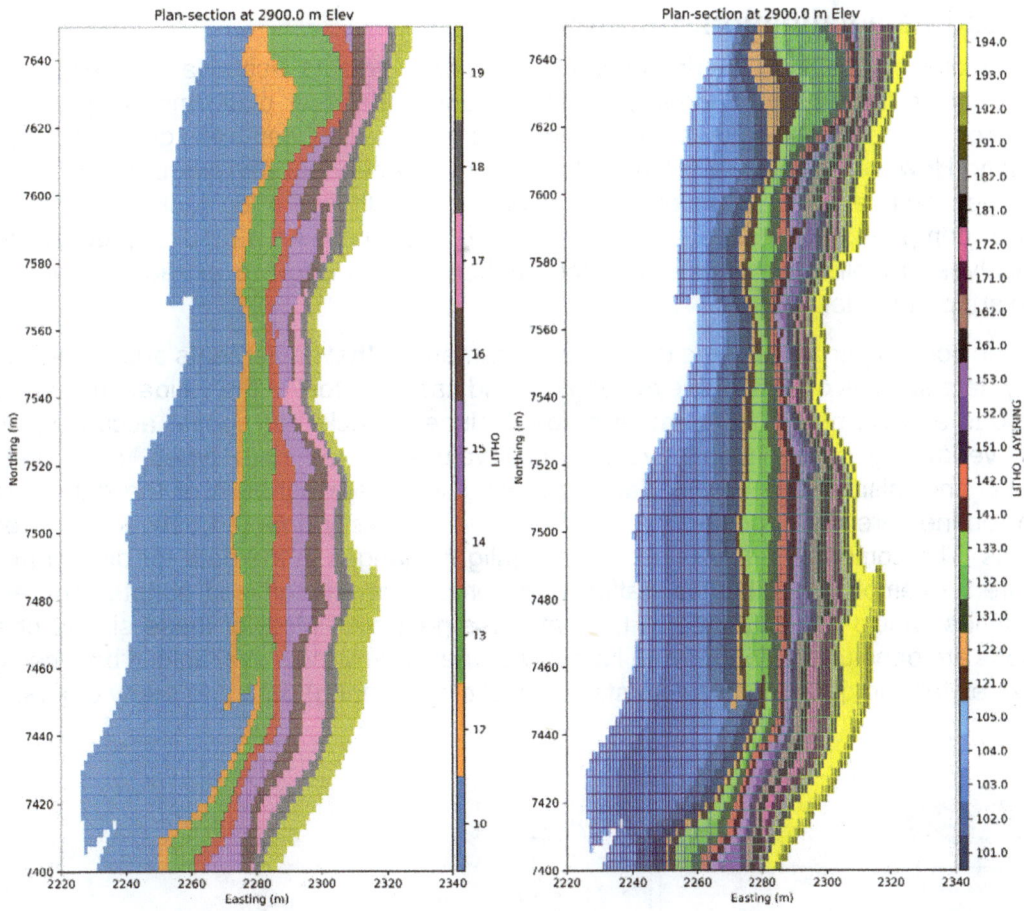

FIG 3 – Layering technique applied to a sediment hosted mineralisation. Note the original lithological units (left) and the sub-layering applied for the validation (right).

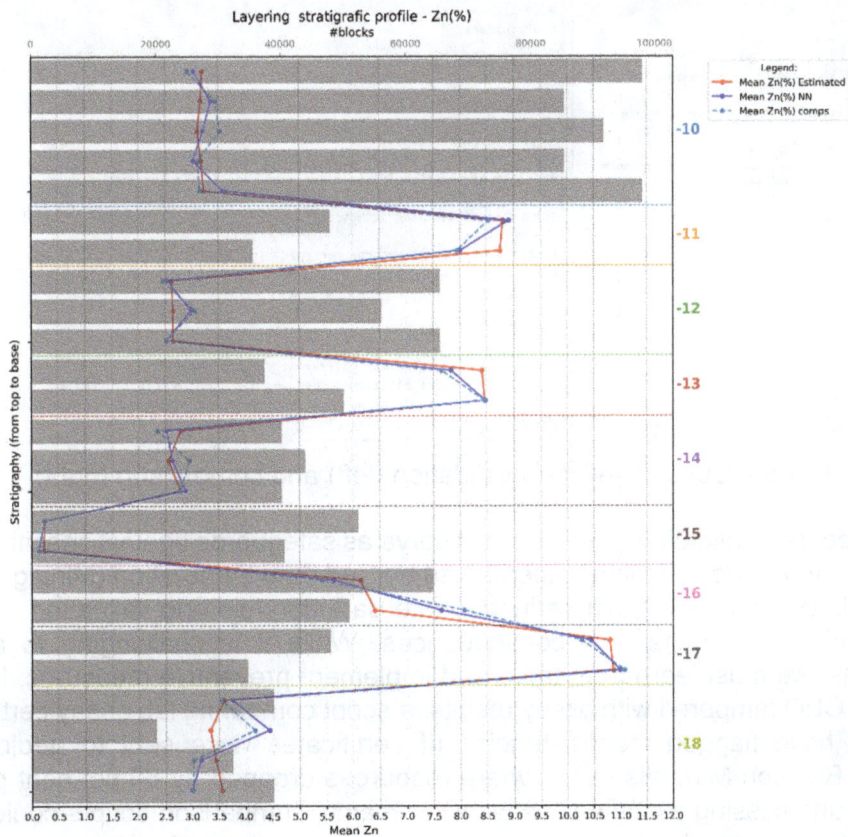

FIG 4 – Stratigraphic grade profile for estimation validation.

Customised data validation routines

In MRE workflows, ensuring the integrity of input data sets is essential for reliable resource estimation and effective operational planning. Many database platforms incorporate built-in validation processes, consisting of predefined checks managed by database geologists, to maintain data integrity. However, it is equally crucial to perform independent reviews of the results to verify their accuracy and reliability. Conventional commercial software often lacks essential tools for validating or comparing databases across different time periods. As a result, modellers frequently find themselves grappling with numerous spreadsheets in an effort to extract critical information needed to support model updates.

Basic Python scripting enables users to develop a framework that implements automated validations across key data such as collars, surveys, logging, and assays. Most of the validations can be carried using Pandas and Numpy. The example below includes checks for spatial accuracy, directional integrity, overlapping or anomalous values, null values, mixed data types, duplicates, invalid coordinates, and mismatches across data sets. As checks are agnostic of the data, customised validation routines are not deposit-dependant and can be easily adapted between different mines and projects. The comparison tool is useful to highlight changes in the data for different periods of time. Figure 5 exemplifies both a validation and comparison customised scripts. Any errors and inconsistencies should be redirected to the database geologist. Most of these simple checks and comparisons are done using Pandas and Numpy libraries. Many tools available in both can generate merged tables for comparison, calculate statistics, slicing (filtering) data and many others.

```
Collar

validCollar(bhid='hole_id',
            xyzcol=['x','y','z'],
            filen=collar_24)

##############################################################
Starting Collar Validation

Validation 1 - There are 0 Zero/Null values

Validation 2 - There are 121 Rounded coordinates

Validation 3 - There are 0 duplicated Hole ID

Validation 4 - There are 423 duplicated coordinates

Validation 5 - There are 0 inverted X and Y

Validation 6 - There are 16 coordinates to be reviewed

Validation 7 - There are 0 Holeid with spaces on last character

Warnings exported to "warnings_collar.csv"
##############################################################
```

```
Litho

ftype="litho"
compData(ftype,
         bhid='hole_id',
         nameo=litho_23,
         namen=litho_24,
         from_i='depth_from',
         to_i='depth_to')

***** OLD LITHO INFO*****
# of lithos: 446663
# of columns: 7

***** NEW LITHO INFO*****
# of lithos: 462382
# of columns: 7

***** COLUMNS NAMES *****

All columns names match!

***** DIFFERENT LITHOS *****
There are 15775 different lithos found.
The list of additional lithos was saved as: "different_lithos.csv"

The total of matching lithos is: 446635

Number of different records per column:
doma = 15214
leaching = 0
litho1 = 793
dyke_selected = 15207

A csv table was saved as: "different_data_litho.csv"
The table contains all lithos that are matching in both old/new files with the respective fields.
It is in binary code, being 0 no difference, and 1 presence of difference.
Whenever a difference exists, both old/new values are shown.
```

FIG 5 – Customised data validation (left) and comparison (right).

These customised data validation routines could serve as safeguards against specific and potentially catastrophic scenarios, such as those experienced by Red Pine Exploration (Mining.com, 2024) and Rubicon Minerals (Koven, 2016), where inadequate data controls and validation processes led to significant financial and reputational consequences. While it is challenging to anticipate such disastrous events, we must learn from them and implement preventive measures. In the Red Pine case, where the CEO tampered with assay results, a script comparing lab assay certificates with the database could have flagged inconsistencies (if certificates were sent to geologists as well). Similarly, in the Rubicon Minerals case, where resources dropped by 88 per cent primarily due to poor domaining and missing assay values within domains, comparison scripts could be developed to verify grade and volumetric statistics as well as the proportions of missing assays within each domain.

An integrated python MRE workflow – George Fisher Mine

The following demonstrates the updated MRE workflow at George Fisher Mine (GFM) stratiform Zn-Pb-Ag deposit (Queensland, Australia). The deposit displays many challenges related to the geometry, structural characteristics, number of estimation domains and variables and processing times. The deposit is divided into two principal mining areas, P49 (south) and L72 (north), both exhibiting comparable stratigraphy, structural features, and mineralisation styles, around 83 estimation domains, separated by multiple fault blocks that can generate up to 700 individual solids and six variables to be estimated.

The updated workflow enhanced the previous geological modelling process by utilising a Python-based script to generate hundreds of thousands of control points, which are then integrated into the Leapfrog Geo workflow to better honour the geometry (1 at Figure 6). These control points are generated within minutes, eliminating manual work that could otherwise take up to four years of a full-time employee (FTE) to complete (Bertossi and Carvalho, 2023). The scripts were developed to address two key challenges in modelling stratiform deposits like GFM: (1) representing pinch-outs and discontinuities where lithological units are absent, and (2) preventing surfaces from crossing other stratigraphic units when drill holes begin or end mid-sequence. These challenges are resolved by generating control points that guide geological surfaces to align with stratigraphy and geological constraints (Figure 7).

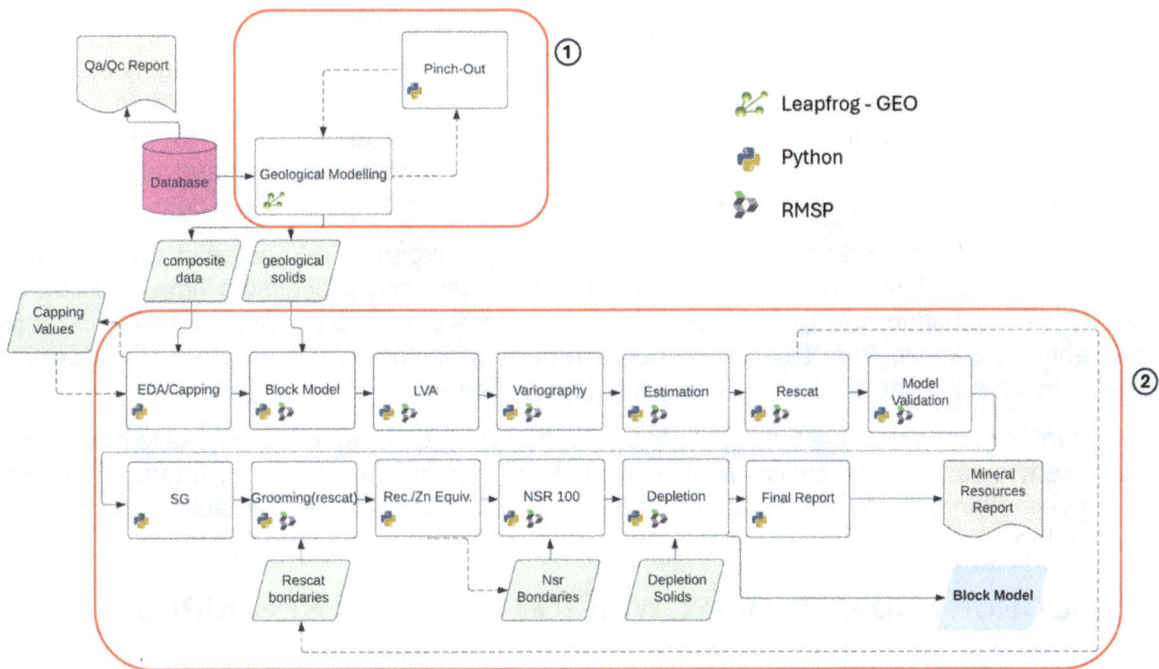

FIG 6 – Flow chart of the Integrated Python MRE workflow at GFM.

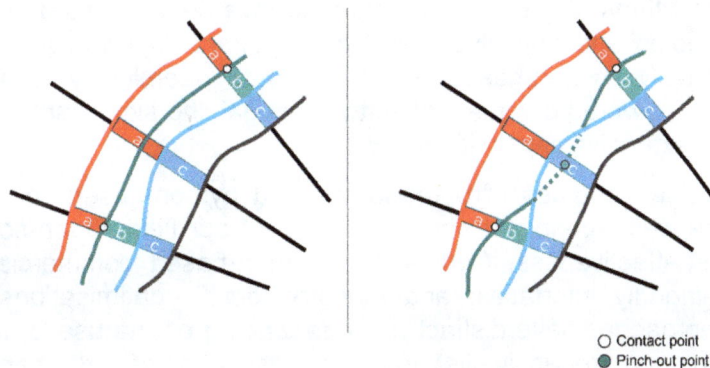

FIG 7 – Schematic representation of original surface/solids interpolation using the deposit tool in a case which unit 'c' cross-cut younger units (left); and surfaces when adding the stacking control points for unit 'c' (right).

Subsequentially, (2 at Figure 6) a fully customised estimation workflow was implemented using Python scripts and RMSP, streamlining processes and integrating multiple validation steps for each task. Each rectangle at Figure 6 represents a specific MRE task, starting from EDA to final resource reporting. Each task is executed within a Jupyter Notebook which contains numerous subtasks. These scripts generate a variety of outputs, including images, sections, graphs, and statistics, which are automatically exported for use in presentations and reports. Any notebook (task) can be run individually or tied together to execute the entire workflow with a single command.

The majority of the notebooks combine Python and RMSP commands and tools. More traditional tasks, such as EDA, block model generation, variography, LVA, grade estimation, and depletion, rely more heavily on the commercial software, while other tasks, including mathematical resource classification, specific gravity regressions, recoveries, and similar processes, are primarily implemented as standalone Python scripts. The integration between Python and the commercial software enables faster processing through multithreading, simplifies the validation of all domains, variables, and fault blocks, and automates the generation of reporting plots, graphs, and statistics.

This modern and flexible workflow replaced an outdated one that relied on multiple scripts, from a conventional software, with more complex syntax and much less robust tasks. The older workflow was inflexible and lacked the ability to generate a comprehensive and streamlined set of validations and tasks. Additionally, manual modifications to the outdated scripts often led to errors that were significantly more difficult to debug compared to Python scripts.

In 2021, an attempt was made to implement a non-scripted workflow using another conventional commercial software. This effort required three months of work by three full-time, highly experienced geologists to manually set-up the 'dynamic' workflow and replicate the previous outdated estimation parameters. At that time, the non-scripted MRE workflow took up to 24 hrs to complete grade estimation for each mine. Additionally, the majority of validations, graphs, and plots had to be generated manually, along with other aspects of the workflow, often requiring weeks to finalise.

In contrast, the updated workflow, implemented by a single experienced geologist over five months (while learning both Python and RMSP), has drastically reduced processing time and increased quality. Grade estimation now takes approximately 20 mins per mine (compared to 24 hrs before), and the entire workflow, from EDA to resource reporting, including the automated generation of graphs, plots, checks, and statistics, takes up to four hrs per mine.

It is important to emphasise that the significant reduction in processing times (weeks to hours) should be redirected towards more critical activities, such as geological modelling, interpreting results, analysing validation plots and statistics, evaluating reconciliation data, and enabling critical thinking and innovation.

OPEN-SOURCE AND PYTHON-POWERED TOOLS FOR RESOURCE MODELLING

When it comes to geostatistics and resource modelling, it is not always necessary to reinvent the wheel. Many current geostatistical tools, whether in commercial or open-source software, are built on well-established algorithms. While these algorithms have been adapted and improved over the years, their core principles and objectives remain consistent. Open-source tools, regardless of whether they are natively Python-based, often integrate seamlessly with Python workflows. Meanwhile, fully Python-powered commercial software is gaining significant traction in the industry due to its flexibility and extensive capabilities.

The advantages and disadvantages of open-source and python-based commercial software for resource modelling are well-recognised and not unique of our field. Open-source software offers transparency and cost-effectiveness. Conversely, Python-based commercial software provides robust support, user-friendly interfaces, and industry-specific optimisations, often at a higher financial cost. Both approaches have distinct roles depending on the user's needs and resources. Below is a selected, non-comprehensive list and short description of both open-source (Python and other languages) and Python-based commercial tools. Due to proprietary considerations, no assumptions or details are provided regarding the source code, algorithms, or other programming languages used in the commercial software descriptions.

Open-source tools

- **GSLIB**: Developed by Deutsch and Journel (1992), is a foundational collection of geostatistical tools written in Fortran, widely regarded as a cornerstone in the field. It provides core functionalities for variogram modelling, kriging, simulation, and post-processing, designed as command-line programs for flexibility in custom workflows. Accompanied by the book GSLIB: Geostatistical Software Library and User's Guide, it is a standard reference in geostatistics, offering detailed instructions and theoretical context. While not user-friendly by modern standards, GSLIB remains highly influential, serving as a benchmark for geostatistical practices. Many commercial software packages have used GSLIB as the basis for their geostatistical algorithms, further solidifying its importance in the industry.

- **GeostatsPy**: developed originally by Michael Pyrcz and expanded through contributions (Pyrcz *et al*, 2021), is a Python-based library designed for teaching and applying geostatistics. It reimplements and expands the established GSLIB algorithms in Python, bringing geostatistical workflows into a modern computational environment. The library provides a comprehensive suite of tools for tasks such as variogram analysis, kriging, simulation, and uncertainty quantification, while leveraging Python's flexibility and integration with many libraries (Pyrcz, nd). Aimed at both students and professionals, GeostatsPy emphasises accessibility and reproducibility, offering detailed examples and workflows that align with industry best practices.

- **PyGSLIB**: A Python package designed originally by Adrian Vargas for mineral resource estimation. While primarily a Python package, it integrates enhanced Fortran code from the original GSLIB to handle computationally intensive geostatistical tasks (Martinez Vargas, nd). Inspired by Datamine Studio macros, it enables workflows such as drill hole compositing, block modelling, and conditional simulations.

- **Pygeostat:** Another Python package for geostatistical modelling, developed by alumni of the Centre for Computational Geostatistics (CCG). It is designed for preparing spatial data, scripting geostatistical workflows, and modelling using tools developed at CCG. Pygeostat also includes many visualisation and plotting capabilities making it a versatile tool for geostatistical analysis (Centre for Computational Geostatistics, nd).

- **SGeMS**: An open-source, GUI-based geostatistical application written in C++ for high-performance computing (Remy, Boucher and Wu, 2009). It provides tools for variogram analysis, kriging, and simulation, enabling users to construct and analyse geostatistical models. SGeMS also integrates with Python, allowing users to automate workflows, extend functionality, and connect with external tools for advanced spatial data analysis.

- **GemPy**: Python-based 3D structural geological modelling software (de la Varga, Schaaf and Wellmann, 2019) that facilitates the implicit creation of complex geological models from interface and orientation data. It supports stochastic modelling to address parameter and model uncertainties, making it a versatile tool for geoscientific applications.

- **GeoStats.jl**: A framework for geostatistical modelling and geospatial data analysis, built in the Julia programming language (JuliaEarth, nd). It provides tools for variogram analysis, kriging, simulation, and visualisation, supporting applications in fields such as mining, hydrogeology, and environmental sciences. Designed for flexibility and performance, it is suitable for both structured and unstructured data workflows.

It is worth noting that these open-source alternatives have their own plotting capabilities, including 2D and 3D visualisation, either within their environments or by leveraging Python libraries such as Matplotlib, Plotly, Mayavi, and PyVista. These libraries enable the creation of interactive and static 3D visualisations, enhancing the analysis and presentation of geological data. An additional recommendation is to use ParaView for interactive 3D visualisation, especially for handling large data sets and creating high-quality visual representations. ParaView's compatibility with Python scripting further enhances its utility allied with the geostatistical libraries above.

Python-powered commercial software

- **RMSP**: a Python-based commercial software designed Resource Modelling Solutions (GeologicAI) for advanced resource modelling and geostatistics. Known for its speed and scalability, RMSP leverages modern algorithms and probabilistic approaches to handle large data sets efficiently. It provides tools for grade estimation, conditional simulation, and uncertainty quantification, enabling streamlined workflows and flexible customisation. RMSP integrates seamlessly with Python's ecosystem, making it an innovative choice for geologists and resource modellers seeking automation, transparency, and reproducibility and integration with artificial intelligence (AI) and machine learning (ML) in their workflows. It can be deployed locally or in the cloud.

- **GeoLime**: a Python library developed by DeepLime for geostatistical modelling and geological data analysis. It offers tools for variogram analysis, kriging, simulation, and visualisation, facilitating resource estimation workflows. GeoLime integrates seamlessly with Python's environment, providing flexibility and efficiency for geoscientists and engineers. Can also be deployed locally or in the cloud.

- **Isatis.Py**: Geostatistical library developed by Geovariances, designed to enhance geostatistical modelling workflows in sectors such as mining, energy, and natural resources. It offers robust algorithms for data preparation, variography, interpolation, estimation with kriging, and uncertainty analysis through conditional simulations. By integrating seamlessly with Python's extensive libraries, Isatis.py enables users to build and automate custom solutions, leveraging data science, artificial intelligence, and visualisation libraries.

INTEGRATION WITH ARTIFICIAL INTELLIGENCE AND MACHINE LEARNING

If there is one group of professionals with the resources, both human and financial, to develop sophisticated GUIs, it is data scientists, along with AI and ML practitioners. Yet, their preferred tool remains scripted environments. As discussed earlier, the preference arises from key advantages such as flexibility, customisation, reproducibility, version control, automation, and scalability. These traits make scripting environments far more suitable for the iterative, complex, and often highly specific workflows characteristic of AI and ML projects. The advantages remain consistent when it comes to resource modelling, whether using traditional or advanced techniques, but they become particularly significant when AI/ML models are employed. There have been numerous advancements in the application of AI/ML techniques within MRE workflows, geostatistics, and geosciences as a whole. A detailed review and discussion of these developments is beyond the scope of this paper, as its intent is not to delve into such advancements and applications. However, some important aspects of its applicability to MRE workflows can be highlighted.

As previously mentioned, due to the lack of full integration with programming languages, popular commercial software offers limited or, in some cases, no support for such techniques. Some of these software solutions may include built-in tools for Principal Component Analysis (PCA), hierarchical clustering or support vector machines, enabling the definition of domains, groups, and lithologies based on spatial continuity and multivariate relationships. Other software solutions may incorporate neural networks and a few ML techniques to generate grade interpolations and geological models. However, these approaches are often seen as less interpretable and may be limited by the range of parameters available within their platform.

There is no alternative: if the MRE practitioner intends to integrate comprehensive and robust ML into their workflow, they must dedicate time to building the model using a programming language. The wide range of potential applications in MRE makes it challenging to implement simplified tools within commercial software. A toolbox approach is recommended when building such models, as demonstrated by AI/ML practitioners, to enable tailored solutions. Given the inherent need for flexibility and modularity, and the relatively gradual pace of AI/ML adoption and implementation popular commercial software, the authors recommend using scripts and programming languages to develop these models.

Python offers a robust ecosystem of machine learning tools, each tailored to specific needs and techniques. A non-comprehensive list and short description is shown below.

- **Scikit-learn**: Widely used for traditional ML tasks, supporting regression techniques like linear and other regressions, and decision trees, as well as classification methods such as support vector machines (SVM), logistic regression, and K-Nearest Neighbours (KNN). It also provides tools for clustering, including K-Means and DBSCAN, and dimensionality reduction techniques like PCA.

- **TensorFlow and PyTorch**: For deep learning, these tools offer capabilities for building neural networks, including convolutional neural networks (CNNs) for image processing, recurrent neural networks (RNNs) for sequence data, and advanced architectures like transformers for natural language processing.

- **TensorFlow**: excels in production environments due to its scalability, while PyTorch is favoured for research and experimentation with its dynamic computation graphs.

- **XGBoost and LightGBM**: specialises in gradient boosted decision trees (GBDTs), which are highly effective for structured data in regression, classification, and ranking tasks, particularly in large-scale applications.

For prototyping some of these models, tools like Orange and KNIME are excellent choices, offering user-friendly, GUI-based environments that allow for rapid experimentation and workflow development without requiring extensive programming expertise.

LEARNING PYTHON

Mastering Python requires time and hands-on experience, much like learning any other software or tool, or developing expertise in geology and resource modelling. While Python's syntax might initially seem complex, it is fundamentally logical and becomes more intuitive with practice. Effective learning often begins with introductory free courses and addressing small, real-world problems, such as tasks typically performed in Excel for structured data processing, and then gradually scaling up to more complex workflows like data analysis, automation and resource modelling. Its extensive libraries and frameworks enable versatility across tasks.

The extensive resources within the Python community, including platforms like Stack Overflow, provide invaluable support as many common questions were likely answered years ago. Additionally, tools such as ChatGPT and other systems based on large language models (LLMs) have the potential to significantly accelerate workflow development, often by factors of 10 or even 100, even for experienced users. However, the outputs must be carefully validated, as inconsistencies in generated code are not uncommon.

CONCLUSIONS

The purpose of this paper is not to persuade readers to abandon conventional commercial software in favour of executing all workflows with Python or Python-based software. Instead, it aims to illustrate the significant innovation, freedom, and flexibility that can be achieved through its use. Additionally, we advocate for stronger integration between software and programming languages. While achieving such integration is no trivial task, it empowers geologists and resource modellers, regardless of their expertise level, to explore and implement innovative ideas.

As noted earlier, many of these results can be achieved using other programming languages; however, Python, as demonstrated, simplifies the process, making it significantly easier and more efficient to implement ideas. These scripted workflows address key limitations of conventional commercial software by allowing customisation, automation, and deeper integration of advanced techniques.

Following the example of pioneers in geostatistics and resource modelling, who relied on scripting to innovate and overcome challenges, adopting these approaches is essential for modernising resource estimation workflows and geology in general. Some practical examples were presented, representing just a small fraction of the virtually limitless possibilities. The paper highlighted numerous tools, both open-source and commercial, that are either Python-based or compatible with it. Likewise, it also reviewed the critical role of Python in machine learning and explored some of the available open-source tools supporting these applications.

Encouraging geologists and modellers to incorporate these tools into their practices will help align the industry with evolving technological standards in other fields and ensure continued progress in geological modelling and resource estimation.

REFERENCES

Bertossi, L and Carvalho, D, 2023. Overcoming implicit modelling software limitations using Python scripting – an innovative geological modelling workflow for George Fisher Mine, Queensland, Australia, in *Proceedings of the Mineral Resource Estimation Conference 2023*, pp 251–265 (The Australasian Institute of Mining and Metallurgy: Melbourne).

Boisvert, J B, 2010. Geostatistics with locally varying anisotropy, Doctoral dissertation, University of Alberta, Edmonton, Alberta.

Cardwell, J and Cartwright, A, 2016. Dynamic unfolding – Complex geology case study of Tenke-Fungurume deposits, *Mining Engineering, Society for Mining, Metallurgy and Exploration*, January:20–27.

Centre for Computational Geostatistics (CCG), nd. Pygeostat: Python package for geostatistical modelling [online] Available from: <https://www.ccgalberta.com/pygeostat/welcome.html> [Accessed: 7 Jan 2025].

de la Varga, M, Schaaf, A and Wellmann, F, 2019. GemPy 1.0: Open-source stochastic geological modeling and inversion, *Geoscientific Model Development*, 12(1):1–32. Available from: <https://www.gempy.org/> [Accessed: 7 Jan 2025].

Deraisme, J, 2004. Recoverable resources estimation: Indicator Kriging or Uniform Conditioning?, presented at the EAGE Conference, Madrid.

Deutsch, C V and Journel, A G, 1992. *GSLIB: Geostatistical software library and user's guide* (Oxford: Oxford University Press).

Goovaerts, P, 2010. Geostatistical software, in *Handbook of Applied Spatial Analysis* (eds: M Fischer and A Getis), (Heidelberg: Springer). https://doi.org/10.1007/978-3-642-03647-7_8

Isaaks, E H and Srivastava, R M, 1989. *An introduction to applied geostatistics* (New York: Oxford University Press).

JuliaEarth, nd. GeoStats.jl: High-performance geostatistics in Julia, JuliaEarth. Available from: <https://juliaearth.github.io/GeoStatsDocs/stable/> [Accessed: 7 Jan 2025].

Koven, P, 2016. Rubicon Minerals Corp shares plunge as miner slashes its gold resources by 88%. *Financial Post*, 11 Jan. Available from: <https://financialpost.com/commodities/mining/rubicon-minerals-corp-shares-plunge-as-miner-slashes-its-gold-resources-by-88> [Accessed: 7 Jan 2025].

Martinez Vargas, A, nd. PyGSLIB: Python Geostatistical Library [online]. Available from: <https://opengeostat.github.io/pygslib/> [Accessed: 7 Jan. 2025].

Mining.com, 2024. Red Pine says former CEO tampered with Wawa gold assays, Mining.com, 10 May. Available from: <https://www.mining.com/red-pine-claims-former-ceo-tampered-with-wawa-gold-assays/> [Accessed: 7 Jan 2025].

Minniakhmetov, I and Clarke, D, 2023. SBRE framework – application to Olympic Dam deposit, in *Proceedings of the Mineral Resource Estimation Conference 2023*, pp 266–275 (The Australasian Institute of Mining and Metallurgy: Melbourne).

Pyrcz, M J, Jo, H, Kupenko, A, Liu, W, Gigliotti, A E, Salomaki, T and Santos, J, 2021. GeostatsPy Python package: Open-source spatial data analytics and geostatistics, *Zenodo*. https://doi.org/10.5281/zenodo.13835444

Pyrcz, M J, nd. My Story, *Professor Michael J Pyrcz*. Available from: <https://michaelpyrcz.com/my-story> [Accessed: 8 Jan 2025].

Remy, N, Boucher, A and Wu, J, 2009. *Applied geostatistics with SGeMS: A user's guide* (Cambridge: Cambridge University Press).

TIOBE, 2025. TIOBE index for January 2025. Available from: <https://www.tiobe.com/tiobe-index/> [Accessed: 7 Jan 2025].

Ward, C, 2023. Rapid resource modelling: New workflows for radically faster and more flexible models, World Mining Congress 2023, Anglo American.

Wilson, C K W, 2019. Reaching for the STARS: AGSTM, SME Conference.

Spatial estimation with machine learning of multiple-fluid phase veining in shear-hosted intrusion-related deposits using geospatial features – a case study from Pogo Mine, Alaska

E Ramos[1], S Sen[2], M Matthews[3] and J Machukera[4]

1. Senior Resource Geologist, Northern Star Resources, Alaska AK 99701.
   Email: eramos@nsrltd.com
2. Senior Resource Geologist, Northern Star Resources, Perth WA 6008. Email: ssen@nsrltd.com
3. Geology Superintendent – Resources, Northern Star Resources, Perth WA 6008.
   Email: mmatthews@nsrltd.com
4. Group Geology Manager – Resources, Northern Star Resources, Perth WA 6008.
   Email: jmachukera@nsrltd.com

ABSTRACT

At Pogo, a shear-hosted, intrusion-related gold deposit in Alaska, the mineralisation is often hosted within 12 m thick quartz veins, exhibiting multiple fluid phases, where gold concentration frequently occurs near the vein margins. The mosaic mineralisation style, when estimated using Ordinary Kriging (OK), fails to reflect the resolution required across the thickness of the vein to manage mining dilution appropriately. Categorical Indicator Kriging (CIK) is utilised to sub-domain grade populations within the same quartz vein. This traditional approach relies on subjective decisions, including the geologist's domain expertise, variogram modelling, cut-off grade selection, and threshold probability determination. In this case study, a machine learning (ML) model was developed without requiring assumptions of stationarity or linearity. The proposed ML model improves performance by applying feature engineering to the original coordinates through polynomial transformation and incorporating a fixed reference point. The initial training utilised several ML algorithms, each contributing to distinct learning patterns. Model performance was further enhanced by employing a stacking approach with diverse base learners and using a meta-learner to ensemble the predictions. The performance of the meta-learner model was evaluated against the CIK model. This case study highlights the ML model's ability to accurately represent the training data while eliminating cognitive bias from human input. It underscores the superior effectiveness of the ML model in estimating this type of mineralisation, positioning it as a compelling alternative to the CIK model.

INTRODUCTION

In geological modelling and resource estimation, stationarity is often assumed, meaning the probability distribution of observations remains consistent over time and space. However, deviations from this assumption can indicate multi-stage processes typical in complex geological systems like gold mineralisation. Gold systems frequently exhibit anisotropy—directional variability—complicating spatial modelling.

Machine learning (ML) models in spatial prediction often introduce grid artifacts, data smoothing, and overfitting near boundaries. While coordinates and their properties are key to spatial data sets, the existing literature inadequately addresses them.

Prior studies relying solely on spatial coordinates as model inputs often overlook critical underlying spatial patterns, leading to inaccurate predictions. Significant improvements have been demonstrated by incorporating distances between observation points (Hengl *et al*, 2015), though this is computationally intensive and difficult to scale for large data sets. Principal Component Analysis (PCA) on distance vectors may reduce computational costs for smaller data sets but its effectiveness in larger data sets remains underexplored and may produce model artifacts (Ahn, Ryu, and Lee, 2020). Recent work (Erten, Deutsch and Yavuz, 2020) demonstrated that rotating coordinates in 5° intervals can reduce artifacts. However, the fixed angle rotation method (Deutsch and Journel, 1998) remains a complex workflow and may not be suitable for mixed-grade populations.

This case study presents an alternative spatial modelling approach by leveraging feature engineering on coordinates within a complex, actively mined, wide-vein gold system. The methodology transforms coordinates using polynomial degrees, introducing higher dimensions that enable the ML model to learn complex interactions. A sensitivity analysis of the CIK model explores varying probability thresholds, 2-Bin and 3-Bin methods, and soft boundary conditions to address model limitations. Comparing ML with the CIK method, this study highlights fundamental differences in how machine learning interprets spatial data, aiming to address a pivotal question: Can an ML model efficiently replicate spatial characteristics observed in OK or CIK models? The proposed methodology demonstrates its potential to provide an effective and efficient method for spatial prediction at both sampled and unmeasured locations using only coordinates and gold values.

GEOLOGICAL BACKGROUND

The Pogo gold deposit is located within the Tintina Gold Province, a metal-rich province spanning northern British Columbia, the Canadian Yukon and much of interior Alaska. It represents one of several Intrusion Related Gold Systems in the central northern Cordillera, spatially and temporally associated with early to mid-Cretaceous magmatic belts.

Mineralisation at Pogo occurs in moderate to flat dipping, stacked quartz veins, hosted within north-west-dipping brittle to semi-ductile fault zones along the regional-scale Liese Fault Zone. Formed in the early to mid-Cretaceous, veins vary in width, locally exceeding 10 m.

Vein geometry has been influenced by dip-slip movement during vein formation, causing duplexing and thinning of veins. Associated Fe-carbonates and sericite alteration suggest late-stage slipping with cooler, likely gold-poor fluids.

Flexures and linkage/relay zones, common in Liese-style mineralisation, often correspond with vein steepening and may be associated with an increase in steep extensional veins. While dilation in these zones can enhance connectivity, creating wide, high-grade zones, excessive dilation may lead to barren quartz infilling, reducing ore grade.

For this study, Lode 1100 – a typical Liese-style mineralised quartz vein, was selected for ML model testing. Ranging from 1 to 12 m in thickness, it exhibits multi-generation veining, creating a mixed-grade, mosaic-style gold distribution. Structural complexity and pervasive alteration prevent effective manual statistical sub-domaining, making Lode 1100 an ideal candidate for advanced modelling techniques.

EXPLORATORY DATA ANALYSIS

Data description

The data set consists of spatial coordinates (Northing, Easting, Elevation) and gold assay values (Au) from 5378 regularised samples, covering a total length of 4071 m. Of these, 47 per cent are underground diamond core drill samples, 51 per cent are underground face channel samples, and the remainder are from exploration surface drilling. The total length of underground diamond drilling is comparable to the total length of the face channel. Most samples are standardised to 0.76 m in length.

Samples are confined within the thick, mineralised quartz-vein Lode 1100, defined by a geological hard boundary during interpretation. Drill spacing averages 4 m, with a maximum of 32 m. Production mining is active, with most mined-out areas concentrated east of the lode. Figure 1 illustrates sample locations.

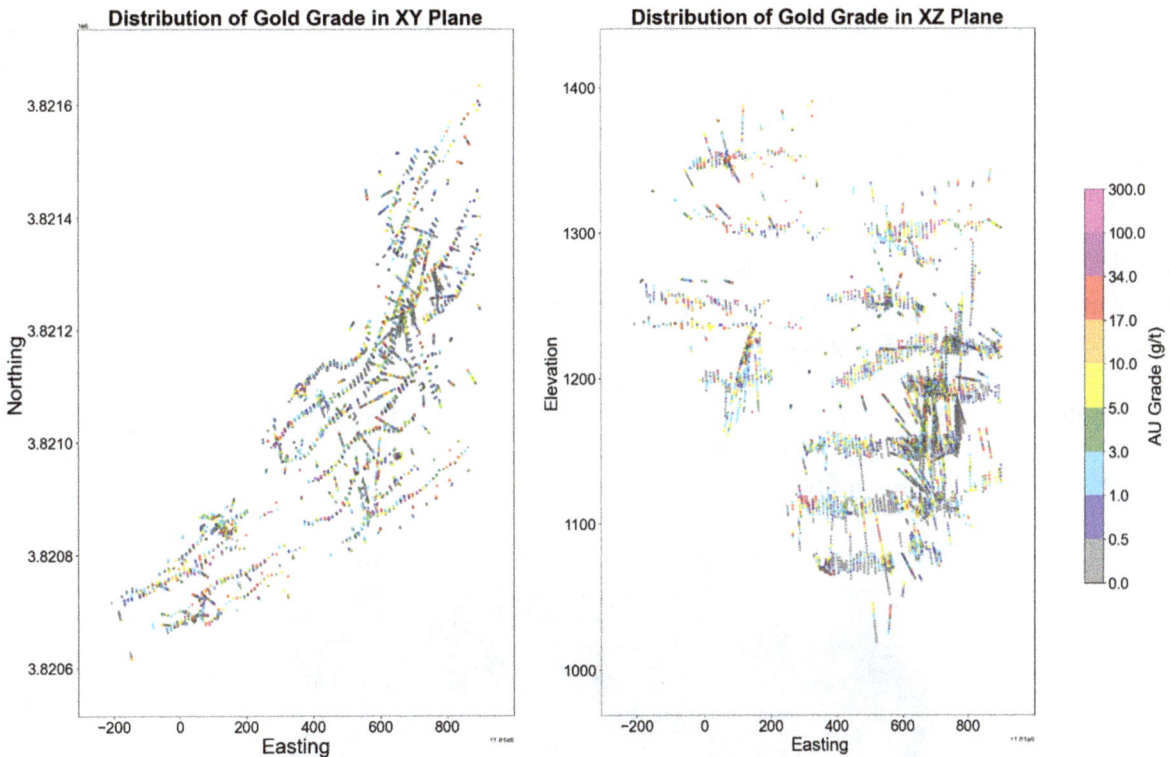

FIG 1 – Location map of samples with gold concentration (Au, g/t) distribution.

Grade distribution

The data set's average gold value is 6.72 g/t, ranging from 0.003 g/t to 130.63 g/t. Figure 2 presents the histogram, basic statistics, and cumulative distribution frequency (CDF) of gold concentration by sample type. Grade distributions for diamond drilling (DD) and face channel (FC) samples are similar and adequately represented across the multivariate space, though FC samples show slightly higher gold concentrations.

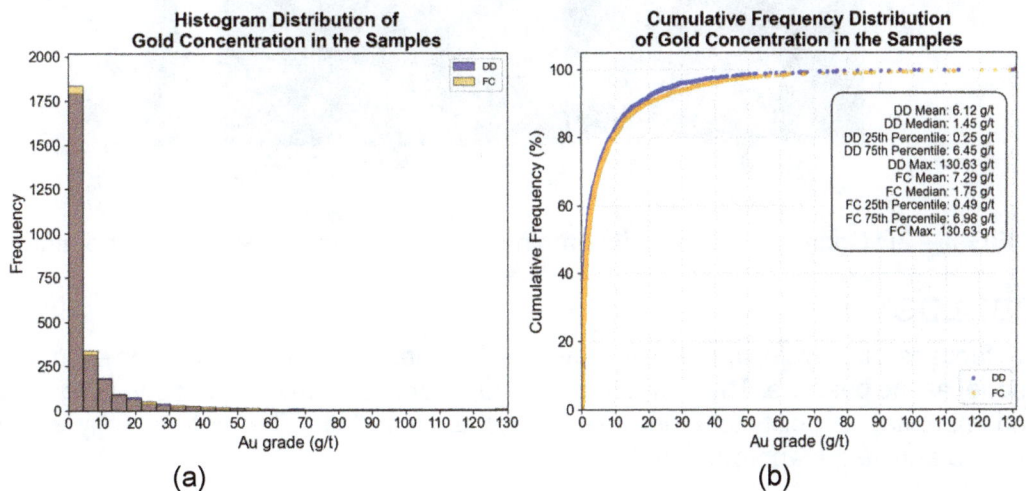

FIG 2 – (a) Histogram; and (b) cumulative frequency distribution of gold concertation in the samples.

The combined gold distribution of DD and FC samples is highly positively skewed, reflecting a non-normal distribution typical of gold deposits. Notably, 37 per cent of samples exceed the economic mining cut-off of 3.43 g/t. The global mean gold value is near the 75th percentile, indicating moderate asymmetry, and the relatively high mean value compared to the median emphasises the presence of high-grade outliers.

Underground drilling constraints can lead to poor drilling orientation, whereas FC samples are always taken perpendicular to the orebody, where the sample intervals are determined by geology: lithology contacts, mineralisation, alteration, and structure. These samples provide greater data density for both models. Although FC samples have been considered less reliable for estimating gold grades and prone to bias in several studies, this case study treats DD and FC as equally valid inputs for both ML and CIK models.

Directional trends in mineralisation

Variogram analyses are commonly used in geostatistics to assess mineralisation's spatial continuity and directionality (plunge). Kernel Density Estimation (KDE) was applied to examine the spatial distribution of gold grades (Figure 3). This validation tool helps identify spatial trends in mineralisation and compare them with the CIK and ML models. Directional smoothing was applied to enhance the visualisation of anisotropy, revealing an anisotropic distribution with higher grade continuity at approximately a 60° angle, expected to be consistent across both models.

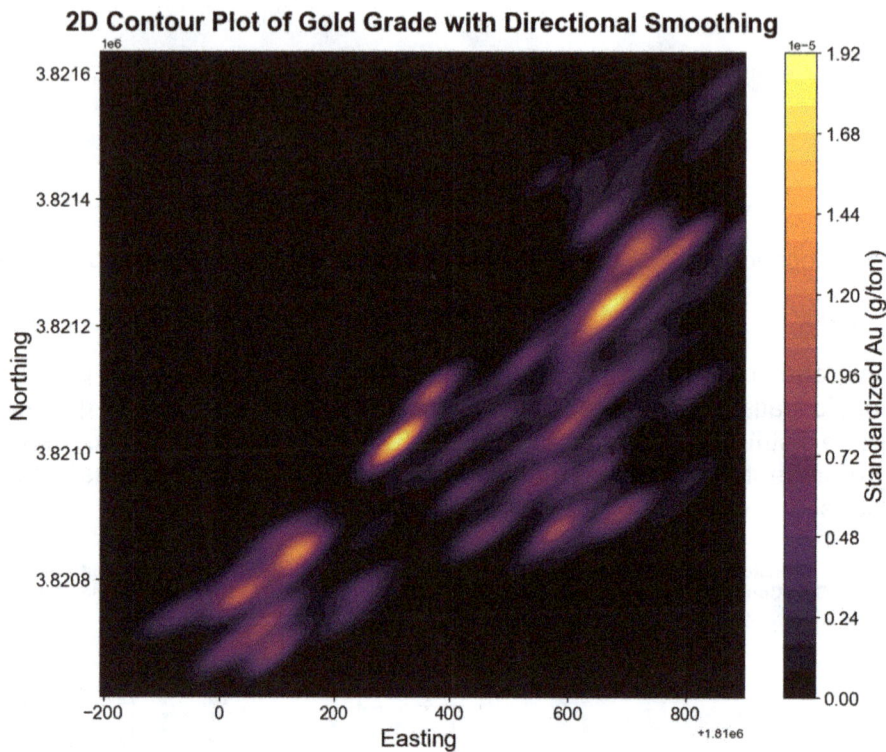

FIG 3 – 2D Contour Plot using a combination of KDE and directional smoothing.

METHODOLOGY

This study focuses on developing an alternate machine learning model (ML) for estimation, using the CIK model as the baseline. The aim is to enhance model performance by optimising input feature design and improving its ability to generalise to new, unseen data. The methodology is divided into key sections to outline the approach taken.

The raw data was pre-processed, including feature engineering and normalisation, ensuring appropriately scaled features for the ML model.

Two experiments were designed to explore several aspects of model performance:

1. Feature Selection: The first experiment focused on selecting key input features for the ML model. The goal was to determine which features would improve the model's accuracy by effectively capturing the underlying patterns in the data.

2. Model generalisation: The second experiment aimed to test the model's ability to generalise to unseen data by training it on 40 per cent, 80 per cent, and 100 per cent of the data set.

Multiple ML algorithms were trained to enhance performance, each capturing unique patterns within the data. The predictions from these models were then combined via a stacking approach where a meta-learner model integrated the outputs from the individual algorithms to improve prediction accuracy. The top-performing ML model was compared against the CIK baseline to evaluate predictive accuracy.

This methodology provides a straightforward, structured approach to training and validating an alternative ML model while ensuring the necessary steps for data preparation, feature selection, and model validation are followed.

Feature engineering

In machine learning, feature engineering converts raw data into meaningful features, enhancing the accuracy of model predictions. To address challenges posed by spatial data, polynomial features were derived from the original undeformed coordinates (Northing, Easting, and Elevation). By introducing higher dimensional components to these coordinates, the model can better capture complex relationships that the original features alone would miss.

Squared terms (X^2, Y^2) help detect non-linear patterns, while interaction terms (XY, XZ, YZ, and XYZ) capture relationships between the features. The Euclidean distance from a defined origin to each sample point was also calculated and combined with the Northing, Easting, and Elevation coordinates as outlined in Equation 1 to generate new features, $X_{distance}$, $Y_{distance}$, and $Z_{distance}$. Table 1 summarises the equation used to calculate each input feature.

$$distance = \sqrt{(X - X_0)^2 + (Y - Y_0)^2 + (Z - Z_0)^2} \tag{1}$$

TABLE 1

Engineered input features derived from raw spatial coordinates for ML model training.

Input feature	Equation
X^2	Easting^2
Y^2	Northing^2
Z^2	Elevation^2
XYZ	Easting × Northing × Elevation
XY	Easting × Northing
XZ	Easting × Elevation
YZ	Northing × Easting
$X_{distance}$	Easting × distance
$Y_{distance}$	Northing × distance
$Z_{distance}$	Elevation × distance

Data preprocessing

To improve the ML model performance, the data set was normalised. Data normalisation is a preprocessing technique that scales or transforms the data, ensuring each feature contributes equally to the training process (Singh and Singh, 2020). This is particularly important in machine learning, as features with more extensive ranges or extreme values can disproportionately impact the model's predictions.

Positively skewed data, such as the gold (Au) variable, can lead to the model overemphasising lower scores during penalisation, resulting in biased predictions in the lower grade range. While some variables in earth sciences may approximate parametric distributions, such as Gaussian or lognormal distributions, no comprehensive theory exists to determine their exact forms (Pyrcz and Deutsch, 2018). In this study, standard score transformation was applied to standardise the gold values, which transforms the data to a mean of zero and a standard deviation of one using Equation 2.

$$normal_{score_{Au}} = \Phi^{-1}\left(\frac{Rank(Au_i)}{N+1}\right) \tag{2}$$

where:

Au_i is the gold value for the i-th sample

N   is the total number of samples

Φ^{-1} is the inverse of the standard normal cumulative distribution function

Additionally, the target Au variable is scaled to a range of [0, 1] to enhance the performance of distance-based algorithms, such as k-nearest neighbours (KNN) and support vector machines (SVR) as defined by Equation 3. This scaling also benefits neural networks (NN) by improving convergence during training.

$$T_{new} = \frac{T-min(T)}{max(T) - min(T)} \tag{3}$$

where:

T   is the target variable for the i-th sample

Data partitioning

Splitting the data into training and testing sets enables the model to be trained on one subset and evaluated on another, providing a more accurate estimate of its real-world performance. In this study, 20 per cent of the data is withheld using Python's train_test_split function from the scikit-learn open-source library. This function ensures random shuffling and keeps the target variable distribution consistent in both sets, creating balanced and spatially representative data sets.

Prediction map

After training, the ML model generates a prediction map on a 0.46 × 0.46 × 0.31 m grid of 1 335 885 centroids aligned with the CIK model's block size. Each centroid represents a spatial unit and serves as a sample point for predicting the target gold value.

Machine learning experimental design

The objective of the ML model is to generate a robust final model that performs well on training and unseen data without introducing spatial artifacts. To achieve this, eight base machine learning models were trained using diverse algorithms, including boosting, bagging, distance-based, kernel-based, non-ensemble, and deep learning approaches. These base learner models were combined into a stacked ensemble using Extreme Gradient Boosting (XGBoost) as the meta-learner, leveraging their combined strengths for improved performance.

To improve accuracy and reduce overfitting, a hyperparameter grid was defined for each base model, and all possible combinations were tested as detailed in Table 2. The same tuning process was applied to the meta-learner, improving the final ensemble's performance. The methodology is summarised in Figure 4, which outlines the sequential steps from training the base learner models to stacking and generating final predictions.

TABLE 2

Hyperparameter values were used during machine learning training, and the corresponding tuned values were obtained from GridSearch.

Base model	Hyperparameter	Best parameter	Search range
XGBoost	n_estimators	1000	[100, 500, 1000, 2000]
	max_depth	12	[8, 10, 12]
	learning_rate	0.01	[0.01, 0.05, 0.08, 0.1]
	reg_lambda	0.1	[0.0, 0.1, 0.5, 1.0]
	max_bin	1000	[None, 400, 500, 1000]
	sub_sample	0.8	[0.8, 0.9, 1.0]
SVM	C	1000	[100, 1000]
	epsilon	0.5	[0.1, 0.5, 0.8, 1]
DT	max_depth	20	[8, 10, 12, 20]
	min_samples_leaf	10	[5, 10, 15, 20]
	min_samples_split	5	[5, 10, 15, 20]
	max_leaf_nodes	None	[None, 10, 20, 50]
	splitter	best	best
RF	n_estimators	1000	[1000,1500,,2000]
	bootstrap	TRUE	TRUE
ADA	n_estimators	1500	[500, 1000, 1500, 2000]
	learning_rate	0.01	[0.01, 0.05, 0.1, 0.5]
GB	max_depth	None	[None, 10, 20, 50]
	learning_rate	0.05	[0.01, 0.05, 0.08, 0.1]
	subsample	0.8	[0.8, 1.0]
KNN	n_neighbors	4	[2, 3, 4, 10]
	weights	distance	['uniform', 'distance']
	algorithm	brute	['auto', 'ball_tree', 'kd_tree', 'brute']
	metric	Euclidean	Euclidean
NN	hidden layers	4	[2, 3, 4, 5]
	neurons number	[64, 64, 32, 16]	[64, 64, 32, 16]
	batch size	32	[16, 32, 64]
	epochs	400	[100, 200, 300, 400, 500]
	learning_rate	0.001	[0.001, 0.1, 0.5, 1]
	optimiser	Adam	Adam

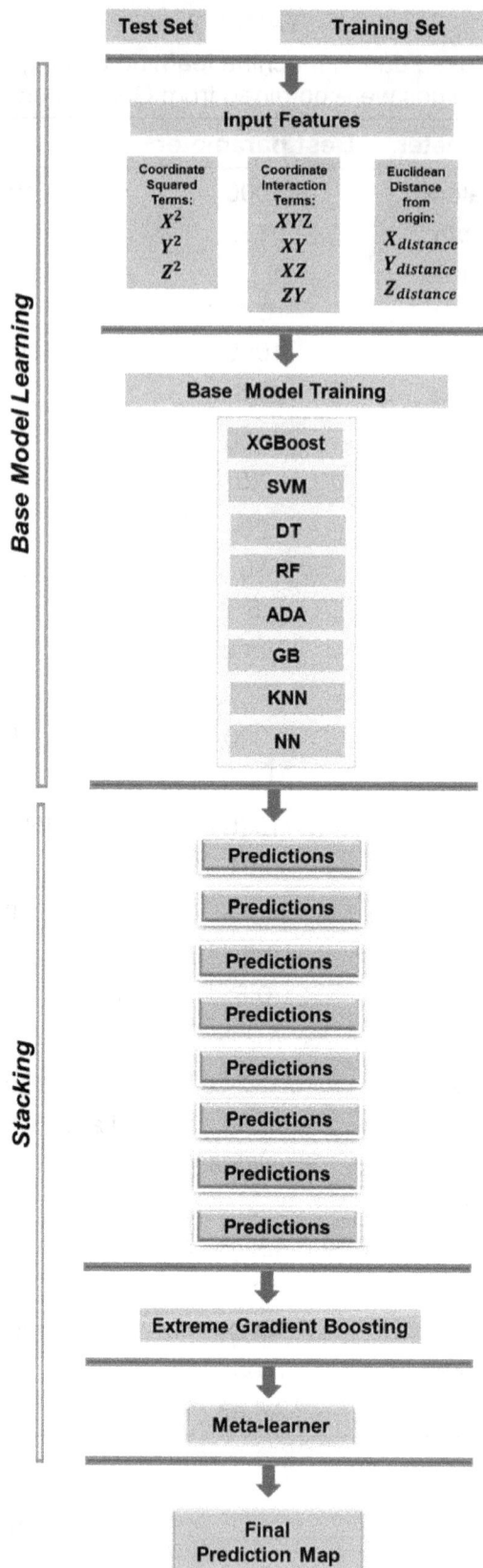

FIG 4 – Flow chart of the proposed methodology (base model and stacking approach).

Machine learning base learners

This section explains the theory behind the eight base learner models used in this study, implemented using Scikit-learn (Pedregosa *et al*, 2011), TensorFlow (Abadi *et al*, 2016), and Keras (Chollet, 2015). Scikit-learn is a popular and user-friendly Python library for machine learning, while TensorFlow and Keras are widely used in deep learning.

Decision tree

A Decision Tree (Quinlan, 1986) is a greedy algorithm that splits data into branches based on locally optimal decisions. While interpretable and easy to visualise, it risks overfitting if unpruned. Decision trees are the foundation for more powerful ensemble methods like Random Forests and Gradient Boosting.

Random forest

Random Forest with Bootstrap Aggregating (Breiman, 2001) is an ensemble learning algorithm that combines multiple decision trees to improve predictive accuracy and reduce overfitting. Each tree is trained on a random subset of the data, selected with replacement through bootstrap sampling. As a result, some data points may appear multiple times in a tree's training set, while others may be excluded. Final predictions are aggregated from all trees, typically using an average or median.

Extreme gradient boosting (XGBoost)

Extreme Gradient Boosting (Chen and Guestrin, 2016) is a decision-tree-based ensemble for classification and regression tasks. It builds sequential decision-tree ensembles to correct errors incrementally and is renowned for its scalability and efficiency, especially with large data sets.

Support vector machine (SVM) with RBF kernal

The Support Vector Machine (SVM) with a Radial Basis Function (RBF) kernel, introduced by Cortes and Vapnik (1995), is a distance-based model designed explicitly for non-linear tasks. The RBF kernel maps input data to a higher-dimensional space, which allows the model to capture complex patterns more effectively. Unlike traditional regression models, Support Vector Regression (SVR) aims to identify a hyperplane where data points within a specified margin around the predicted function are error-free. Only those points outside this margin contribute to the loss function, which helps minimise the risk of overfitting.

AdaBoost

AdaBoost (Freund and Schapire, 1996) sequentially trains multiple weak learners, often decision trees. After each learner is trained, the algorithm updates the weights of the data points based on their residual errors, assigning greater emphasis to incorrectly predicted instances. This iterative process ensures the subsequent learners focus on harder-to-predict instances. The final prediction is a weighted average of all learners, with more accurate models receiving higher weights.

Gradient boosting

Gradient Boosting (Friedman, 1999) is an ensemble learning algorithm that constructs a strong predictive model by sequentially training weak learners. Unlike AdaBoost, which emphasises misclassified points, Gradient Boosting minimises a specified loss function using gradient descent. Each iteration optimises the model by correcting errors in the direction of the steepest descent of the loss function, reducing prediction error at every step.

K-nearest neighbours (KNN)

K-Nearest Neighbour (Cover and Hart, 1967) is a simple, non-parametric, and instance-based learning algorithm that predicts outcomes based on the closest training examples. Its performance depends on key factors, including the choice of the number of neighbours (K), distance metrics, and feature scaling.

Neural networks

Neural Networks (Hinton, LeCun and Bengio, 1986) are a powerful class of machine learning models inspired by the brain's architecture. They comprise three main layers: input, hidden, and output. The training relies on the backpropagation algorithm, enabling weight adjustments to minimise prediction errors. Known for their flexibility, neural networks form the basis of many modern AI advancements.

Hyper-parameter optimisation and cross-validation

In machine learning, hyperparameters are parameters set before the learning process that define the model's structure and training process. Correct selection of hyperparameters significantly enhances model performance. To prevent underfitting or overfitting, the exhaustive method of Python's GridSearchCV function from the sci-kit-learn library was employed to identify the optimal combination of hyperparameters.

To ensure the model is neither underfitting nor overfitting, cross-validation is used in combination with the GridSearchCV function. Cross-validation divides the data set into multiple subsets (folds). For each fold, the model is trained on a portion of the data and validated on the remaining fold. This process is repeated for each fold, and the model's performance is averaged across all folds.

In this study, five-fold cross-validation was applied during hyperparameter optimisation for all base models. Table 2 outlines the hyperparameter search space and the final configurations selected for each base model.

Ensemble model

Stacking is an ensemble technique that integrates the predictions of several models through a final model, known as the meta-learner, to produce the final prediction. The base learner models are trained on the same data set, and their predictions serve as inputs for the meta-learner. This method is especially effective when combining models with different biases.

Stacking enhances generalisation, improves accuracy, and helps mitigate challenges such as overfitting and model variance. This study used the XGBoost algorithm to generate predictions from seven base learner models. This choice was made due to its speed and robust regularisation capabilities, effectively reducing the risk of overfitting.

Case study – MI model optimisation

This study features two key experiments to enhance model performance. The first experiment focuses on selecting the best features for training to optimise both computational complexity and dimensionality. Various forms of coordinate features, as detailed in Table 3, are tested, and each resulting ML model is evaluated.

TABLE 3

Experiment 1: List of coordinate Input types used for base model training and meta-learner model.

Meta-learner model	Coordinate input type	Applying Euclidean distance from the origin?
ML-Coord	X, Y, Z	No
ML-Coord-Dist	X, Y, Z	Yes
ML-Dist	None	Yes
ML-Poly	X^2, Y^2, Z^2, XYZ, XY, XZ, YZ	No
ML-Poly-Dist	X^2, Y^2, Z^2, XYZ, XY, XZ, YZ	Yes

The second experiment uses the selected features from the first experiment to train the model and evaluate its ability to predict unseen data accurately. This is achieved by varying the amount of training data as outlined in Table 4.

TABLE 4

Experiment 2 – List the percentage of training samples used for the final model design (ML-Poly-Dist).

Meta-learner model	Percentage of training samples	Using diamond core samples?	Using face chip samples?
ML-40	40%	Yes	No
ML-80	80%	Yes	Yes
ML-100	100%	Yes	Yes

Together, these experiments aim to enhance the model's accuracy and robustness. The best-performing model from both experiments is then compared to the CIK model.

RESULTS

Model performance

The ML model's performance was assessed using a withheld 20 per cent test data set. Common evaluation metrics included root mean squared error (RMSE), coefficient of determination (R^2), descriptive statistics, probability density plots, grade bin volume, spatial artifact evaluation, and geologic explainability.

The performance of the base learner models and meta-learner in predicting gold (Au) values on the prediction map was evaluated against the 20 per cent test data, with results shown in Table 5. Despite improvements from hyperparameter tuning, none of the models achieved an R^2 greater than 60 per cent. This suggests that the base learner models still struggle to capture the data's complexity. Overall, the meta-learner outperformed the individual models, demonstrating the effectiveness of the ensemble approach.

TABLE 5

Performance comparison of base learner models and the meta-learner model trained on 80 per cent of the data.

	TRUTH	XGBoost	SVM	DT	RF	ADA	GB	KNN	NN	Meta-learner
RMSE	-	19.56	6.96	6.84	2.50	5.59	5.75	19.38	6.50	1.60
R-squared	-	49%	25%	26%	51%	47%	49%	45%	33%	99%
Mean	4.69	3.63	2.85	4.25	3.46	3.43	3.94	4.42	3.29	4.50
SD	7.26	3.77	2.02	5.04	3.26	3.12	4.42	5.38	3.12	7.25
CV	1.55	0.96	1.41	0.84	1.06	1.10	0.89	0.82	1.05	1.61
Minimum	0.00	0.14	0.03	0.14	0.14	0.21	0.10	0.10	0.03	0.11
25th percentile	0.37	1.20	3.70	0.96	1.27	1.27	1.06	0.89	1.23	1.00
Median	1.43	2.37	2.33	2.19	2.33	2.40	2.33	2.33	2.37	2.25
75th percentile	5.75	4.73	3.70	5.90	4.59	4.59	5.14	6.00	4.53	4.66
Maximum	39.07	49.68	19.54	25.85	41.69	28.56	63.33	53.07	50.78	63.03

Feature optimisation

Experiment 1 revealed minor statistical differences but notable variations in probability plots and spatial maps (Figure 5). The best feature set (ML-Poly-Dist) for ML training combined higher-dimensional coordinates with a Euclidean distance vector, enhancing predictive accuracy. For five different combinations of input features, all meta-learner scenarios achieve high R^2 values, as presented in Table 6.

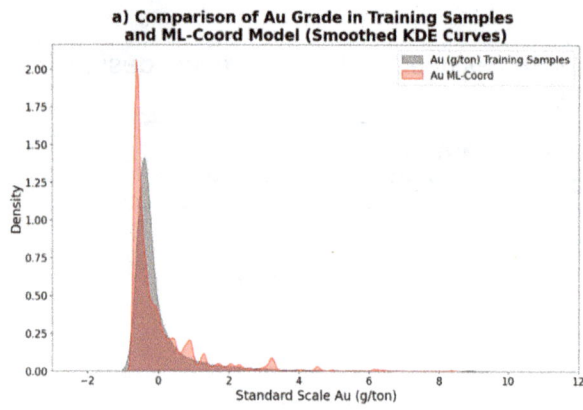

a) Comparison of Au Grade in Training Samples and ML-Coord Model (Smoothed KDE Curves)

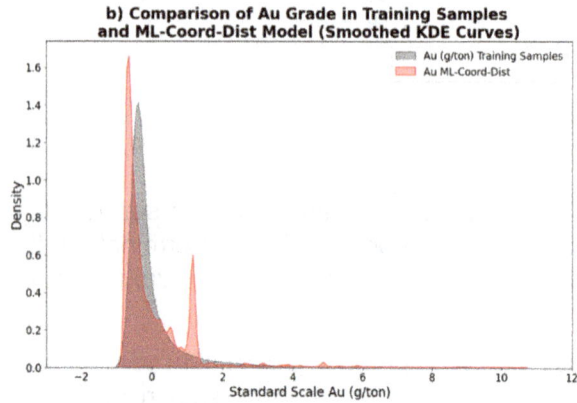

b) Comparison of Au Grade in Training Samples and ML-Coord-Dist Model (Smoothed KDE Curves)

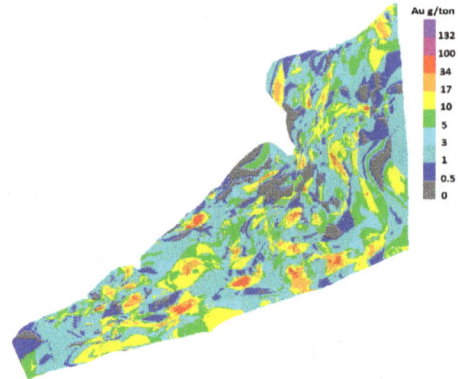

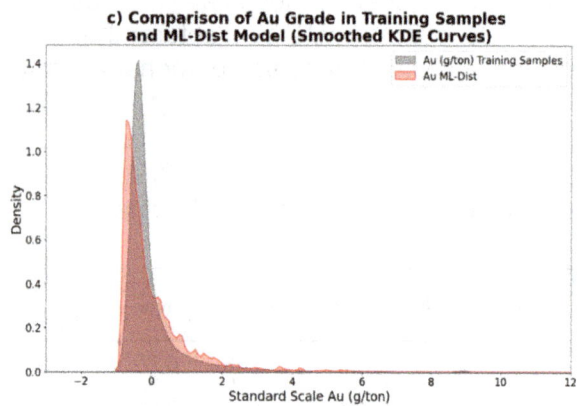

c) Comparison of Au Grade in Training Samples and ML-Dist Model (Smoothed KDE Curves)

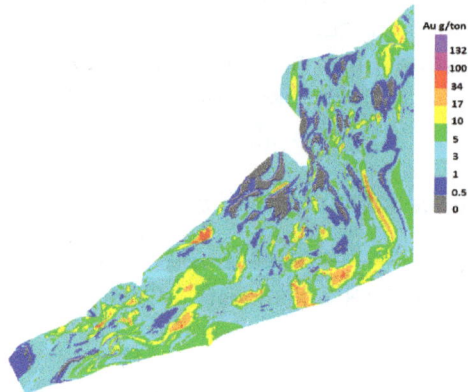

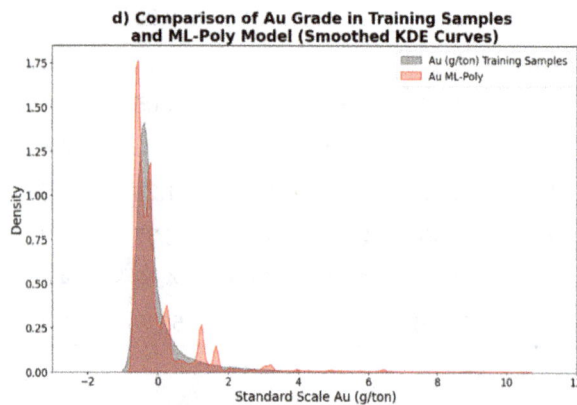

d) Comparison of Au Grade in Training Samples and ML-Poly Model (Smoothed KDE Curves)

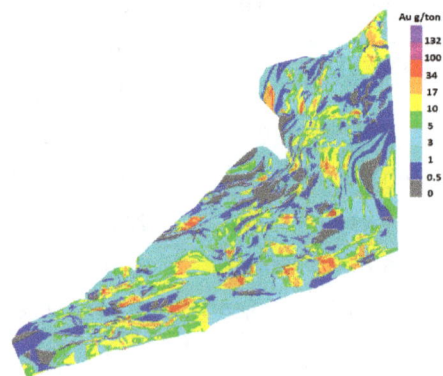

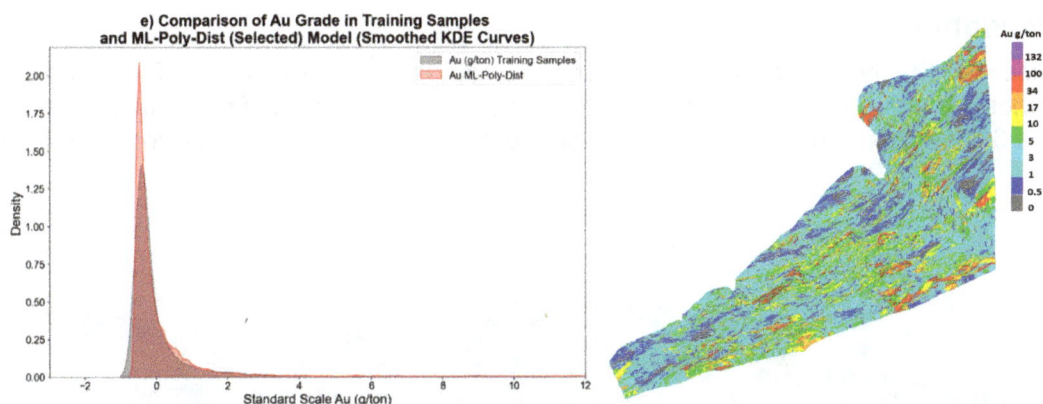

FIG 5 – Probability density plot of normalised Au values and spatial map for experiment 1: (a) ML-Coord, (b) ML-Coord-Dist, (c) ML-Dist, (d) ML-Poly, and (e) ML-Poly_Dist.

TABLE 6

Performance of different coordinate input types for base and meta-learner models.

	TRUTH	ML-Coord	ML-Coord-Dist	ML-Dist	ML-Poly	ML-Poly-Dist (selected)
RMSE	-	1.60	1.60	1.60	1.60	1.60
R-squared	-	99%	99%	99%	99%	99%
Mean	4.69	5.07	4.55	3.90	4.03	4.50
SD	7.26	7.27	5.80	4.46	5.86	7.25
CV	1.55	1.43	1.28	1.14	1.45	1.61
Minimum	0.00	0.11	0.11	0.11	0.11	0.11
25th percentile	0.37	0.75	0.86	1.13	0.79	1.00
Median	1.43	2.24	2.25	2.36	2.24	2.25
75th percentile	5.75	6.29	6.55	5.07	4.67	4.66
Maximum	39.07	65.89	65.89	65.88	65.89	63.03

A probability density plot was used to assess model alignment with the grade distribution of the training data. Gold grades were standardised using Z-score normalisation, using Equation 4 to adjust for sample size differences:

$$z = \frac{X_i - \mu}{\sigma} \tag{4}$$

where:

z   is the standardised value(z-score)

X_i is the individual point

μ   is the mean of the population

σ   is the standard deviation of the population

Figure 5 shows consistent mineralisation continuity across ML model scenarios, although spatial artifacts and smoothing are more apparent in Figure 5a–5d. The best model from Experiment 1, combining polynomial coordinates with a Euclidean distance vector from the origin, significantly improved grade precision in both in-plane and cross-sectional prediction maps. The selected ML model (ML-Poly-Dist) also best reflected the underlying distribution of the training data.

Model generalisability

Experiment 2 assessed the model's ability to predict unseen data after optimisation in Experiment 1. ML model scenarios are trained on 40 per cent, 80 per cent, and 100 per cent of the data set. Cross-sections in Figure 6 show that all models capture the spatial patterns with varying amounts of training data.

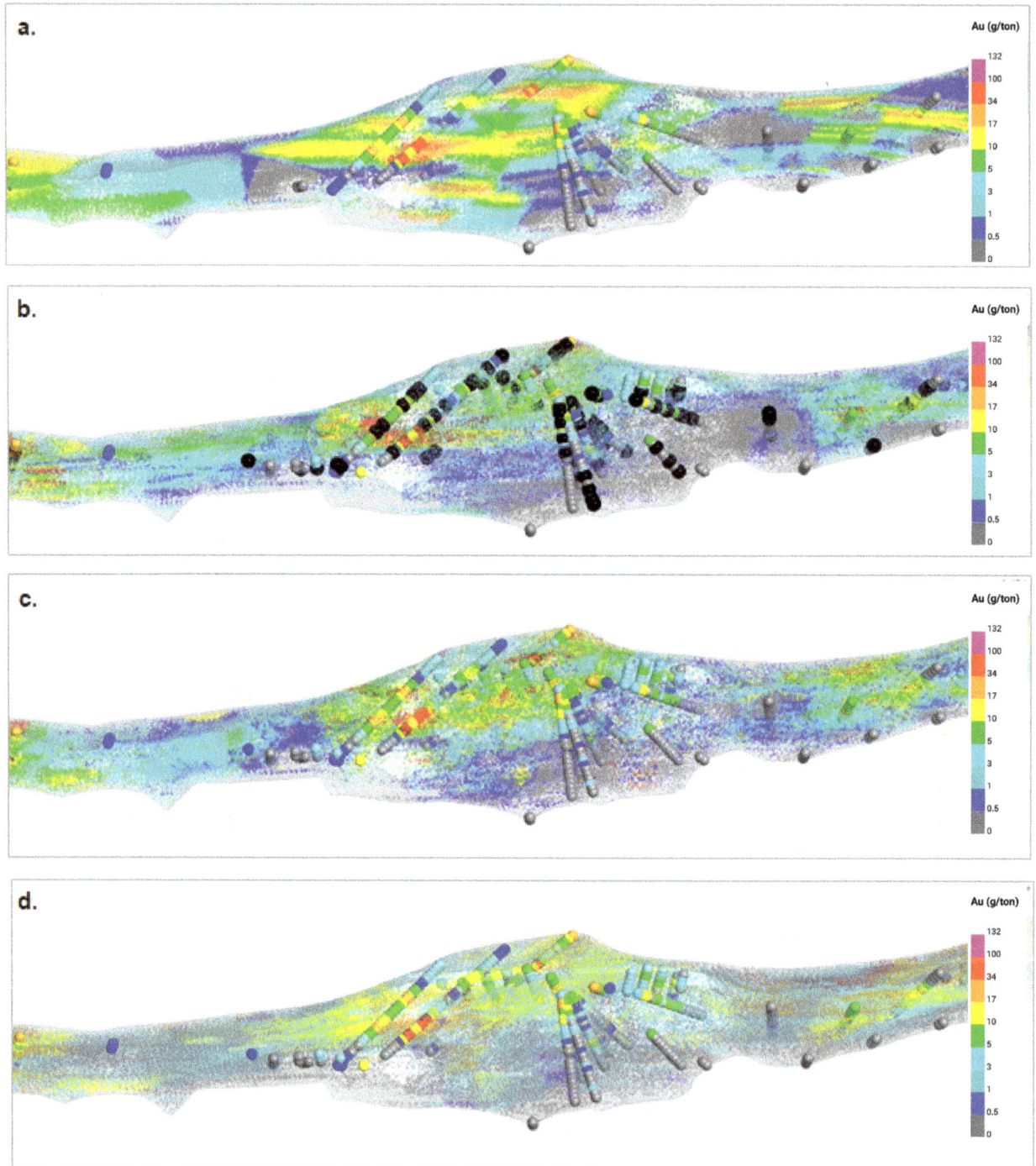

FIG 6 – Cross-section in SE: (a) ML-40, (b) ML-80 (black coloured samples are test samples not used in training), (c) ML-100, and (d) CIK model.

80 per cent trained ML model

The 80 per cent trained ML model produced a highly accurate spatial prediction map aligning well with the sample data. In Figure 6b, black-coloured samples represent the test data excluded from the training process.

100 per cent trained ML model

The 100 per cent trained ML model improved precision, achieving enhanced model performance by including all training samples.

40 per cent trained ML model

The 40 per cent trained model, which excluded face channel samples, successfully identified the primary direction of mineralisation. Despite training on a smaller data set, the model revealed similar high-grade enrichment patterns as models trained on larger data sets. Notably, excluding face channel samples resulted in greater spatial continuity of high-grade areas.

A twinning analysis of 37 pairs of drill core and face channel samples located within 1 m showed that face channel samples slightly underestimated gold grades compared to nearby drill core samples (Figure 7). This analysis offers insights into the spatial variability of mineralisation and highlights potential biases arising from different sampling methods. However, it is beyond the scope of this study to determine whether this difference reflects an actual bias inherent to the face channel sampling method or is related to nugget effects in the deposit.

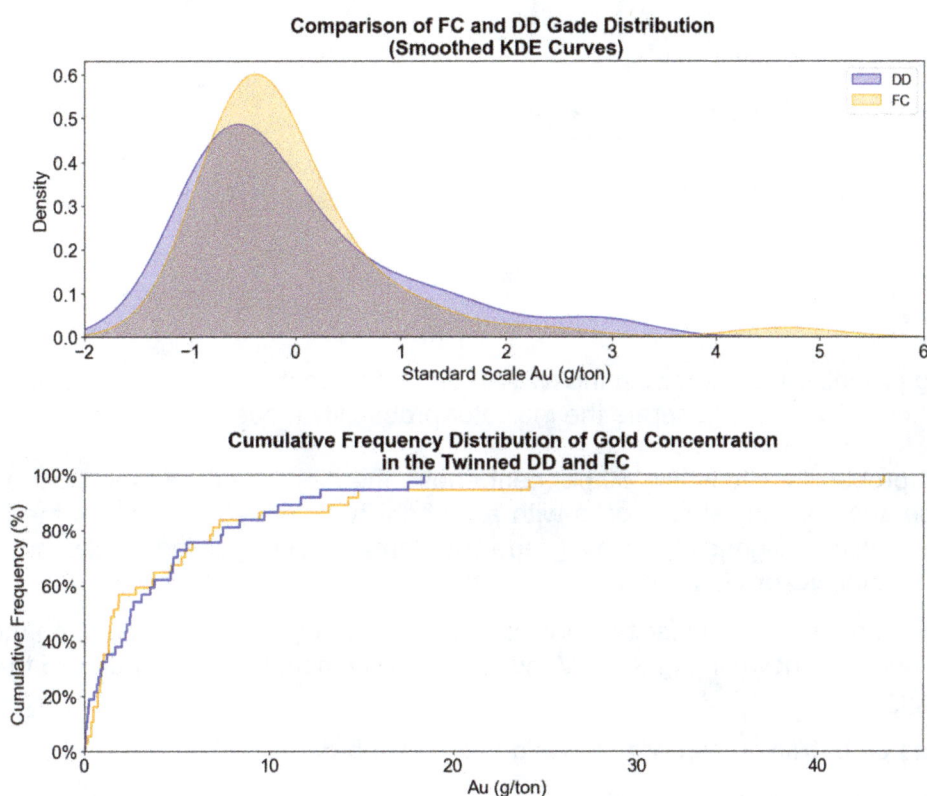

FIG 7 – Twinning analysis of diamond core and face chips within one metre proximity.

While the 40 per cent trained model is less detailed than those trained on 80 per cent and 100 per cent ML, it is a reliable interim model for drilling targeting and planning.

CIK MODEL

A categorical indicator model was applied to Lode 1100, to address the multi-modal grade populations and the complexity of controls on high-grade mineralisation emplacement. Grade cut-off-based indicator estimation was used to replicate the different gold populations within the lode.

Two grade thresholds were applied to define three distinct grade populations within Lode 1100, with an indicator probability model generated for each grade threshold. These thresholds were used to classify the lodes into three sub-domains, High-Grade (HG), Medium-Grade (MG) and Low-Grade (LG), representing the varying grade populations (Figure 8).

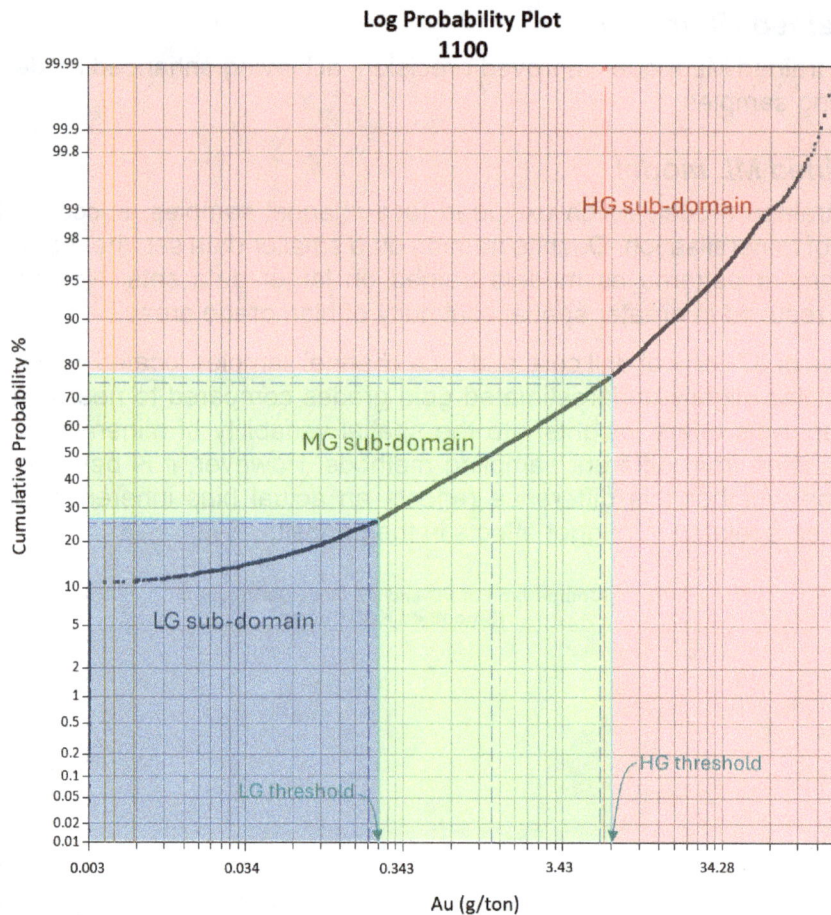

FIG 8 – Log probability plot showing the two Au grade thresholds, 0.24 g/t and 6.85 g/t, used to generate the indicator probability model.

Blocks with a probability exceeding 60 per cent above the high-grade threshold were assigned to the high-grade sub-domain, while blocks with a probability below 40 per cent of meeting the low-grade threshold were assigned to the low-grade sub-domain. All remaining blocks that did not meet the criteria for either HG or LG were allocated to the medium-grade sub-domain.

Despite its utility, the CIK model faces significant challenges in accurately capturing the underlying sample population as shown in Figure 9. Various sensitivity models were conducted to optimise the model, including:

- Adjusting probability thresholds for assignment of sub-domains.

- Evaluating two grade bins indicator models.

- Exploring soft boundary conditions using three bins.

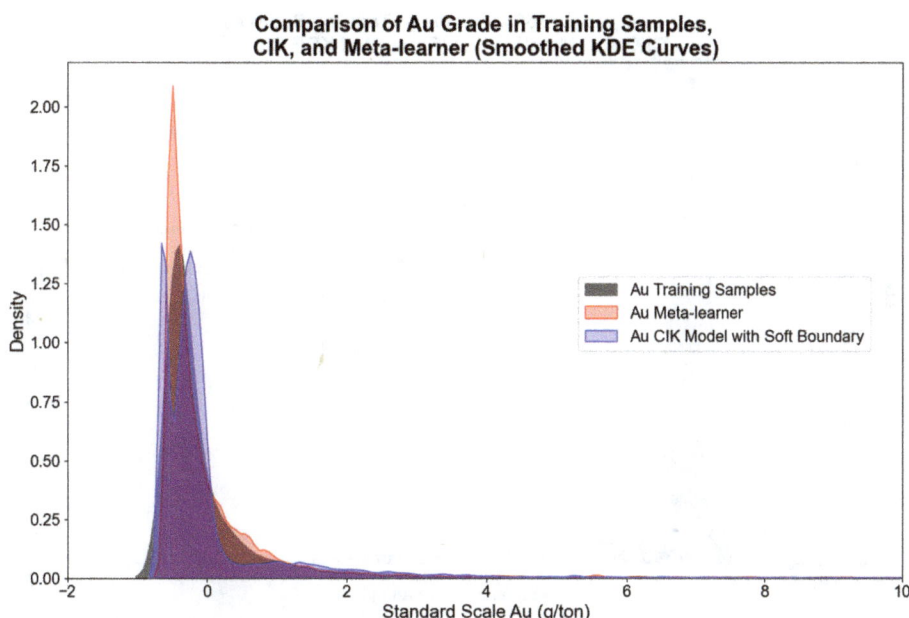

FIG 9 – Kernel Density Estimate (KDE) plot comparing the distribution of Au values across the composite samples, the ML model, and the CIK with soft boundary.

The results indicated that a hard boundary CIK model with three grade bins provided the most representative output among all tested configurations. This is the CIK model used as benchmark for comparison with the ML model.

MODEL VALIDATION

The best model selected from Experiments 1 and 2 was compared to the CIK model using the same evaluation metrics outlined in the previous section. Additionally, the strengths and limitations of both models are discussed to provide further insights into their predictive capabilities.

Global statistics

As shown in Table 7 both the CIK and Meta-learner models have slightly lower means compared to the training data set, which is considered the true value. This phenomenon is common in estimation. However, the Meta-learner exhibits a wider predicted grade range than the CIK model. Figure 10 illustrates that the CIK model has a significant gap above its mean, indicating that it does not accurately represent the higher-grade population within the samples.

TABLE 7

Summary statistics for training data, CIK model, and meta-learner model.

	TRUTH	CIK	Meta-learner 100% training data
Number of samples	5378	5378	5378
Mean	6.72	5.48	5.80
SD	13.90	8.79	9.86
CV	2.07	1.61	1.70
Minimum	0.00	0.00	0.11
25th percentile	0.38	0.18	1.04
Median	1.60	1.88	2.57
75th percentile	6.68	3.03	6.51
Maximum	130.63	62.76	130.63

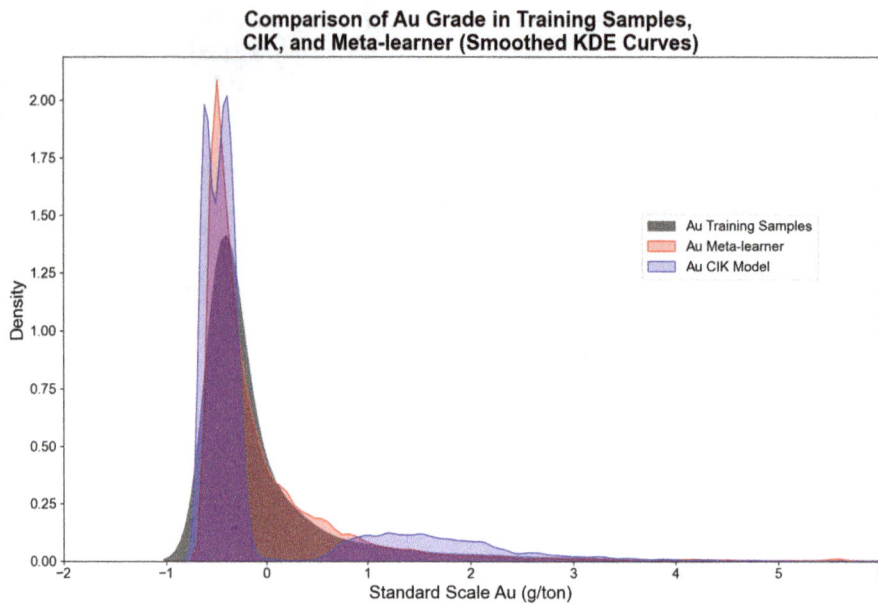

FIG 10 – Kernel Density Estimate (KDE) plot comparing the distribution of Au values across three data sets.

The CIK model has a smaller standard deviation (SD) and coefficient of variation (CV), resulting in predicted grades that are more closely clustered together. However, these predictions are biased, either too low or too high. In contrast, the Meta-learner model more effectively captures the upper-middle values present in the training data.

ML residual analysis

Residuals, calculated as the difference between predicted and actual Au values, are most effectively visualised using a scatter plot and histograms. As shown in Figure 11, the residuals of the meta-learner model are randomly distributed around zero, indicating a good model fit and homoscedasticity, with no evident bias across different grade ranges.

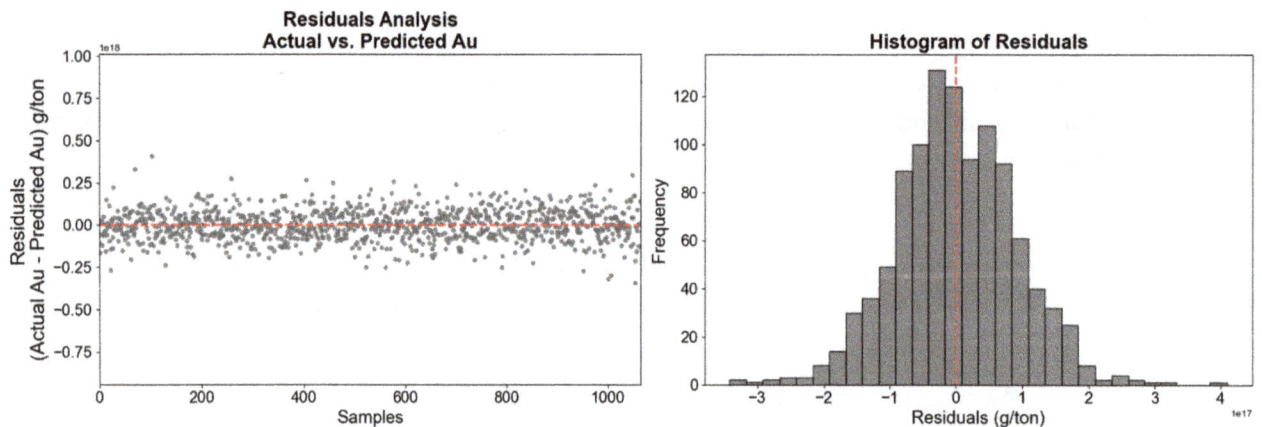

FIG 11 – Residuals analysis for Meta-Learner Model trained on 80 per cent data: actual versus predicted Au.

The histogram of residuals suggests a normal distribution, supporting the assumption of random prediction errors. Overall, the residual analysis reveals no significant biases, demonstrating the model's ability to generalise to new data and its effectiveness in making predictions.

Local scale performance

Examining the ML model's performance compared to the CIK at a local scale is important for understanding how well the model captures regional variations. For this, actual samples within 1 m of the predicted grade are paired and analysed.

The CIK model, as shown in Figure 12, provides limited estimates typically ranging from 3.5 to 10 g/t. It often misclassifies grades by either underestimating or overestimating them. In contrast, the Meta-learner model demonstrates improved accuracy in predicting actual sample values at specific locations. It effectively manages data variance, making it a more reliable tool for estimating gold (Au) values at a local scale.

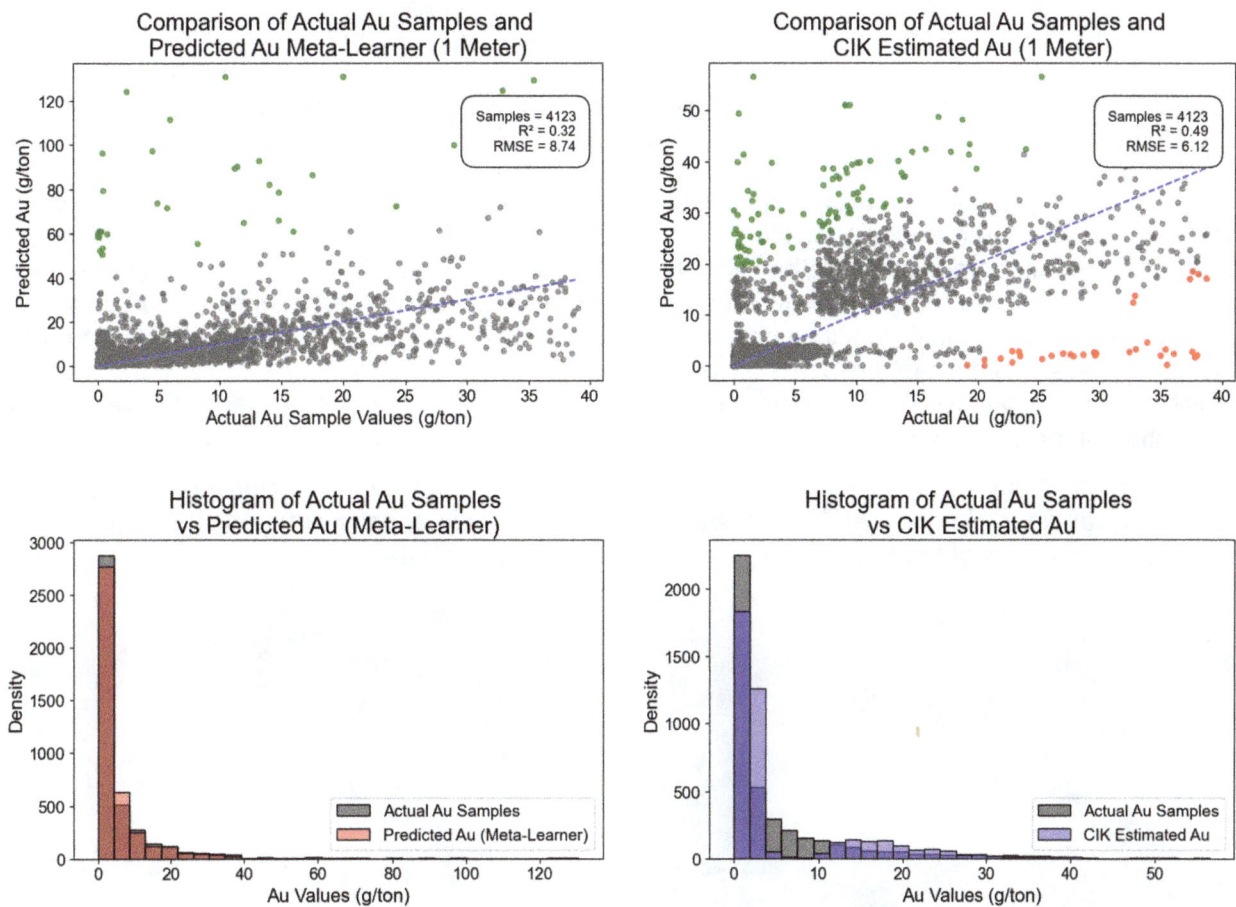

FIG 12 – Scatter plot comparing actual gold values to predicted values from the CIK and Meta-Learner models for paired samples within one metre (top). Histograms showing the distribution of actual gold values versus predicted values for both models (bottom).

Spatial grids

The prediction maps in Figure 13 for both the CIK and Meta-learner models capture the main trend of mineralisation as modelled in the variogram and Figure 3. Both models show no artifacts, but the Meta-learner model identifies secondary trends that reveal geological complexities not captured by the CIK model.

Consistent with statistical validation discussed previously, the CIK model tends to underestimate gold grades within the middle bin (5–10 g/t range), resulting in consistently low predictions (3–5 g/t), as illustrated in the spatial map. In contrast, the Meta-learner model better captures the upper-middle grade range and displays a smoother transition in grade values across the study area.

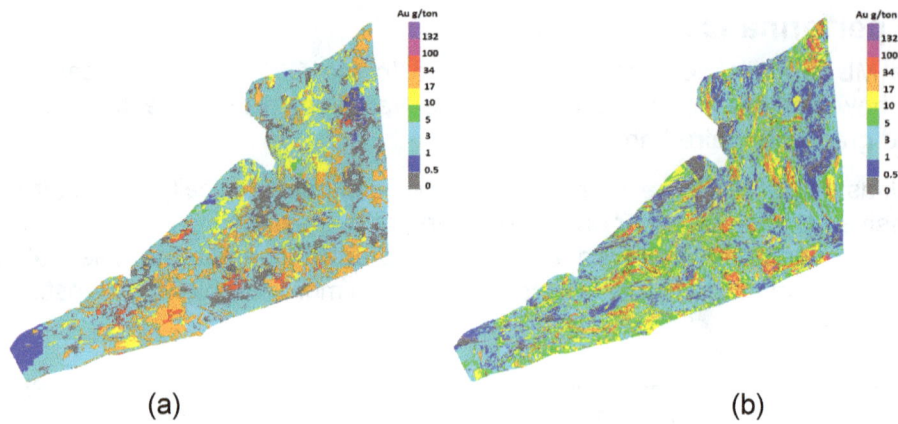

FIG 13 – Prediction maps of gold grade in plan view for (a) CIK model; and (b) Meta-learner model trained on 100 per cent of the data.

The CIK model maintains a narrower grade range (17–34 g/t) within the higher-grade zones and has difficulty accurately estimating the volume of bins exceeding 34 g/t. Efforts to adjust the probability thresholds and utilise a two-bin method with soft boundary conditions for the middle and high-grade bins did not lead to significant improvements. This limitation is due to the Ordinary Kriging method utilised by the CIK model, which tends to predict values close to the average of sub-domains, restricting its ability to capture complex grade distributions. In contrast, the Meta-learner model provides a more nuanced understanding of grade variability and best represents the underlying population of the training samples.

In the cross-sections shown in Figure 14 to Figure 16, the ML model captures more complex patterns with finer granularity. In contrast, both models predict high-grade enrichment along the hanging wall and footwall margins of the thick mineralised vein, consistent with the vein's geology.

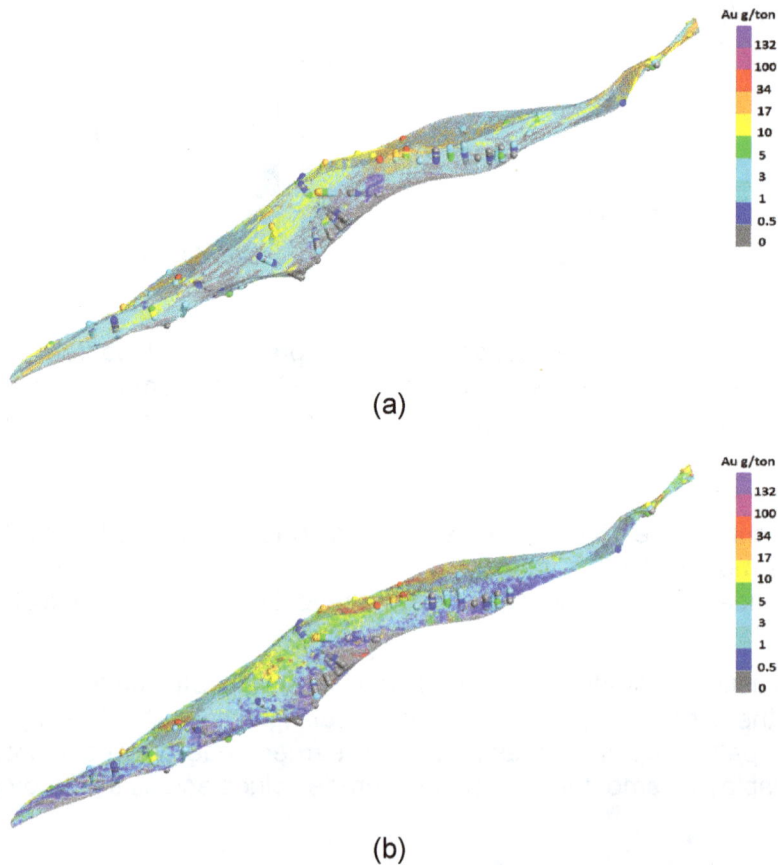

(a)

(b)

FIG 14 – Prediction maps of gold grade in cross-section view looking NE for (a) CIK model; and (b) Meta-learner model trained on 100 per cent of the data.

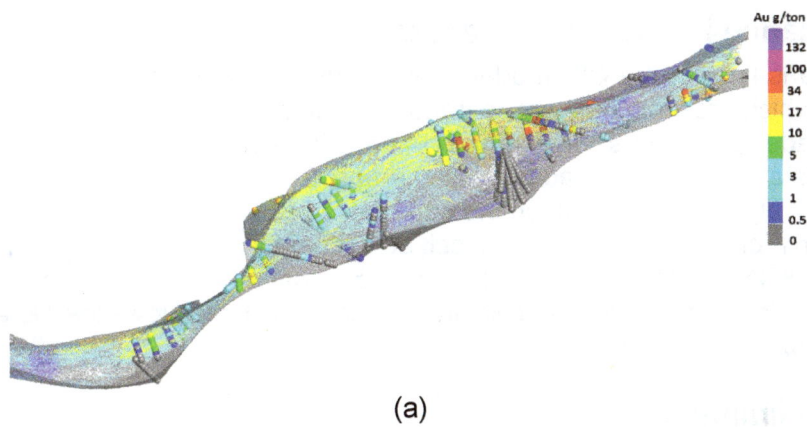

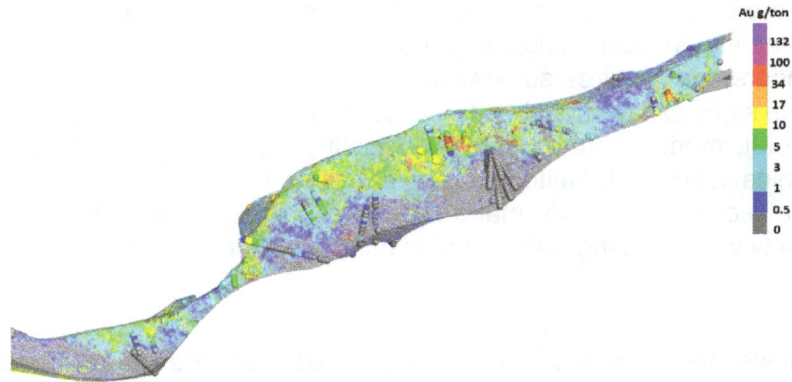

FIG 15 – Prediction maps of gold grade in cross-section view looking NW for (a) CIK model; and (b) Meta-learner model trained on 100 per cent of the data.

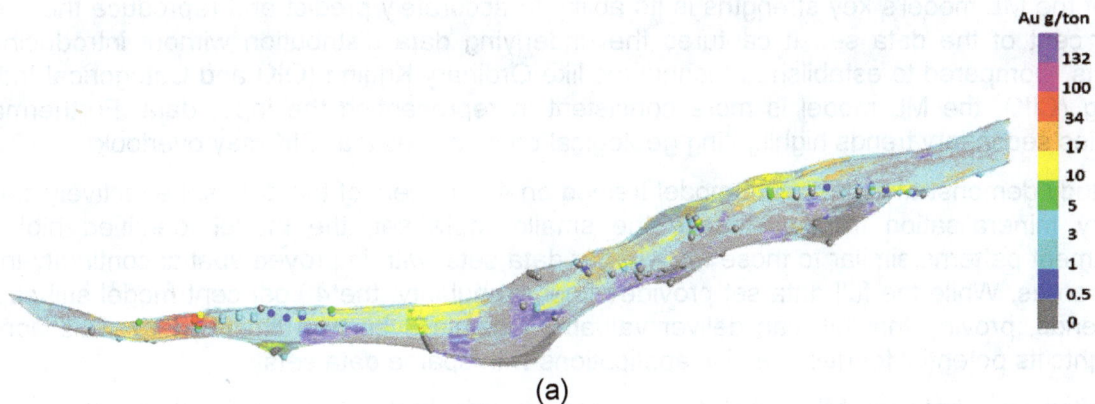

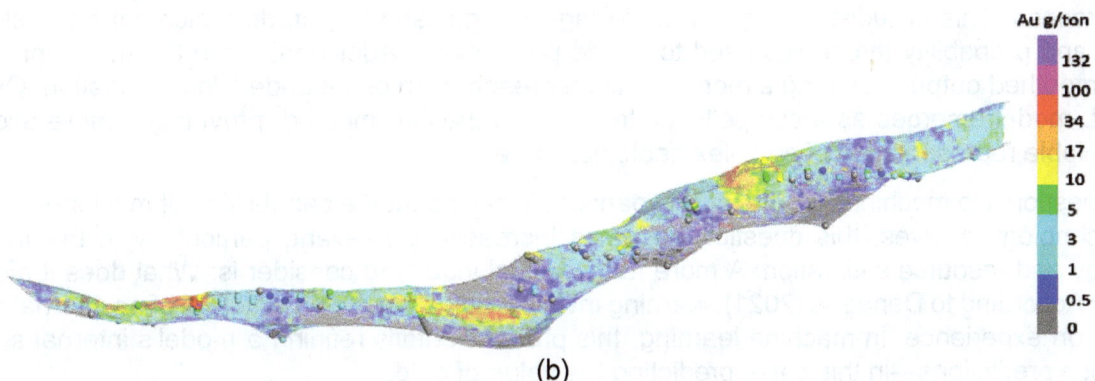

FIG 16 – Prediction maps of gold grade in cross-section view looking NW for (a) CIK model; and (b) Meta-learner model trained on 100 per cent of the data.

Model performance in sparse data areas

When drilling data is limited, the CIK model often maintains spatial continuity along the preferred plunge direction, which preserves the grade trend, even in areas with sparse data. This can lead to oversimplified grade predictions. In contrast, the Meta-learner model is more effective at accommodating data variations in sparsely drilled areas. By combining multiple base learners, each leveraging different algorithms, the ensemble captures a broader range of patterns in the data. For example, Random Forest and XGBoost excel at capturing nonlinear relationships and feature interactions, while KNN is well-suited for identifying local patterns, and SVM is effective at defining boundaries in high-dimensional spaces. This diversity in algorithms allows the Meta-learner to adapt to sparse data sets.

Computational limitations

A key limitation of the CIK process is its high computational demand. Depending on data set size and software implementation, estimating categories into subblocks using alternative software for CIK can take several hours and requires substantially greater storage space than the ML model. This long processing time significantly impacts efficiency, especially when dealing with large data sets. In contrast, while the ML model requires extensive testing and tuning, once established, it runs in approximately 30 mins, taking only limited storage capacity. The ML model's faster computation time and lower resource requirements may make it a more practical choice for large-scale geological modelling, particularly when dealing with complex mineralisation patterns.

DISCUSSIONS

This case study shows that an ML model can efficiently estimate the spatial characteristics of multi-phase fluid shear-hosted vein mineralisation, proving to be a strong alternative to traditional methods. The ML model performed exceptionally well across all evaluation metrics, including root mean squared error (RMSE), coefficient of determination (R^2), descriptive statistics, probability density plots, grade bin volumes, spatial artifact evaluation, and geological explainability.

One of the ML model's key strengths is its ability to accurately predict and reproduce the withheld 20 per cent of the data set. It captures the underlying data distribution without introducing any artifacts. Compared to established techniques like Ordinary Kriging (OK) and Categorical Indicator Kriging (CIK), the ML model is more consistent in representing the input data. Furthermore, it identifies secondary trends highlighting geological complexities that CIK may overlook.

The study demonstrates that a ML model trained on 40 per cent of the data set effectively captures primary mineralisation trends. Despite the smaller data set, the model identified high-grade enrichment patterns similar to those from larger data sets, with improved spatial continuity in high-grade areas. While the full data set provides finer granularity, the 40 per cent model still captured key trends, proving that ML can deliver valuable insights even with limited data. This approach highlights its potential for geoscientific applications with sparse data sets.

The differences between ML and CIK outcomes are largely due to the subjectivity inherent in the CIK process. This includes variogram modelling for high- and low-grade indicators and selecting grade and probability thresholds used to define populations. Additionally, the ML model produces less smoothed outputs, offering a more precise representation of the underlying population. Overall, the ML model emerges as a compelling alternative to the CIK method, providing a more accurate and reliable representation of complex geological data.

The question 'Do machines learn?' often sparks curiosity about the capabilities of machine learning. As technology evolves, this question becomes increasingly relevant, particularly in the fields of geology and resource estimation. A more fundamental inquiry to consider is: What does it mean to learn? According to Dehaene (2021), learning involves adjusting the parameters of an internal model based on experience. In machine learning, this process entails refining a model's internal state to enhance predictions—in this case, predicting the value of gold.

The machine learning model has learned to explore a vast search space using combinatorial combinations during hyperparameter tuning. It also minimises prediction errors during the training process. This learning is further enhanced by employing multiple base learners, introducing randomness and diversity into the model. Additionally, incorporating varied amounts of training data sets adds another layer of exploration, enabling the model to establish robust connections across different scenarios and learn from a wider range of data.

CONCLUSIONS

This case study demonstrates that the ML model effectively estimates the spatial characteristics of multiple-fluid-phase vein mineralisation at the Pogo Mine. Unlike traditional methods, the ML model achieves accurate results without relying on assumptions of stationarity or linearity. It emerges as a compelling alternative to the Categorical Indicator Kriging (CIK) technique, overcoming the inherent subjectivity associated with the CIK model.

The ML model provides a more accurate representation of the input data distribution. This case study's design and feature engineering, tailored for a highly complex and extensive mineralised vein gold deposit, successfully captured the underlying sample population. Importantly, the spatial grids generated by the ML model are devoid of grade artifacts, and the primary plunge of mineralisation is accurately represented.

This case study overcomes the limitations of previous research, which often faced computational inefficiencies and artifacts. It introduces innovative feature engineering by utilising polynomial coordinates and a fixed reference point as input features, providing a robust solution for complex geological modelling. The ML model's adaptability is particularly valuable for data sets that only include coordinates and gold values, where traditional variogram modelling proves impractical.

Future research will broaden the training data set to include a larger mining area, extending beyond specific geological domains. This approach aims to tackle challenges in complex geological settings where it is difficult to parameterise the Categorical Indicator Kriging (CIK) model. It offers a practical, efficient, and reliable alternative for geostatistical modelling.

ACKNOWLEDGEMENTS

The authors thank Northern Star Resources Ltd. for granting access to the Pogo database.

REFERENCES

Abadi, M, Barham, P, Chen, J, Chen, Z, Davis, A, Dean, J, Devin, M, Ghemawat, S, Irving, G, Isard, M, Kudlur, M, Levenberg, J, Monga, R, Moore, S, Murray, D, Steiner, B, Tucker, P, Vasudevan, V, Warden, P, Wicke, M, Yu, Y and Zheng, X, 2016. TensorFlow: A System for Large-Scale Machine Learning, in *Proceedings of the 12th USENIX Symposium on Operating Systems Design and Implementation (OSDI)*, pp 265–283.

Ahn, S, Ryu, D-W and Lee, S, 2020. A machine learning-based approach for spatial estimation using the spatial features of coordinate information, *International Journal of Geo-Information*, 18.

Breiman, L, 2001. *Random Forests, Machine Learning*, 45(1):5–32.

Chen, T and Guestrin, C, 2016. XGBoost: A Scalable Tree Boosting System, in *Proceedings of the 22nd ACM SIGKDD International Conference on Knowledge Discovery and Data Mining*, pp 785–794.

Chollet, F, 2015. Keras, GitHub repository. Available from: <https://github.com/keras-team/keras> [Accessed: 3 January 2025].

Cortes, C and Vapnik, V, 1995. Support-Vector Networks, *Machine Learning*, 20(3):273–297.

Cover, T M and Hart, P E, 1967. Nearest Neighbor Pattern Classification, *IEEE Transactions on Information Theory*, 13(1):21–27.

Dehaene, S, 2021. How We Learn: Why Brains Learn Better Than Any Machine. for Now, Viking.

Deutsch, C V and Journel, A G, 1998. *GSLIB: Geostatistical software library and user's guide*, 2nd ed (New York, NY: Oxford University Press).

Erten, G E, Deutsch, C V and Yavuz, M, 2020. Managing estimation artifacts in machine learning spatial estimation, CCG Annual Report 22, Paper 105.

Freund, Y and Schapire, R E, 1996. Experiments with a New Boosting Algorithm, in *Proceedings of the 13th International Conference on Machine Learning (ICML)*, pp 148–156.

Friedman, J H, 1999. Greedy Function Approximation: A Gradient Boosting Machine, in *Proceedings of the 13th International Conference on Neural Information Processing Systems (NIPS)*, pp 118–126.

Hengl, T, Heuvelink, G B M, Kempen, B, Leenaars, J G B, Walsh, M G, Shepherd, K D, Sila, A, MacMillan, R A, de Jesus, J M, Tamene, L and Tondoh, J E, 2015. Mapping soil properties of Africa at 250 m resolution: Random forests significantly improve current predictions, *PLoS ONE*, 10:e0125814. https://doi.org/10.1371/journal.pone.0125814

Hinton, G E, LeCun, Y and Bengio, Y, 1986. Learning Representations by Backpropagating Errors, *Nature*, 3236088, pp 533–536.

Pedregosa, F, Varoquaux, G, Gramfort, A, Michel, V, Thirion, B, Grisel, O, Blondel, M, Prettenhofer, P, Weiss, R, Dubourg, V, Vanderplas, J, Passos, A, Cournapeau, D, Brucher, M, Perrot, M and Duchesnay, É, 2011. Scikit-learn: Machine Learning in Python, *Journal of Machine Learning Research*, 12:2825–2830.

Pyrcz, M J and Deutsch, C V, 2018. Transforming Data to a Gaussian Distribution, GeostatisticalLessons.com.

Quinlan, J R, 1986. Induction of Decision Trees, *Machine Learning*, 1(1):81–106.

Singh, D and Singh, B, 2020. *Investigating the Impact of Data Normalization on Classification Performance* (Elsevier Applied Soft Computing).

Estimating recoverable resources – is it still hopeless?

O Rondón[1] and W Assibey-Bonsu[2]

1. Principal Geostatistician, Datamine, Perth WA 6000.
   Email: oscar.rondon@dataminesoftware.com
2. Principal: Geostatistics and Assurance, Gold Fields, Perth WA 6000.
   Email: winfred.assibeybonsu@goldfields.com

ABSTRACT

The mining industry relies on recoverable resources as a key component for resource evaluation, mine planning, and economic decision-making. The ability to estimate recoverable resources has profound implications for project feasibility, financial forecasts, and profitability. From global to local techniques, the evolution of methodologies for estimating recoverable resources reflects the industry's need for practical and effective approaches to assess recoverable resources in mining operations. This paper provides an overview of these methodologies, exploring their underlying theories, principles, and applications. Case studies from the geostatistics literature that demonstrate the advantages and limitations of various recoverable resource techniques are discussed.

By bridging the past and present, this paper offers an assessment of recoverable resource estimation techniques, providing insights into their evolution and their impact on the mining industry's ability to evaluate resources with confidence and required precision.

INTRODUCTION

The mining industry requires very capital-intensive investments. For example, recent acquisitions of the Newcrest assets by Newmont (Businesswire, 2023) cost US$16.8 billion, and BHP's partnership with Lundin Mining in the acquisition of Filo Corp., a South American copper miner, cost approximately US$3.0 billion (BHP, 2025). In 2007, Rio Tinto's acquisition of Alcan, a Canadian aluminium company, cost US$38.1 billion (Financial Review, 2007). Gold Fields Limited acquired the South Deep Gold Mine in South Africa at a cost of US$2.5 billion (Reed, 2006).

Mineral Resources and reserves are the fundamental assets of mining companies, serving as the basis for significant capital investments. The strategic objective is to explore, acquire, develop, and mine these assets. However, a critical risk lies in the uncertainty inherent in resource and reserve estimation. Inaccurate estimation can lead to billions of dollars in financial losses if the expected resources and reserves are later found to have been inefficiently estimated or valued. Effective evaluation of mineral resource assets requires key foundational elements, including data integrity, an appropriate geological model, and optimal geostatistical techniques. Geostatistics provides technical solutions to mitigate in part both technical and financial risks during the evaluation of these fundamental assets.

Geostatistical techniques for recoverable resource estimation involve assessing the tonnage, grade, and metal content above a relevant economic cut-off that could be extracted at the time of mining. This process is more than a technical exercise; it directly influences critical aspects such as effective mine planning, risk assessment, informed decision-making, financial forecasting, and investment viability. Unfortunately, over the past 30 years, geostatistical advances addressing the critical problem of estimating recoverable resources have been limited. While the theoretical foundations laid decades ago remain valuable, few transformative innovations have emerged to address the evolving challenges in this domain. This lack of progress can be attributed to several factors. Firstly, the inherent complexities of geological variability and mining selectivity continue to challenge existing change of support models, requiring not only refinements to current techniques, but also the development of entirely new methodologies. Secondly, many advances in geostatistics, though commendable, have concentrated on areas such as machine learning and big data integration, often neglecting the specialised needs of recoverable resource estimation techniques. Lastly, while the adoption of conditional simulation has been significant, its potential has often not been fully realised due to practical implementation constraints and a lack of widespread professional training.

Building on the foundational contribution of Ross (1950), Krige (1951), Matheron (1963, 1965), Rossi and Parker (1994), this paper explores the theories, principles, and practices underlying the assessment of recoverable resources. It examines the evolution of techniques, from traditional methods to contemporary approaches, such as conditional simulation tools. By presenting practical examples, this paper demonstrates how these methodologies address the challenges of the recoverable resource estimation problem and underscores the critical importance it has in modern mining.

THE PROBLEM OF ESTIMATING RECOVERABLE RESOURCES

During the early stages of a mining project, drill hole data are often on relatively large grids, which are adequate for estimating tonnage and grade using large blocks or panels whose dimensions reasonably conform to the drill hole spacing. However, this spacing is inadequate for estimating recoverable resources, as the estimation of smaller blocks at the scale required for future selective mining will be oversmoothed. As a result, such estimates fail to provide an efficient and unbiased assessment of the corresponding tonnage and grade. In particular, the average grade above cut-off is consistently understated. This is illustrated in Case Study 1.

Recoverable resource estimation techniques address the inevitable smoothing effect problem by deriving the grade distribution of selective mining units (SMUs) required for efficient mine planning and financial forecasts at the early stages of mining projects. This approach enables the estimation of the expected grade-tonnage curve at the required scale for respective cut-offs. The primary data available for deriving the block grade distribution is the composited drill hole data, also known as point support data. This presents a challenge because the distribution of actual selective mining units (SMUs) is largely unknown when starting from the point support distribution. What is known, however, is that the mean of the block distribution matches that of the point support distribution, and that the variance of the block distribution is lower due to the reduced grade variability that occurs when transitioning from smaller to larger supports. These known statistical relationships are insufficient to fully characterise the block or SMU distribution. Even if the point support distribution is assumed to be known, determining the corresponding block support distribution remains a challenge (Lantuejoul, 1988).

The block grade distribution can be determined at two levels: global and local. A global block distribution represents the grade of blocks across the entire deposit or domain, but does not consider grade variations within smaller areas or individual block grades. In contrast, a local block distribution focuses on grades within specific areas, such as SMUs within a larger panel, providing more detailed information for selective mining.

The choice between global and local approaches depends on the scale of decision-making. Global distributions are typically used for high-level global recoverable resource assessments, also for global project resource model validations against the corresponding global estimates derived from aggregated local distribution/model estimates (when local estimates are available). Local distributions are useful for long-term planning, and, in particular, for detailed mine design and short-term production forecasting, also for equipment selection.

Several techniques can be employed to derive these distributions. Global block distributions are often obtained through methods like histogram transformation or global change of support models, which translate point support data (eg composited drill hole grades) to block support. Local block distributions, on the other hand, rely on spatially-informed methods such as ordinary kriging, simple kriging, or conditional simulation, which incorporate spatial variability and account for local grade patterns.

A critical step in deriving any block or SMU grade distribution is determining its variance based on composited drill hole data. This process relies on the additivity property of dispersion variances, commonly referred to as Krige's relationship (Krige, 1951).

The following sections present Krige's relationship and explore the principles of global and local block distributions, the techniques thereof, and their respective roles in estimating recoverable resources.

Recoverable resources – Krige's relationship

The concept of *support* and its corresponding variances, which is fundamental to geostatistics, was first covered by Ross (1950) and further developed empirically by Krige (1951, 1960, 1966), and Krige and Ueckermann (1963) including Krige's variance-size of area relationship. These variances, computed for different block sizes (or supports) and often referred to as dispersion variances in the geostatistical literature, are closely linked to the point support variogram model (Journel and Huijbregts, 1978). This relationship allows the total variance σ^2 of the point support data to be decomposed into the sum of the variance of the block grade distribution σ_{block}^2 and the variance of points inside the block $\bar{\gamma}_{block}$. from which the desired block variance is obtained.

$$\sigma_{block}^2 = \sigma^2 - \bar{\gamma}_{block}$$

The term $\bar{\gamma}_{block}$ is computed using a discretisation of the block based on the point variogram. Krige's relationship highlights an important property, that block average grades always have less variability than drill hole data.

It is important to note that the calculation of the block variance is significantly influenced by the variogram model parameters. An incorrect specification of any of these parameters can result in a block variance estimate that does not accurately capture the underlying variability of block distributions in the orebody. For example, if the variogram range is small compared to the block size, the points within the block will appear less correlated, leading to higher values of $\bar{\gamma}_{block}$ and correspondingly lower block variances. Similarly, a large nugget effect relative to the total variogram sill increases $\bar{\gamma}_{block}$, reducing the block variance. Additionally, setting the variogram sill either higher or lower than the point support variance will respectively overstate or understate the block variance. During production, follow-up (reconciliation) should always be carried out to validate the geostatistics parameters (including block variances) and the recoverable resource estimates.

Recoverable resources – global techniques

Several techniques have been developed to derive global block distributions and estimate recoverable resources, each based on distinct theoretical frameworks. This session focuses on the methods commonly used in practice or implemented in commercial software, including the affine correction, the lognormal and indirect lognormal corrections, and the discrete Gaussian method. Chilès and Delfiner (2012) provide a comprehensive review of these and other techniques.

In the following sections, $Z(x)$ represents the composited point support data with a mean value of m, $Z(v)$ denotes the block support (SMU) data, and $Z(V)$ represents the panel support data.

Affine correction

The affine correction relies on the assumption of permanence of distribution, meaning that the point-support and block-support distributions have the same shape, differing only in variance. This assumption holds reasonably well when the difference in support is small, but becomes invalid as the support increases, as the block-support distribution tends to become more symmetric.

The correction applies an affine transformation to the point support distribution. Specifically, it assumes that $Z(v)$ and $\sqrt{f}(Z(x) - m) + m$ have the same distribution. Here, f is the factor that ensures the block distribution has the expected block variance σ_{block}^2. It is defined as $f = \sigma_{block}^2/\sigma^2$ and is commonly referred to as the variance correction factor in the geostatistical literature. Journel and Huijbregts (1978) suggested that the affine correction is appropriate when $f > 0.70$. This threshold implies that the difference in variance between the block and point-support distributions is small enough for the affine transformation to be consistent with the permanence of distribution assumption. Additionally, the proportion of blocks above a cut-off z is restricted to cut-off values z greater than $m(1 - \sqrt{f})$.

Lognormal correction

The lognormal correction is designed for point-support data that follows a lognormal distribution, assuming that the block-support variable also follows a lognormal distribution with the same mean but a reduced variance. This approach enables the explicit derivation of the block-support distribution

(Journel and Huijbregts, 1978; Isaaks and Srivastava, 1989). The distribution of $Z(v)$ is equal to the distribution of $aZ(x)^b$ where a and b are functions of the variance correction factor f, the mean value m, and the coefficient of variation $CV = \sigma/m$ (Isaaks and Srivastava, 1989).

The lognormal correction has significant limitations, primarily because it relies heavily on the assumption that the underlying data follows a lognormal distribution. In practice, many data sets deviate from this strict lognormality, even in cases of highly positively-skewed distributions. Deviation from the lognormal distribution is likely to affect the reproduction of the point-support mean value m. Moreover, from a theoretical standpoint, block grades cannot strictly maintain a lognormal distribution.

Indirect lognormal correction

The indirect lognormal correction (Isaaks and Srivastava, 1989) is a two-step process. In the first step, a lognormal correction is applied, even if the point-support distribution is not strictly lognormal. Since this step can alter the mean value m, the second step involves an additional rescaling to ensure that both the point-support and block-support distributions have the same mean value m. As a result, the distribution of $Z(v)$ is equal to the distribution of $cZ'(v)$ where $Z'(v)$ is the distribution obtained after the lognormal correction with $c = m'/m$. Here m' is the mean value of $Z'(v)$.

This rescaling also affects the reproduction of the target block variance σ^2_{block}. Isaaks and Srivastava (1989) argue that, in practice, since the block mean is typically better known than the variance, prioritising an exact match for the mean is considered more important. However, Emery (2004) proposed a distinct approach to calculating the parameter b using explicit analytical equations, allowing for the simultaneous reproduction of both the prescribed mean and block-support variance.

Discrete Gaussian method

The discrete Gaussian model (DGM), introduced by Matheron (1976a), assumes that $Z(x)$ and $Z(v)$ can be expressed as functions of two standard Gaussian variables $Y(x)$ and Y_v. These functions are given by $Z(x) = \Phi(Y(x))$ and $Z(v) = \Phi_v(Y_v)$ with Φ and Φ_v being the point support and block support anamorphosis functions, respectively.

The point support anamorphosis is derived from the drill hole data using a Hermite polynomial expansion. This expansion enables the explicit computation of the block-support anamorphosis, Φ_v as:

$$\Phi_v(y) = \sum_{n \geq 0} \phi_n r^n H_n(y)$$

with H_n the Hermite polynomials, ϕ_n the corresponding Hermite coefficients (Rivoirard, 1994) and r being a coefficient that accounts for the reduction in variance of the block support distribution.

The coefficient r is specifically determined to ensure that the anamorphosis function defining the block-support grades reproduces the expected variance σ^2_{block}, satisfying Krige's relationship:

$$\sigma^2_{block} = \sigma^2 - \bar{\gamma}_{block} = \sum_{n \geq 1} \phi_n^2 r^{2n}$$

The DGM offers several advantages over traditional methods. It provides a mathematically rigorous framework for deriving the block-support distribution and accounts for the block symmetrisation effect, explicitly modelling the tendency of the block-support distribution to become more symmetric as the block size increases (Chilès and Delfiner, 2012). Furthermore, the DGM can easily incorporate the information effect, that is the consideration that block selection above a cut-off is based on estimates rather than actual block grades when calculating tonnage, grade, and metal content at the block-support level (Roth and Deraisme, 2001).

All the previously discussed methods are affected by the zero effect, defined as the inability of a transformation to adequately handle values at or near zero in a data set. However, by modelling the Gaussian anamorphosis using the approach proposed by Deraisme and Rivoirard (2009), the Discrete Gaussian Model (DGM) can be effectively utilised.

Recoverable resources – local techniques

A common approach across various techniques for estimating the distribution of blocks or SMUs within a larger panel is to use local conditions to guide the estimation. This allows to account for spatial grade variations within each panel, enabling the localised estimation of the block distributions. In contrast to global techniques, which generate a single grade-tonnage curve for the entire deposit or domain, local techniques produce grade-tonnage curves at block support for each individual panel. This provides more detailed and practical information for mine planning and selective mining.

The local estimation of recoverable resources has long been a challenging problem in geostatistics. Over time, several techniques have been developed to address this challenge, each offering unique strengths and limitations. This section examines five widely used methods for locally estimating recoverable resources: MultiGaussian Kriging (MGK), Multiple Indicator Kriging (MIK), Uniform Conditioning (UC), Direct and Indirect approaches and Conditional Simulation (CS). Disjunctive Kriging (Matheron, 1976b), once regarded as a promising geostatistical method, is rarely utilised nowadays in modern geostatistical practice, due mainly to stationarity issues and will not be included in the description of techniques presented in this section.

MultiGaussian Kriging

MGK relies on a key property of the multiGaussian distribution, which enables the derivation of the conditional distribution at an unknown location using only its simple Kriging estimate and the associated Kriging variance (Verly, 1983). In practice, the drill hole data $Z(x)$ is transformed into Gaussian data $Y(x)$, typically through the anamorphosis function $Z(x) = \Phi(Y(x))$. Following this transformation, it is assumed that values $Y(x)$ have a multiGaussian distribution.

Using this transformed data, a Simple Kriging estimation is performed using a search neighbourhood that collects nearby data around the target location $Y(x_0)$. The Simple Kriging $Y_{SK}^*(x_0)$ and its variance σ_{SK}^2 are then obtained through the kriging formalism.

The probability of $Z(x_0)$ being below a cut-off z, given the nearby data, is given by:

$$P(Z(x_0) \leq z/nearby\ samples\ Z) = P(\Phi(Y(x_0)) \leq y/nearby\ samples\ Y) = G\left(\frac{y - Y_{SK}^*(x_0)}{\sigma_{SK}}\right)$$

where y is the equivalent Gaussian cut-off $z = \Phi(y)$ and G is the Gaussian distribution function. (Chilès and Delfiner, 2012; Verly, 1983).

Despite its simplicity, MGK encountered limitations in practice. Firstly, the block conditional distribution could not be derived analytically and resorting to numerical approaches was required. Secondly, all conditioning is concentrated in the Simple Kriging estimate $Y_{SK}^*(x_0)$, which works under the assumption of a known mean of zero and variance one. Using Ordinary Kriging to re-estimate the mean at each location is not suitable for this framework (Chilès and Delfiner, 2012).

Emery (2005) elegantly addressed these challenges by introducing analytical expressions to derive the conditional distributions for both point and block support and by generalising the approach to account for local stationarity using Ordinary Kriging. Unfortunately, these advancements are not yet readily available in commercial software.

Multiple Indicator Kriging

MIK is a non-parametric geostatistical technique widely used in the mining industry, particularly for estimation in the presence of highly skewed data. MIK partitions the input drill hole data distribution into multiple thresholds using a set of indicators, which are modelled and kriged independently for each panel V (Journel, 1983). Unlike traditional kriging methods, the output of MIK is not a direct grade estimate but an estimate of the point-support distribution or cumulative distribution function (CDF) at each location. This distribution function, informed by the data, enables the estimation of the probability that any point within the panel V has a grade no greater than a specified threshold (Figure 1).

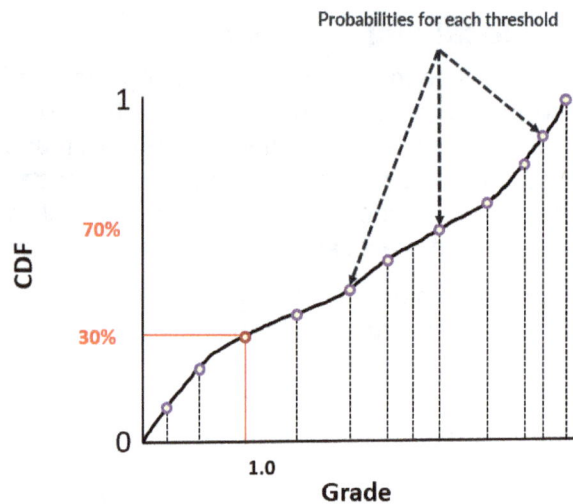

FIG 1 – Example of a local CDF estimation using MIK.

MIK gained widespread adoption in the mining industry due to its early advantages. These included the absence of distributional assumptions and its offering to model different spatial continuity directions for each threshold. However, inconsistencies with the method became evident. Notably, estimating the indicators independently led to inconsistencies in the estimated probabilities, known as order relations problems, where the estimated probabilities fail to maintain logical ordering across thresholds. These inconsistencies must be corrected a posteriori before the distribution function can be used effectively. In practice, order relation problems can be minimised or avoided by using the same variogram model for all indicators or by gradually-changing models across the indicators. However, having a single variogram model contradicts MIK's initial offering of modelling different spatial continuities for each threshold. Despite its complexity and associated challenges, MIK has proven to be a valuable tool in numerous practical applications within the mining industry, particularly in complex deposits or geological settings, where classical techniques yield unsatisfactory results.

MIK can be applied for both grade estimation, using a probability-weighted average known as an E-type estimator, and recoverable resource estimation. A critical step in applying MIK to recoverable resources is deriving the block-support distribution from the calculated point-support distribution (Vann, Guibal and Harley, 2000). The affine correction or indirect lognormal correction methods are typically employed to adjust the point-support distribution using a quantile identification approach on a block-by-block basis. Rossi and Parker (1994) present illustrative examples that demonstrate the process and highlight its practical implications.

An important aspect of this procedure is selecting the appropriate mean for the correction. Rossi and Parker (1994) demonstrated that using the panel E-type estimate can introduce biases, leading to overestimation of grades above the global mean and underestimation below it. Unfortunately, this practice continues to date, more than 30 years after this issue was first identified.

Additionally, Chilès and Delfiner (2012) highlight the potential for inconsistencies between local and global grade-tonnage curves when using MIK. These discrepancies may arise when aggregated local estimates do not fully align with the global distribution.

When compared to MGK, Vincent and Deutsch (2019) report findings that do not indicate any significant advantage of MIK over MGK, even within the scenario where MIK is expected to be superior.

Uniform Conditioning

UC is a non-linear technique for estimating recoverable resources inside a mining panel V using the estimated panel grade (Rivoirard, 1994). This grade is typically derived from ordinary kriging (OK) or simple kriging (SK) with a local mean to account for potential non-stationarity. The technique is a sound approach for assessing recoverable resources because it is well known that when drilling data is sparse, estimation into panels provides more reliable results than directly estimating into small blocks. The mining industry's acceptance of UC has been apparent for several years, and a good reconciliation is generally found between UC medium-to long-term estimates and production data.

A critical aspect of UC is the need for a robust, conditionally unbiased estimate of the panel grade (see Case Study 3), supported by a model that relates grades at different supports: $Z(x)$ at point support, $Z(v)$ at SMU support, and the estimated grades $Z^*(V)$ at panel support. UC employs the DGM and incorporates the additional assumption that the panel grade can also be expressed using an anamorphosis function as:

$$Z^*(V) = \Phi_V(Y_V^*)$$

where Y_V^* is a standard Gaussian variable and the panel anamorphosis function Φ_V is given by:

$$\Phi_V(y) = \sum_{n \geq 0} \phi_n s^n H_n(y)$$

The correction factor s is selected to respect the variance $Z^*(V)$ and satisfies $0 \leq s \leq r$. This ensures the variance at panel support is less than the variance at SMU support. If it is further assumed that Y_v and Y_V^* have a joint Gaussian distribution with correlation R, then $R = s/r$ and the conditional distribution of the Gaussian equivalent SMU grades Y_v given the Gaussian equivalent panel grade Y_V^* is known. For a panel with $Y_V^* = y_V$ this distribution is Gaussian with mean Ry_V and variance $1 - R^2$. This key result allows computing the recoverable resources inside the panel in the form of panel-specific grade-tonnage curves (Rivoirard, 1994).

A key distinction between MIK and UC lies in their processes. MIK follows a two-step approach: it first derives the point support distribution and then applies a support correction to transform it into a block support distribution. In contrast, UC integrates support modelling directly within its process, using panel estimates to derive the block support distribution. This embedded approach simplifies the transition between supports and ensures the internal consistency of the model. However, using the panel estimate to condition the results comes with a limitation: the distribution of SMUs within the panel depends entirely on the estimated panel grade, regardless of the drill hole data surrounding the panel.

A common feature of local recoverable resource techniques presented here is the use of a local condition combined with a global variance correction factor. Emery (2008) examines the local change of support problem and demonstrates that the variance correction factor between point and block-support local distributions depends on the specific block and is generally smaller than the global variance correction factor. As a result, applying the global variance reduction factor to derive local block grade distributions tends to underestimate the impact of change of support at the local scale, potentially leading to optimistic evaluations at the selective mining scale.

Emery's approach provides a practical solution to this issue and can be easily implemented in current commercial software. Unfortunately, this capability is not yet available in such tools. The only limitation of this method is the requirement that the discrete Gaussian model (DGM) is suitable for the specific deposit being modelled, as the validity of the results depends on this assumption.

UC and, in general, all indirect methods, allow the derivation of the distribution of SMUs grades within local panels, but not their individual spatial locations within the panel. Abzalov (2006) proposed a localised uniform conditioning (LUC) technique to spatially locate selective mining unit grades that have been derived using uniform conditioning for the assessment of recoverable resources. The technique has the advantage of producing selective mining unit estimates conforming to the uniform conditioning panel-specific grade-tonnage curve while introducing spatial information at the scale of the selective mining units.

Direct and Indirect Localised Conditioning methods

The techniques as proposed by Assibey-Bonsu and Krige (1999) and Assibey-Bonsu et al (2015) use direct and indirect distributions of selective mining units.

The indirect approach, like UC, involves deriving the unknown SMU grade distribution within relatively large blocks. Specifically, this approach estimates the grade distribution at the SMU scale within a panel, conditioned on the estimated panel grade. These panel grades are typically derived using ordinary kriging (OK) or simple kriging (SK) with a local mean. In this method, the post-processing of the local probability distributions of SMUs within large planning blocks holds that they are assumed to follow a lognormal distribution.

The direct approach, on the other hand, estimates recoverable functions based directly on individual untransformed SMU grades, regardless of the inevitable smoothing effect caused by the estimation process. This method replaces the smoothed kriged SMU estimates with probability distributions that represent the expected 'actual' grades to be observed during mining. The post-processing of the local probability distributions of the SMUs assumes a lognormal distribution.

In all cases, block estimates, whether for large panels or SMUs, must be conditionally unbiased and adjusted to account for the expected average grade improvements and tonnage reductions based on the additional information that will become available during the production stage. These adjustments are essential prerequisites for producing efficient and reliable estimates of local recoverable resources. This stands in contrast to the practice of some estimators who, during feasibility mine planning, assume perfect information at the production stage. Such an approach overlooks the information effect, which reflects the variability and uncertainty inherent in SMU estimates at the final production stage.

Conditional Simulation

Conditional simulation can be used for assessing recoverable resources, offering a flexible framework that does not rely on a specific change of support model. The approach involves simulating block grades, either by reblocking point-support simulations or through direct block simulation (Emery and Ortiz, 2011; Deraisme and Assibey-Bonsu, 2012), which are then used to derive the local grade distribution of SMUs within a larger panel (Figure 2). An added advantage of conditional simulation is its ability to assess uncertainty at both the local and global scale (Journel and Kyriakidis, 2004). This approach provides not only the expected recoverable quantities but also the uncertainty associated with them, enabling more informed decision-making in mine planning and resource evaluation.

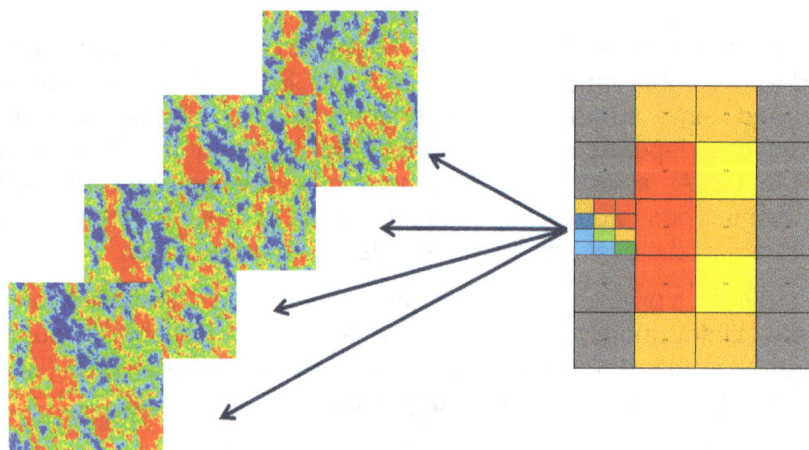

FIG 2 – Example of using conditional simulations to assess recoverable resources.

Conditional simulation is a well-established and mature topic in geostatistics, with decades of development and substantial practical applications. While advances in technology, including more powerful desktops, laptops, and cloud computing platforms, have significantly enhanced computational capabilities, the application of conditional simulation remains computationally intensive. Generating multiple realisations of spatial grade distributions requires substantial memory, storage, and processing power, along with significant time for set-up, execution, and analysis. Mine planning using several equi-probable realisations also presents a challenge; selection of some of the simulations for mine planning has been proposed as a practical solution, which is still not exhaustive. These resource demands might explain why many practitioners often lean toward simpler techniques, such as UC or MIK, for estimating recoverable resources. These simpler methods, while less robust in capturing uncertainty, are faster to implement and analyse, making them more practical in time-sensitive or resource-constrained situations.

CASE STUDIES

This section provides a summary of case studies from the geostatistics literature that demonstrate the advantages and limitations of various recoverable resource techniques. Additional details, including spatial modelling parameters, are available in the references for each case study.

Case study 1 – recoverable versus non-recoverable resource estimates for mine planning

This case study is based on various recoverable resource modelling techniques, derived using typical feasibility study or new mine drilling data configurations sourced from a comprehensive production database (Assibey-Bonsu et al, 2024). The database originates from the Gold Fields Tarkwa Gold Mine, which has approximately 5 Moz in Mineral Reserves and a 12-year Life-of-Mine. The operation exploits narrow auriferous conglomerates, like those in South Africa's Witwatersrand Basin, and currently mines from four open pits. The study discusses the resource modelling, mine planning, and financial risks.

For this case study, the original SK estimated panels (KRIG) of 50 m × 50 m × 3 m based on local means were retained as the non-recoverable estimates for the study. The SK estimated panels were further subjected to post-processing as per the indirect (INDLN) and direct (LIC) recoverable resource methodologies (Assibey-Bonsu and Krige, 1999). The methodology incorporates the information effect and change of support correction for the relevant SMUs of 10 m × 10 m × 3 m. A final production grade-control grid of 25 m × 25 m has been assumed in deriving the recoverable resources, which is based on the expected final production grade control drilling on the mine. The post-processed output provides recoverable tonnages, grades, and metal content estimates above respective cut-offs per panel. The tonnages and metals represented by the grade tonnage curves estimated by the indirect technique were decomposed and distributed into the SMUs within respective panels according to a ranking of the main element grade estimate of the SMUs to derive the localised or direct recoverable resource (LIC) model following the localisation process proposed by Abzalov (2006). These indirect (INDLN) and indirect but subsequently localised (LIC) recoverable as well as the non-recoverable estimates were analysed using data from sections of the 3D gold deposit. Sets of resource drilling data, equivalent to a typical Indicated class Resource drilling data configurations on an approximate 200 m × 100 m, were selected to represent the data at the early stage of a new mining project.

The INDLN and LIC recoverable Whittle outputs, as well as the non-recoverable (KRIG) outputs were compared to the final production Grade Control model to determine the efficiency of the respective approaches. The efficiencies of the Whittle optimisation physicals and the corresponding financial profiles are measured on the basis of the spreads of percentage errors. Figure 3 shows the grade-tonnage curves (GTCs) for the post-processed (INDLN) and non-post-processed results compared with the corresponding 25 m × 25 m production grade control-data (GCM). These are the *in situ* resource models which were subsequently diluted and used for the Whittle optimisation analysis. The GTCs are for all the reefs in the area.

KRIG vs INDLN vs GCM SK 10103 (Undiluted)

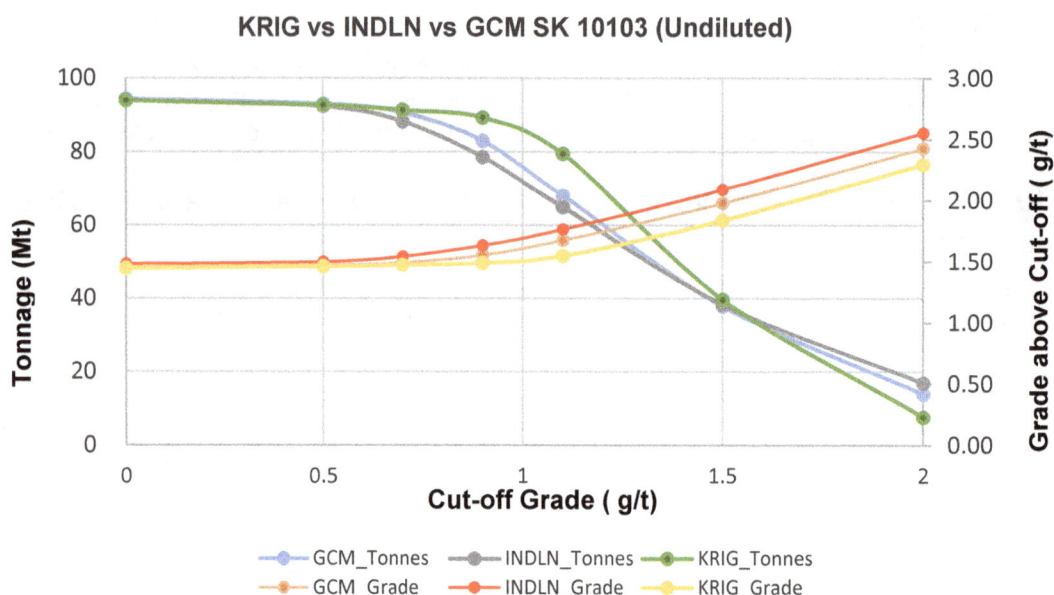

FIG 3 – *In situ* comparison between: (1) Indirect Postprocessed (INDLN) model, versus (2) Non-post-processed Kriged panels (KRIG), versus (3) Grade Control model (GCM).

It is worth noting that:

- At lower cut-offs there are no material differences.

- For higher cut-offs >= ~1 g/t required for high-grade blending/scheduling, the kriged (non-post-processed/non-recoverable) model shows materially higher tonnes (+17 per cent for non-post-processed) with lower grades (-8 per cent for non-post-processed).

- At ~1 g/t cut-off, the post-processed results show non-material underestimation of tonnes (-5 per cent) and grade variances (+5 per cent) translating to similar ounces as the GCM.

The analysis using the open pit Whittle optimisation process applied dilution factors to both the INDLN and non-recoverable models prior to the application of techno-economic parameters for optimisation. These techno-economic parameters were then applied to both resource models using similar processes to prepare them for the Whittle optimisation process (Assibey-Bonsu *et al*, 2024). The results are demonstrated in Figure 4 at an economic evaluation cut-off of 1.1 g/t. Based on the GCM, ~70 per cent of Reserves within the Whittle pit shell are above a 1.1 g/t cut-off. Figure 4 shows that the results for the non-post-processed Whittle outputs indicate material differences when compared to the production Grade Control (GCM) results. For example, the non-post-processed Whittle outputs 'ore tonnes mined' shows overestimation of 12.6 million tonnes (+19 per cent). Similarly, total waste mined, and the total cost (mining, processing, and selling) show material variances of -44 per cent and +14 per cent respectively for the non-post-processed model. The non-post-processed Reserve ounces and revenue are overstated by (10 per cent) respectively. This translates to overestimation of revenue by US$443 million, which reflects significant financial risks. The corresponding post-processed revenue is within 2 per cent variance, demonstrating significant risk mitigation when the post-processing model is used to underpin the mine planning. The post-processed recoverable results show lower variances in all cases.

The Whittle optimisation results of the smoothed non-post-processed kriged model illustrate the typical 'problem of the vanishing tonnes' as documented in the literature (eg David, 1977). As observed in this case study, non-post-processed kriged models tend to overestimate the recoverable tonnages, underestimate mined grades and significantly underestimate the waste to be mined (-44 per cent), leading to material mine planning inefficiencies and material financial risks. Figure 4 demonstrates the significant variances and inefficiencies of the non-recoverable models when used to underpin mine planning. Whittle optimisation sensitivity analyses based on higher gold prices (US$1750/oz) and higher total cost (mining, processing and selling costs) showed similar results.

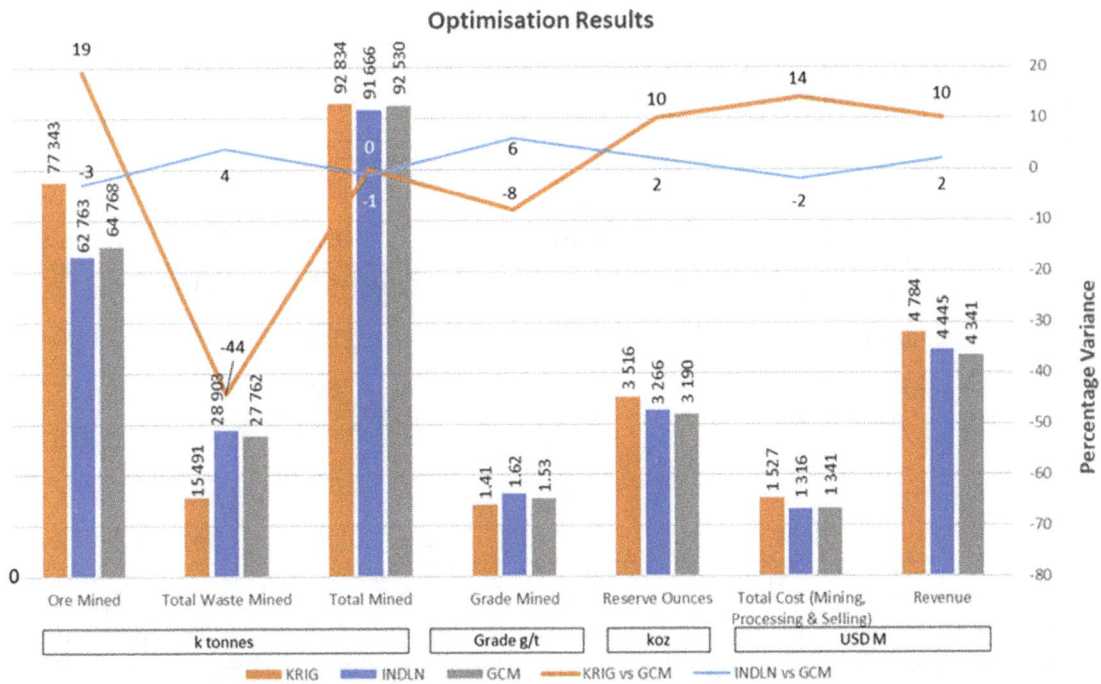

FIG 4 – Whittle optimisation results showing physicals and financials for indirect recoverable (INDLN) and non-recoverable Resource Models plus variances. Economic evaluation at 1.1 g/t cut-off (constitutes ~70 per cent of Reserves within the Whittle pit shell based on the GCM).

The findings showed that the indirect recoverable modelling approach can serve as a practical and effective tool for computing recoverable resources for mine planning and financial forecasts. The study also showed that the localised approach presents comparable results to the currently used indirect method. Additionally, Assibey-Bonsu *et al* (2024) highlight that relying on smoothed non-recoverable estimates for capital-intensive mining projects or mine extensions can misrepresent the true economic value of the project or operation.

Case study 2 – direct and indirect distributions of SMUs for estimation of recoverable resources and reserves for new mining projects

This case study utilises the indirect and direct distributions as discussed in Case Study 1, but in addition it includes UC modelling.

This case study is based on the results obtained by Assibey-Bonsu and Krige (1999) where conditional simulation using the Turning Bands Method (Matheron, 1973) was used to simulate a lognormal variable on a 5 × 5 × 5 m grid to represent the closely-spaced data available at the final mining stage. From this data, the equivalent of exploration or feasibility borehole values on a 20 × 20 m have been selected to represent the data at the early stage of a new mining project for a hydrothermal deposit. UC and the direct and indirect techniques were applied to the 20 × 20 m grid data for recoverable resource estimates which are usually required for feasibility mine planning. The results for each approach were then compared with the corresponding 'actual' values based on the available 5 × 5 × 5 m grid to determine the efficiency of each approach.

Table 1 summarises the results of aggregating recoverable resources from individual block estimates across the global area. While the total metal content estimates are satisfactory, there is a slight overestimation of grade for all three techniques. However, these discrepancies are not considered significant for practical purposes. Sensitivity analysis of the variogram parameters indicates that these overestimations disappear with minor adjustments to the variogram. This occurs because, with these changes, the theoretical SMU block variance calculated from the variogram aligns closely with the observed SMU block variance. As previously noted, accurate modelling of the variogram parameters is crucial for the success of all techniques.

TABLE 1

Global aggregate of direct and indirect distributions of SMUs for recoverable resource estimates compared with actuals. Uniform conditioning (UC), Indirect SMU (INDLN) and Direct SMU (DLN) distributions.

Cut-off	Mean grade above cut-off				Metal quantity			
	UC	INDLN	DLN	Actuals	UC	INDLN	DLN	Actuals
0.0	1.02	1.02	1.02	1.01	1.019	1.019	1.016	1.015
0.3	1.07	1.09	1.08	1.06	1.00	0.999	0.997	1.000
0.5	1.23	1.26	1.25	1.20	0.933	0.928	0.925	0.931
0.7	1.42	1.46	1.46	1.40	0.825	0.824	0.818	0.818
0.8	1.53	1.57	1.57	1.48	0.766	0.768	0.761	0.767
1.0	1.74	1.80	1.80	1.68	0.651	0.659	0.645	0.652

The study demonstrates that the approaches compared can effectively be used to compute recoverable resources for new mining projects. It also highlights that the direct, more straightforward technique, when applied efficiently, can serve as an equally practical tool for estimating recoverable resources in feasibility mine planning. Here 'efficient' emphasises the importance of ensuring that all block estimates are conditionally unbiased using an adequate search routine. In situations where this is impractical, the use of simple kriging with a more localised mean is recommended. As such, the practice of artificially increasing dispersion variance by limiting search routines, employed by some practitioners, is not justified (Krige, 1997; Assibey-Bonsu and Krige, 1999; Magri *et al*, 2003).

Case study 3 – limitations in accepting localised conditioning recoverable resource estimates for medium-term, long-term, and feasibility-stage mining projects

Assibey-Bonsu and Muller (2014) used LUC with typical feasibility or new mining drilling data configurations on 40 m × 20 m drawn from a massive database from a mined-out area in a hydrothermal gold deposit. The results were then compared with the corresponding 8 m × 5 m grid grade-control data to determine the goodness of the approach and the validity of the recoverable resource estimates for mine planning and financial forecasts. Figure 5 demonstrates that LUC results are systematically biased, with a slope of regression of approximately 0.1. This is the result of the invariable limited resource drilling data available for the LUC estimates, typical of most exploration and new mining projects.

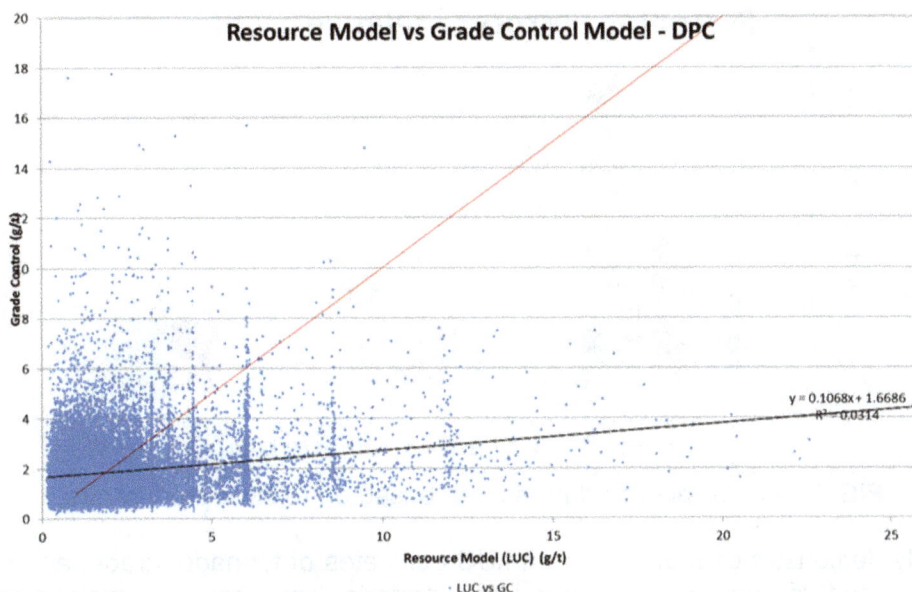

FIG 5 – Scatterplot of Grade Control block grades versus the corresponding LUC estimates.

Further analyses demonstrated that:

- LUC estimates with grades of 12 g/t ended up on average just above 3 g/t, based on the grade control block values.

- LUC estimates of 0.3 g/t classified as waste based on a 0.6 g/t cut-off, ended up on average just about 2 g/t, indicating that they have been significantly underestimated and misclassified.

- The percentage of errors of ore mined (ie tonnes and grades above relevant economic cut-offs) over six months, one year and three years are serious and show that the LUC estimates provide a misleading pattern of mine production and financial forecasts for short to medium-term plans of the Mine (Table 2).

- The percentage errors associated with the profit profiles for the production periods analysed are very significant. In the short-term production period of 6 months, the percentage errors range from 36–74 per cent for respective cut-offs. Although the percentage errors decrease as the production volumes increase, the percentage errors are still substantial, and any financial forecast based on these estimates can be misleading (Figure 6).

TABLE 2

Percentage errors of ore mined (grade control) over respective time periods compared to LUC predictions.

Cut-off (g/t)	6 Months		1 Year		3 Years	
	tonnes (%)	grade (%)	tonnes (%)	grade (%)	tonnes (%)	grade (%)
0.6	-15%	31%	-15%	20%	-15%	16%
0.7	-14%	31%	-16%	22%	-16%	18%
1.0	-3%	27%	-16%	24%	-17%	19%

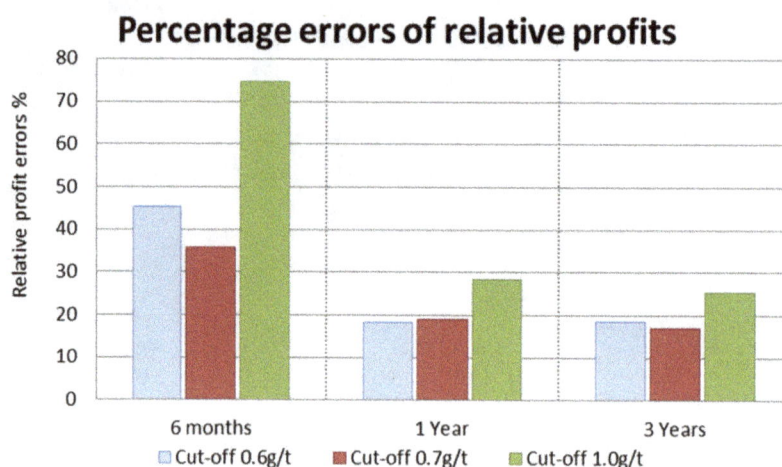

Percentage errors of relative profits

FIG 6 – Financial impact measured based on relative profit errors.

This case study demonstrates that relying on LUC estimates of tonnage, grade, and profit figures, particularly for short to medium-term production periods, can result in misleading outcomes, especially when the data set is limited. This occurs because the smaller data set available for individual block and LUC estimates through Ordinary Kriging, as well as for the sections to be mined during these periods, may prevent the generation of efficient and reasonable estimates for planning purposes. Additionally, such estimates may suffer from significant conditional biases, undermining the accuracy of financial forecasts and potentially invalidating feasibility assessments for new projects or extensions of existing mines.

Case study 4 – production reconciliation of a multivariate recoverable estimate

Deraisme and Assibey-Bonsu (2011) and Assibey-Bonsu *et al* (2015) present a production reconciliation case study based on a porphyry copper-gold deposit in Peru. The reconciliation study compares the long-term mineral resource model to the corresponding production blasthole-based grade control model, as well as the final plant production.

Localised Multivariate Uniform Conditioning (LMUC) (Deraisme and Assibey-Bonsu, 2011) was applied to a relatively large grid of 100 m × 50 m to develop the long-term model. The estimation process involved co-kriging gold and copper on panels with dimensions of 40 m × 40 m × 10 m, assuming SMUs of 10 m × 10 m × 10 m. Significant conditional biases were observed for ordinary panel kriging estimates, which, as previously demonstrated, have a negative impact on ore and waste selection for mine planning, as well as financial planning. To mitigate this issue, simple co-kriging with local means was used for panel conditioning in all cases.

Table 3 summarises the percentage errors for tonnes, copper grade, and gold grade on quarterly, six-month, and annual bases. The mine reports these production results to shareholders quarterly.

TABLE 3

Distribution of percentage errors between resource model and plant production over various production periods.

Period	Tonnes	Grade	
		Gold	Copper
Quarterly	6%	2%	-7%
Six monthly	7%	5%	-2%
Annually	-1%	3%	-1%

The production reconciliation results show the overall advantage gained by using localised multivariate uniform conditioning (LMUC) estimates based on simple co-kriging as demonstrated by the narrow spreads of the monthly percentage errors. The central 80 per cent confidence limits of the monthly production errors were -12 per cent/+10 per cent, -6 per cent/+14 per cent, and -8 per cent/+8 per cent for tonnes, gold, and copper grades respectively. The case study also showed percentage errors of +6 per cent/+2 per cent/-7 per cent on a quarterly basis for tonnes, gold, and copper grades respectively (Table 3). The narrowing of the observed confidence limits is also observed as shown by the reduced observed average percentage errors of -1 per cent/ +3 per cent for the plant production reconciliations on a macro or long-term production basis.

Practitioners should exercise caution regarding some drawbacks associated with LMUC (Rondon and Talebi, 2023). The current implementation of LMUC simplifies a multivariate problem to a bivariate one by designating a 'master' or anchor attribute to identify SMUs above a specified cut-off. While this approach may appear reasonable for multivariate uniform conditioning applications, localising the results introduces two significant artifacts. Firstly, LMUC results may exhibit linear-like patterns that are inconsistent with the characteristics of the input data. This issue is evident in the scatterplots of localised attributes within a panel. Secondly and more critically, the correlation between localised attributes differs drastically between the global scale (across the entire mineralised domain) and the local scale (within individual mining panels). These artifacts stem from the reliance on the master attribute to guide the localisation process.

These limitations highlight the need for careful consideration and validation. These issues need to be checked, and the potential negative impacts assessed.

Case study 5 – block simulation and localised multivariate UC

Deraisme and Assibey-Bonsu (2012) compared the use of localised UC in the multivariate case with localised block simulations for estimating recoverable resources. The latter involves deriving local grade-tonnage curves using multivariate simulations and localising the grades of the SMUs. The paper provides a brief review of Multivariate Uniform Conditioning (MUC), Direct Block Simulation (DBS), and their corresponding localised techniques, LMUC and LDBS. Additionally, the authors present a comparative case study based on a porphyry copper-gold deposit in Peru (see Case Study 4). The favourable production reconciliation results observed in Case Study 4 serve as a benchmark for assessing the efficiency of the multivariate block simulation technique. The multivariate block simulation has an advantage in design, as it incorporates the correlations between both the main and secondary elements, whereas the MUC approach does not explicitly consider these correlations between secondary variables.

In the LDBS process, the tonnages and metals represented by the grade tonnage curves simulated through 50 multivariate block simulations are used to derive the local grade tonnage curves. These are then decomposed and distributed into the SMUs within respective panels using Abzalov's approach (2006).

The results depicted in Figures 7 and 8 show that globally, the grade tonnage curves derived from LMUC and LDBS are similar as well as the scatter diagrams between localised SMU grades assigned by both approaches. This demonstrates that two different approaches based on MUC and DBS, compared on a real-life case study, show similar results. Since Case Study 4 for MUC showed good production reconciliation results, this work demonstrates that the block simulation approach, when applied appropriately, will deliver efficient recoverable resource estimates for mine planning. Simulations also provide the additional dimension of the quantification of uncertainty, allowing the selection of scenarios corresponding to different risk levels or the building of localised confidence intervals to assist with scheduling priorities.

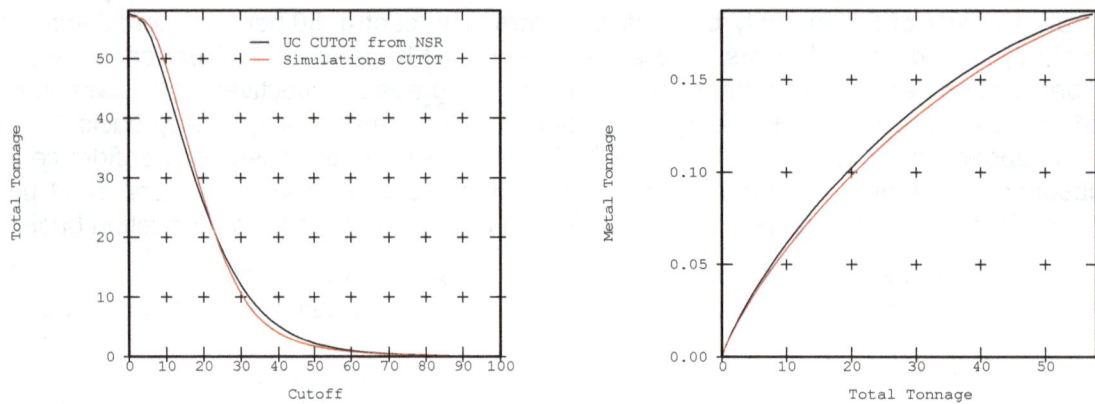

FIG 7 – Global grade tonnage curves on SMUs calculated from MUC and LDBS.

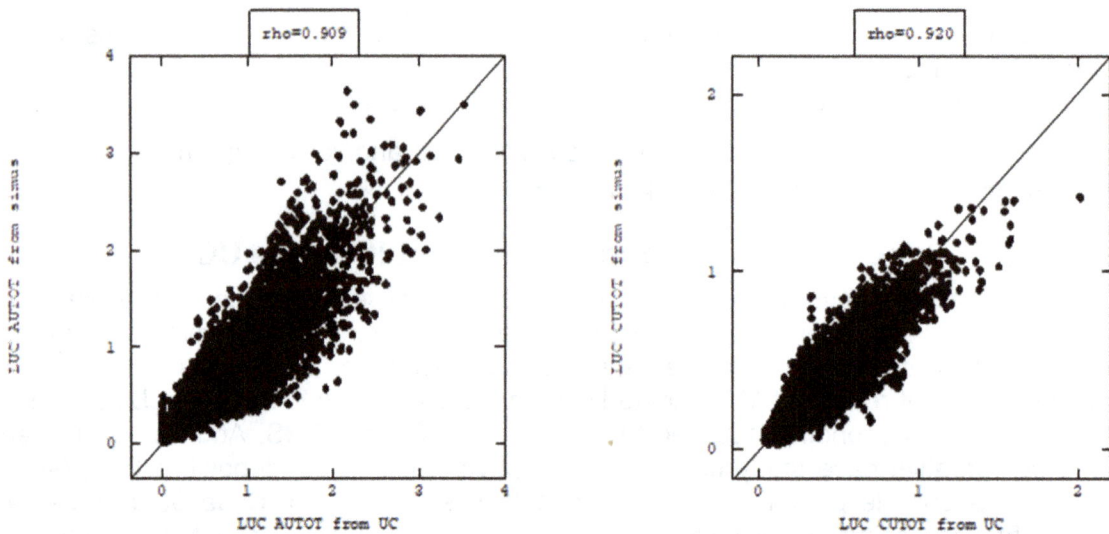

FIG 8 – Scatter diagrams of SMU grades AUTOT and CUTOT assigned from LMUC and LDBS.

CONCLUSIONS

This paper highlights the ongoing challenges and advancements in estimating recoverable resources, a critical component of mining operations and financial planning. The evolution of methodologies from traditional to localised approaches, together with the use of conditional simulations, demonstrates the industry's commitment to improving resource estimation accuracy. However, significant risks and limitations persist, particularly in the context of conditional biases, the information effect, and the appropriate/optimal implementation of geostatistical techniques. These issues underscore the importance of rigorous validation and follow-up reconciliations against actual production, wherever possible.

The comparison of direct and indirect techniques in case studies reveals that both can effectively address the complexities of recoverable resource estimation when applied efficiently. Yet, caution is warranted. The paper also demonstrates that incorporating post-processing techniques and leveraging the strengths of conditional simulation can mitigate risks and enhance decision-making in feasibility studies and production forecasting.

While advancements such as Localised Multivariate Uniform Conditioning (LMUC) show promise, their practical limitations, including the potential for artifacts and inconsistent correlations, require careful consideration. The reconciliation of production data with resource models, as demonstrated in the case studies, highlights the value of continuous refinement and adaptation of geostatistical methods.

Looking forward, the integration of modern computational tools, improved data acquisition techniques, and advanced geostatistical modelling will be essential to address remaining challenges. This paper reaffirms the importance of robust and adaptable methodologies in enabling mining companies to optimise resource utilisation, manage financial risks, and ensure sustainable operations.

The authors believe that it is correct to say that the estimation of recoverable resources is not hopeless. While the paper acknowledges the challenges and limitations inherent in the process, it also highlights the advances in methodologies. The case studies presented in the paper demonstrate that accurate and reliable recoverable resource estimates can be achieved; moreover, the integration of modern computational tools and advanced geostatistical techniques provides a promising path forward for refining and optimising these estimates.

In summary, while challenges remain, the paper clearly shows that recoverable resource estimation is a manageable and solvable problem with the right methodologies and careful application.

it is also important that the industry researches and develops new geostatistical techniques for recoverable resource estimates and the application thereof, but there is a critical need to validate new techniques by way of (follow-up) checks to confirm the absence of biases and advantages to be gained when they are applied in practice.

REFERENCES

Abzalov, M Z, 2006. Localised Uniform Conditioning (LUC): A New Approach to Direct modelling of Small Blocks, *Mathematical Geology*, 38(4):393–411.

Assibey-Bonsu, W and Krige, D G, 1999. Use of direct and indirect distributions of selective mining units for estimation of recoverable resource/reserves for new mining projects, in Proceedings of APCOM'99 Symposium, Colorado School of Mines, Golden, Co.

Assibey-Bonsu, W and Muller, C, 2014. Limitations in accepting localised conditioning recoverable resource estimates for medium term, long-term and feasibility-stage mining projects, particularly for sections of an ore deposit, *Journal of the Southern African Institute of Mining and Metallurgy*, 114(8):619–624.

Assibey-Bonsu, W, Aboagye, M, Appau, K and Muller, C J, 2024. Orebody and mine planning assessment based on alternative recoverable and non-recoverable resource modelling techniques, in 12th International Geostatistics Congress.

Assibey-Bonsu, W, Deraisme, J, Garcia, E, Gomez, P and Rios, H, 2015. Production reconciliation of a multivariate uniform conditioning technique for mineral resource modelling of a porphyry copper gold deposit, in *The Danie Krige Geostatistics Conference*, Southern African Institute of Mining and Metallurgy, Symposium Series S84, pp 231–243.

BHP, 2025. BHP and Lundin Mining complete the acquisition of Filo Corp, BHP News. Available from: <https://www.bhp.com/news/media-centre/releases/2025/01/bhp-and-lundin-mining-complete-the-acquisition-of-filo-corp>

Businesswire, 2023. Newmont Acquires Newcrest, Successfully Creating World's Leading Gold Mining Business. Available from: <https://www.businesswire.com/news/home/20231106048474/en/>

Chilès, J-P and Delfiner, P, 2012. *Geostatistics: Modelling Spatial Uncertainty*, 2nd ed (Wiley).

David, M, 1977. *Geostatistical Ore Reserve Estimation*, 364 p (Elsevier: Amsterdam).

Deraisme, J and Assibey-Bonsu, W, 2011. Localised uniform conditioning in the multivariate case, an application to a porphyry copper gold deposit, in Proceedings of the 35th APCOM Conference, Wollongong.

Deraisme, J and Assibey-Bonsu, W, 2012. Comparative study of Localized Block Simulations and Localized Uniform Conditioning in the Multivariate case, in Proceedings of the International Geostatistics Conference 2012, pp 309–320.

Deraisme, J and Rivoirard, J, 2009 Histogram Modelling and Simulations in the Case of Skewed Distributions with a 0-Effect: Issues and New Developments, IAMG 2009.

Emery, X and Ortiz, J, 2011. Two approaches to direct block-support conditional co-simulation, *Computers and Geosciences,* 37(8).

Emery, X, 2004. On the Consistency of the Indirect Lognormal correction, *Stochastic Environmental Research and Risk Assessment,* 18:258–264.

Emery, X, 2005. Simple and Ordinary Multigaussian Kriging for Estimating Recoverable Reserves, *Mathematical Geology,* 37:295–319.

Emery, X, 2008. Change of Support for Estimating Local Block Grade Distributions, *Mathematical Geology,* 40:671–688.

Financial Review, 2007. Rio swallows Alcan for $44bn. Available from: <https://www.afr.com/politics/rio-swallows-alcan-for-44bn-20070713-jkdo0>

Isaaks, E and Srivastava, M, 1989. An Introduction to Applied Geostatistics (Oxford University Press: New York).

Journel, A G and Huijbregts, C J, 1978. *Mining Geostatistics* (Academic Press: London).

Journel, A G and Kyriakidis, P, 2004. *Evaluation of Mineral Reserves: A Simulation Approach* (Applied Geostatistics Series: Oxford University Press).

Journel, A G, 1983. Nonparametric estimation of spatial distributions, *Mathematical Geology*, 15(3):445–446.

Krige, D G and Ueckermann, H J, 1963. Value contours and improved regression techniques for ore reserve valuations, *Journal of the South African Institute of Mining and Metallurgy*, pp 429–452.

Krige, D G, 1951. A statistical approach to some basic mine valuation problems on the Witwatersrand, *Journal of the Chemical, Metallurgical and Mining Society of South Africa*, 52(6):119–139.

Krige, D G, 1960. On the departure of ore value distributions from the lognormal model in South African gold mines, *Journal of the South African Institute of Mining and Metallurgy*, 61:231–244.

Krige, D G, 1966. Two-dimensional weighted moving average trend surfaces for ore valuations, in *Proceedings of the Symposium on Mathematical, Statistics and Computer Applications in Ore Valuation*, pp 13–38 (South African Institute of Mining and Metallurgy).

Krige, D G, 1997. Block Kriging and the fallacy of endeavouring to reduce or eliminate smoothing, Regional APCOM Conference, Moscow.

Lantuejoul, C, 1988. On the Importance of Choosing a Change of Support Model for Global Reserves Estimation, *Mathematical Geology*, 20:1001–1019.

Magri, E, Gonzalez, M, Couble, A and Emery, X, 2003. The influence of conditional bias in optimum ultimate pit planning APCOM 2003 Conference.

Matheron, G, 1963. Principles of Geostatistics, *Economic Geology*, 58:1246–1266.

Matheron, G, 1965. Les variables regionalisées et leur éstimation, Masson et Cie, Paris.

Matheron, G, 1973. The Intrinsic Random Functions and Their Applications, *Advances in Applied Probability*, 5(3):439–468.

Matheron, G, 1976a. Forecasting block grade distributions: The transfer functions, in *Advanced Geostatistics in the Mining Industry* (eds: M Guarasico, M David and C Huijbregts).

Matheron, G, 1976b. A simple substitute for conditional expectation, The disjuntive kriging, in *Advanced Geostatistics in the Mining Industry* (eds: M Guarasico, M David and C Huijbregts).

Reed, J, 2006. Gold Fields in $2.5bn South Deep deal, Financial Times. Available from <https://www.ft.com/content/dd22b0b8-416e-11db-b4ab-0000779e2340>

Rivoirard, J, 1994. *Introduction to Disjunctive Kriging and non-linear Geostatistics* (Clarendon Press Oxford).

Rondon, O and Talebi, H, 2023. Artifacts in Localised Multivariate Uniform Conditioning: A Case Study, in *Geostatistics Toronto, GEOSTATS 2021* (eds: S A Avalos Sotomayor, J M Ortiz and R M Srivastava), (Springer Proceedings in Earth and Environmental Sciences, Springer).

Ross, F W, 1950. The development and some practical applications of a statistical value distribution theory for the Witwatersrand auriferous deposits, MSc(Eng) thesis, University of Witwatersrand.

Rossi, M E and Parker, H M, 1994. Estimating Recoverable Reserves: Is it Hopeless?, in *Geostatistics for the Next Century, Quantitative Geology and Geostatistics* (ed: R Dimitrakopoulos), p 6 (Springer: Dordrecht).

Roth, C and Deraisme, J, 2001. The information effect and estimating recoverable reserves, in *Proceedings Geostatistics 2000* (eds: W J Kleingeld and D G Krige) pp 776–778 (Geostatistical Association of Southern Africa: Johannesburg).

Vann, J, Guibal, D and Harley, M, 2000. Multiple Indicator Kriging – Is it suited to my deposit?, in Fourth International Mining Geology Conference.

Verly, G, 1983. The multiGaussian approach and its applications to the estimation of local reserves, *Mathematical Geology*, 15:259–286.

Vincent, J and Deutsch, C, 2019. Multiple Indicator Kriging - A Review of Good Practice, CCG Annual Report 21, paper 112.

Preg-robbing gold ores – an example of a non-linear estimation workflow when dealing with non-additive geometallurgical variables

M Samson[1], G Benthen[2], J M Clark[3], C Le Cornu[4] and D Conn[5]

1. Manager, Resource Geology, Nevada Gold Mines, Elko, Nevada, 89801.
   Email: matthew.samson@nevadagoldmines.com
2. Technical Specialist, Metallurgy, Nevada Gold Mines, Elko, Nevada, 89801.
   Email: george.benthen@nevadagoldmines.com
3. Chief Geologist, Nevada Gold Mines, Elko, Nevada, 89801.
   Email: jesse.clark@nevadagoldmines.com
4. Manager, Resource Geology, Nevada Gold Mines, Elko, Nevada, 89801.
   Email: christopher.lecornu@nevadagoldmines.com
5. Senior Mine Geologist, Nevada Gold Mines, Elko, Nevada, 89801.
   Email: dconn@nevadagoldmines.com

ABSTRACT

Carlin-type mineral systems exhibit a unique combination of geometallurgical material types characterised as oxide, single refractory, and double refractory ores. Oxide ores are the simplest to process and have been the dominant ore source since bulk open pit mining began in the mid-1960s in the Carlin Trend. Though most carlin ores are refractory, that is nanoscale gold inclusions in the crystal lattice of hydrothermal arsenian pyrite rims. Autoclave technology has been utilised since the 1990s to process refractory ores, which breaks down the pyrite crystal structure using moderate heat under high-pressure with oxidising agents. The residue is then treated with carbon-in-leach during cyanidation that uses activated carbon to adsorb free gold particles as the primary recovery method.

The problem is that most high-grade carlin ores exhibit high total carbonaceous matter from graphite to organic carbon that may offer a higher affinity for gold to adsorb to, essentially robbing gold from the pregnant solution, significantly reducing gold recoveries ('preg-robbing'). These ore types are referred to as 'double' refractory and require energy-intensive roasting at high-temperatures for effective processing. The challenges this poses are multi-faceted. Preg-robbing material is a contaminant to carbon-in-leach processing thus it is not the concentration rather the presence. Not all preg-robbing ores behave alike either, with some carbonaceous matter may significantly impact gold recoveries versus others that essentially 'preg-borrow' ie only a partial loss of recovery.

Robust resource estimation is therefore critical to ensuring accurate geometallurgical characterisation of ore types for economic, blending and routing considerations. The misclassification of oxide as refractory material has material impacts to the total potential recovered gold ounces produced in any given year as processing facilities are spatially dispersed across Nevada. Sites compete for roast material to be processed with those located furthest away required to meet a higher-grade threshold often resulting in material to be stockpiled.

Most geometallurgical variables are non-additive meaning they are the product of a calculation of multiple variables eg from a regression, that do not average linearly yet traditional linear estimation techniques like inverse weighted distance is commonly used. To calculate the final geometallurgical variable usually requires each variable comprising the underlying equation to be estimated but only if consistent data exists. Preg-robbing analyses involve a gold spike, which is subject to change over time or with different laboratories that is not reliably documented. To reduce estimation error, this study explores the use of a non-linear estimation approach utilising a Truncated Pluri-Gaussian (TPG) simulation that categorises preg-robbing data using a threshold of 40 per cent. This approach utilised carbon logging that improved geologic domaining aiding in higher resolution zones of potentially preg-robbing zones that were otherwise poorly sampled.

This study highlights the importance of integrating advanced geostatistical methods, geological interpretations, and robust assay frameworks for addressing complex geometallurgical challenges.

INTRODUCTION

Carlin-type gold deposits are renowned for their complex geometallurgical variability that presents unique challenges in ore control classification and metallurgical processing. These deposits are typically categorised into three material types: i) oxide, ii) single refractory, and iii) double refractory, each requiring distinct treatment methodologies. Historically, oxide ores were prioritised for economic extraction due to their compatibility with heap leaching and oxide milling processes. Though, as mining transitions to deeper deposits, the prevalence of refractory ores, characterised by complex pregnant-robbing (preg-robbing) behaviours, has significantly increased.

Preg-robbing is a phenomenon where organic carbon sequesters gold from cyanide solutions, which complicates the recovery of gold in refractory ores. This behaviour varies based on the carbon's structural characteristics, with more disordered forms posing higher risks. The emergence of double refractory ores, which combine preg-robbing behaviour with finely disseminated gold in sulfides, necessitates advanced and energy-intensive processing techniques such as roasting. In this context, accurate geometallurgical modelling is critical to minimising production losses and optimising resource utilisation.

This paper provides a technical review of the challenges and solutions associated with preg-robbing estimation and its impact on the Crossroads Life-of-mine (LOM) oxide and refractory ore classification. Historical metallurgical practices utilised traditional linear estimation methods such as inverse distance weighting (IDW), kriging, and nearest neighbour (NN) techniques that led to significant risks in ore misclassification. Preg-robbing estimations rely on the non-additive ratio of cyanide-soluble assays that are not suitable for linear estimations, resulting in the overestimation of oxide material and material production discrepancies. Preliminary analyses project a total recoverable gold loss of approximately 12 per cent of total Mineral Reserve ounces from Q2 2024 to present due to miss routing, highlighting the urgency for methodological improvements.

To address these issues, this study explores a Truncated Pluri-Gaussian (TPG) simulation framework, treating preg-robbing data categorically and integrating geological insights such as carbon logging. By refining the estimation process and leveraging advanced probabilistic geostatistical techniques, this approach aims to better differentiate oxide and refractory ores, ensuring reliable classification and reconciliation with production data. Additionally, the implementation of TPG simulations demonstrates improved alignment with ore control classifications and enhanced reliability in capturing the spatial variability of preg-robbing behaviour.

By transitioning to TPG simulations, it led to sustained improvements, evidenced by sensitivity analyses of preg-robbing probability thresholds and continuous integration of geological data to refine classification accuracy and mitigate production risks. Through active open pit case studies, this paper highlights the efficacy of these methodologies in reducing estimation variability and aligning resource models with operational requirements.

PROCESSING

Effective geostatistical modelling requires a clear understanding of the purpose of the variables being modelled and their physical behaviour within the processes they aim to represent. A rudimentary understanding of the various available processing facilities at the Nevada Gold Mines (NGM) complex need to be understood. At Crossroads there are three processing streams for ore: i) oxide leach, ii) oxide mill, and iii) double refractory. Oxide leach material is typically low-grade ore ($\geq$0.14 g/t Au) that exhibits no preg-robbing behaviour or gold sulfide encapsulation, oxide mill is higher grade oxide ore ($\geq$0.69 g/t Au) that exhibits no preg-robbing behaviour or gold sulfide encapsulation also known as free milling ore, and double refractory ore exhibits both sulfide encapsulation and preg-robbing. Due to the scale of material movement across NGM, double-refractory material at Crossroads must be >5.5–6 g/t Au to compete with closer ore sources in the Carlin Trend.

Each processing facility has a different processing cost, recoveries and geometallurgical processing constraints. Generally, the processing facilities costs can be ranked from least to most expensive: leach pads as the cheapest option, to an oxide mill, followed by the roaster. All oxide material can be successfully processed in any of the processing streams, whereas refractory material can be processed in both a roaster and autoclave, with only double refractory processed in a roaster. Thus, estimating the gold grade accurately is not the single most important variable, as the intense processing cost differential between processing types means that predicting the appropriate material type is imperative to underlying economics.

The cut-off grade of gold for the Oxide Leach Pad, Oxide Mill, and Roaster are shown in Table 1. The Leach Pad is the most cost-effective option, with the Oxide Mill having a processing cost five times higher, and the Roaster costing 15 times more than the Leach Pad.

TABLE 1

Relative cog grades between the different processing option at Crossroads.

Process	Relative cut-off grade
Oxide leach pad	1
Oxide mill	5
Roaster	15

Oxide leach pad

Low-grade oxide ore is heap leached with cyanide on lined leach pads. Ore is stacked in lifts, typically 6 m tall. Cyanide solution is applied to the heap, which dissolves the gold into a pregnant solution. Once leaching of one lift is completed, fresh ore is stacked on top of the previously leached bench. The cyanide solution is then applied to the fresh lift. This solution passes through the entire leach pad height and is collected at the bottom. A mature leach pad may contain very many lifts.

The gold bearing solution which is collected at the bottom of the pad in a pregnant leach solution pond and passed through carbon-in-column vessels. Gold is adsorbed onto the activated carbon in the carbon-in-columns, and barren solution after this process is sent back to the leach pad for further leaching of the ore. Make-up cyanide and water is needed to replenish losses from evaporation and reagents consumed by the ore. When the carbon in the carbon-in-columns reaches a high gold loading, this carbon is removed and sent to an elution circuit where the gold is stripped and passed through electrowinning cells. Stripped carbon is then regenerated and returned to the process. The gold precipitate from the electrowinning cells is then retorted to remove mercury, fluxed, melted and poured into doré bars.

Preg-robbing material on a leach pad is highly undesirable as the carbonaceous material will contaminate the leach pad. Despite recoveries in a leach pad being typically lower than in the other processes, this processing route is typically by far the cheapest, and most suitable for low-grade oxide gold ore.

Oxide mill

Higher grade oxide gold ore, which can pay for the crushing and grinding costs is typically sent to an oxide mill, which can obtain much higher recoveries than a ROM leach pad. The oxide mill starts with a crushing circuit where ROM ore is broken down to 5 inches in size and then fed a comminution circuit for further size reduction. A wet grinding circuit typically consists of a semi-autogenous grinding (SAG) mill and ball mill, working in closed circuit with hydrocyclones to classify the ground material. Water is added to the ore to create a slurry. Steel balls are used in the rotating mills to grind the material to a very fine size. Ground slurry is passed through the hydrocyclones which classify the particles by removing the finer ones which can be sent to downstream processing, while sending the coarser particles back to the mills for further grinding.

Once size reduction is complete, the ground slurry typically has too low of a density. The density is increased bypassing the slurry through a thickener, where excess water is removed. Thickened slurry is then sent for cyanidation in agitated and aerated carbon-in-leach (CIL) tanks which contain activated carbon to adsorb the gold as it is leaching. Typically, this process takes 24 hrs. This is a continuous process, with fresh slurry being added to the first tank, then passing via gravity through a series of tanks (typically to 6–10 total) until the tailings are discharged from the last tank.

As the cyanide concentration can still be high in the tailings, the slurry then passed through a cyanide destruction (Detox) process, before it can be released into a tailings storage facility (TSF). Loaded carbon is then eluted, and undergoes the same steps as outlined in the 'Oxide Leach Pad' section.

As we can see, there are many more steps involved in recovering gold via an oxide mill compared to a leach pad, and overall cost is substantially higher than heap leaching (mostly due to the cost of comminution), also this process requires cyanide detoxification and a TSF which further add costs. Thus, generally higher-grade oxide ore is processed via this route, when the increased gold recovery offsets the additional cost.

It should be noted that if preg-robbing ore is misrouted to the oxide process plant it will negatively impact gold recovery on the material that is being processed at that time, as the retention time in the plant is typically 24 hrs. Once the preg-robbing material is processed it is disposed of at the TSF and will not negatively impact gold recovery on material that follows.

Though it must be stress that preg-robbing material contaminates oxide ore, meaning recovery on blocks that are non-preg-robbing will be negatively impacted if blended with preg-robbing material. It should be clearly stated that preg-robbing cannot be effectively 'blended out'. This applies to autoclave processing as well, which is described below.

Autoclave

Single refractory ore requires oxidative pre-treatment before the gold can be recovered by means of conventional cyanidation. This is due to the gold being physically encapsulated within sulfide minerals such as pyrite and arsenian pyrite. Cyanide is unable to contact the gold, thus recoveries via conventional oxide processing are very low.

An effective way of oxidising sulfide minerals is pressure oxidation, which is carried out in autoclaves. The extreme conditions (temperature and pressure) in this process quickly dissolve the sulfide minerals and release the gold particles contained within. Once these particles are released, the material can be treated via conventional cyanidation.

An autoclave facility has a standard comminution circuit, followed by an acidulation stage where sulfuric acid is added to dissolve carbonate minerals before the slurry can be fed to the autoclave. Dissolving carbonate minerals is highly important, because if they are present in sufficient quantities inside the pressure oxidation process – they generate carbon dioxide which competes with oxygen, effectively slowing down or even stopping the oxidation reactions inside the pressure vessel.

Once the slurry is oxidised in the autoclave, it is sent to a neutralisation stage, where the acidity generated by the dissolution of sulfides (sulfides when oxidised generate sulfuric acid) is neutralised with slaked lime to bring the pH to 10.5, which allows downstream cyanidation to take place.

The downstream processing (leaching and carbon handling) after neutralisation at an autoclave facility is the same as in an oxide mill, as described above.

Whole ore autoclaving is typically expensive due to additional costs for acid, steam, and lime which are added to the baseline oxide mill processing costs.

A downside of autoclaving is that the conditions in process are not sufficient to oxidise carbonaceous (preg-robbing) material, hence this process is generally unsuitable for double refractory ores.

Roaster

Double refractory ores are very challenging if not impossible to efficiently process by conventional means. This is due to the gold not only being encapsulated in sulfides, but also due to the presence of carbonaceous material, which adsorbs leached gold, effectively competing with activated carbon in the CIL thus significantly lowering plant gold recovery.

Roasting is a very efficient process to deal with this dual challenge. Roasting oxidises the sulfides in the first stage and oxidises the carbonaceous material in the hotter second stage. This effectively eliminates both problems, and the product can be successfully leached via conventional cyanidation.

Roasting has some differences from conventional gold mineral processing techniques, namely dry grinding in air classified double rotator mills (which are more common in the cement industry), as well as very complex gas cleaning systems which are required to meet modern stringent environmental regulations. The gas cleaning systems are designed to handle sulfur dioxide, mercury, NOx gases, and carbon monoxide.

The roasted calcine is then quenched to reduce its temperature, followed by neutralisation and thickening of the slurry. Once the slurry is thickened it sent to a conventional CIL cyanidation circuit for gold recovery.

Sampling and analytical methods

The limitations of the assay method of preg-robbing must be understood to appropriately model variables in geostatistics. Preg-rob is a two-part assay consisting of a cyanide soluble atomic absorption spectrometry assay (AuCN) where the sub-sample is pre-treated with cyanide for one hour in a shake test to determine the cyanide soluble gold fraction. The resulting sub-sample then has its gold concentration in solution measured using atomic absorption spectroscopy. The second portion of the preg-rob assay is to spike a parallel sub-sample with a gold solution where a concentration is known, then preform an AuCN shake test and assay to obtain the value of gold that was preg-robbed from the spike solution (AuPr). The AuPr is the residual concentration of gold in solution assay, thus, the lower the value – the more gold was preg-robbed by the carbonaceous material in the ore, the higher the preg-rob percentage. The AuPr is compared with AuCN, to understand if the material has preg-robbing potential. The spike solution concentration is 1.7 g/t, and it is added at a 2:1 solution to solids ratio, hence bringing the spike to a total of 3.4 g/t on a solids basis. All results are reported on a solids g/t basis. From the above assays the preg-rob percent can be calculated using Equation 1.

$$\text{Preg} - \text{rob}_\% = \frac{(AuCn+spike)-AuPr}{AuCn+spike} * 100 \tag{1}$$

It should be noted that the test described above is the Barrick standard preg-rob test. Other companies have slight differences in test parameters and calculations, as well as format of reporting the results.

Another value used to classify and route the ore is the AuCN/AuFA ratio. This test compares the AuCN value to the total gold content in the sample, measured via fire assay (AuFA). This ratio is a good indicator for determining whether the gold is oxide or free milling, as a high AuCN/AuFA ratio indicates a high proportion of oxide gold.

To accurately classify the three distinct material types—preg-robbing, sulfide-encapsulated, and oxide—all three assay methods, along with the spike, are necessary to determine the final material type. The AuFA assay provides the total gold concentration within the sample, while the AuCN assay quantifies the cyanide-soluble portion of that gold. Combined with Equation 1, the AuPr assay enables the determination of whether the sample exhibits oxide, sulfide-encapsulated, or preg-robbing metallurgical characteristics. Without the inclusion of the spike in the AuPr assay, it would not be possible to distinguish between sulfide-encapsulated and preg-robbing material. Additionally, since preg-robbing is expressed as a percentage, the spike serves to reduce assay noise near detection limit thresholds, which could otherwise misclassify samples with potential economic value for leach pad processing.

The preg-robbing assay has notable limitations, as it does not fully characterise the behaviour of preg-robbing material. One key limitation is that the assay does not account for preg-robbing strength of the bond. With the current methodology, it remains unclear whether activated carbon introduced into the processing stream would exhibit a stronger bond strength for gold compared to the naturally occurring carbonaceous material in the sample. This phenomenon could suggest less severe recovery losses within an oxide mill or autoclave making it difficult to quantify a true recovery impact. Because a finite amount of gold is introduced into the assay as a spike, it cannot accurately quantify the preg-robbing capacity of the carbonaceous material if the assay result indicates 100 per cent preg-robbing. In many cases, the sample may exhibit preg-robbing capacity far exceeding this threshold. As a result, preg-robbing must be treated as a contaminant and cannot be averaged linearly.

Across NGM deposits the percent preg-rob is considered and refractory material differs; however, typically ranges are between 20–60 per cent. This is because the current assay method does not explain the preg-robbing bond strength. The current preg-rob cut-offs are defined site by site based on historical performance in the processing facilities. At Crossroads the preg-rob cut-off is 40 per cent; however, additional work should be done to better account for preg-robbing bond strength. Equation 2 is used to convert preg-rob to a categorical indicator.

$$\begin{cases} 0 \; if \; \mathrm{Preg-rob_{\%}} < 40\% \\ 1 \; if \; \mathrm{Preg-rob_{\%}} \geq 40\% \end{cases} \tag{2}$$

Figure 1 illustrates Equation 2 graphically, it is interesting to note that a majority of the material typically assays as strong preg-robbing (preg-rob$_{\%}$ ≥ 80 per cent) or non-preg-robbing (preg-rob$_{\%}$ ≤ 20 per cent) the transitional material (preg-rob$_{\%}$ < 80 per cent and preg-rob$_{\%}$ > 20 per cent) represents a very small proportion of the total samples. The sample portion suggests that the transitional zone is typically very discrete and may not be practical to model based on the current mining limitations. Processing transitional preg-rob ore at a reduced recovery in the oxide and single refractory facilities should be recognised and quantified as a potential opportunity.

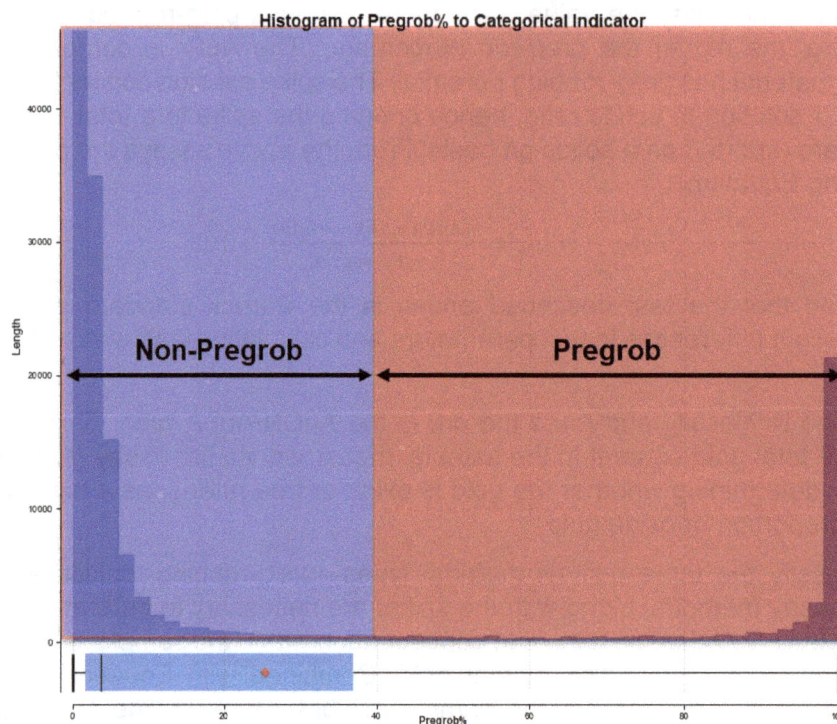

FIG 1 – Preg-rob indicator conversion.

CROSSROADS GEOLOGY

Litho-structural

Gold mineralisation at the Crossroads deposit is dominantly hosted within the Lower Plate stratigraphic package, which is unconformably overlain by a Quaternary alluvium unit (Blamey, Campbell and Heizler, 2017). Throughout Northern Nevada, the Lower Plate package consists of a series of silty, calcareous mudstones interbedded with varying degrees of limestones which are interpreted to have been deposited along a carbonate slope platform. At Crossroads, these lower plate units can be divided into the Devonian Wenban formation, which overlies the Silurian Roberts Mountain, defined by varying abundances of siltstone, mudstone and limestone. These rocks are interpreted to have been deposited in anoxic to euxinic conditions, resulting in the preservation of organic matter (Large, Bull and Maslennikov, 2011).

Lower Plate rocks throughout Northern Nevada have subsequently undergone a poly-phase history of deformation, summarised by Rhys *et al* (2015), including:

- thick skinned compression during the Devonian
- thin-skinned compression during the Jurassic
- far-field contraction during the Cretaceous and associated magmatism
- Eocene extension associated with magmatism and Carlin gold mineralisation.

This tectonic history therefore resulted in a complex structural fabric consisting of low angle fold and thrusts, cross-cut by later extensional faults, all of which contribute to the geometry of gold mineralisation.

Carbon remobilisation

Organic carbon content at Crossroads has been impacted by the deformation and magmatic events highlighted above, resulting in hard to resolve geometries in areas of low data density. As observed in Figure 2, preg-rob assay data and organic carbon have strong spatial relationships with specific sedimentary units such as the Silurian Roberts Mountain, however, they also follow smaller scale trends not thought to be associated with stratigraphy. Although remobilisation of organic carbon is not well documented in Carlin-type systems, it is a well-documented phenomenon in other geologic settings (Simoneit, 1994).

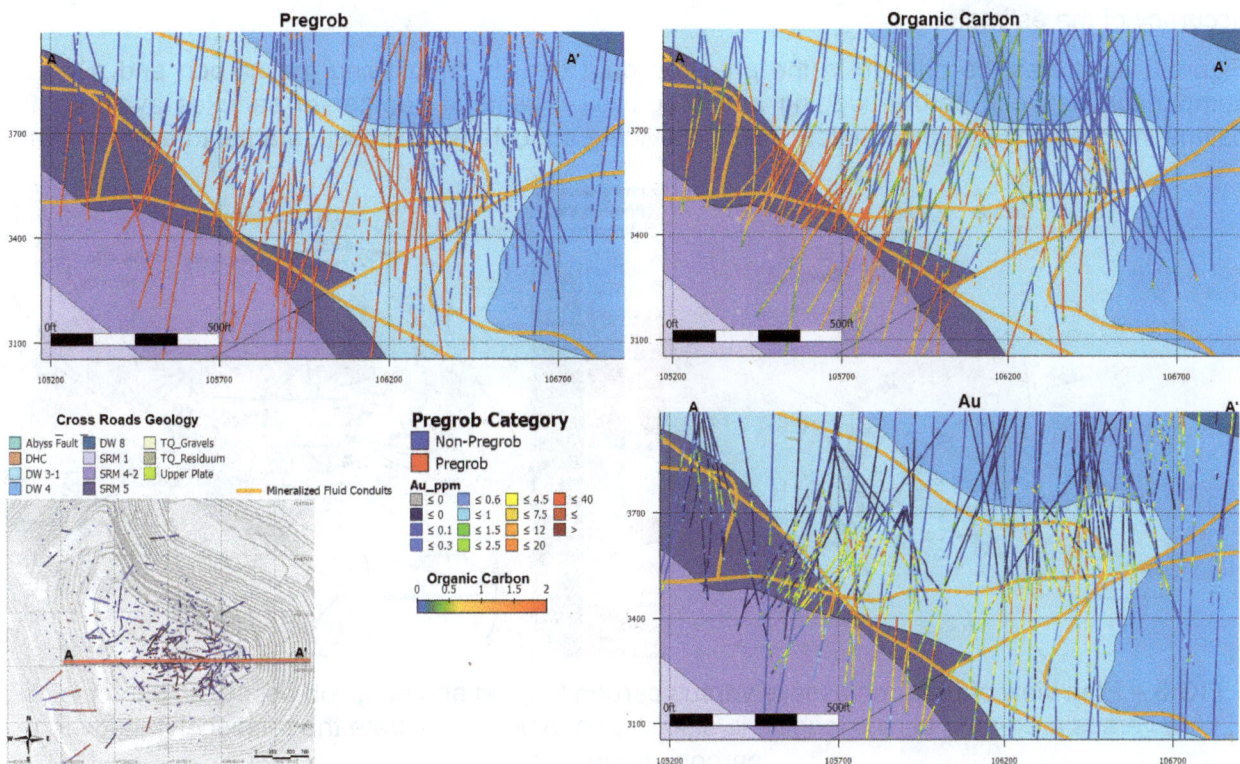

FIG 2 – Litho-structural model showing preg-rob indicators, organic carbon, gold assays, and currently understood mineralisation trends.

Metasomatism associated with the emplacement of intrusions appears to impact the organic carbon content within surrounding host rocks, as organic carbon is often observed to be depleted near intrusive features. The Crossroads open pit is located approximately 2.5 km to the south-east of the Cretaceous Gold Acres stock, thought to be responsible for marbelisation of limestone units within the same area. The proximity of this deposit to the Gold Acres stock is inferred to be the main reason for the unpredictable nature of the organic carbon observed at Crossroads, however, further studies are ongoing to define these local controls.

Preg-rob and carbon alteration correlation

It was recognised that the number of preg-rob assays was insufficient to accurately define the boundary between preg-rob and non-preg-rob material. A significant amount of carbon logging data was available. Further investigation revealed a strong correlation between carbon logging and preg-rob behaviour, as shown in Table 2.

TABLE 2

Carbon logging versus preg-rob statistics.

Name	Count	Mean	Minimum	Lower quartile	Median	Upper quartile	Maximum
S	2927	**91**	0	**98**	**100**	100	100
M	5578	**80**	0	**76**	**99**	100	100
W	9651	**64**	0	14	**90**	100	100
na	34 599	16	0	2	3	7	100

In Table 2, where both preg-rob assay and carbon logging data are available, samples labelled as strong ('S'), medium ('M'), or weak ('W') generally exhibit preg-robbing characteristics. Conversely, samples without carbon logging data are typically classified as non-preg-rob. While carbon logging is not a perfect match to preg-rob behaviour, the additional information substantially improves the accuracy of the estimates.

Figure 3 illustrates the change in the preg-rob data density by utilising the carbon and a visual comparison between carbon logging and preg-rob%. The total number of samples available for estimating preg-rob is approximately tripled providing a much more robust data set to estimate with.

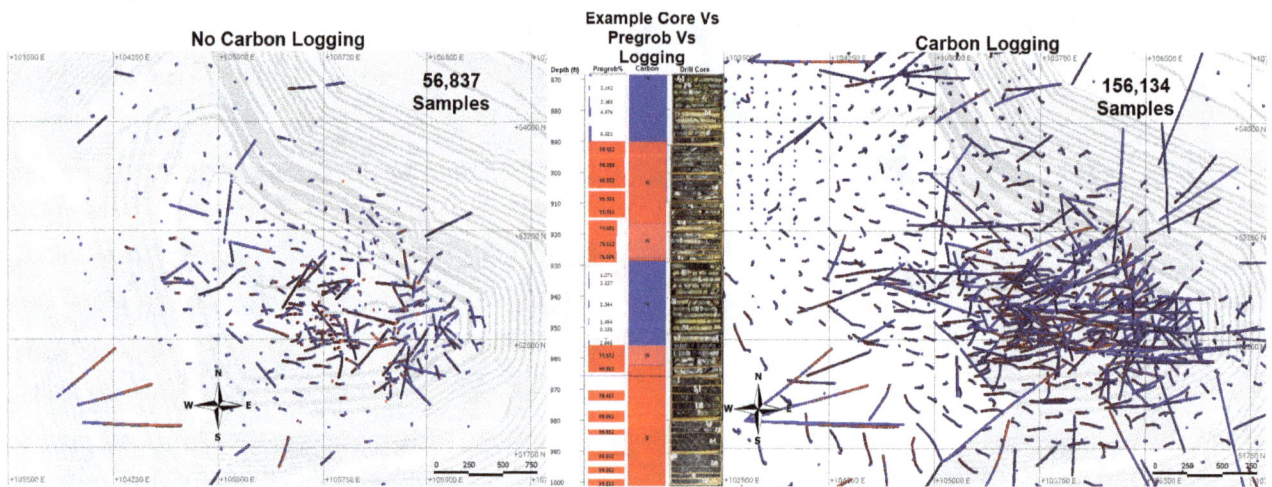

FIG 3 – Plan view of preg-rob assays versus carbon logging and preg-rob assay. an example of core versus preg-rob% versus logging is also shown to better illustrate the relationship between carbon and preg-rob.

CROSSROADS MINING

Crossroads utilises a Hitachi EX 5500 hydraulic excavator (35 yd³ bucket capacity), a P&H 4100 (approximately 60 yd³ bucket capacity), and a P&H 2800 (approximately 40 yd³ bucket capacity). The truck fleet primarily consists of Komatsu 930E trucks (~320 ton capacity) and CAT 795 trucks (372 ton capacity). The minimum selective mining unit (SMU) that Crossroads applies is 40 ft × 40 ft × 50 ft based on the current mining fleet, although further work is needed to confirm this assumption. Studies have shown that using 50 ft benches instead of 25 ft benches results in more efficient load times and significantly improves the total tons per hr that can be moved from the Crossroads pits. Considering SMU is critical because the selectivity of the equipment could lead to preg-rob contamination of non-preg-rob material.

GEOSTATISTICAL MODELLING

The purpose of geostatistic models is to analyse and estimate spatial data that has not been directly measured. Geostatistical models are estimated at unsampled locations by utilising the spatial correlation at sampled locations derived from the sample location to create a continuous map. It is of particular importance to understand the physical behaviour of the variables being modelled and ensure that the behaviour reflects in the resultant estimate. For example, grade variables typically exhibit linearly properties, where blending high-grade material with low-grade material will result in medium grade material. The choice between linear and non-linear estimation techniques can have a significant impact on the resultant geostatistical modelling and must be considered carefully.

Linear estimation techniques

At NGM the most common methods of linear resource estimation being utilised are Ordinary Kriging (OK) and Inverse Distance Weighting (IDW).

The ordinary kriging system can be written as:

$$Z^*(\boldsymbol{u}) = \sum_{i=1}^{N} \lambda_i Z(u_i) \tag{3}$$

where:

$Z^*(\boldsymbol{u})$ unsampled location to be estimated

λ_i is the weight assigned to the sampled location $Z(u_i)$

N is the number of samples utilised for estimation

The ordinary kriging weights can be solved as such:

$$\sum_{j=1}^{N} \lambda_j \gamma(\mathbf{z_i}, \mathbf{z_j}) + \mu = \gamma(\mathbf{z_i}, \mathbf{z_0}), \quad i = 1, 2, \dots, N, \tag{4}$$

$$\sum_{j=1}^{N} \lambda_j = 1$$

Where:

$\gamma(\mathbf{z_i}, \mathbf{z_j})$ represents the spatial semi-variogram between two known sample locations

         represents the spatial semi-variogram between the estimation location and a known sample location

μ is the LaGrange multiplier

Additionally, information on Ordinary kriging can be found in Journel (1989).

The IDW equation can be written as:

$$Z^*(\mathbf{u}) = \frac{\sum_{i=1}^{N} \frac{Z(\mathbf{u_i})}{d(\mathbf{u_i})^w}}{\sum_{i=1}^{N} \frac{1}{d(\mathbf{u_i})^w}} \quad \alpha = 1, 2, \dots, K \tag{5}$$

Where:

$d(\mathbf{u_i})$ is the distance from the estimation location to the sampled location $\mathbf{u_i}$.

w   is the estimation power.

The IDW equation can be simplified to:

$$Z^*(\boldsymbol{u}) = \sum_{i=1}^{N} \lambda_i Z(\mathbf{u_i}) \quad \alpha = 1, 2, \dots, K \tag{6}$$

By setting:

$$\lambda_i = \frac{\dfrac{1}{d(\mathbf{u_i})^w}}{\sum_{i=1}^{N} \dfrac{1}{d(\mathbf{u_i})^w}} \tag{7}$$

The OK and IDW equations both act as a linear averaging system based on a distance based weighting scheme. From the processing and assaying method sections above, we know that preg-rob does not average linearly; thus, OK or IDW will not work to estimate preg-rob. Two issues arise:

1. The spiked concentration of the assay is not always known as it has changed throughout the years and was not always recorded, or slightly different methods were employed by different assay labs.

2. Due to the AuPr assay not producing a true preg-rob capacity means the assay is incomplete.

In other words, a AuPr equal to 3.4 g/t and a AuPr equal to 34 g/t could have the same preg-rob capacity and would likely result in a lower AuPr estimate between the assays. For these reasons a categorical estimation method based on the preg-rob estimate was utilised. Additionally, a simulation method was utilised to ensure the point distribution was accurately reproduced, prior to upscaling to the mining they are upscaled to the relevant SMU. The simulation method also allows additional risk management and mitigation in future years of mining.

Non-linear estimation techniques

In recent years the development of machine learning techniques has led to a multitude of new ways of dealing with non-linear variables (Silva, 2024); however, for the purpose of this paper the focus will be on utilising simulation and pre-processing data into an indicator framework. Simulation provides a suitable framework for dealing with non-linear variables as it avoids concerns with averaging, loss of variability and potential bias. Common methods of simulation Sequential Gaussian Simulation (SGS), Indicator Simulation, and Turning Bands Simulation (TBS). The simulation method implemented in this workflow utilises TBS embed into a Truncated Pluri-Gaussian (TPG) workflow.

The TPG can be summarised as:

- Design the truncation rules.
  - The truncation rule should take into consideration the relationship between the categorical variables. Geological expertise and transition probabilities can help design the truncation rules.
- Define the required inputs.
  - High resolution trend model of all categories
  - Indicator variograms of all categories.
- Infer the required parameters for the TPG Simulation
  - Variograms of the latent Gaussian variable
  - Infer the data. Impute multiple realisation of the data in Gaussian space.
- Apply the TPG.
  - Multi-Gaussian Simulation
  - Apply Truncation Rules

More on TPG can be found in Silva and Deutsch (2018).

CROSSROADS IMPLEMENTATION

In this section the TPG workflow implemented at Crossroads will be discussed. The new workflow will be compared to the previous IDW workflow with visual validation and relative impacts to the final ore routing.

TPG workflow

The selection of composite length and compositing method plays a crucial role in the accuracy of the estimate. Practitioners should consider factors such as the typical assay length, block size, and the appropriate points to split composites. Generally, it is recommended to use a composite length that is 50 per cent to 100 per cent of the block size and limit the number of samples to 2 or 3 per holes to minimise the impact of the string effect (Markvoort and Deutsch, 2024).

Block sizes should also be determined based on the data spacing. A spacing of approximately 1 to 1.25 times the block size is typically ideal to achieve the desired level of smoothness. However, in a simulation workflow, the sample distribution is reproduced directly, and smoothing is not applied. In addition to length-based splitting, composites may also be divided into additional categories if necessary for the analysis.

For the TPG workflow, a composite size of 10 feet was selected at Crossroads, as this matches the most common assay length at the site. To ensure consistency, the sample-scale data distribution was reproduced within a simulation framework, resulting in a 10 ft × 10 ft × 10 ft block size to align with the composite length. This is particularly important for preg-rob estimation, as preg-rob does not average linearly.

After compositing a trend model (Figure 4a) is generated of the categorical data and the residuals (Equation 8) used to calculate the variogram (Figure 4b) used in the TPG global proportion dictionary (Figure 4c). Because this is a binary implementation of TPG there is only one way to set-up the truncation tree which is shown in Figure 4c. From the truncation tree a latent variable is generated. The variogram of the latent variogram is then fit (in Figure 4d) and used to imputed. In areas where data is sparse and trend modelling will be unreliable and results in artifacts the estimate will default to preg-rob rob; however, most of these area are not within current LOM designs.

$$Pregrob_{residual} = Pregrob_{Indicator} - Pregrob_{trend} \tag{8}$$

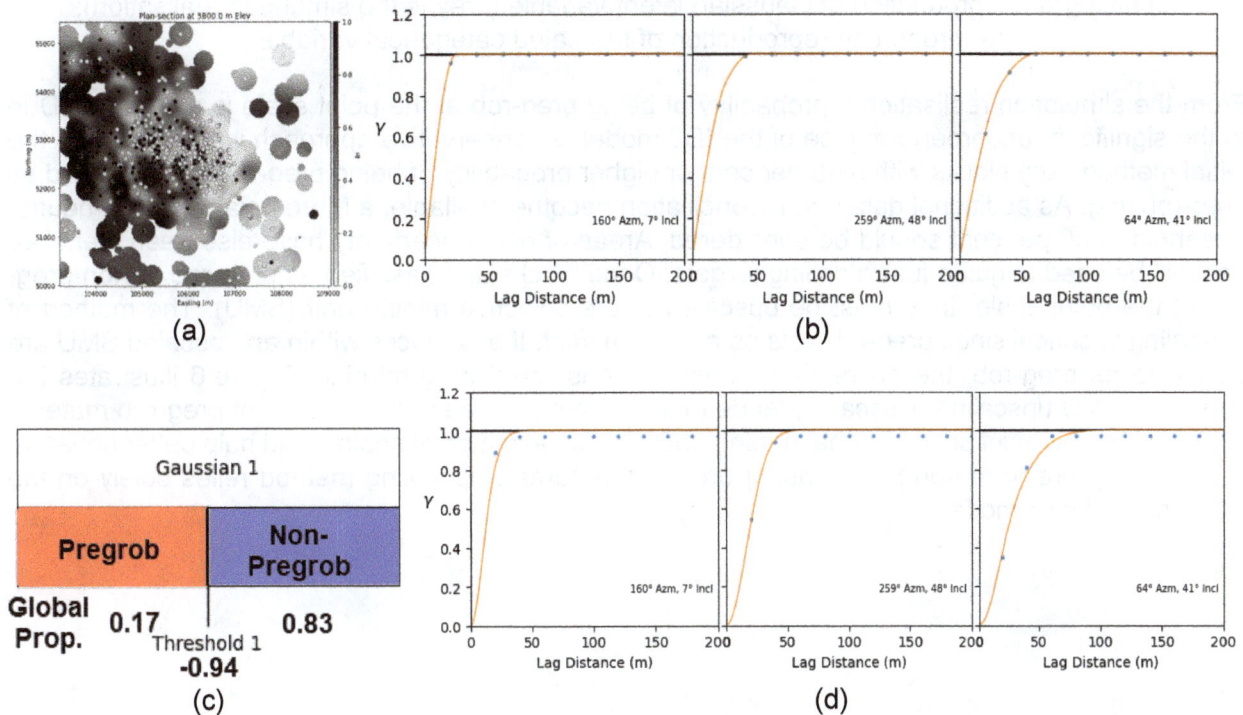

FIG 4 – Crossroads trend model a different elevations: (a) preg-rob trend model; (b) variogram of the preg-rob residuals; (c) truncation tree; (d) variogram of the latent variable.

The distribution of the latent variable imputation is checked to ensure that it still reflects the original distribution and is then used for simulation (Figure 5a). The simulation is checked to ensure that the variograms are reproduced (Figure 5b), the latent variable distribution is reproduced (Figure 5c), and the original proportions of the of the categorical preg-rob indicator are reproduced (Figure 5d). All validations seem reasonable, except we see some reduction of continuity in the 3rd direction of the variogram.

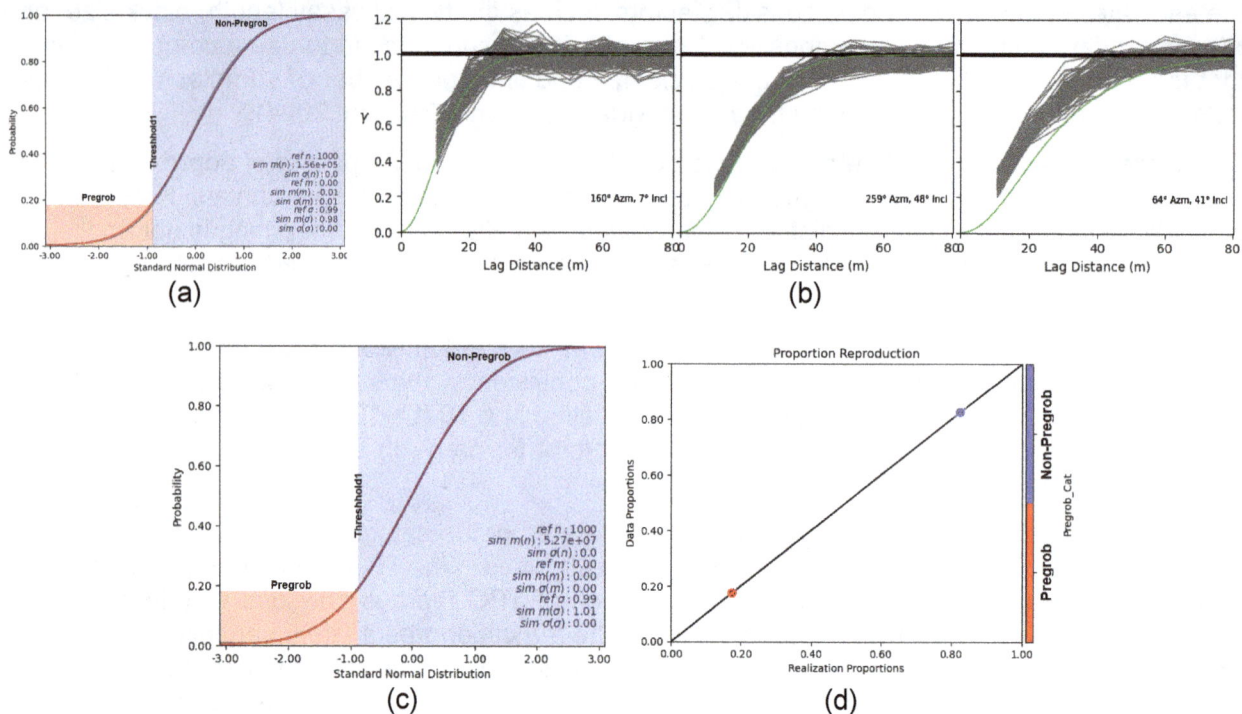

FIG 5 – TPG validation plots: (a) imputation of the Gaussian latent variable; (b) variogram reproduction of the simulated Gaussian latent variable (grey is the simulated realisation); (c) histogram reproduction of Gaussian latent variable (grey is the simulated realisation); (d) proportion reproduction of truncated categorical variable.

From the simulation realisation a probability of being preg-rob at the point scale is generated. Due to the significant underperformance of the ID2 model, a conservative approach was chosen as the initial method. Any blocks with a 40 per cent or higher probability of being preg-rob are classified as preg-robbing. As additional data and reconciliation become available, a more balanced, risk-neutral threshold of 50 per cent should be considered. Areas of high uncertainty have also been identified and will be used to guide future drilling targets. Once blocks are classified as preg-rob or non-preg-rob at the point scale, they must be upscaled to the selective mining unit (SMU). The method of upscaling is critical since preg-rob acts as a contaminant. If any blocks within an upscaled SMU are classified as preg-rob, the entire SMU must be considered preg-robbing. Figure 6 illustrates the impact of SMU upscaling. Upscaling tends to significantly increase the volume of preg-rob material compared to the point scale. Further mining optimisation at the point scale could help better separate preg-robbing ore from non-preg-robbing ore, as the current upscaling method relies solely on the origin of the block model.

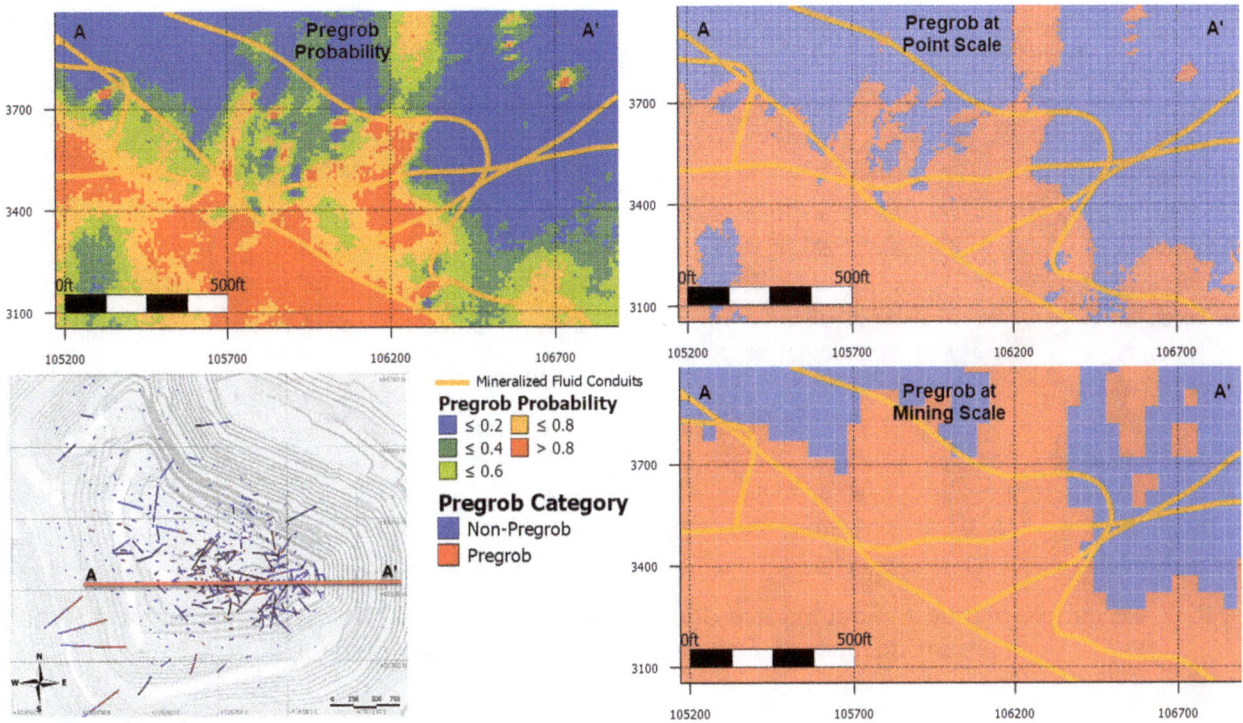

FIG 6 – Section showing the probability to be preg-robbing versus point scale preg-rob versus mining scale preg-rob.

Visual comparison and impacts

Figure 7 shows a plan-section example highlighting blocks identified as preg-rob by both the ID2 method and the TPG method, compared to core samples. In the core, a layering of preg-rob and non-preg-rob material is observed. Treating compositing and estimation linearly can significantly underrepresent preg-rob material, as seen in the comparison between the ID2 block model and the TPG block model.

The section marked with a red circle contains a mix of preg-rob and non-preg-rob material in proximity. Using the ID2 method results in a significant portion of the material being classified as non-preg-rob, despite sporadic occurrences of preg-rob in the area. In contrast, the TPG method flags the material as preg-rob, primarily because it represents the estimate at a point scale rather than averaging preg-rob material linearly.

A similar result is evident in the cross-section in Figure 8, where the ID2 method significantly underestimates preg-rob material, while the TPG method avoids misrepresentation. In both Figures 7 and 8, misclassification typically occurs in zones where the assays indicate a mix of preg-rob and non-preg-rob material.

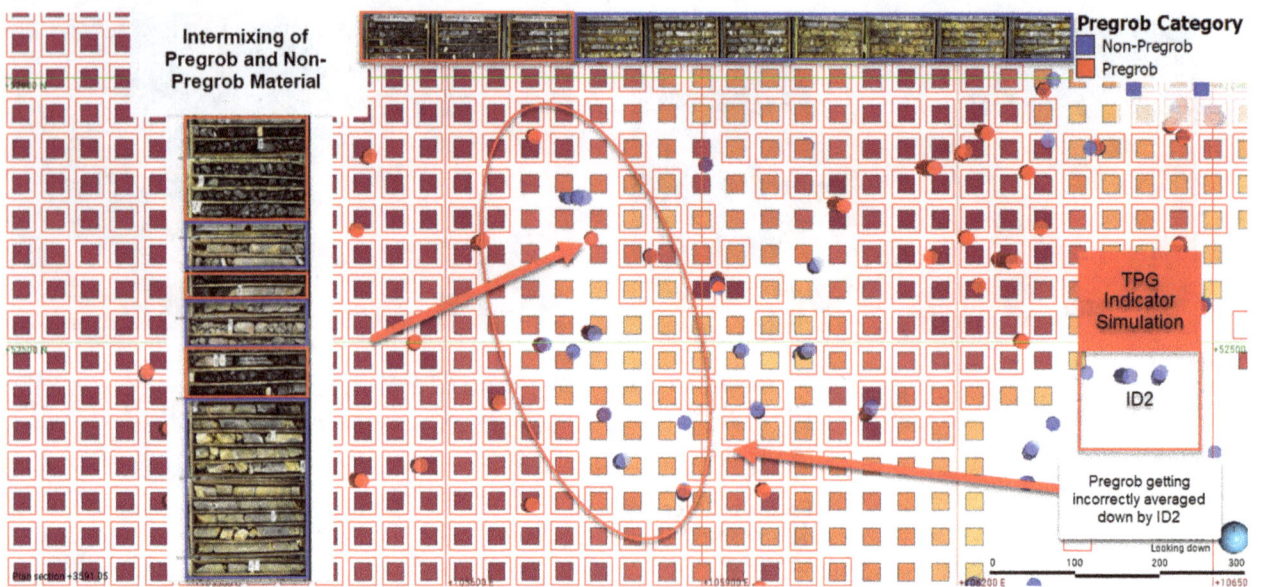

FIG 7 – Example of preg-rob getting averaged down by ID2 versus TPG Simulation showing lens of preg-robbing core. Showing only blocks that are preg-robbing.

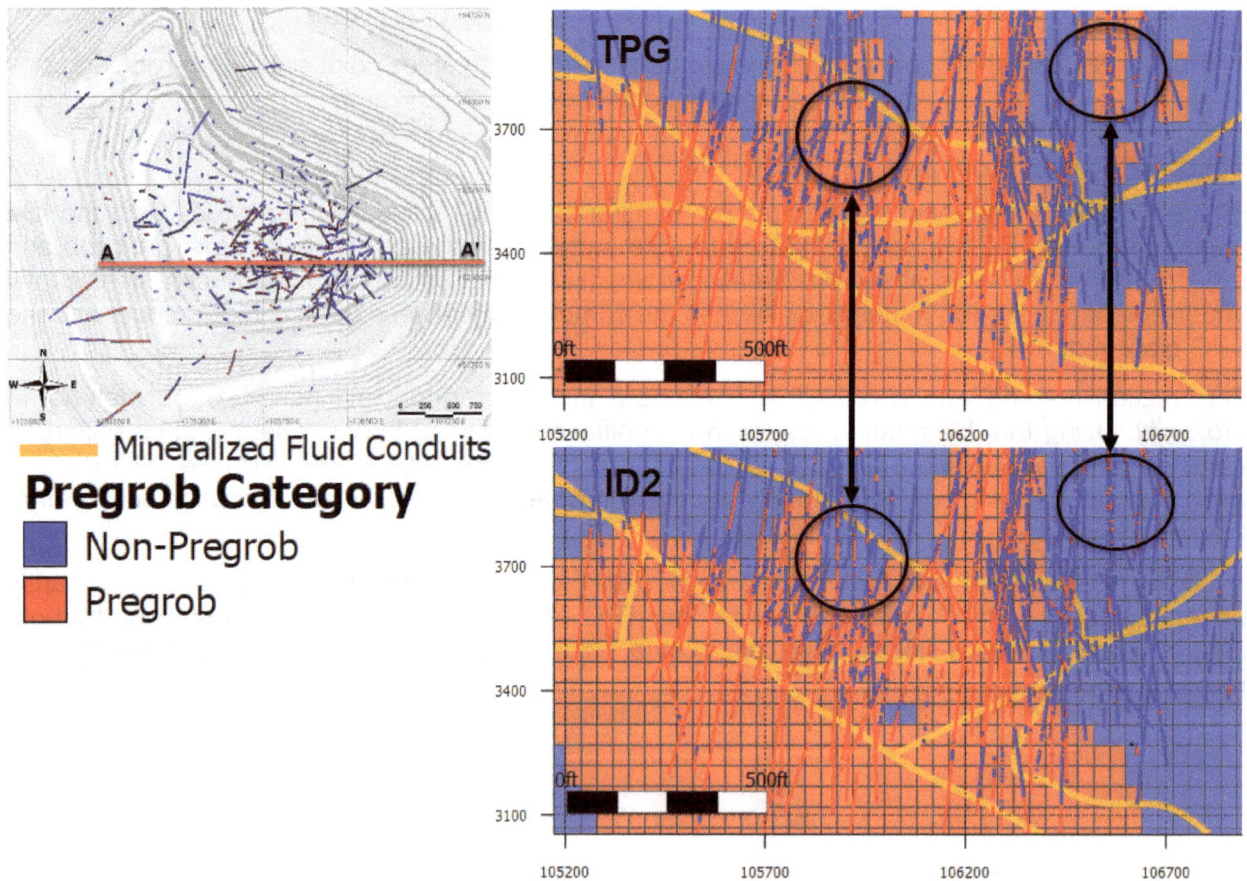

FIG 8 – Cross-section of TPG versus preg-rob.

Figure 9 compares the ID2 and TPG estimates to production data (blastholes) and production mining blockouts. Like earlier observations, while the ID2 method generally represents preg-rob and non-preg-rob data reasonably well, there are areas where it is incorrectly averaged down. Additionally, relying solely on the preg-rob assay lacks the resolution needed to match the production data that indicates preg-rob. Including carbon logging is essential, as it highlights preg-rob in certain areas where the preg-rob assays were not taken.

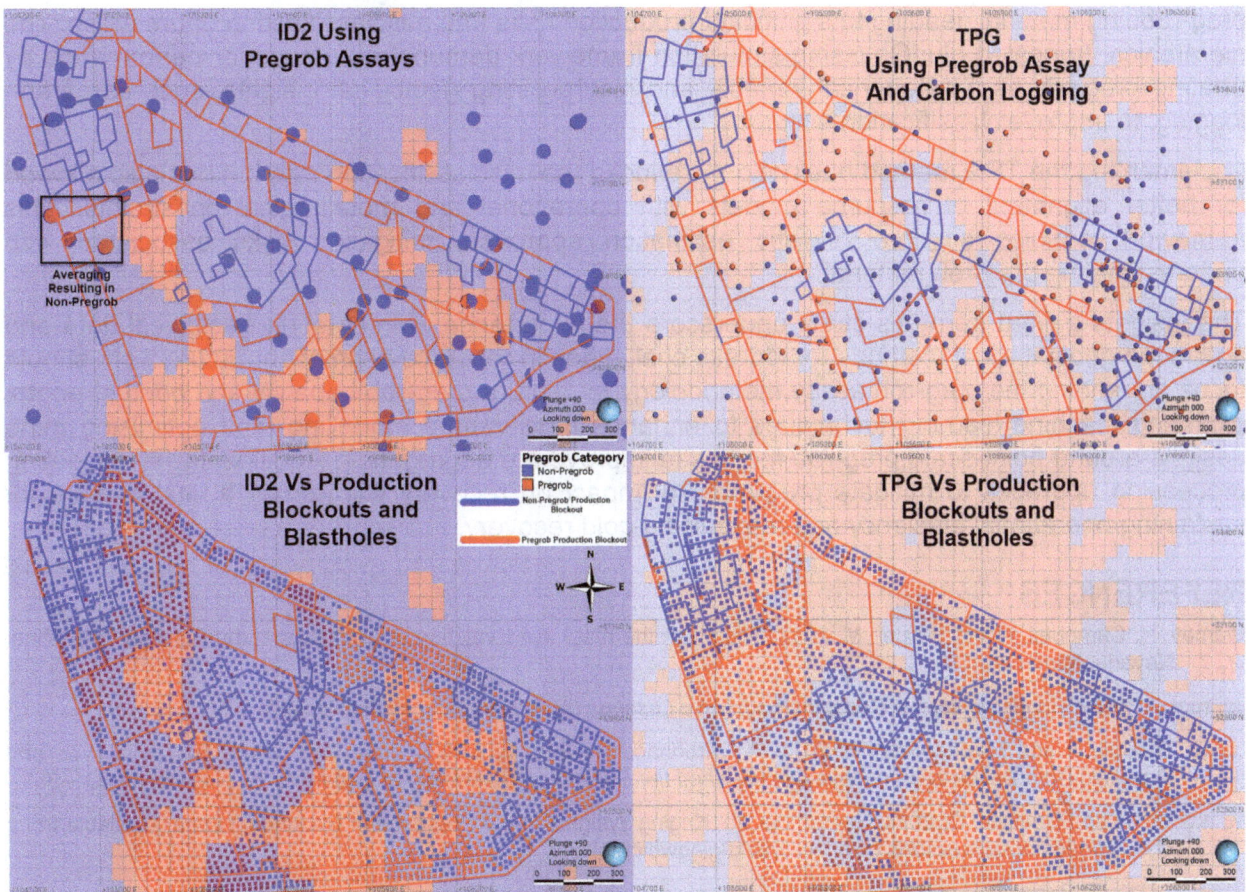

FIG 9 – Raw blasthole comparison to ID2 versus TPG preg-rob model.

It is also noteworthy to consider the impact of minimum SMU on production blockouts. In some areas designated as preg-rob, a small number of samples indicating preg-rob are contaminating the entire blockout. This occurs due to the processing constraints associated with preg-rob and the severe negative impact even a small amount of preg-rob can have on overall gold recovery.

Figure 9 clearly show that the TPG method utilising carbon logging estimates a significantly larger proportion of preg-rob material. When the TPG method was implemented, approximately 12 per cent of the total reserve ounces were flagged as misclassified. Due to the higher relative cost of a roaster process stream compared to an oxide process stream, most of these misclassified ounces are now considered waste. Table 3 summarises the preg-rob reconciliation for 2024. The new TPG estimate is slightly conservative, predicting more preg-rob tons than were present. In contrast, the ID2 estimate was overly optimistic, predicting significantly fewer preg-rob tons that are associate with a lower COG process. While a risk-neutral model typically aims for balance, the overall impact on production is less severe when preg-rob is overestimated rather than underestimated, as there remains an opportunity to consider these tons as ore.

TABLE 3

Preg-rob ton reconciliation 2024.

	TPG	ID2	Production	TPG % delta to production	ID2 % delta to production
Preg-rob tons	21 019 163	13 432 644	19 105 435	10%	-30%

CONCLUSION AND FUTURE WORKS

The study presented here emphasises the critical need for accurate and robust modelling techniques when dealing with complex geometallurgical variables like preg-rob. Traditional linear estimation methods, such as ID2, have proven to be inadequate in accurately capturing the spatial variability of preg-robbing material, leading to significant production risks and misclassification of ore types. The hierarchical truncated pluriGaussian simulation framework demonstrated superior performance by treating preg-rob as a categorical variable and incorporating geological insights, such as carbon logging, to improve classification accuracy.

Implementing the TPG method has not only reduced the misclassification rate but has also allowed for better alignment of resource models with operational and metallurgical constraints. This framework enables a more realistic estimation approach, mitigating risks associated with underestimating preg-rob material.

The insights gained from this study underscore the importance of integrating geological data and advanced geostatistical methods to address challenges in resource estimation. Future work should focus on refining the preg-rob assay methodology to capture preg-rob capacity and bond strength, exploring machine learning techniques to enhance multi-variate modelling, and optimising mine designs around point-scale preg-rob models to improve ore segregation. Additionally, investigating processing upgrades to mitigate preg-robbing impact on recovery will remain a vital avenue for sustaining operational efficiency and maximising gold recovery.

REFERENCES

Blamey, N, Campbell, A and Heizler, M, 2017. The Hydrothermal Fluid Evolution of Vein Sets at the Pipeline Gold Mine, Nevada, *Minerals*, 100.

Journel, A, 1989. *Fundamentals of Geostatistics in Five Lessons* (American Geophysical Union).

Large, R, Bull, S and Maslennikov, V, 2011. A Carbonaceous Sedimentary Source-Rock Model for Carlin-Type and Orogenic Gold Deposits, *Economic Geology and the Bulletin of the Society of Economic Geologists*, pp 331–358.

Markvoort, H and Deutsch, C, 2024, September 10. Kriging Weights in the Presence of Redundant Data. Available from: <https://geostatisticslessons.com/lessons/krigingweights>

Rhys, D, Valli, F, Burgess, R and Heitt, D, 2015. Controls of fault and fold geometry on the distribution of gold mineralisation on the Carlin trend, in *New concepts and discoveries, Geological Society of Nevada Symposium* (eds: W M Pennell and L J Garside), pp 333–389.

Silva, D and Deutsch, C, 2018. Guide to Categorical Modeling With HTPG, Edmonton Alberta: Centre for Computational Geostatistics.

Silva, D, 2024, December. Spatial Modeling of Geometallurgical, Resource Modeling Solutions.

Simoneit, B, 1994. *Organic matter alteration and fluid migration in hydrothermal systems,* pp 261–274 (Geological Society Special Publications: London).

Exploring tailings reprocessing options – from characterisation to preliminary block modelling of the Mount Isa Mines tailings storage facility

L Stefoni[1], A Wilhelmsen[2] and D Ambach[3]

1. Senior Geostatistical Engineer, Glencore, Dorval H9P 2N4, Canada. Email: lucia.stefoni@glencore.ca
2. Principal Geology, Glencore, Mount Isa Qld 4825. Email: aaron.wilhelmsen@glencore.com.au
3. Principal Projects, Glencore, Brisbane Qld 4000. Email: daisy.ambach@glencore.com.au

ABSTRACT

Mine waste reprocessing can offer a dual benefit of unlocking additional revenue while mitigating long-term environmental impacts. Mount Isa Mines is in the process of investigating reprocessing options for its tailings storage facility (MIM TSF), operational since 1958. The facility has accumulated tailings from multiple ore-types processed through various concentrators, creating a complex deposit. Spanning 1500 hectares, the MIM TSF presents significant opportunities due to the volume of material but also poses challenges for characterisation and modelling.

MIM TSF faces challenges due to sparse legacy data similar to many historical facilities. Although metallurgical accounting data exists, there are unknowns in the spatial distribution of metals and key material properties such as density, mineralogy, and metallurgical performance critical to evaluating reprocessing opportunities. To address these gaps, a phased drilling and characterisation program was developed, aligned with the exploratory nature of reprocessing investigations. This paper outlines the program, from sampling to resource estimation, and highlights key methodologies and insights.

The staged sampling program along with Geographic Information System (GIS) work package – integrating historical contour maps and satellite imagery with metallurgical data – resulted in a preliminary block model for the MIM TSF, integrating geochemical, geometallurgical, and acid-rock-drainage characterisation. The 3D model remains internal given the infancy of the project and is therefore not intended for reporting.

While this early-stage model provides a strong foundation for understanding the facility's holistic properties, further sampling and testing will be necessary to eventually develop a resource estimate that would comply with the requirements for reporting under the JORC Code (2012). Drill hole Spacing Analysis (DHSA) was implemented to understand the required sampling configuration for future Mineral Resource Reporting should potential economic extraction be demonstrated. The insights and diverse techniques from this study demonstrate how comprehensive, multi-disciplinary approaches to material characterisation are integral to assessing the potential for latent metal recovery and improving environmental sustainability at legacy mine sites.

INTRODUCTION

Mine waste reprocessing at any stage during the life of a facility has the potential to bring environmental and economic benefits. In the last century, advances in metallurgical processing technologies continue to allow for more efficient metal recovery in complex minerals or newly valued geochemical elements. Opportunities emerge with legacy waste facilities, sometimes hosting significant residual metal content which was not recovered in the initial mineral extraction process. As well, specific targeting of sulfide minerals during waste reprocessing could reduce acid mine drainage risks thereby achieving environmental improvements as well as enhance closure outcomes and costs.

The Mount Isa Mines tailings storage facility (MIM TSF) is located near the town of Mount Isa in North-West Queensland, Australia (Figure 1). MIM TSF has been operational since 1958, with ongoing deposition occurring and planned until the site's Life-of-mine. Tailings has been generated through two separate concentrators – copper, and zinc-lead, processing a mix of massive sulfide, copper-enriched supergene, and disseminated minerals. Spanning 1500 hectares (Glencore,

2024b), the MIM TSF is one of Australia's largest tailings storage facilities by footprint and will require extensive earthworks to construct a suitable cover system which can allow the land to be returned to native bushland at closure. On the upside, its sheer size presents an opportunity for reprocessing due to the large volume of material containing low-grade residual metals.

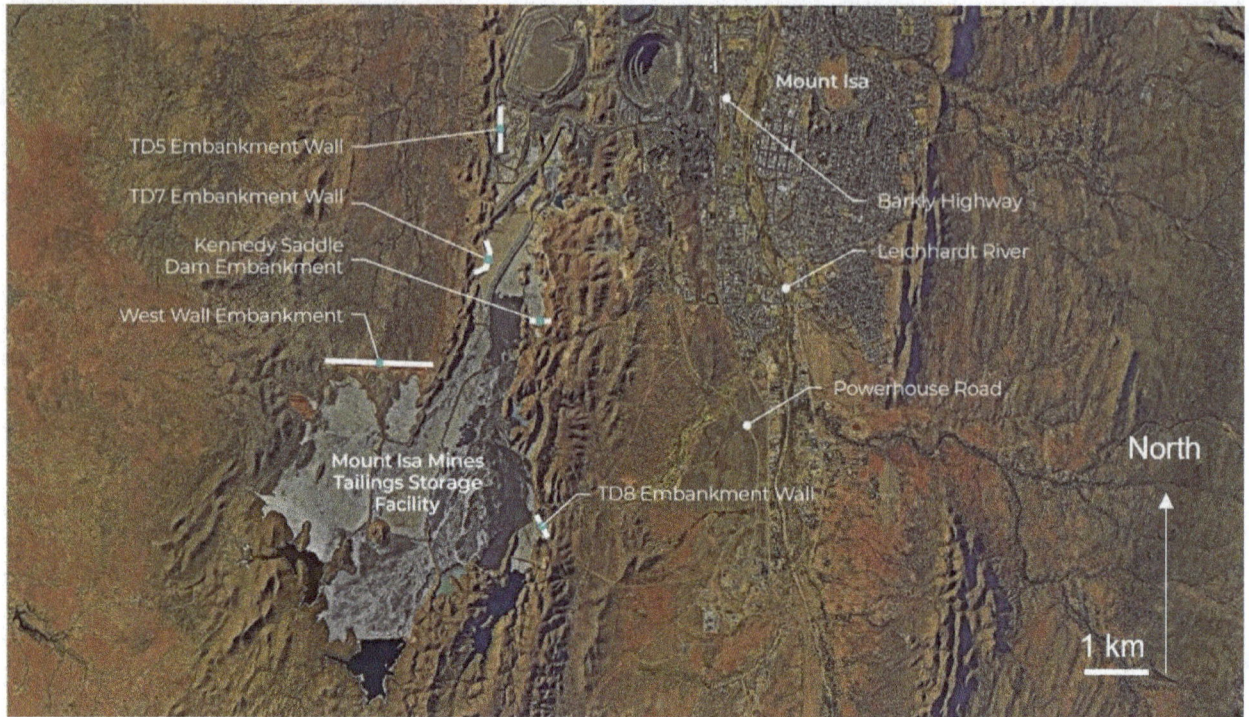

FIG 1 – Planview of Mount Isa Tailings Location relative to Mount Isa City, showing embankments. Image source: Glencore's website. Mount Isa Mines Tailings Storage Facility (Glencore, 2024a).

Cognisant of the opportunity to improve sustainability outcomes as well as recover residual metal value, Glencore initiated a concept study investigating the potential to reprocess tailings from the MIM TSF. At the onset of the study, there were significant data gaps with respect to critical geochemical and metallurgical characteristics of the tailings, and their spatial distribution, inhibiting the project team from estimating the value of the resource and exploring pathways for mineral recovery.

The objective of this paper is to provide an overview of key methodologies, from sampling to preliminary block modelling, used to address the challenges of characterising tailings at a large-scale, long-standing site with incomplete historical records. It also shares techniques for developing an exploratory tailings characterisation and resource estimating program, offering a foundation for other sites facing similar challenges in investigating mine waste repurposing or reprocessing opportunities.

LITERATURE REVIEW

The following literature review covers the core concepts developed in this publication. First, the review includes papers discussing common strategic targets such as reprocessing of tailings material with the objective to improve economic and environmental aspects of legacy mine sites. Second, articles about tailings characterisation such as sampling, geochemistry and mineralogy are reviewed. Third, literature related to block modelling of geochemical and geometallurgical variables for preliminary resource estimation is presented. Fourth and last, the review includes selected industry applications with learnings on best practices in tailings reprocessing and characterisation.

Improved strategies of tailings management

Dold and Fontboté (2001) made a case study of successful mine waste reprocessing projects involving complex tailings deposits. In this publication a comparison of three tailings from Andina, El Teniente and El Salvador was performed. The geochemical, mineralogical, and microbiological compositions with focus on the transition zone between oxides and sulfidic flotation tailings was studied. All three tails are low-sulfide and low-carbonate.

Jackson and Parbhakar-Fox's (2016) study examines the effectiveness of a water cover in mitigating acid mine drainage (AMD) at the Old Tailings Dam in Australia. One key finding is that water covers effectively reduce sulfide oxidation and AMD generation. The study highlights the importance of integrated mineralogical and geochemical analysis for effective tailings management.

Other works include Siqueros Valencia *et al* (2022) who performed a general review of mine waste reprocessing, discussing economic and environmental benefits. Kossoff *et al* (2014) examine the risks associated with mine tailings dams, including their potential for failure and the severe environmental consequences. Kinnunen *et al* (2022) review circular economy strategies for mine tailings in Finland.

Tailings characterisation

Juutinen, Seitsari and Pietilä (2023) provide detailed sampling techniques and analytical methods chosen for the geochemical and mineralogical characterisation of the Rautuvaara mine site tailings inferring its environmental conditions and its resource potential.

Other works by Smith *et al* (2006) focus on statistically based, cost-effective sampling strategy for mine waste materials. James *et al* (2011) investigated the dynamic properties of gold mine tailings in Quebec. Wang *et al* (2018) investigated the rheological properties of tailings from different mines.

Block modelling for preliminary estimation

Mineral resource estimation comprises several aspects such as data analysis using geostatistical techniques, material quantification using interpolation and block modelling techniques, and resource classification through risk assessment techniques. Geostatistical methods have proven useful for spatial interpolation and uncertainty assessment of mining projects and operations with an increasing number of applications at legacy sites and tailings facilities.

Wilson *et al* (2021) covered a case study from a tailing's facility located in Taltal in Northern Chile in which tailings are revalourised as input sands in the cement manufacturing chain. Composition and properties of tailings can vary significantly throughout the dam; an assessment was performed on geochemical variability of the tailing's materials using advanced geostatistical methods such as discrete event simulation (DES) to understand the risk in the output cement quality.

Blannin *et al* (2023) investigated the potential for resource recovery and environmental impact mitigation in the Davidschacht tailings storage facility through 3D geostatistical modelling. Universal sequential Gaussian simulation aimed to account for the strong trends and heterogeneity in chemical distribution within the Davidschacht TSF. Findings suggest the method provides robust tonnage estimates even with a small number of drill holes. The MIM Tailings Project focused on Ordinary Kriging instead, a traditional estimation methodology, given trends and heterogeneity were not as prevalent in its early data sets.

The paper also provides an insightful discussion on the application of industry reporting guidelines published under umbrella of Committee for Mineral Reserves International Reporting Standards (CRIRSCO, 2024) and the unique challenges presented in applying standard reporting codes to resource estimating for tailings facilities. This is due to significant differences between anthropogenic and geogenic deposits in terms of resource and reserve definitions, as well as the distinct re-mining and re-processing methods required. Innovative approaches are needed to effectively adapt these codes to the specific characteristics of tailings deposits.

Industry applications

The following refers to studies specific to legacy mine sites and complex tailings deposits:

Minera Valle Central and Amerigo Resources published public reports (NI 43–101) for their Initial Study (Robert and Henderson, 2014), Concept Study (Robert and Henderson, 2018) and Pre-feasibility Study (Global Resource Engineering Ltd, 2022) for the reprocessing of the Cauquenes sulfide tailings facility located in central Chile. A blended strategy is used with fresh feed from the operating El Teniente Mine. Cauquenes is a 100-year-old legacy facility that does not meet current environmental regulations and must be dismantled. MIM Tailings variography modelling parameters were benchmarked against Cauquenes for comparison at each study level.

Other works include Surrette *et al* (2024) investigating the heterogeneity of the Cantung Mine tungsten tailings located in the Northwest Territories in Canada assessing its reprocessing potential for value recovery and remediation purposes. Dinis *et al* (2020) investigate the characterisation of ancient mine wastes from the Cabeço do Pião tailings impoundment in Portugal, highlighting the importance of thorough characterisation of ancient mine wastes to inform effective environmental management and potential resource recovery strategies.

TAILINGS CHARACTERISATION METHODOLOGY

Several work packages were implemented to be able to characterise the physical, geochemical and geometallurgical aspects of the MIM tailings facility requiring a multidisciplinary team. First, the overall volume of the facility was defined and later split in sub-domains that align with changes in metallurgical processing attributes. Second, two staged drilling campaigns were deployed to understand its geochemistry and geometallurgy. Third, physical attributes such as porosity, moisture and bulk density were investigated through various methodologies. Fourth, mineralogy assessment to define the existence of complex speciation, mineral associations and particle liberation to assess reprocessing opportunities.

GIS geological and topographical mapping

An integrated GIS scope was fundamental to the entire characterisation and modelling program. SLR Consulting, an international sustainability consultancy, were engaged to complete this scope. Their methodology for volume modelling was intuitive, stacking surfaces from the oldest at the bottom (pre-deposition) to the youngest upwards, until reaching most current up-to-date topography at the very top. Its main objectives were to:

- Refine the pre-deposition surface, from available legacy data sets.

- Generate metallurgical domains, useful in the metallurgical validation of drilling and block model results.

- Develop a deposition volume to be used in the block modelling stage.

A significant focus of the work developed by SLR Consulting was on reviewing and refining pre-deposition and historic surface topographies, particularly given the importance in generating geometallurgically controlled tailings volumes for preliminary resource estimation. Efforts were made to align deposition locations with provided concentrator and mine production profiles to improve understanding of what material was deposited in which areas of the MIM TSF.

First, SLR completed a review of historical aerial imagery, survey, constructions records, digital elevation data and orthophotos. A data rating system was applied to each data set to assess the data quality and its ability to be translated into a surface extent that aligns with the temporal extents of the metallurgical domains.

The existing pre-deposition surface was developed from topographic survey contours that were digitised from georeferenced PDFs and converted from imperial to metric measurements. The digitised contours were interpolated into digital surface models (DSM) with a 10 m resolution on the horizontal Planview – XY axis – and 1 m resolution on the elevation – Z axis. As with any model, the accuracy is contingent upon the quality of the underlying data.

Given the age of the facility (with deposition commencing in 1958), there may be inherent inaccuracies with using historic spatial data, due to the resolution of data and changes in survey practices; as such, historical data can be a significant source of error, directly impacting the accuracy of final tonnage estimations. SLR Consulting was able to include additional legacy data sets in the following formats: preliminary 10 m resolution DXF surface, drawings, contours, and LiDAR; and with the A2 drilling campaign intersecting the basement, SLR was able to further refine the pre-deposition surface model DSM (Figure 2).

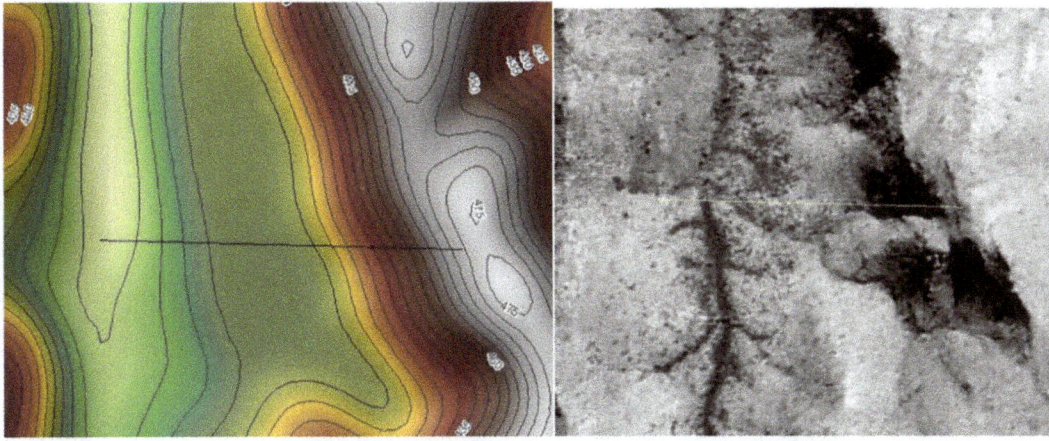

FIG 2 – Pre-deposition surface and 1956 imagery.

Drilling and sampling

Overview of phased-sampling campaigns

To assess the potential resource within the MIM TSF, a two-phased characterisation program was implemented to investigate the geochemical, mineralogical and physical properties of the tailings, as well as their respective spatial distributions.

To develop the initial phase of the program, the Mine Waste Transformation Through Characterisation (MIWATCH) Group from the University of Queensland (UQ) were engaged due to well renowned experience on tailings characterisation. This phase, aptly named Program A1, included sampling of 21 holes which were drilled to a depth of approximately 10 m using hand augers. These samples were collected in May 2023 and were sent for geochemical and mineralogical analysis in a program designed and implemented by the MIWATCH group.

Due to the depth limitations of hand augering, a subsequent phase of traditional core drilling took place (with modified drill bit, basket lifters and drilling techniques). Program A2, was conducted in December 2023 with a total of 254 m drilled across ten holes. The objective of the A2 program was to obtain undisturbed representative samples to bedrock across the entire facility and to improve upon certain logistics and methodology aspects of the A1 sampling. Both A1 and A2 had an additional objective to assess short-distance *in situ* sampling variability, by duplicate and triplicate sampling comparisons.

The most recent 2023 topography along with collars, coloured by sampling campaign are shown from different views in Figures 3 and 4. Most samples are located along the main access road due to constraints with sampling at a facility with active ongoing deposition. At the time of sampling, deposition was occurring in several areas of the TSF. As a result, certain areas were considered unsafe for personnel and drilling equipment to access, as its surface was not fully consolidated. Additionally, decant water was present in certain areas of the TSF. Access to other areas of the facility could have been facilitated but would require construction of access roads to ensure safety of personnel. A vertical exaggeration of five was used to highlight the terrain features, collar's elevations range between 384 to 418 m above sea level.

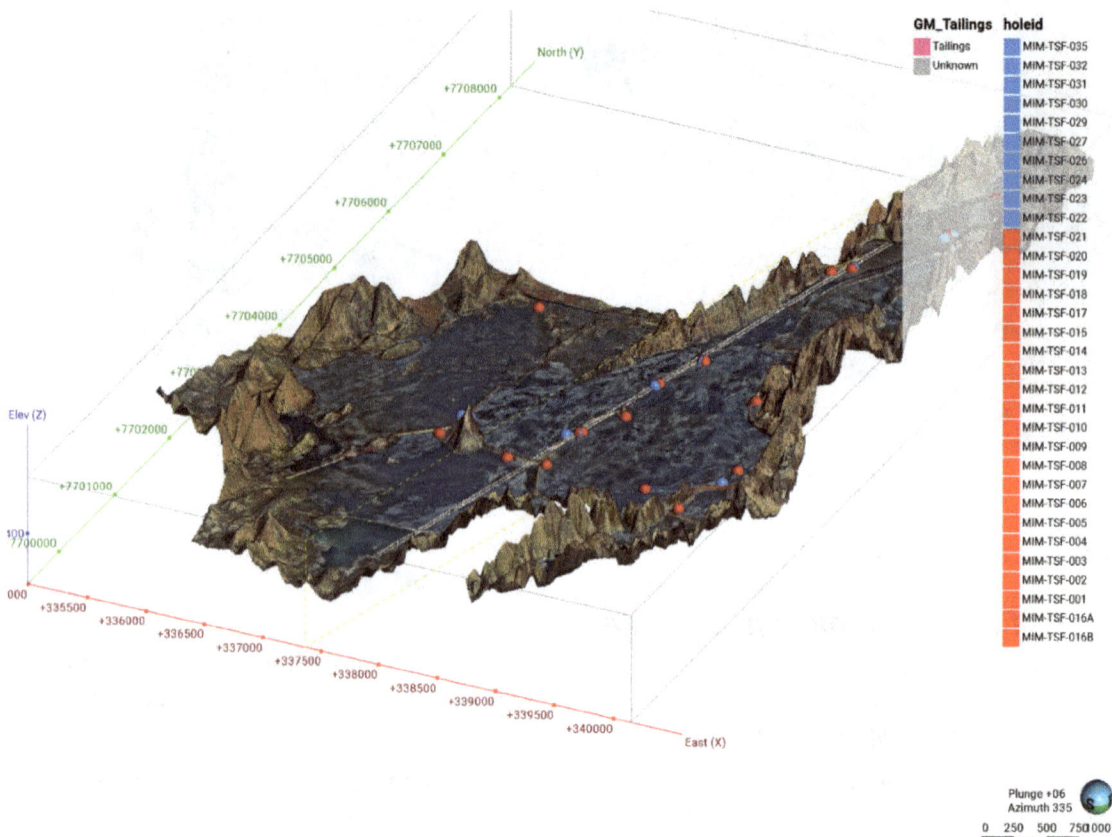

FIG 3 – 3D view of 2023 topography and collars of samples coloured by sampling campaign (red A1 and blue A2). Vertical exaggeration of five.

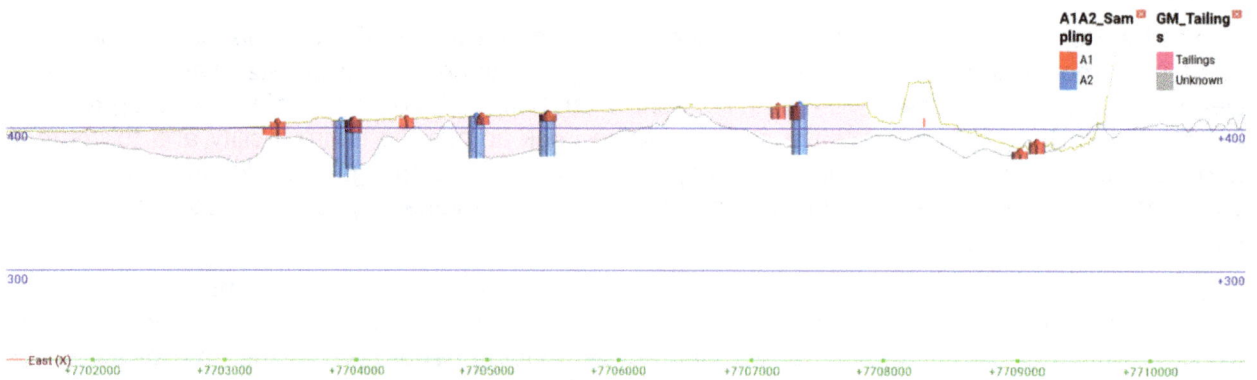

FIG 4 – Section view along the road, displaying topography and sample collars coloured by sampling campaign (red A1 and blue A2) and vertical traces in black. Vertical exaggeration of five to highlight end of hole feature. North points to the right of the above image.

Program A2 – modified drilling technique

Due to the continuous, active deposition within the MIM TSF, the sediments have not fully consolidated, resulting in high moisture content and uncertain crust thickness. Consequently, through the risk assessment process it was decided that drill pads would need to be constructed on the tailings surface to provide a stable platform for drilling to eliminate the hazard of liquefaction (see Figure 5 for safety drill pad).

FIG 5 – DE710 diamond drill rig and support truck on MIM-TSF-022 drill pad looking NNW.

Due to the variably wet and unconsolidated nature of the TSF material, a tailored conventional drilling method was implemented for achieving suitably representative samples. The modified drilling approach adopted minor modifications to the drill-bit cutting face to improve penetration and sample recovery. Drilling involved pushing through the tailing's material with no rotation, relying on the weight on the bit for penetration. No drill fluids were used to minimise impacts on moisture content or potential wash out of the sample. Modifications made to the drill bit and use of wire basket lifter are illustrated in Figure 6.

FIG 6 – Modified diamond drill bit on drill rod and wire basket lifter.

These drilling modifications are adopted from those implemented for the Century Tailings Deposit; further information on the drilling approach, feasibility study results and restart plans for the Century Mine can be found in the New Century Resources ASX announcement (2017).

Assay result comparison A1 versus A2

A1 and A2 assays were conducted by ALS Geochemistry, a well-regarded certified laboratory used extensively by Glencore. ALS has facilities in Mount Isa, allowing for cold storage prior preparation and analysis. Given the exploratory nature of the campaign and the need to evaluate both major metals and trace elements, a comprehensive characterisation package was chosen for both programs.

Figure 7 shows there is a strong correlation between assays for A1 and A2 sampling campaigns at upper levels. There is limited variance between the grades observed between the two campaigns at the same depths; the top 10 m of A2 depositional grades and characteristics align with those observed from the A1 Campaign. Global differences are due to lower grades at deeper depths observed in A2. Grades from A2 show a stable behaviour with slightly lower grades at depth for copper, zinc, lead and silver (Cu%, Zn%, Pb%, and Ag g/t).

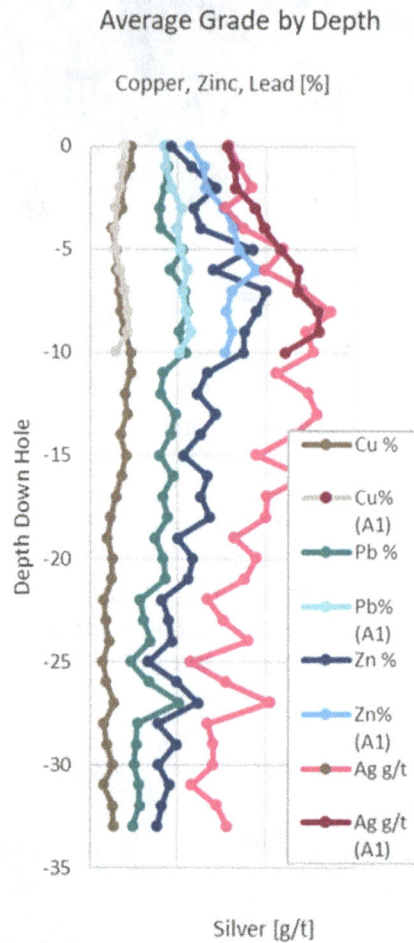

FIG 7 – Mean grade for each metal versus relative downhole depth.

Bulk density

Given the transient and evolving properties of tailings material, accurately measuring *in situ* bulk density presents considerable challenges and risks for projects considering material recovery. This issue is particularly significant for the MIM TSF given the magnitude of the ~1500 ha footprint. Even a minor variance in density, such as a change of 0.1 g/cm³, could result in an addition or subtraction of approximately ~20 million tonnes of material. Therefore, bulk density emerged as a critical focus of the characterisation program.

A bulk density assessment was completed for campaign A1 but as it was not an initial focus of this campaign, the sampling methodology had limitations in respect to bulk density measurement. Long sample retrieval times translated to long exposure of the samples in open conditions, leading to humidity losses which directly impact moisture, porosity and bulk density calculations. Furthermore, hand augering does not have enough control on core volume to enable volumetric bulk density calculation.

Learnings from Program A1 on bulk density were implemented to improve the accuracy of the bulk density understanding for Program A2. The final value for bulk density was determined by taking the average value for samples determined to be of high quality. This determination is based on alignment between two methods of bulk density calculation, Volumetric and Saturation. Each method calculates the upper and lower limits of bulk density with truly representative samples selected as those where the variance between both methods is minimal. This implies that there has been no significant loss of mass or moisture and/or open space porosity that cannot be easily measured.

The difference between both methods is how the total volume is calculated. Volumetric bulk density assumes a fixed volume of HQ3 and corrected for sample recovery, this represents the lower limit for bulk density. In volumetric bulk density, volume is not reliant upon full recovery but is calculated assuming full saturation, calculating the volume of solids, from dry mass and particle density and volume of water from the mass of water and assuming water density of 1.

As a result, the average dry bulk density for campaign A2 was 1.59 g/cm³ with a porosity of 0.45 and average moisture of 20 per cent. Figure 8 shows that MIM's bulk density and porosity compares favourably to similar deposits documented in the literature (Bhanbhro, 2014).

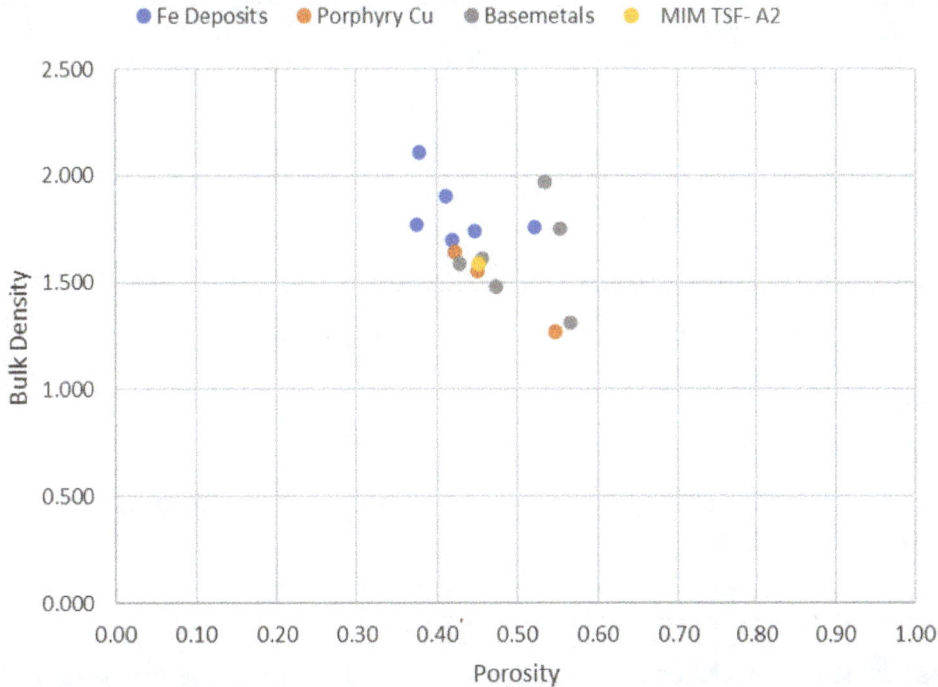

FIG 8 – MIM TSF A2 Sampling campaign dry bulk density and porosity plot in comparison to global TSFs (Bhanbhro, 2014).

Mineralogy

Half-core samples from the A1 and A2 sampling campaigns were sent for mineralogical characterisation by the UQ at its laboratory facilities. Blannin *et al* (2021) proposed the use of the bootstrap resampling technique to quantify statistical uncertainties in automated mineralogical studies and was implemented by the team at the UQ to support the mineralogical assessment.

A brief overview of the main modal mineralogy and its relevance to geology and geometallurgy is provided below.

Modal mineralogy

Figure 9 shows the modal mineralogy for zinc and lead (Zn-Pb) samples presented in order of global modal importance in the following stacked-bar plot. Key takeaways relevant for modal mineralogy are:

- Gangue minerals (carbonates and silicates) are in order of modal importance: quartz, dolomite, orthoclase, dolomite-silicate, pyroxene, chlorite, ankerite, olivine, muscovite, biotite, silicate mix, talc, gypsum anhydrite, calcite.

- Sulfide minerals in order of modal importance: pyrite, pyrrhotite, sphalerite, siderite, chalcopyrite, galena.

- Oxide minerals in order of modal importance: oxide iron, iron zinc oxide.

- Deleterious minerals in order of modal importance: pyrrhotite, arsenopyrite.

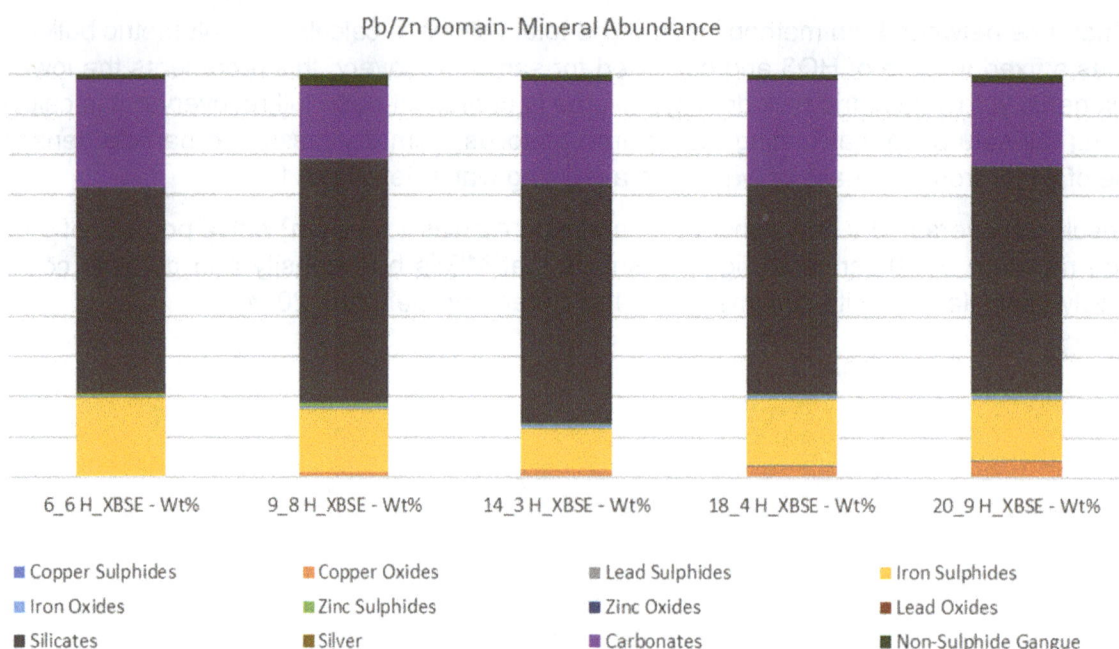

Pb/Zn Domain- Mineral Abundance

Legend:
- Copper Sulphides
- Copper Oxides
- Lead Sulphides
- Iron Sulphides
- Iron Oxides
- Zinc Sulphides
- Zinc Oxides
- Lead Oxides
- Silicates
- Silver
- Carbonates
- Non-Sulphide Gangue

X-axis categories: 6_6 H_XBSE - Wt%, 9_8 H_XBSE - Wt%, 14_3 H_XBSE - Wt%, 18_4 H_XBSE - Wt%, 20_9 H_XBSE - Wt%

FIG 9 – Modal mineralogy of select tailings sampled.

Mineral abundances

Key takeaways for mineral abundances are:

- Main zinc mineral associations in relative order of modal importance are sphalerite, zinc oxide, smithsonite, tetrahedrite.

- Main copper mineral associations in relative order of modal importance are chalcopyrite, bornite, copper iron oxide, tetrahedrite, freibergite.

- Main lead mineral associations in relative order of modal importance are galena and anglesite.

- Main silver mineral associations in relative order of modal importance are freibergite, tetrahedrite and pyrargyrite.

Sulfur abundance and iron deportment

Relevant for geology, metallurgical flotation and environmental impact are sulfide iron minerals such as pyrite, sphalerite, galena, pyrrhotite and chalcopyrite. Sulfur minerals and Iron in pyrite are of particular importance because of its relevance in the potential acid generation index. Key insights for sulfur abundance and iron deportment are:

- Main sulfur mineral associations in relative order of modal abundance are pyrite, pyrrhotite, gypsum-anhydrite, sphalerite, chalcopyrite, barite, cobaltite, bornite, galena, anglesite, arsenopyrite, tetrahedrite, alunite, freibergite.

- Iron is carried in the following minerals, in order of modal importance pyrite, pyroxene, olivine, pyrrhotite, chlorite, iron oxide, ankerite, biotite, siderite, zinc iron oxide, chalcopyrite.

Particle size distribution

Particle size distribution (PSD) has a strong influence in grade distribution as well as in metallurgical recovery, how well liberated particles are, and their relative size has a strong influence in the flotation circuits. Furthermore, re-processing cost is highly dependent on re-grinding requirements for complex particle associations as milling is the highest cost of the processing plant. Figure 10 shows element (copper, zinc and lead) by particle size distribution, for MIM the highest metal content is in the ultrafine particle size.

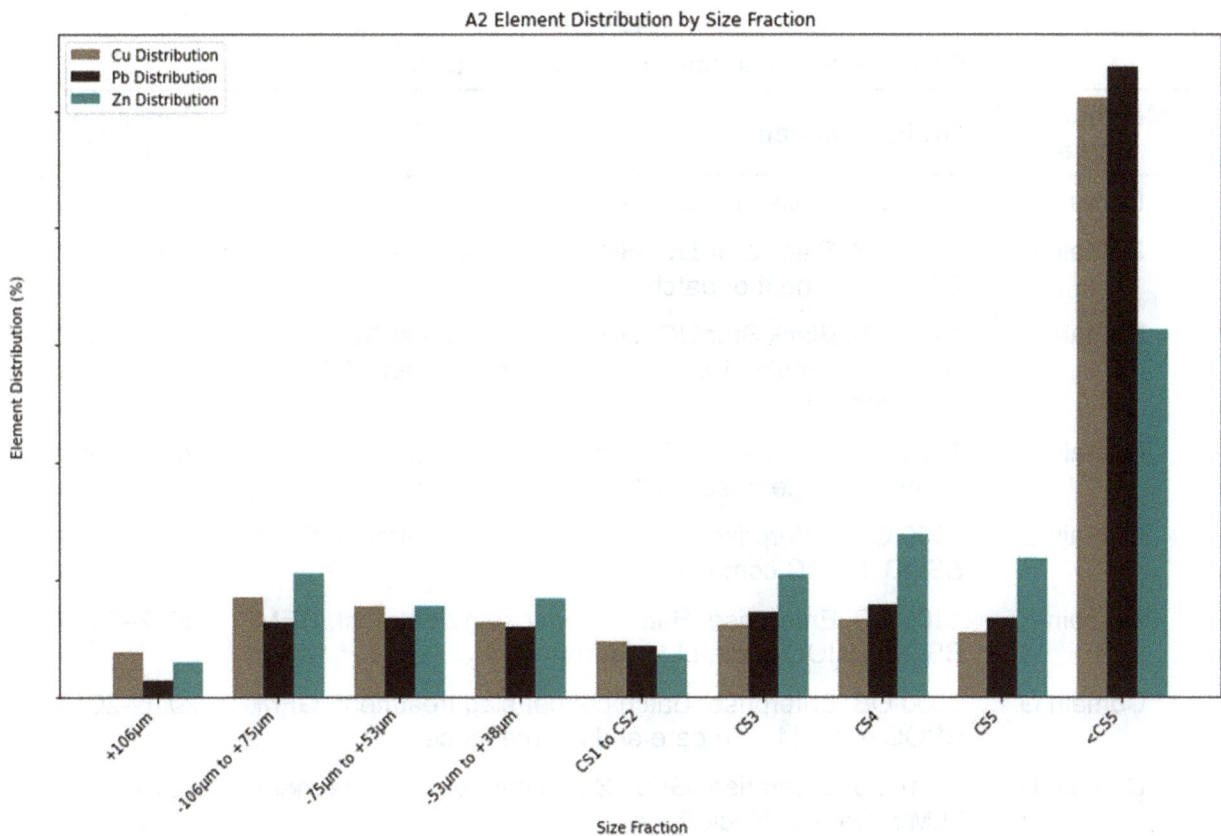

FIG 10 – Element distribution for copper, zinc and lead (Cu%, Pb%, Zn%) by particle size distribution fraction.

Key mineralogy takeaways

Mineralogy seems to show variability in speciation, which is relevant for metallurgical recovery:

- Copper species such as sulfide, oxide and transitional (chalcopyrite, chalcocite) play a strong role in metallurgical recovery.

- Zinc, lead and silver species are less diverse (sphalerite, galena) than for copper but still show binary and tertiary associations that are more difficult to recover.

- Iron shows a wide range of mineral species as well, however sulfide species such as pyrite might present subtypes that make a difficult targeted recovery.

Metallurgical domains

Metallurgical accounting and contributing mines

To determine the nature of the tailings at the main MIM TSF, metal accounting records were reviewed for two reasons. The first being to understand the orebody feeds to the mill to obtain insights on the mineralogical characteristics of the tailings; and the second to understand tailings grades at the time of deposition. Table 1 summarises the contributing mined sources from available mine records. Early production profiles were richer in sulfide zinc and lead mineralisation and the later mining fronts were richer in sulfide copper mineralisation. As a result, opposite trends are expected in the tailing's deposition sequence (copper enrichment near surface, zinc and lead enrichment at depth).

TABLE 1

Contributing mined sources since 1967 to the MIM TSF.

Metallurgical domain	Orebody mined*	Deposition period
Domain A	1100 OB Begins, Black Star UG	1967–1986
Domain B	1100 OB, Black Star UG, Hilton ore treated at No.2, Commencement of batch copper slag treatment	1987–1997
Domain C	1100 OB, Black Star UG, Hilton ore treated at No.2, Batch copper slag treatment, Enterprise Begins, GFM Commences	1998–2002
Domain D	1100 OB, Black Star UG (ends 2005), Batch copper slag treatment, Enterprise, GFM, BSOC commences	2003–2007
Domain E	1100 OB, Enterprise, Batch copper slag treatment, GFM, BSOC, HHOC commences	2008–2011
Domain F	1100 OB, Enterprise, Batch copper slag treatment, GFM, BSOC, HHOC ends, LLM Starts	2012–2015
Domain G	1100 OB, Enterprise, Batch copper slag treatment, GFM, BSOC ends, LLM in care and maintenance	2016–2018
Domain H	1100 OB, Enterprise, GFM, Batch copper slag treatment, LLM restarted, Black Rock	2019–2022

*Main contributing mines are 1100 Orebody, Black Star Underground (BSUG), Hilton, George Fisher Mine North (GFM N), Enterprise, Lady Loretta Mine (LLM), Black Star Open Cut (BSOC), copper slag, Handlebar Hill Open Cut (HHOC).

GIS metallurgical domains

To identify the spatial location and depth of each of the abovementioned metallurgical domains, a GIS work package was initiated to review historical survey and imagery data to determine 3D shapes for each of these domains. The main objective was to improve the overall understanding about what/when each material was deposited in the MIM Tailings Storage Facility.

GIS Domains respect a sedimentary nature of deposition with emphasis on nearly horizontal controls. Domain A is the oldest and deepest, followed by domains B, C, D, E, F, G, H higher up. Upper and younger domains were added to continue to fill up to current topographic level. As an example, the pre-deposition surface (year 1958) was used as a starting bottom surface for the oldest deposited material (Domain A). The final year of deposition for this domain was 1986, for which a digital surface model was developed. Using the bottom and top DSM's for each domain, and a contour line defining its extents, a three-dimensional volume could be created to represent the tailings volume distribution for each domain. Digital terrain models (DTM's) were not always 100 per cent compatible with required time periods of metallurgical domains; in this case, an exemption was made to select the closest time-period required by the domain.

The digital volume created for each domain was then validated with the metallurgical accounting record. For example, the digital volume created for the Domain A should have a volume equal to the equivalent volume noted in the metallurgical balance for this period. This validation showed good alignment between the digital volumes created and deposited tailings volumes from metallurgical accounting records.

An overview of the 3D shapes for each of the metallurgical domains within the MIM TSF are illustrated in Figure 11.

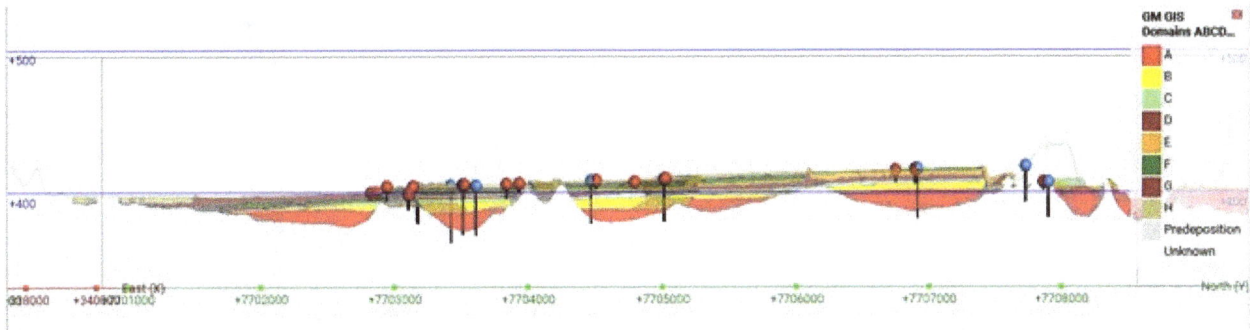

FIG 11 – 3D metallurgical domains and characterisation programs A1 and A2- cross-section along the access road (Section North-South, North points to the right of the image above).

PRELIMINARY BLOCK MODELLING

A preliminary resource estimation workflow was implemented following characterisation sampling campaigns A1 and A2 within the Mount Isa Mines Tailings Storage Facility. The intent of the model is to inform an order of magnitude assessment and planning for future drill programs. It will **not** be used for mineral resource reporting.

The slow depositional nature of the fine tailings particles allows for nearly horizontal continuity of the residual materials being stored in an upwards-younging sequence. Most samples are located along the main access road; the southern area is covered with water and no sampling was feasible.

A preliminary block model estimate was obtained for the following main elements, grouped into three categories:

1. Metals associations, potentially economic:
   - Zn and Pb and Ag: zinc, lead and silver.
   - Cu and Co and In: copper, cobalt, and indium.
2. Deleterious/concentrate penalties:
   - As, Cd, Sb, SiO_2: arsenic, cadmium, antimony, and silica.
3. Acid generating and/or neutralising:
   - S and Fe: sulfur and iron.
   - Ca and Mg and C: calcium, magnesium and carbon.

Metallurgical domains could be used in future project stages provided additional sampling campaigns are performed. Block model estimation was based on a global merged domain selected due to limited data available within each unique metallurgical domain. Figure 12 shows a 3D image for the preliminary block estimation model along A1 and A2 samples for zinc.

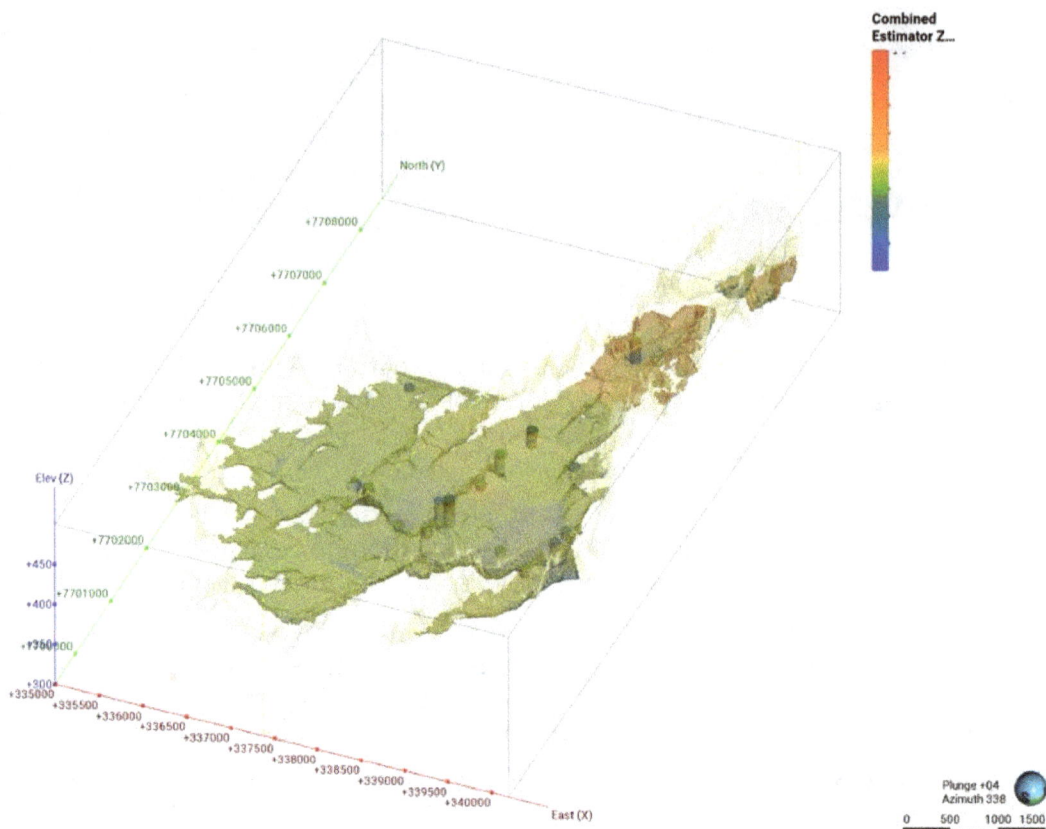

FIG 12 – 3D view of A1 and A2 zinc samples along preliminary block model estimates for zinc within the MIM TSF.

Exploratory data analysis

Since low-grade residual materials are deposited in the tailing facilities, most of the chemical elements (zinc Zn[%], lead Pb[%], silver Ag[g/t], copper Cu[%]) associated with sulfide minerals being floated in the processing plant; have grade distributions with histograms showing nearly bell-shaped curves (such as Normal or Gaussian), this means very few high-grade samples or outliers.

Nugget effect or down the hole variability

Grade variability was a concern in the early stages of the characterisation study given the numerous, diverse contributing sources over such a long period of time. Down-the-hole sample variability is also known as nugget effect and reflects the field elements variability at close distances, typically 1–2 m or sample length (1.5 m).

However, the nature of the materials yielded lower than expected variances, hinting for more homogeneity than expected; A2 samples confirmed again very low nugget effects (variance between samples that are closely located in space).

Figure 13 shows a spider plot with the downhole grade variability for main sulfide minerals, deleterious and penalties elements. The A1 and A2 combined sampling campaigns show even lower than expected nugget effects for all elements combined.

All elements show down-the-hole sample variability ranges within lower-than-normal values for *in situ* resources (1–15 per cent compared to 20–50 per cent). Metal associations of copper, cobalt and indium (Cu, Co, In with max 15 per cent) show higher nugget effect than zinc, lead, silver (Zn, Pb, Ag with max 5 per cent). Acid neutralisation elements such as calcium, magnesium, silica and carbon (Ca, Mg, SiO_2, C) have low variability at 5 per cent. Deleterious and penalties elements such as arsenic cadmium, antimony and Iron (As, Cd, Sb, Fe_2O_3) show variability ranging within expected values (max 10 per cent). Indium shows a higher-than-average variability (15 per cent), mostly due to outliers.

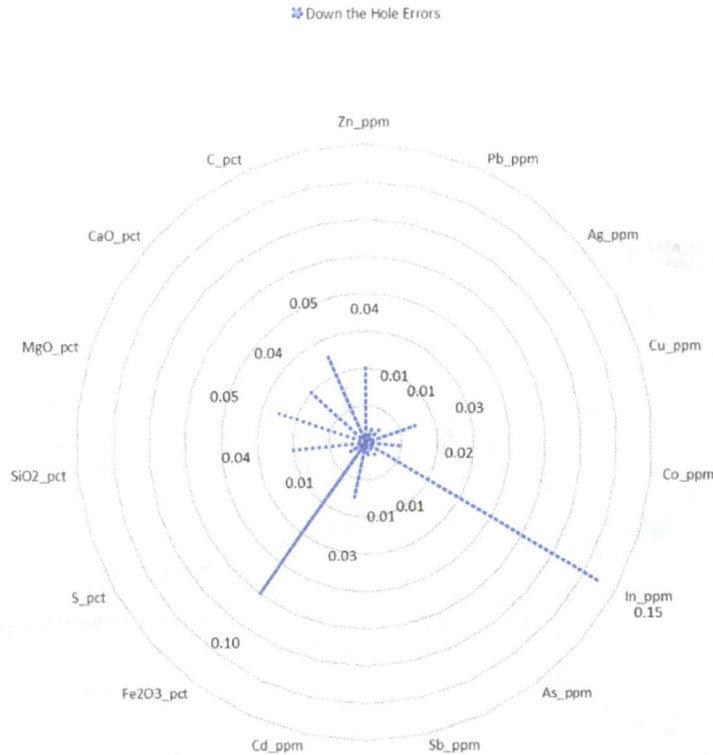

FIG 13 – Spider plot for nugget effect variability of chemical elements A1 and A2 combined.

Benchmark on Cauquenes variography

A benchmarking exercise was done with A1 samples against a similar tailing's deposition such as Cauquenes; a historic tailings facility belonging to El Teniente porphyry copper underground mine in Chile (the study report was also at a similar project stage – before pre-feasibility). Cauquenes tailing is located high in a mountain valley in the Andes, its footprint is similar with a 1200 hectares and slightly deeper deposition, having the same global volume as the MIM TSF. Cauquenes Tailings Facility shows a similar variogram behaviour, with longer ranges in the horizontal and shorter ranges in the vertical direction. Cauquenes public mineral resource and mineral reserves reports for concept and pre-feasibility stages are stated in Robert and Henderson (2014) and Global Resource Engineering Ltd (2022).

- Cauquenes Concept: 30 drill holes at 400 m spacing.

- Cauquenes Pre-feasibility: 83 drill holes at 100 m spacing.

Long ranges for horizontal variograms from A1 (~1 km) are uncertain at concept stage because of the low number of samples which are separated at large spacings (320 m on average); a shorter range would mean a smaller portion of the volume can be estimated with certainty. Verification of variogram ranges is needed with more drilling data at closer sample spacings. For reference, Cauquenes pre-feasibility data set spacing is ~100 m:

- MIM TSF horizontal variability range for total copper CuT is 350 m compared to 800 m for Cauquenes at pre-feasibility Stage.

- MIM TSF vertical variability range for total copper CuT is 10 m compared to 15 m for Cauquenes at pre-Feasibility Stage.

Vertical variability in A2 showed a decreased nugget effect but shorter ranges from A1. Horizontal variability ranges shortened after incorporating additional A2 sampling information. The horizontal variogram range decreased from 800 m to 350 m when increasing the data set from A1 to A2. Figure 14 contains the copper raw variogram for its main directions (horizontal range in red and vertical range in blue).

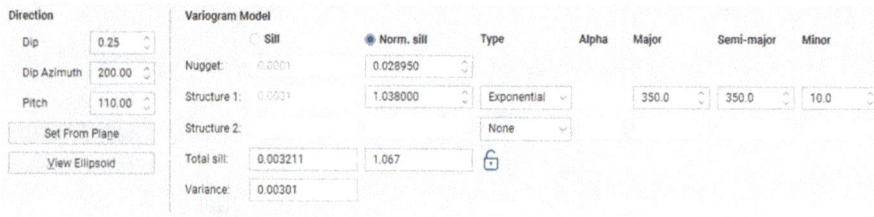

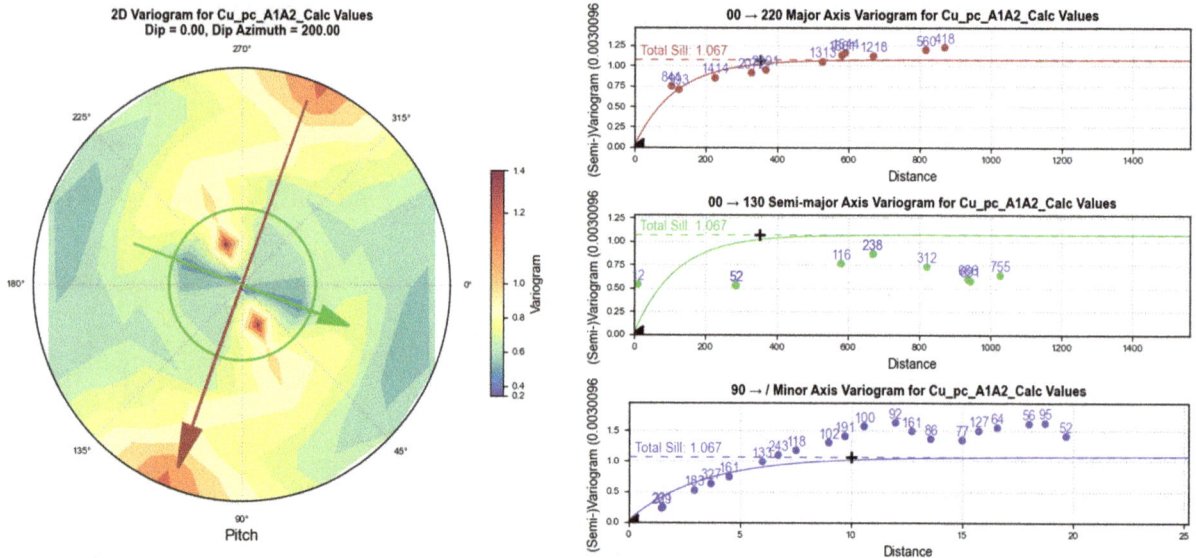

FIG 14 – Copper raw variogram MIM TSF – Concept stage A1A2 data set.

Acid classification calculations

Acid generation relies on the existence of sulfide minerals capable of oxidising and leaching acid and is contingent on the presence of neutralising carbonate minerals. The most acid-relevant sulfide mineral is pyrite (FeS_2). The relative proportion of pyrite to calcite ($CaCO_3$) is the foundation basis for the acid generation categories.

For acid mine drainage (AMD), the Fe:S molar ratio in pyritic material can provide insight into the potential for acid generation. A ratio near 2:1 suggests pure pyrite, which can generate significant acidity upon oxidation.

Table 2 exposes the acid classification system implemented in the block model; it uses inferred Net Acid Production Potential (NAPP) as a basis for classifying potentially acid forming materials in the MIM TSF developed by SLR Consulting for the project. The acid classification system includes the following categories: AC: acid consuming, NAF: non-acid forming, NAF-LC: non-acid forming low capacity, PAF-LC: potential acid forming low capacity, PAF: potential acid forming, PAF-HC: potential acid forming high capacity.

TABLE 2

Acid classification system with acid generating categories used by the project.

SLR Classification derived using Inferred NAPP	Inferred NAPP$_{FESulfide}$ kg H$_2$SO$_4$/tonne	Inferred NPR
AC	< -100	> 1
NAF	> -100	> 1
NAF-LC	Negative	1
PAF-LC	Positive	1
PAF	<= 30	<1
PAF-HC	> 30	<1

Figure 15 shows as section view of the acid classification system. The intent of integrating these calculations into the block model was to enable the model to act as a screening tool for predicting areas where potentially acid forming material could be placed. Making of it a useful tool to support with closure planning activities for the site, as well as production planning from a tailings reprocessing perspective.

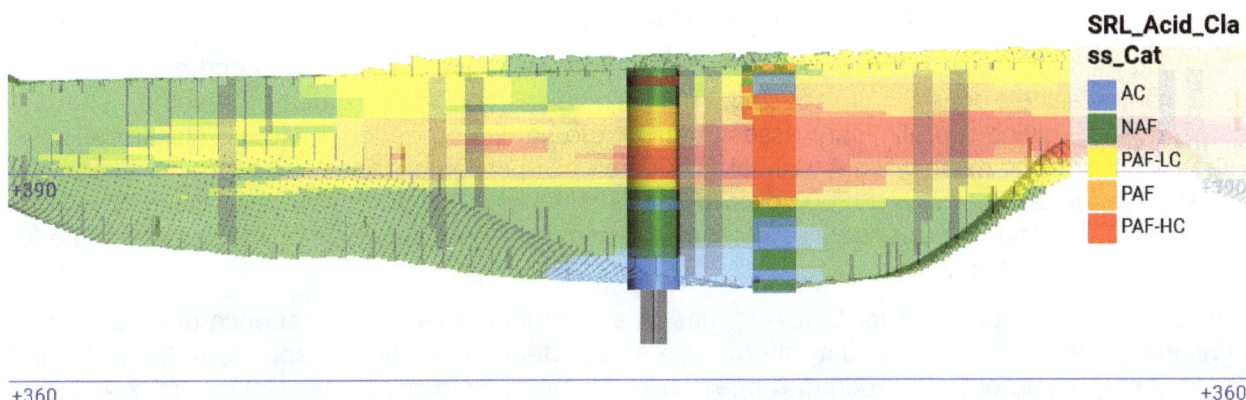

FIG 15 – Cross-section view in Leapfrog with acid generating classification block model and program A2 drill holes at the southern end of the facility (looking east along access road; north is pointing to the right of the image).

There are opportunities to further refine the acid classification system, which is based on data from static analytical methods. Due to the expense and duration of kinetic leach methods, these tests were not performed on the samples collected in this program. Future stages of characterisation should look to complete further kinetic leach tests, enabling refining of the acid classification framework to determine acid neutralisation capacity (ANC) depletion rates.

Drill hole spacing analysis – DHSA

In this concept study, the mineral resource risk classification level is high as there is insufficient information to declare mineral resources with confidence. Compounded factors for the inability to declare mineral resources at this stage are mainly a result of insufficient sample density and legacy topographical data for the pre-deposition surface. Inherently, it is not possible to demonstrate reasonable prospects for eventual economic extraction, a fundamental criterion for reporting under the JORC Code (2012). Higher confidence in Mineral Resources can be achieved through additional sampling campaigns, which increase the quantity and quality of geological data, thereby reducing the associated risk levels. Subsequent study stages such as pre-feasibility (PFS) and feasibility (FS) will need a better definition for each resource classification category and further low risk conversion as the project seeks execution.

DHSA is a geostatistical technique that supports the mathematical assessment of a sampling mesh for each Mineral Resource category conversion (example: conversion of Inferred to Indicated). Gaussian simulation for DHSA was conducted to support with planning future drill hole campaigns.

DHSA was run twice at MIM tailings facility for zinc grades: first with MIM sampling campaign A1 results and second with sampling results from A1 and A2 campaigns combined. The results showed strong alignment between the two runs.

Preliminary DHSA results supported by the above-mentioned parameters could be used for preliminary resource classification. Also, DHSA results can be a helpful guide for sample spacing and a sampling pattern for future mineral resource conversion drilling campaigns or assisting in designing additional drilling campaigns aimed at informing geotechnical or geometallurgical attributes.

Current DHSA results indicate that a 400 m drill spacing may not provide the data density required for confident Mineral Resource classification at the PFS or FS stage, where detailed assessments of resource continuity and grade are necessary. However, this spacing may be sufficient for preliminary geotechnical investigations at the concept study stage.

At the concept level stage, variography does not yet yield reliable results in continuity and anisotropy. Reliable variography results help in determining the search distances and anisotropy directions for first, second and third passes for kriging estimation. Kriging passes or runs are closely related to confidence levels, upon further smoothing and operational factors.

Sampling meshes were tested to enhance confidence, beginning with a 500 m grid and narrowing in 50 m intervals to a denser 50 m spacing. The drill hole spacing analysis method assumes a square sampling grid, giving a preliminary assessment of the samples needed for future classification. A wide 500 m spacing mesh will need about 20–30 sampling collar locations and a tight 50 m spacing mesh will need approximately 1700 collar locations. Neither the wide nor the tight mesh are suited for PFS of FS resource classification, instead, an intermediate mesh would be more appropriate for the future classification.

See plots below (Figure 16) for DHSA results of a combined A1-A2 run. Variance decreases to a minimum plateau at around 250 m, then stays at a plateau. The 250 m happens to be within the order of magnitude of the variogram ranges for all elements combined (min 120 m – max 400 m in the nearly horizontal direction).

- A 250 m drilling mesh will require about 70 drill hole collar locations (680 samples/10–20 samples per hole).

- Reasonably low variability for a sampling spacing is achieved at 300 m. There would be a need of around 25–45 drilling locations (450 samples/10–20 samples per hole).

- There is less variability for a sample spacing of 400 m. There would be a need for 13–26 drill hole locations (260 samples/10–20 samples per hole).

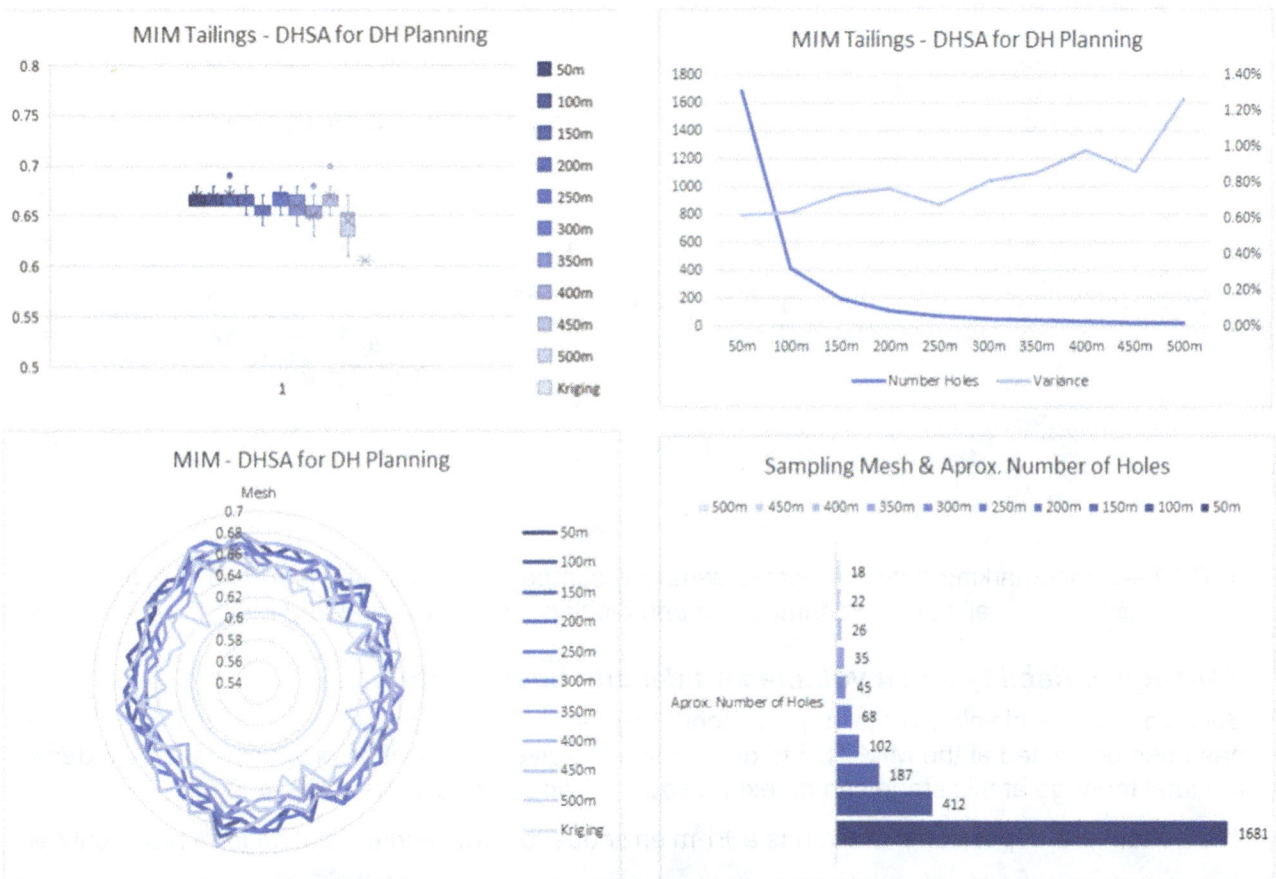

FIG 16 – Drill hole spacing analysis (DHSA) results indicate minimum variance at 250 m drill hole spacing.

Drilling costs related to these proposed drilling campaigns might indicate it is more cost-effective to step up from 13 to 25, or 45 to 70 drilling collar locations, pending data variability performance.

Grade variability by successive sampling campaigns

Assays results from Program A1 are compared to tailings plant data from 1967–2022; BML's test work (Brisbane Met Labs in 2020) and preliminary block models 2023 and 2024 from A1 and A1-A2 sampling campaigns.

A box-and-whisker plot in Figure 17 shows the histogram of grades in a summarised profile where only six statistics are highlighted: Minimum, Maximum, Median, Mean, Q25–25% Quartile and Q75–75% Quartile. Box-and-whisker plot statistics by successive sampling campaigns (BML, A1, A2) are consistent with each other.

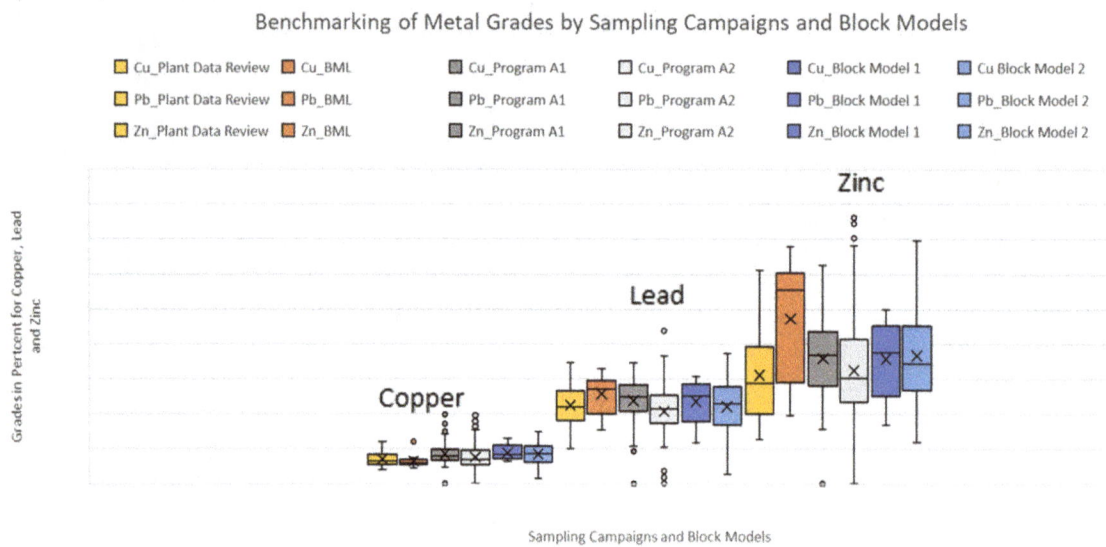

FIG 17 – Benchmarking exercise across sampling campaigns for copper, lead and zinc grades for all sampling campaigns and block models 2023 and 2024.

Tonnage variability given volume and density uncertainty

Based on accessible plant metallurgical accounting records, it is expected that over 300 Mt of tailings has been deposited at the MIM TSF to date, however; given variabilities in volume and bulk density the total tonnage at this stage can be expressed in a range, for example 300–400 Mt.

The bottom pre-deposition surface has a ±1 m error due to surface interpolation from historical data.

A sensitivity analysis was performed on the global volume provided by SLR. The volume was increased by moving the pre-deposition surface down by 1 m (minus) and decreased by moving surface up 1 m (plus). This sensitivity analysis allows for understanding of tonnage uncertainty.

Bulk Density (BD) on A1 and A2 samples shows variability, which was used to understand variability on tonnage. A density range of 1.5–1.65 g/cm^3 was applied to understand volume variability given bulk density variations. A fixed value of 1.59 g/cm^3 (Dry Bulk Density) was retained for preliminary tonnage reporting for subsequent order of magnitude studies. Further work is required for more reliable density measurements to refine the bulk density to a narrower range, even variations of 0.1 in bulk density can cause a material change on the order of 18 Mt in estimated tonnages.

The Box and Whisker plot below (Figure 18) presents the pre-depositional surface and density tonnage sensitivity analyses. The outcome of the analyses presents a final tonnage range. The true tonnage is likely closer to the total metallurgical accounting calculation.

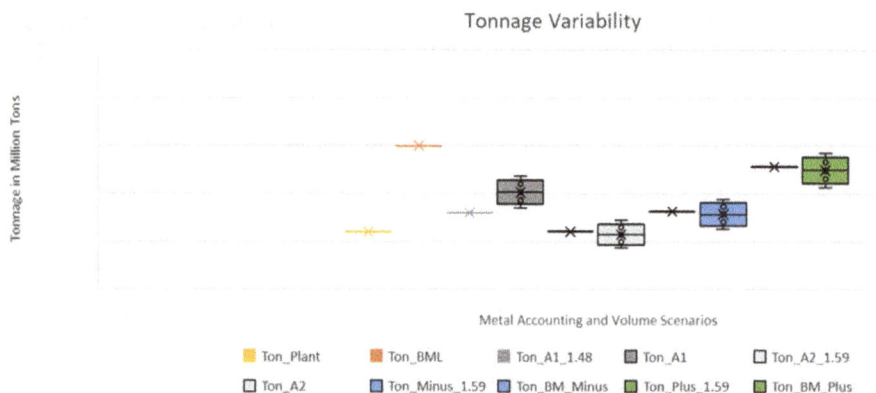

FIG 18 – Benchmarking exercise for tonnage, assessing tonnage variability due to volume and density variability compounded.

Block model validation – swath plots by geometallurgical units

Figure 19 show swath plots by geometallurgical units for copper and zinc which compare the metal tonnages and grades between the metallurgical accounting data, the characterisation data and estimates from the block model. Globally, sample statistics by domains respect the expected global trends from historical metallurgical accounting; for example: copper grades have a downward trend consistent with declining grades of the orebody. Zinc grades are higher when more complex multi-sourced feed was in place. Geometallurgical swath plots show a reasonable correspondence of zinc and copper grades by geometallurgical domain between the three data sets. Block model grades present smoothing which is expected given the small data set and the extensively interpolated area covered by the block model.

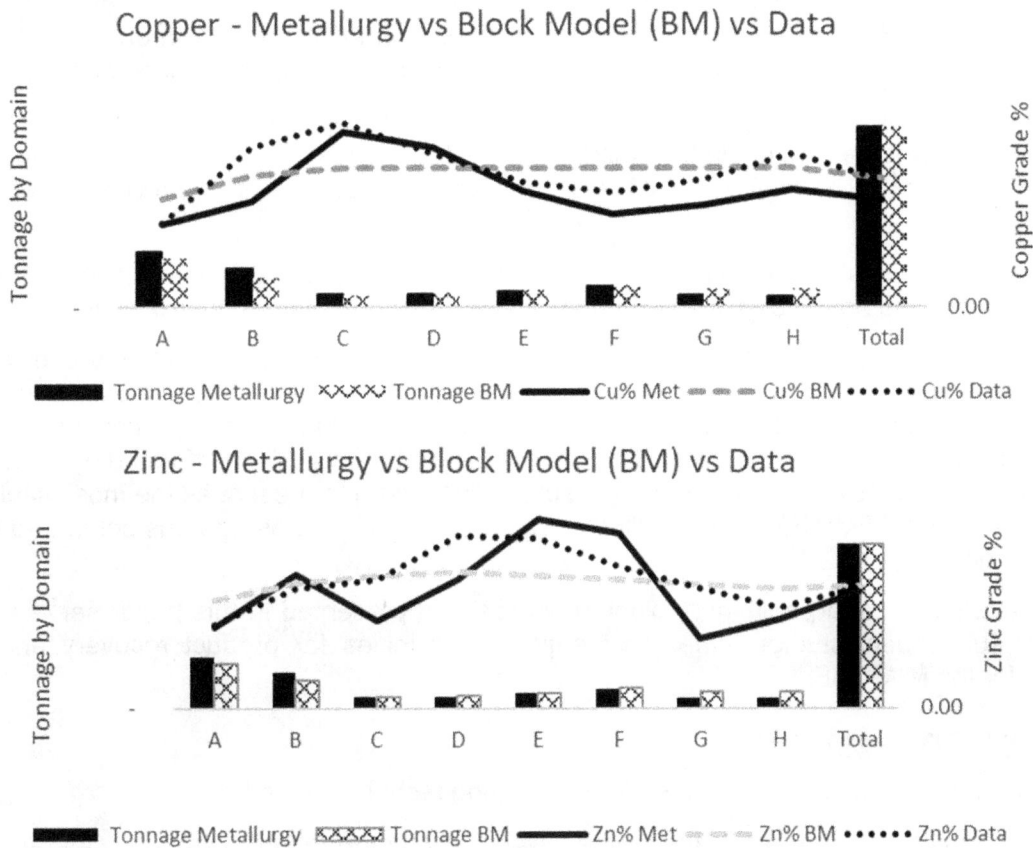

FIG 19 – Swath plots for Zinc and Copper grades for data set versus metallurgy versus block model 2024.

CONCLUSIONS

To support MIM in exploring reprocessing options for its largest TSF, a comprehensive characterisation and resource estimating program was required to set a foundation for subsequent metallurgical and environmental studies. With the large-scale, longstanding and ongoing operation of the MIM TSF, this presented multiple challenges for which a staged characterisation program was developed to address the objectives and preliminary nature of the project.

A TSF is different to a traditional orebody given the anthropogenic nature of the deposited resource. A traditional geostatistical block modelling workflow can be applied to the geochemical data set from a tailings sampling campaign, however, input from multi-disciplinary experts is material to this workflow for tailings deposits. The accomplished work required input from specialists in fields such as geology for sampling and characterisation; GIS spatial experts for volume modelling of geometallurgical domains; metallurgist for understanding MLA analysis results and its geometallurgical variables; geo-statisticians needed for resource estimation and preliminary resource classification; environmental chemistry experts for acid accounting classification; risk management for safety procedures; and last, project management experts to enable the development and implementation of a targeted characterisation program which allows exploratory assessment of potential reprocessing and closure liability reduction opportunities.

One key learning applicable for other legacy tailings facilities is the staged approach to sampling and drilling TSFs with large footprints. Rather than implementing an extensive and costly drilling campaign, a phased approach ensures that characterisation efforts are in line with the business' current stage in evaluating prospectivity opportunities from tailings. In the case of MIM, samples were required to initiate exploratory metallurgical and closure improvement assessments.

Due to the preliminary nature of the resource model, multiple variables are not fully known or understood and should be subject for further focus if future drilling campaigns progress:

- Volume uncertainty resulting from the use of legacy data for the generation of the pre-deposition surface, which has an error in the elevation location of ±1 m due to interpolation. Future drilling campaigns should seek to improve confidence on the ground surface.

- Density measurements prove sensitive to sampling techniques with only a handful of measurements being used in the calculation of a fixed value of 1.59 g/cm^3 (Dry Bulk Density). Further efforts will be required to refine dry bulk density estimates, including potential for spatial variability.

- Geochemical spatial correlations not fully understood. Variograms are not fully informed in their range and anisotropy due to the current wide data spacing of approximately 300 m.

With the unique challenges of tailings deposits, one significant unknown is what is considered an acceptable level of resource confidence to support a financial investment decision on tailings reprocessing projects. This is important to inform the design of future drilling campaigns to mature the resource estimate. One observation from the project is that this will be informed by target metals or minerals. For example, if environmental desulfurisation were to present as the most viable option for tailings from the MIM TSF, future drilling campaigns may not be as rigorous compared to metal recovery options.

The characterisation and preliminary block model results presented in this paper serve as critical inputs for subsequent studies aimed to identify opportunities for product recovery and liability reduction for the MIM TSF.

ACKNOWLEDGEMENTS

We would like to highlight contributions by the following technical experts:

- Anita Parbhakar-Fox and the Mine Waste Transformation Through Characterisation (MIWATCH) Group at the University of Queensland (UQ) for their expertise and contributions to the sampling and lab testing program.

- Nathan Turner and his Geospatial team at SLR Consulting for developing the 3D geometallurgical domains from sparse spatial records.

- Greg Maddocks and his team at SLR Consulting for the contributions of their geochemistry expertise to enabling acid mine drainage prediction capabilities in the preliminary block model.

- Roxanne O'Donnell, Selwyn Orwe and Sameer Morar from Glencore Zinc's Capital Studies Metallurgy Team for their guidance and support on the geometallurgical analysis.

- Allan Huard from Glencore Canada Zinc's Technical Team for his contributions to the analysis of bulk density, porosity and moisture.

- Amanda Landriault from Glencore Canada Zinc Technical Team for her encouragement and involved peer review in the publication of this paper.

- Alastair Grubb from the Glencore Zinc's Capital Studies Australian Team, for his guidance and mentorship in meeting strategic needs of the MIM tailings reprocessing project.

REFERENCES

Bhanbhro, R, 2014. Mechanical Properties of Tailings: Basic Description of a Tailings Material from Sweden Licentiate dissertation, Luleå Tekniska Universitet. Available from: <https://www.researchgate.net/publication/265296679_Mechanical_Properties_of_Tailings_Basic_Description_of_a_Tailings_Material_from_Sweden>

Blannin, R, Frenzel, M, Tolosana-Delgado, R, Büttner, P and Gutzmer, J, 2023. 3D Geostatistical Modelling of a Tailings Storage Facility: Resource Potential and Environmental Implications, *Ore Geology Reviews*, 154(2023):105337. https://doi.org/10.1016/j.oregeorev.2023.105337

Blannin, R, Frenzel, M, Tuşa, L, Birtel, S, Ivăşcanu, P, Baker, T and Gutzmer, J, 2021. Uncertainties in quantitative mineralogical studies using scanning electron microscope-based image analysis, *Minerals Engineering*, 167:106836. https://doi.org/10.1016/j.mineng.2021.106836

Committee for Mineral Reserves International Reporting Standards (CRIRSCO), 2024. International Reporting Template. Available from: <https://crirsco.com/the-international-reporting-template/> [Accessed: 21 Nov 2024].

Dinis, M D, Fiúza, A, Futuro, A, Leite, A, Martins, D, Figueiredo, J, Góis, J and Vila, M C, 2020. Characterization of a mine legacy site: an approach for environmental management and metals recovery, *Environmental Science and Pollution Research International*, 27. https://doi.org/10.1007/s11356-019-06987-x

Dold, B and Fontboté, L, 2001. Element cycling and secondary mineralogy in porphyry copper tailings as a function of climate, primary mineralogy and mineral processing, *Journal of Geochemical Exploration*, 74(2001):3–55. https://doi.org/10.1016/S0375-6742(01)00174-1

Glencore, 2024a. Mount Isa Mines Tailings Storage Facility, Glencore. Available from: <https://www.glencore.com.au/operations-and-projects/qld-metals/operations/mount-isa-mines/mount-isa-mines-tailings-storage-facility> [Accessed: 21 Nov 2024].

Glencore, 2024b. Mount Isa Mines Tailings Storage Facility – Fact Sheet, Glencore. Available from: <https://www.glencore.com.au/.rest/api/v1/documents/fbe7f96f7e84134aefc7b2218a0d9122/Tailings+Storage+Facility+Fact+Sheet_MIM_Feb+2023_v2.pdf> [Accessed: 21 Nov 2024].

Global Resource Engineering Ltd, 2022. Technical Report for Pre-Feasibility Study – Amerigo Resources Ltd, NI 43–101, Mineral Resource Estimate Technical Report Minera Valle Central Operation Rancagua, Region VI, Chile, Section 14.0 Mineral Resource Estimates, pp 109–133. Available from: <https://www.amerigoresources.com/_resources/reports/Technical-Report-20220330.pdf> [Accessed: 21 Nov 2024].

Jackson, L M and Parbhakar-Fox, A, 2016. Mineralogical and geochemical characterization of the Old Tailings Dam, Australia: Evaluating the effectiveness of a water cover for long-term AMD control, *Applied Geochemistry*, 68:64–78. https://doi.org/10.1016/j.apgeochem.2016.03.009

James, M, Aubertin, M, Wijewickreme, D and Wilson, W, 2011. A laboratory investigation of the dynamic properties of tailings, *Canadian Geotechnical Journal*, 48(11):1587–1600. https://doi.org/10.1139/t11-060

JORC, 2012. Australasian Code for Reporting of Exploration Results, Mineral Resources and Ore Reserves (The JORC Code) [online]. Available from: <http://www.jorc.org> (The Joint Ore Reserves Committee of The Australasian Institute of Mining and Metallurgy, Australian Institute of Geoscientists and Minerals Council of Australia).

Juutinen, S, Seitsari, J and Pietilä, J, 2023. Geochemical and mineralogical characterization of mine tailings at the Rautuvaara mine site and aspects to environmental conditions and resource potential, *Suomen Geologinen Seura*, 95(1):5–22. https://doi.org/10.17741/bgsf/95.1.005

Kinnunen, P, Karhu, M, Yli-Rantala, E, Kivikytö-Reponen, P and Mäkinen, J, 2022. A review of circular economy strategies for mine tailings, *Cleaner Engineering and Technology*, 8:100499. https://doi.org/10.1016/j.clet.2022.100499

Kossoff, D, Dubbin, W E, Alfredsson, M, Edwards, S J, Macklin, M G and Hudson-Edwards, K A, 2014. Mine tailings dams: Characteristics, failure, environmental impacts and remediation, *Applied Geochemistry*, 55:243–260. https://doi.org/10.1016/j.apgeochem.2014.09.010

New Century Resources, 2017. New Century Reports Outstanding Feasibility Results that Confirm a Highly Profitable, Large-Scale Production and Low-Cost Operation for the Century Mine Restart, ASX Announcement, 28 Nov 2017. Available from: <https://hotcopper.com.au/data/announcements/ASX/6A862558_NCZ.pdf> [Accessed: 21 Nov 2024].

Robert, D and Henderson, P, 2014. Technical Report for Scoping Study – Amerigo Resources Ltd, Minera Valle Central Operation Rancagua, Region VI, Chile, NI 43–101, Technical Report, Section 14.0 Mineral Resource Estimates, pp 52–67. Available from: <https://www.amerigoresources.com/_resources/reports/tech_report_april_8_2014.pdf> [Accessed: 21 Nov 2024].

Robert, D and Henderson, P, 2018. MVC Technical Report for Concept Study – Minera Valle Central Operation Rancagua, Region VI, Chile NI 43–101 Technical Report, Section 14.0 Mineral Resource Estimates, pp 54–71. Available from: <https://minedocs.com/20/MVC-TR-12-31-2018.pdf> [Accessed: 21 Nov 2024].

Siqueros Valencia, J R, Loredo Portales, R, Moreno Rodríguez, V and Del Rio Salas, R, 2022. Use of biomaterials as an alternative for remediation of mine tailings, Uso de Bio-materiales Como Alternativa Para La Remediación De Jales Mineros, *EPISTEMUS*, 15(31). https://doi.org/10.36790/epistemus.v15i31.208

Smith, K, Hageman, P, Ramsey, C, Wilderman, T and Ranville, J, 2006. Reconnaissance sampling and characterization of mine-waste material, Proceedings of the 2006 Mine Waste Technology Symposium. Available from: <https://www.clu-in.org/conf/tio/r10hardrock3_030513/smith-epa-hardrock-2006-proceed-paper.pdf>

Surrette, A, Dobosz, A, Lambiv Dzemua, G, Falck, H and Jamieson, H E, 2024. Geochemical and mineralogical heterogeneity of the Cantung mine tailings: implications for remediation and reprocessing, *Frontiers in Geochemistry*, 2:1392021. https://doi.org/10.3389/fgeoc.2024.1392021

Wang, X, Wei, Z, Li, Q and Chen, Y, 2018. Experimental research on the rheological properties of tailings and its effect factors, *Environ Sci Pollut Res Int*, 25(35):35738–35747. https://doi.org/10.1007/s11356-018-3481-1

Wilson, R, Toro, N, Naranjo, O, Emery, X and Navarra, A, 2021. Integration of geostatistical modeling into discrete event simulation for development of tailings dam retreatment applications, *Minerals Engineering*, 164:106814. https://doi.org/10.1016/j.mineng.2021.106814

Quantifying, integrating and modelling

Assessment of revenue uncertainty in life-of-mine planning at George Fisher deposit – Queensland, Australia

B C Afonseca[1]

1. MAusIMM(CP), Principal Geologist, Glencore, Brisbane Qld 4000.
   Email: bruno.dedeusafonseca@glencore.au

ABSTRACT

Uncertainty in mineral resources may arise from various factors, including statistical fluctuations, data acquisition, and geological interpretation. In mining, the multi-source nature of uncertainty is categorised into major resource categories to convey the overall risk associated with estimates. However, conventional classification approaches may fail to breakdown uncertainty into its components, particularly when evaluating risks related to financial outcomes or economic variables. Challenges arise when financial functions depend on indirect or non-geological variables that are frequently overlooked in conventional resource classification frameworks such as revenue functions, metallurgical inputs, and deleterious elements. In this context, this paper explores the impact of large-scale spatial uncertainty on net smelter return, at the George Fisher polymetallic deposit. The workflow integrates both modern and traditional multivariate geostatistical techniques, including simulation, multivariate data imputation, data decorrelation, and histogram uncertainty analysis, to assess revenue risk over the life-of-mine. In the presented workflow, profit-related variables are simulated individually in a multivariate fashion to preserve the high-order statistics of the original data in the simulated realisations. Revenue uncertainty is evaluated by processing all realisations through technological and financial functions within the context of production volumes. In the deposit, the net smelter return is mainly influenced by three key variables: lead, zinc, and silver. Additionally, metallurgical recovery, expressed as a linear combination of primary and secondary variables such as sulfur, iron, and copper, also contributes to determining the NSR. Overall, the revenue uncertainty generally aligns well with the current resource classification. However, there are instances where higher risk is observed, driven by the variability of secondary variables, the spatial configuration of mining stopes, and the scale of production across the years. The uncertainty in mine planning from 2025 to 2036 is quantitatively assessed at both point and production scales, with an evaluation of the nature of uncertainty and the influence of each variable. This case study demonstrates that a multivariate approach to evaluating revenue uncertainty provides a more comprehensive assessment of financial risk. It offers valuable insights for decision-making by identifying sources of uncertainty that are often overlooked in traditional classification methods.

INTRODUCTION

For clarity, the terms 'risk' and 'uncertainty' will be used interchangeably, as they convey the same idea in the context of the study.

Research and practical mining examples demonstrate the contributions of geostatistics to uncertainty modelling. Stochastic methods such as Monte Carlo simulations have long been used as a quantitative metric to measure risk. The so-called '90 per cent – 15 per cent rule' is a popular criterion adopted by many mining companies to bring in a degree of objectiveness when communicating risk and classifying mineral resources. This approach defines that, at the 90 per cent confidence level, the material with a deviation from the mean below 15 per cent for a quarterly production scale is classified as measured, while regions with a deviation below 15 per cent at an annual production scale are assigned as indicated (Parker and Dohm, 2014).

Commonly, conditional simulation applied for drill hole spacing optimisation (Englund and Heravi, 1993; Boucher, Dimitrakopoulos and Vargas-Guzman, 2005; Martínez-Vargas, 2017) and for classification predominantly uses only the primary grade variable. When this occurs, other aspects contributing to the total uncertainty are not addressed, which can lead to an understated and less realistic perception of it. Regarding financial return, quantifying uncertainty solely based on the primary variable does not fully represent the risk, as it also depends on other variables, such as deleterious elements or technological factors.

The use of stochastic planning that considers uncertainty in profit variables is not new to mining, although it may still be somewhat distant from becoming a common practice. Handling multiple scenarios can be a laborious and time-consuming task. Nevertheless, various studies demonstrate how stochastic mine planning has been valuable in assessing the uncertainty of economic functions. Dowd (1994) and Dimitrakopoulos and Ramazan (2004) were perhaps among the first researchers to consider the risk of financial and geological variables in resource and reserve assessment and mine planning. The use of sequential simulations for similar purposes continues to be applied for transferring geological uncertainty to the planning strategy. Recent research by Jelvez *et al* (2023), Maleki, Madani and Jelvez (2021) and Acorn, Boisvert and Leuangthong (2020) provides contributions on profit maximisations given the risk of geological variables. However, the bibliographic survey conducted for this work suggests that there is a significant predominance of research focused on the stochastic approach for open pit operations, with much fewer applications of similar nature for underground operations.

Likewise, the purpose of this paper is to quantify the uncertainty of the financial return across the life-of-mine (LOM) in an underground operation. The work started from deterministic sequenced stopes, optimised from the last mineral resource estimate. Within the workflow illustrated in Figure 1, the uncertainty on profit, as Net Smelter Return (NSR), is quantified based on grades and recovery functions. Due to the nature of the variables involved, the uncertainty modelling considered a multivariate fashion to honour the multivariate spatial and statistical features of the variables.

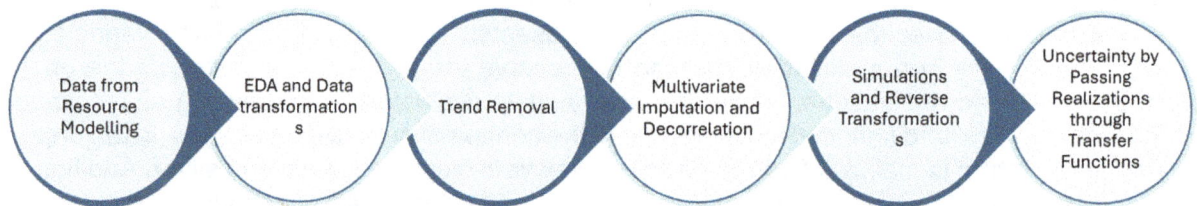

FIG 1 – Simplified workflow adopted for NSR uncertainty calculation within mining volumes.

The case study applied the methodology to the George Fisher Mine (GFM) polymetallic deposit (Queensland, Australia). Due to confidentiality concerns, some information might be omitted or distorted if deemed sensitive by the company. Recent studies conducted on an analogous deposit within the same geological context suggest that uncertainty in tonnes, grade, and metal is not significantly sensitive to domaining uncertainty. Based on these studies, the methodology was applied to deterministic domains. However, re-evaluating the study within a stochastic framework is part of future works.

GFM CONTEXT

The GFM polymetallic deposit, located in Queensland, Australia, is divided into two primary orebodies based on their geographical positions: the L72 orebody in the north and the P49 orebody in the south. This study focuses exclusively on the P49 mine. GFM is situated within the north–south Western Fold Belt of the Mount Isa Inlier, a remnant of an ancient intracontinental extensional basin. The chronostratigraphic sequence of the Western Fold Belt comprises four unconformable sequences, each corresponding to distinct rifting and volcanic events. The Zn-Pb-Ag mineralisation in the region typically occurs along a north–south strike and dips westward within the stratigraphic sequence (Figure 2). The P49 deposit consists of 13 stacked ore lenses distributed within a 250 m-thick stratigraphic interval. A barren dolomitic siltstone unit separates the mineralised sequence into two parts: the sulfide-rich hanging wall orebodies and the footwall orebodies (Chapman, 2004; Bertossi and Carvalho, 2023).

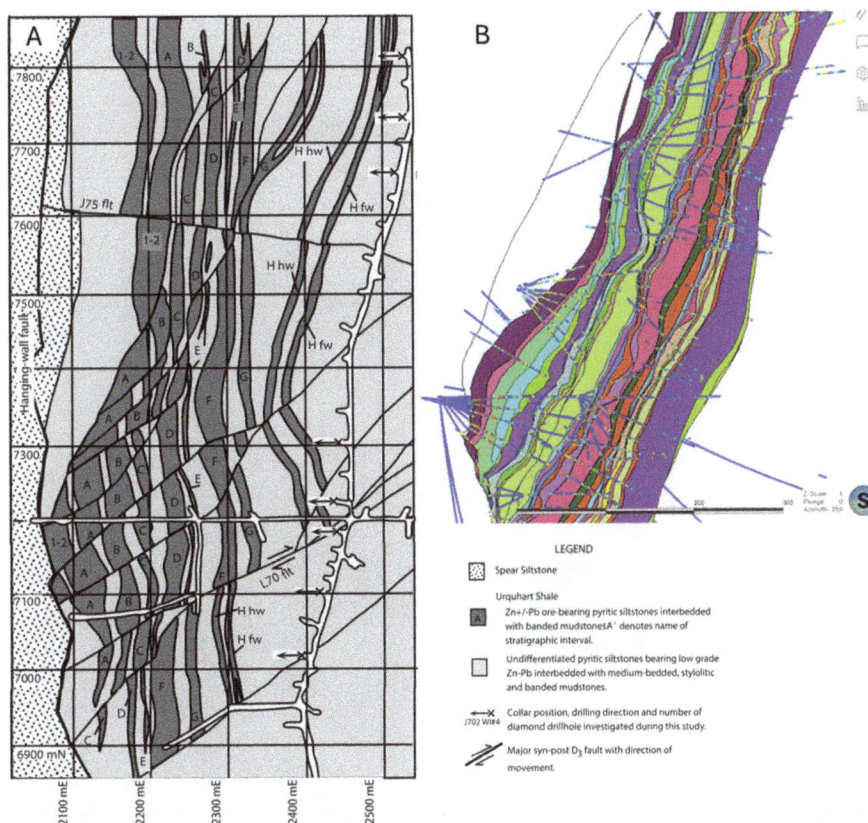

FIG 2 – A simplified plan view (A) of George Fisher North mineralised lenses (modified from Chapman, 2004). An east–west cross-section showing the modelled domains (B).

At the time, the P49 deposit count with 3620 drill holes, arranged in a semi-regular grid with spacing varying between 100 to 20 m, predominantly drilled from underground. The ore lenses were modelled as stratigraphic sequences using the Leapfrog software, encompassing 35 distinct stacked domains. The stratigraphic sequence, from base to top, includes dark shale (n), banded shales (8d, 8cm, 8b, and 8am), siltstone/shales (7c, 7bm, and 7a), calcitic shales/siltstones (6f, 6el, 6dm, 6c, 6bl, and 6am), shaley siltstones (5cl, 5bm, and 5bl), pale siltstones/shale (4g, 4fm, 4e, 4dl, 4cm, 4b, and 4al), dark siltstone/shale (3em, 3dl, 3cm, 3b, and 3a), chloritic siltstone (2dm, 2c, 2bm, and 2al), chloritic shale (1m), and hanging wall upper shale/siltstone (hwl). Each stratigraphic domain (Figure 2b) is estimated independently, with its own composites, variograms, and search parameters.

The case study uses the geological interpretation and composites from the 2023 year-end mineral resource prepared by Glencore's Technical Services department. Therefore, topics such as the decision of stationarity, capping definitions, and other aspects essential to mineral resource estimates are beyond the scope of this paper. Instead, the focus is on assessing NSR uncertainty that reflects the current practices on data management, domaining, classification, and mine planning at GFM.

METHODOLOGY

Scope framing

Domain's wireframes cover a much larger area compared to the stopes for which NSR uncertainty is assessed. This makes simulation workflows both time-consuming and computationally expensive. For this reason, the study's scope was limited to the mine design zones, plus a reasonable buffer to prevent artifacts at the borders. The composites were used in their original spatial form. To further reduce processing time, domains were ranked based on their contribution to the total planned tonnage. Out of the initial 35 domains, 13 were excluded from the analysis as each represented less than 1 per cent of the total material being mined. The final scope now covers approximately

95 per cent of the LOM tonnes. Figure 3 provides a breakdown of the proportion each domain contributes to mine planning.

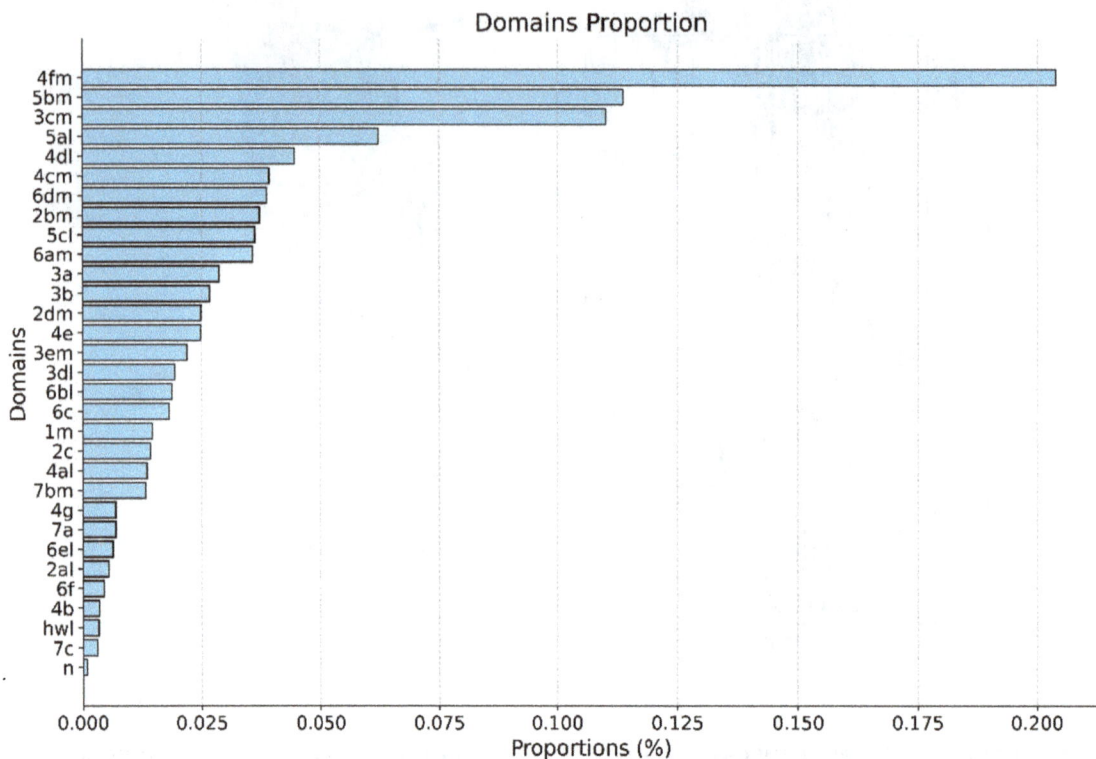

FIG 3 – Proportions of each domain's contribution to total mass over the life-of-mine.

The workflow was applied to a high-resolution unrotated grid with 2 × 2 × 2 m spacing. This decision was made to align with the data and domain geometry, ensuring the grid was consistent with the composite support and that the wireframes were adequately filled. The model specifications are outlined in Table 1.

TABLE 1

Modelling specifications.

	Easting	**Northing**	**Elevation**
Minimum (m)	1591.69	4559.26	2565.92
Maximum (m)	3263.69	5919.26	3253.92
Spacing (m)	2.0	2.0	2.0
Number	838	682	346

Data properties

The study simulated six variables: Pb, Zn, and Ag as the primary ore components, and S, Fe, and Cu as variables influencing metallurgical recovery. To ensure conciseness and better readability, the demonstrations and analyses in the following sections will focus on the three most representative domains (4fm, 5bm, and 3cm).

The data is analysed in both univariate and multivariate dimensions. Accessing the actual data properties, such as statistics and correlations, is crucial, especially in a multivariate context, since the realisations are not intended to reproduce solely the univariate but also any complex multivariate spatial features (Leuangthong, McLennan and Deutsch, 2004). To this end, the exploratory data analysis included statistical tables, cross-plots, boxplots, cumulative distribution functions, correlation matrixes (Figure 4), and variograms for all variables and domains. Declustering weights were determined using an inverse distance method with a power of 2. Weights are different for each element as they reflect its variables' heterotopic characteristics. Going forward in the workflow, the next steps consider the N-Scores of the variables transformed using declustering weights. Given the number of variables and domains, a total of 136 variogram models were required. To streamline this task, an automatic fitting function available on dedicated geostatistical packages was used to fit two to three exponential structures depending on the variable. However, constraints were applied to prevent the introduction of artificial or unrealistic structures by the auto-fitting. The principal variogram directions were aligned with those established in the 2023 mineral resource estimates. The principal orientations are 360°/-20°, 47°/72°, and 277°/19°. The first two define the reference plane, while the third represents the orebody thickness. The nugget contribution inference used omnidirectional variograms, and short ranges were extrapolated to determine the variogram at zero lag. Figure 5 illustrates the principal directions of the Pb and Zn de-trended variogram models for domain '4fm'. The accuracy of local uncertainty determined by the fitted variogram models was checked using accuracy plots, as illustrated in Figure 6.

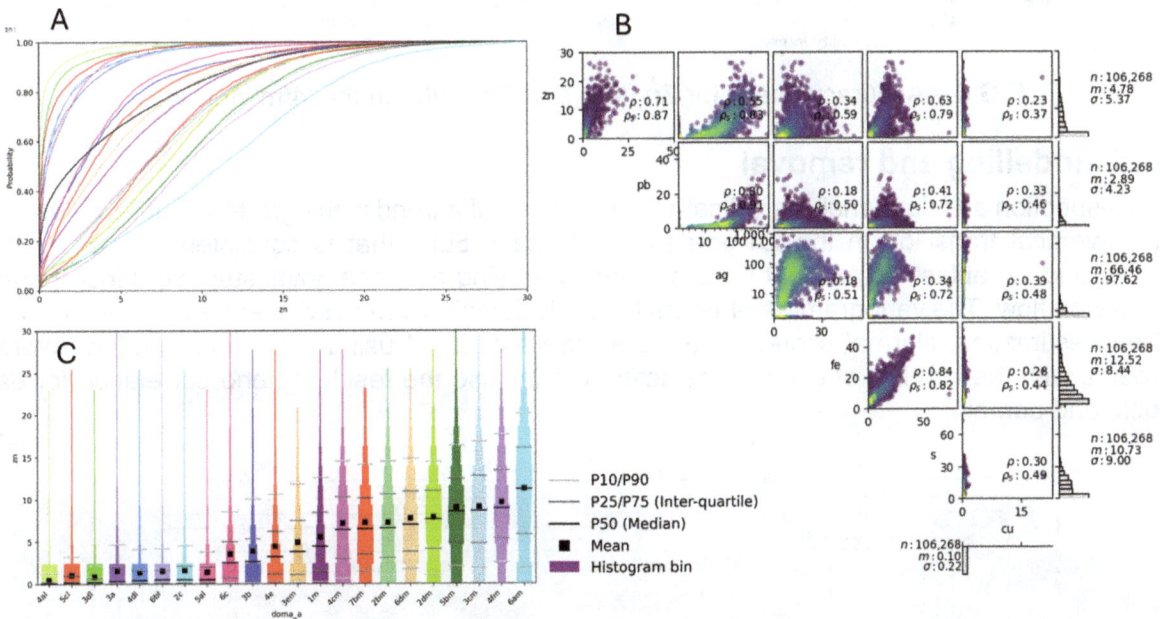

FIG 4 – Outputs of exploratory data analysis. Cumulative distribution (A) of Zn, global cross-plots (B), and Zn boxplots (C) by domain.

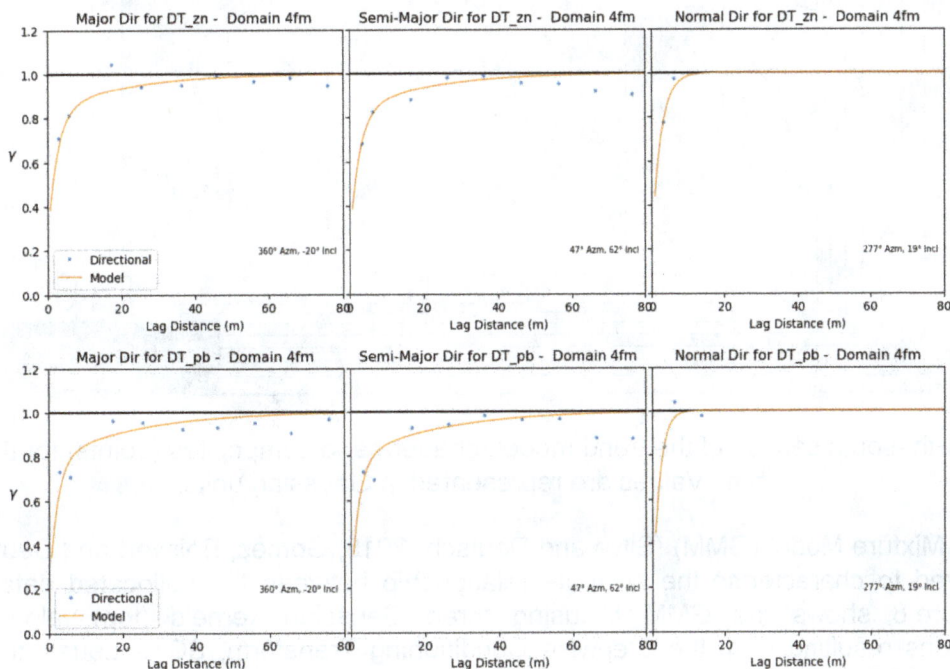

FIG 5 – Directional variogram models of de-trended Zn (top) and Pb (bottom) for domain '4fm'.

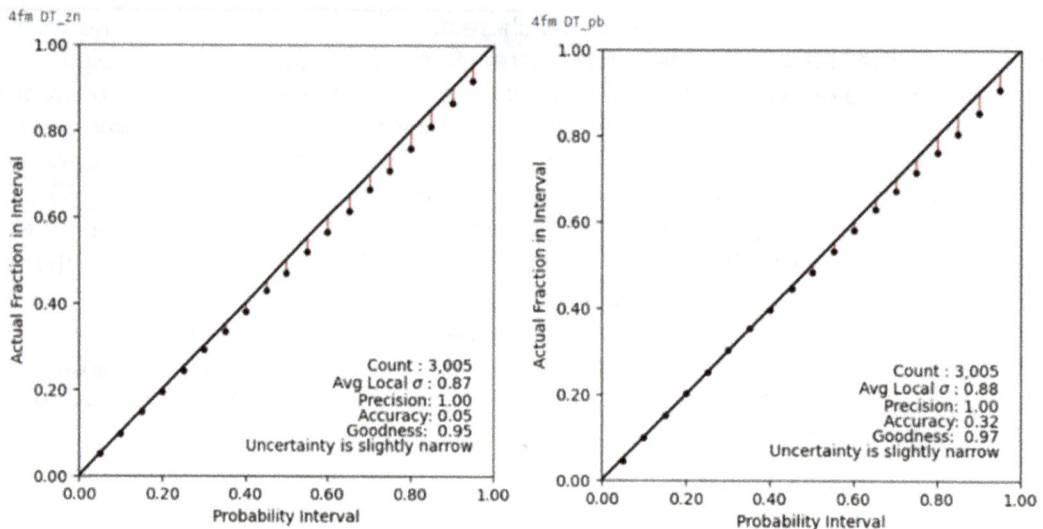

FIG 6 – Accuracy check for Zn (left) and Pb (right) in the '4fm' domain.

Trend modelling and removal

Visual inspection and data analysis indicate the presence of a trend in the grades. Figure 7 illustrates a quasi-vertical transition in the Zn grades for Domain '5bm' that is consistent across multiple domains and variables. To account for this, trend modelling and a removal step were incorporated into the workflow. The variogram parameters for each domain were used to establish the anisotropy for trend estimation. Normal scores data were then estimated using a Moving Window Average approach, with the number of samples selected to minimise the residual-trend correlation for each variable and domain.

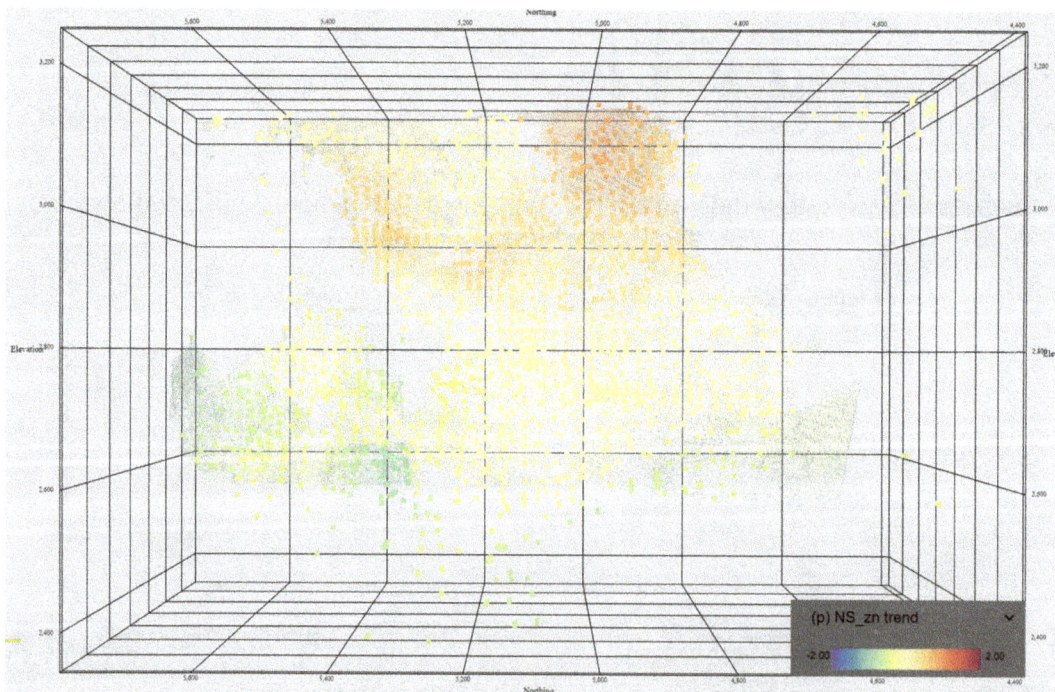

FIG 7 – North–south section of the trend model (shaded) and composites (points) for the domain '5bm'. Values are represented in Gaussian units.

A Gaussian Mixture Model (GMM) (Silva and Deutsch, 2015; Gomes, Boisvert and Deutsch, 2022) was employed to characterise the bivariate relationship between the collocated data and trend points. Figure 8 shows the GMM fit using three Gaussian kernels (top), along with the transformations resulting from the Stepwise Conditioning Transform, SCT, (Leuangthong, 2003) (bottom). The SCT process transforms the data conditioned to the trend distribution fit by the GMM, resulting in a perfect uncorrelated pairing. This step is necessary because the trend and residuals are correlated. Estimating or simulating the residuals independently of the trend could lead to poor reproduction of this feature. The SCT allows for the independent modelling of residuals, and the original dependence is restored by re-introducing the trend at the end.

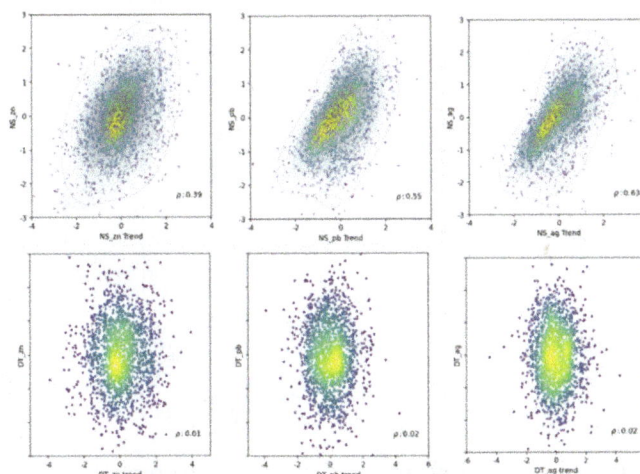

FIG 8 – GMM (top) + SCT (bottom) for data-trend decorrelation. The variables Zn (left), Pb (middle) and Ag (right) within the domain '4fm'.

Data imputation

GFM is a mature operation, with exploratory work beginning in the 1940s and production in the early 1980s. The drilling and sampling data from the deposit results from a combination of historical and modern campaigns, each with distinct purposes and lab protocols. This has resulted in a highly heterotopic database nowadays. However, in a multivariate fashion, some techniques employed in the workflow require the data to be isotopic. To tackle this, one simple solution would be to enforce isotopy by omitting the missing locations. However, this approach was found to be inadequate for the objectives of the study. The secondary variables are less sampled than the primary ones. Dropping data would result in a much sparser data set for Zn, Pb, and Ag, thereby not accurately reflecting the current uncertainty associated with these variables. The second approach, which was employed in this study, involved generating realisations for the missing data conditioned on the known values of other related elements. The multiple isotopic scenarios are used to condition the simulations in a manner that the uncertainty in the inputted values is transferred to the geostatistical realisations.

When working with heterotopic data, it is important to characterise the missingness mechanism, as this influences the decision of the imputation approach. There are three main forms that describe the nature of missingness (Rubin, 1976; da Silva, 2019). The term MCAR (Missing Completely at Random) refers to a situation where data is missed in an entirely random manner, with no dependence between the absent and observed data. This is often uncommon in geological data due to the systematic employed in mining sampling and data acquisition. The MAR (Missing at Random) mechanism does not align exactly with its acronym. Here, the probability of missing data depends on the observed data's occurrence (but not the value). The last mechanism, MNAR (Missing Not at Random), is likely the most prevalent in geological data imputation. In MNAR cases, the probability of data being missing is related to the value of the missing data itself. For example, in the case of a gold deposit where sampling occurs only in mineralised areas, the missing data reflects the expectation of values of gold. The last two mechanisms characterise most of the missing values in the GFM mine; there is a bias of +10 per cent, +12 per cent, and +10 per cent in the averages of Zn, Pb, and Ag, respectively, when data is made isotopic. This confirms that missing data are predominantly present in lower-grade regions.

Multiple imputation for continuous variables employs a Bayesian Updating formalism. This technique combines a prior probability distribution with a likelihood function to produce an updated probability for the variable being imputed. The prior distribution is obtained through simple kriging with Gaussian units. In a multivariate case, the likelihood's mean and variance represent a linear combination of the n conditioning variables, with weights accounting for the correlations and redundancies. Ultimately, the updated distribution can be sampled stochastically to generate realisations of the variables at missing locations. This approach is powerful for imputing data in multivariate problems, as it incorporates both spatial and multivariate dependencies between variables. Figure 9 illustrates the imputed distributions for the three secondary variables in domain '5bm'.

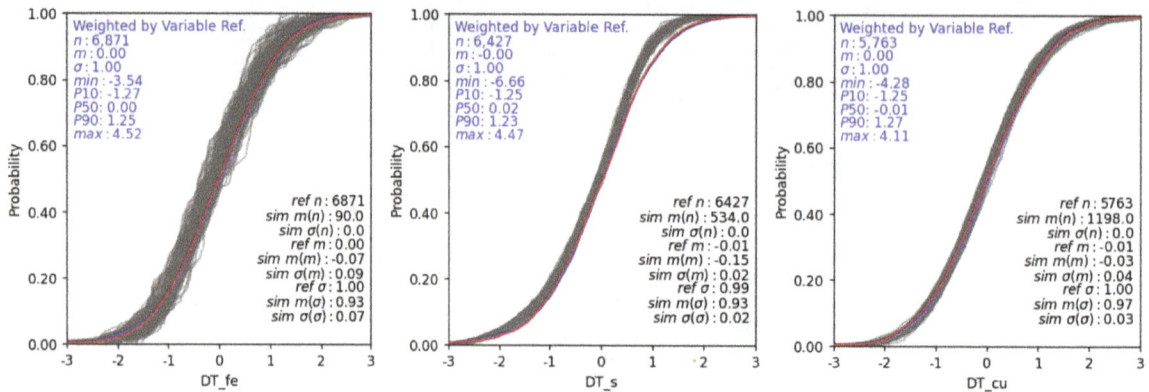

FIG 9 – Imputed realisations of the secondary variables in domain '5bm'. Iron (left), Sulfur (centre), Copper (right).

The lower variance and averages observed in the imputed realisations (in grey) reflect the MAR nature of the missingness. If the missing data followed a completely random pattern, the imputed realisations would be expected to align more closely with the original reference distributions. The imputed values were also evaluated for their multivariate relationships and fairness in accordance with the fitted likelihood GMMs (Figure 10).

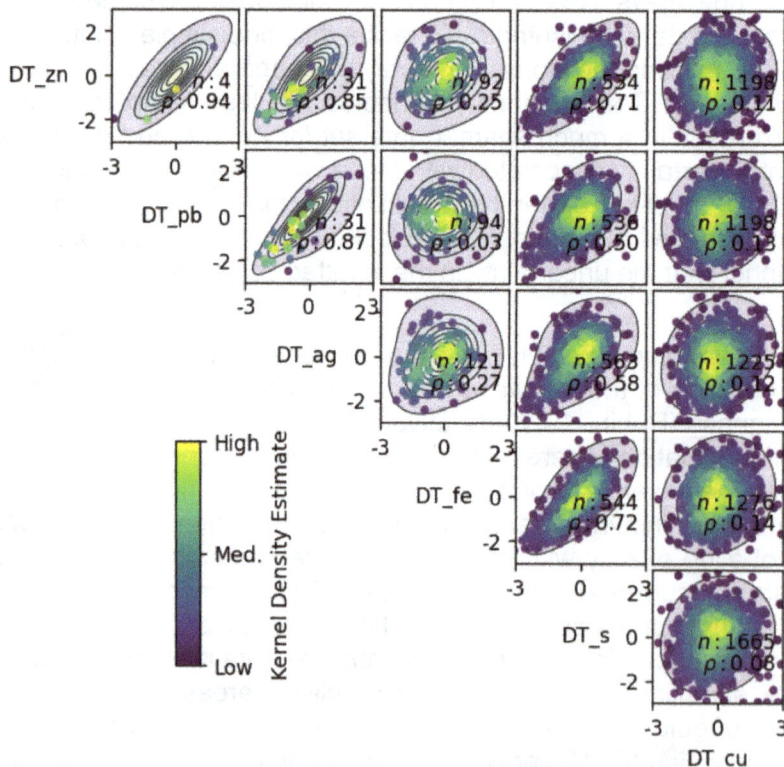

FIG 10 – Realisation 0 of the imputed data for domain '5bm' and fitted bivariate GMMs.

Histogram uncertainty

When evaluating uncertainty on a large scale, such as for annual production volumes, the impact of uncertainty can be partially attenuated, leading to a misleading perception that the overall uncertainty is smaller than it truly is. To address the issue, uncertainty can be more accurately characterised by generating multiple realisations of the data and using these reference realisations in the simulation workflow. This technique, known as Spatial Bootstrap (Deutsch, 2004; Rezvandehy, 2016), ensures that the uncertainty in the global distribution is properly transferred to the final simulated realisations honouring the underlying spatial correlation as informed by the variograms. However, the process could be extremely computationally intensive if employed on inputted realisation as each inputted realisation from the previous step must be resampled once. The Spatial Bootstrap process involves performing non-conditional simulation at the original data locations, followed by stochastic resampling to generate multiple equiprobable realisations from the reference distributions (Figure 11).

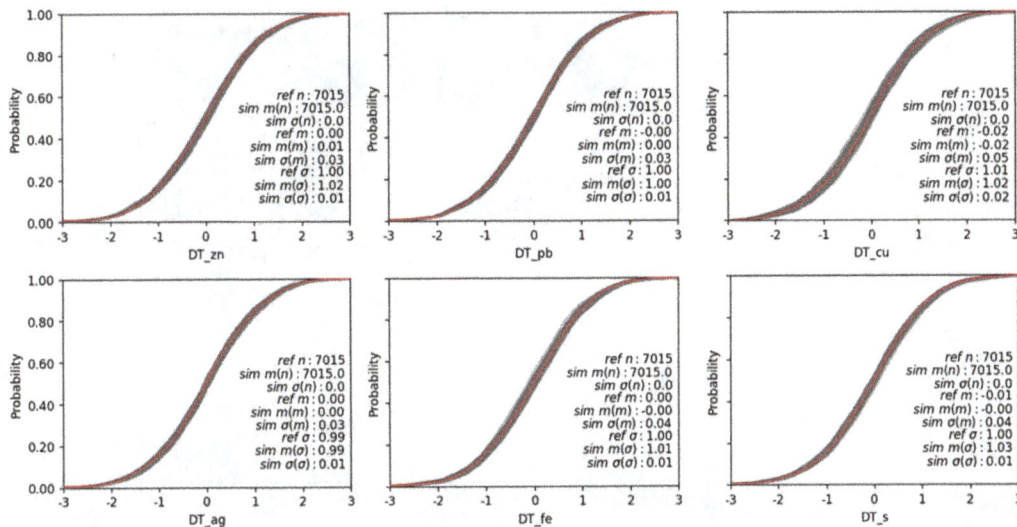

FIG 11 – Spatial Bootstrap distributions for the first imputed realisation in domain '3cm'. Histogram realisations are in grey, and the reference inputted realisation is in red.

Each one of the histogram realisations is transformed using declustered normal scores transformations prior to decorrelation.

Decorrelation

To simulate the data independently while preserving the back-transformed multivariate relationships, the data was transformed to ensure uncorrelated variables prior to modelling. A Projection Pursuit Multivariate Transformation (PPMT) (Barnett and Deutsch, 2015; Barnett, Manchuk and Deutsch, 2014; Barnett, 2017) was employed to achieve this, converting the variables into a multivariate Gaussian form. PPMT operates on the principle that if data is multivariate Gaussian, any projection of the data onto a vector will also be Gaussian. This is a fundamental concept, as the PPMT algorithm searches for a vector in the multidimensional space that, when data is projected onto it, yields the most non-Gaussian distribution. The values along that vector are then transformed into normal scores. This process iterates, repeatedly searching for the next maximum non-Gaussianity and performing transformations on the projections until the data is completely uncorrelated or a stopping criterion achieved.

To verify the reduction of non-Gaussianity, the projection index is plotted against the number of iterations. The projection index can be seen as a statistical measure of the degree of non-Gaussianity in the projected data (Friedman, 1987). A high projection index indicates a high degree of non-normality, while a decrease in the index across iterations indicates progress toward normality. After sufficient iterations, the projection index should stabilise. Figure 12 illustrates the results of the PPMT and the normality check after 100 iterations (top) and the resulting uncorrelated bivariate distributions (bottom).

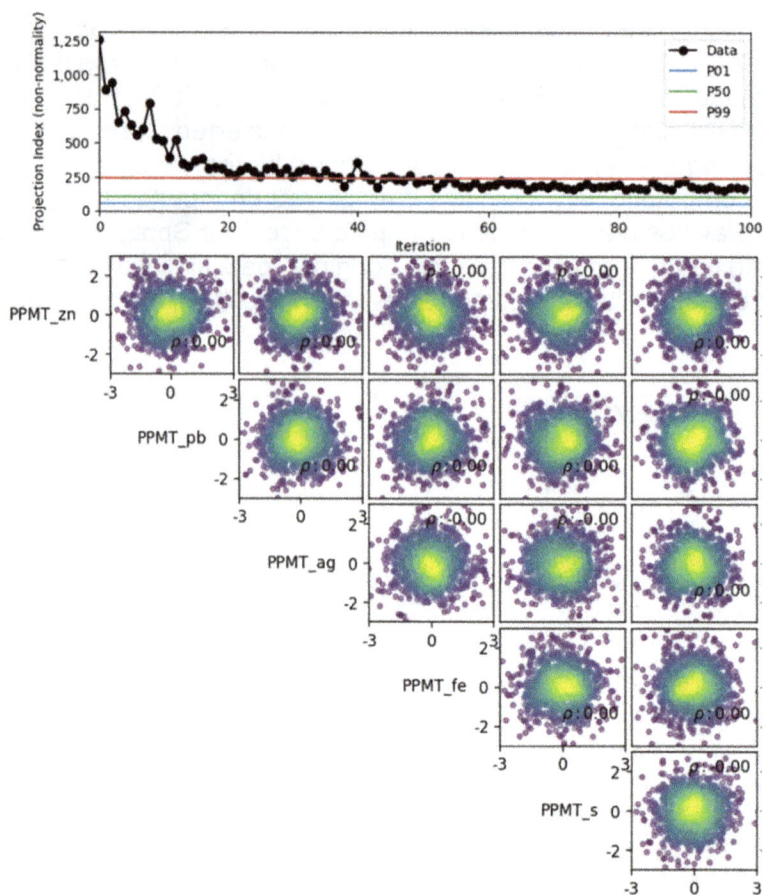

FIG 12 – PPMT decorrelation bivariate plots for domain '4fm' (below). Projection index demonstrating the decay of non-normality by iteration (above).

Simulation and uncertainty

The PPMT variables were simulated using the Turning Bands (TB) method (Matheron, 1973; Journel, 1974). The TB method was the first simulation technique developed in geostatistics. Fundamentally, it is based on simulations along independent radial lines using 1D covariance functions to generate 3D simulations. Typically, moving averages are employed for the initial 1D simulations, and the simulated values are assigned to all discretisation points within each 1D band. The simulated realisations in 3D are defined as the averages of all orthogonal projections onto the bands. Conditioning of TB simulations is achieved by simple kriging (SK) with the original data and unconditional realisations.

In this case study, the decision to use the TB method over other traditional simulation techniques was based on the efficiency of the algorithm. Given that methods, such as Sequential Gaussian Simulation (Isaaks, 1990), would be impractical for handling six variables across 22 domains, TB was chosen for its computational efficiency. Furthermore, the TB method requires less disk space and operates significantly faster than other techniques because it doesn't rely on previously simulated values stored in a structured grid. This makes it a better choice for larger domains and multiple variables. In this study, a hundred TB realisations employed 1000 bands and 2000 discretisation points. The appropriateness of the selection of the number of bands and discretisation was visually verified to ensure that undesired spatial patterns did not occur. In addition to enhancing computational efficiency, the scope framing assisted in preventing artifacts by ensuring the simulated grid is reasonably aligned with the composite's spatial configuration.

The simulated realisations were transformed back to the original units by sequentially reversing the transformations applied during the preprocessing stage. It is important to ensure that all transformations are reversed in the correct order, from the last applied to the first: (1) PPMT multivariate decorrelation, (2) Spatial Bootstrap normal scores transformation, (3) SCT for reintroducing the trend, and finally (4) the initial normal scores transformation. This systematic reversal ensured accurate reproduction of the declustered grade distributions, which was verified for all variables and domains. Figure 13b illustrates the successful histogram reproduction of the simulated realisations for Zn, Pb, and Ag, while Figure 13a displays the reproduction of the Gaussian variograms.

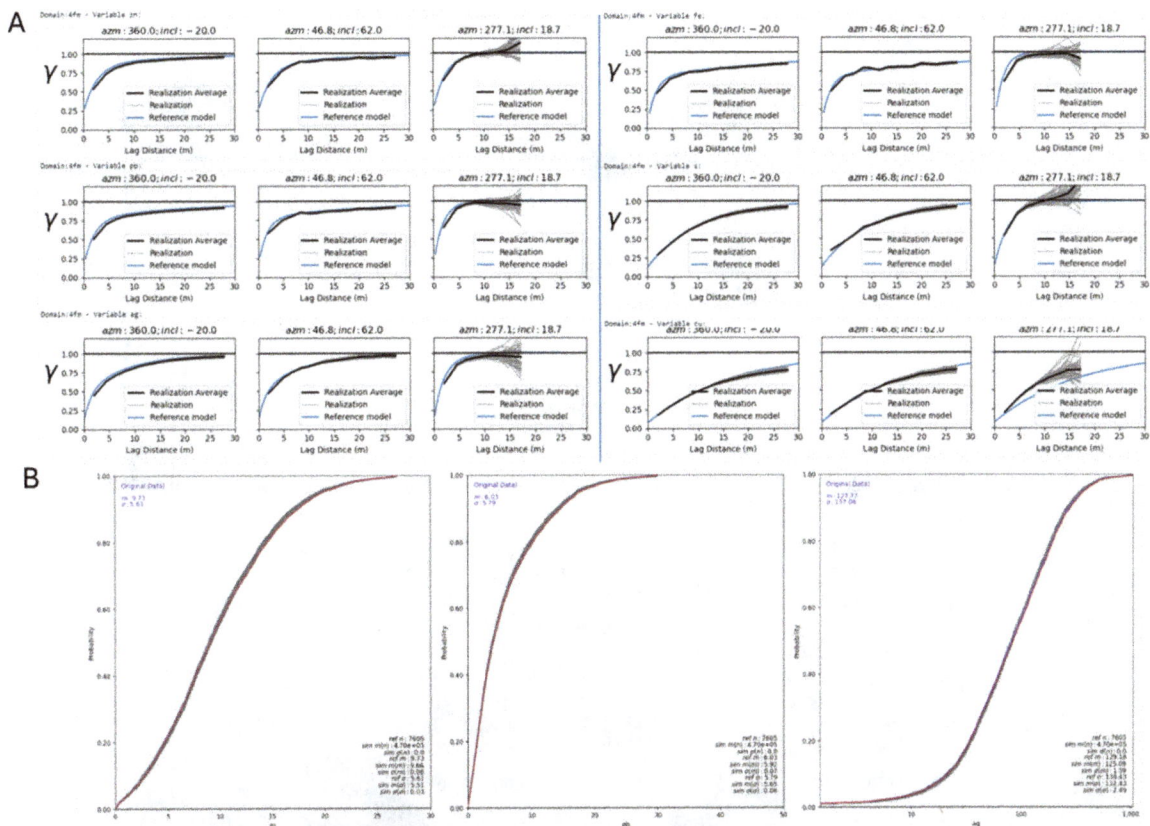

FIG 13 – Variogram reproduction in Gaussian units for domain '4fm' (A). B shows the reproduction of Zn (left) Pb (centre) and Ag (right) grades for domain '4fm'.

Simulated realisations in original units were also validated by inspecting their multivariate statistical properties. The scatter plots in Figure 14 show that the first simulated realisation effectively captures most of the key bivariate features of the reference data. The matrix (Figure 15) represents the bivariate correlations across all realisations by averaging the coefficients from each one. Overall, the examination of multivariate relationships demonstrates an adequate reproduction of the original multivariate properties of the data.

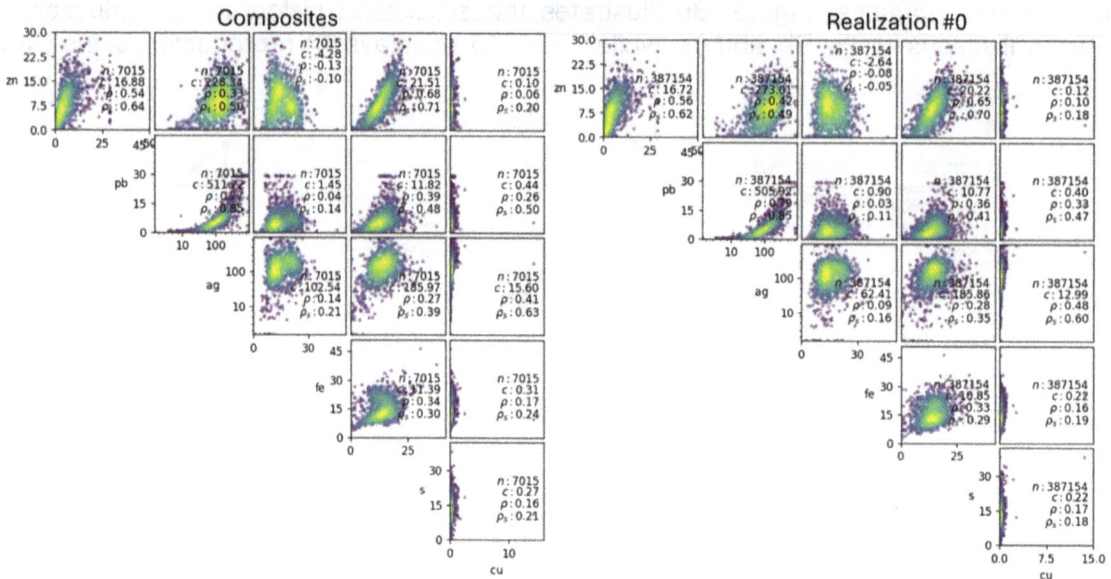

FIG 14 – Bivariate relations on composites and first realisation for domain '3cm'.

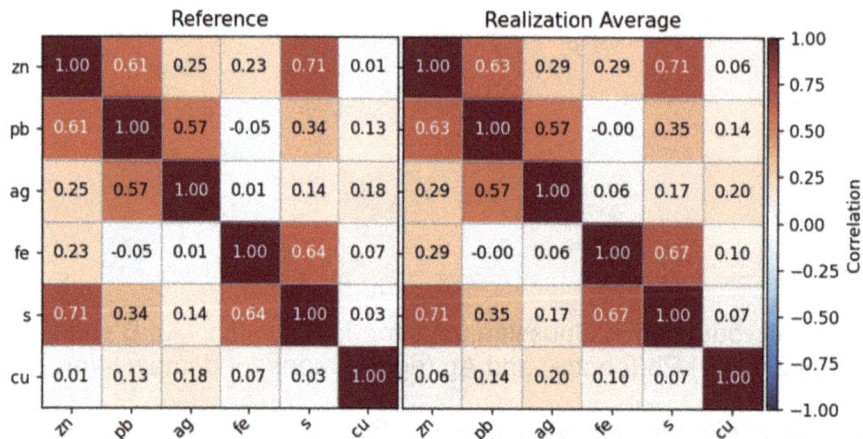

FIG 15 – Matrix comparing variables correlations in the data and simulated realisations.

The e-type estimate was also used to assess TB conditioning and data reproduction. Global reproduction across all domains was evaluated using the e-type model collocated with the actual composited data (Figure 16). The results were within expectations, evidenced by a low deviation in the mean and a strong correlation for all variables/domains. Additionally, swath plots (Figure 17) comparing the e-type and composites in 25 m slices further demonstrates adequate conditioning at a larger scale.

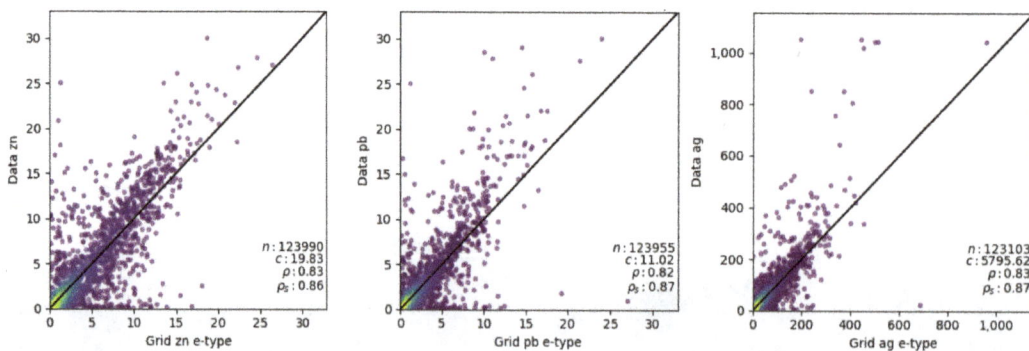

FIG 16 – Conditioning checking at data-realisation collocated points for all domains.

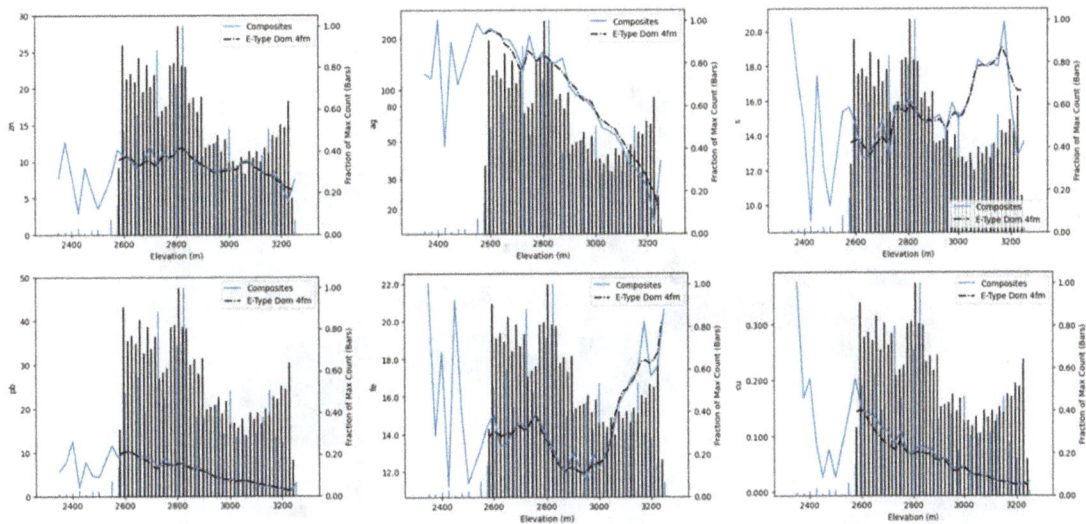

FIG 17 – Swath plots of E-type and composites for the six variables in domain '4fm'.

The uncertainty in expected revenue within stope designs was evaluated bypassing all realisations through NSR and metallurgical functions. To achieve this, simulated points were upscaled to the stopes dimensions for each year, spanning from 2024 to 2036. The upscaling process involved averaging point-scale realisations within each production volume. A cross-section of a single point-scale realisation and the corresponding yearly production volumes is illustrated in Figure 18. The simplified economic functions are as follows:

$$NSR = a_1[Zn * Zn_{rec}] + a_2[Pb * Pb_{rec}] + a_3[Ag * Ag_{rec}] + a_4$$

Where:

$$Zn_{rec} = a_1 - a_2 \left[\frac{Pb}{Zn}\right] - a_3[(b_1Fe + b_2S - b_3Cu)/Zn]$$

$$Pb_{rec} = a_1Pb - a_2Zn - a_3 \left[\frac{(b_1Fe + b_2S - b_3Cu)}{Pb}\right] + a_4$$

$$Ag_{rec} = Pb_{rec}|Zn_{rec} - a_1 \left[\frac{(b_1Fe + b_2S - b_3Cu)}{Ag}\right] + a_2$$

The coefficients a and b are experimentally determined values derived from geometallurgical and economic assumptions historically employed in mineral resource reporting and mine planning at GFM.

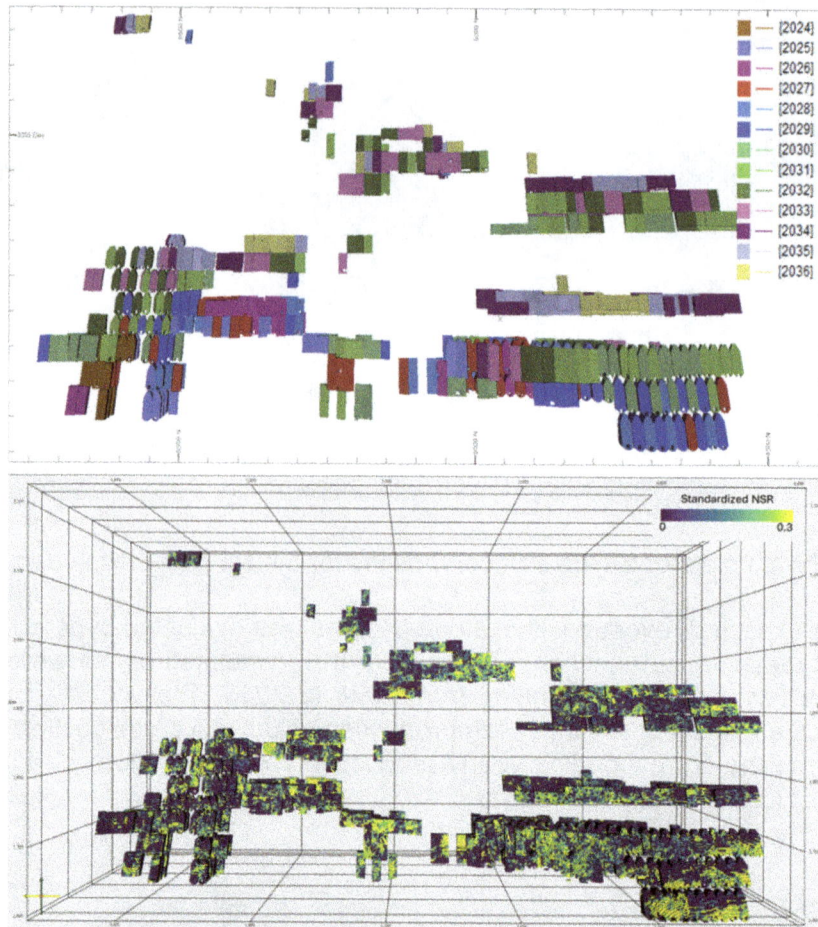

FIG 18 – North South cross-section showing the production plan (above) and the first realisation of NSR at point scale (below). NSR values are presented on a standardised scale.

RESULTS AND DISCUSSION

The validation steps introduced at each stage of the workflow (Figure 1), including the final checks on the simulated realisations, confirm that the models are adequate for assessing the uncertainty of the variables themselves and the relevant transfer functions.

NSR uncertainty is presented for each mining stage, covering the period from 2024 to 2036. Revenue distributions were generated by processing all realisations through the recovery and NSR functions. The resulting distributions were then averaged within the mining volumes to reflect uncertainty at the appropriate scale for each production year. This process resulted in 100 NSR values per annum, from which a probability distribution could be derived, and uncertainty inferred. Table 2 presents the uncertainty as the variability around the mean at the 90th confidence interval. This metric quantifies the percentage variation around the mean necessary to encompass 90 of the 100 realisations. Additionally, the table includes the average drill hole spacing calculation, estimated by using three samples from different drill holes with a search-oriented nearly orthogonal to the overall drilling direction. The average distance (radius) was approximated to the side of a square polygon.

TABLE 2

NSR Uncertainty at production scale.

Year	P5	P25	P75	P95	Mean	90th Confidence interval [%]	Average drill hole spacing
2024	148.48	156.28	169.89	178.70	163.08	9.27	22.67
2025	171.79	178.09	184.58	191.09	181.34	5.32	21.52
2026	208.79	214.13	222.85	228.03	218.46	4.40	23.02
2027	208.88	214.58	222.19	228.92	218.58	4.58	24.59
2028	242.25	246.72	254.21	259.98	250.64	3.54	26.30
2029	214.60	218.98	227.94	235.09	224.09	4.57	30.46
2030	178.55	183.32	188.15	192.81	185.89	3.84	29.95
2031	172.15	176.50	182.03	186.85	179.15	4.10	38.43
2032	169.92	172.88	181.00	185.34	177.25	4.35	30.04
2033	163.04	168.20	175.90	182.67	172.23	5.70	27.18
2034	168.44	171.45	177.71	182.93	174.72	4.15	31.84
2035	167.09	170.69	178.68	182.65	174.95	4.45	24.82
2036	168.90	174.60	181.82	187.33	178.27	5.17	23.09

We observe that uncertainty does not necessarily reflect the average data availability across production years. When pairing the 90th percentile confidence interval with the average drill hole spacing for each year, no meaningful correlation becomes evident. In fact, the drill hole coverage is reasonably well distributed across the entire deposit, suggesting that it is not a primary driver of uncertainty in NSR at an annual scale.

However, when interpreting the results, it is important to consider that the estimated uncertainty is scale-dependent. As point realisations are averaged to match production volumes, the NSR distribution narrows around the mean, and the degree of smoothing reflects the disparity between the source and target supports when upscaling. Consequently, the observed post-processed uncertainty encompasses not only statistical variance but also the differences in production volumes across years. Figure 19 illustrates how each year contributes to the overall production throughout the LOM.

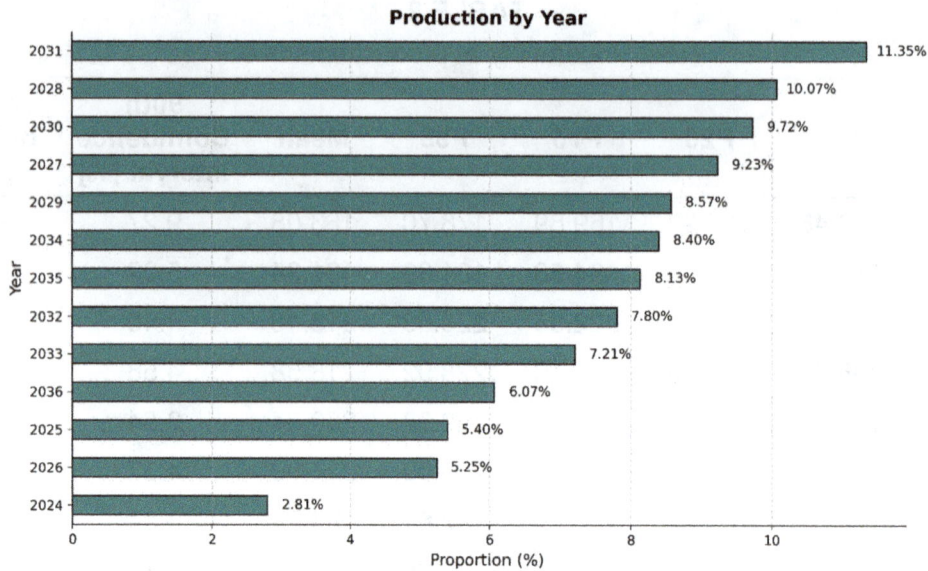

FIG 19 – Contribution of each year to the LOM production.

The highest uncertainty observed in 2024 is partially attributed to the smaller mass involved, as averaging over larger volumes tends to smooth out variability among realisations. However, this is not the only factor contributing to the higher uncertainty within the planned stopes for 2024. To isolate the effect of volume, uncertainty was also calculated as the probability of being within 15 per cent of the mean on a point scale (Figure 20).

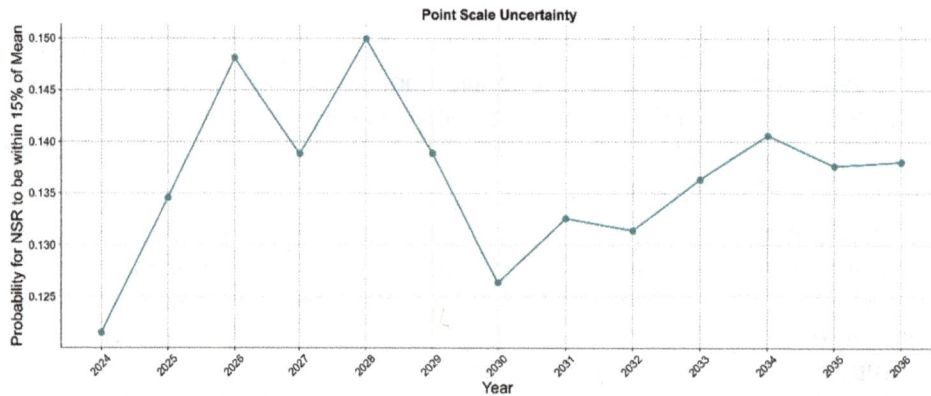

FIG 20 – Point-scale uncertainty represented as the probability for NSR being within 15 per cent of mean.

As shown in Figure 18, the stopes planned for 2024 (brownish solids at the bottom left) are clustered in a localised portion of the deposit, whereas the stopes for subsequent years are more broadly distributed across the ore strike. This suggests that geographical diversification in mining helps reduce uncertainty in annual production as the project expands into more dispersed areas of the deposit, thereby attenuating the variability in NSR estimates. This analysis reveals that the 2024 region exhibits the greatest NSR variability, while the 2026 regions present some of the lowest uncertainties on a point scale despite having one of the smallest production volumes. This observation reinforces the earlier finding.

A sensitivity analysis of the NSR inputs was also evaluated. Feature importance is a concept used in machine learning applications to quantify the relevance of different features in regression problems. In algorithms such as stochastic gradient boosting (GB) (Friedman, 2001) or Random Forest (RF) (Ho, 1995), feature importance is determined by assessing how much each feature contributes to the predictive power of the model. In this case study, the relevance of Zn, Pb, Ag, and recovery in explaining the NSR was analysed. This approach helps identify which variables most significantly influence the NSR calculation, allowing for a better interpretation of the modelled uncertainty. With this purpose, a GB regressor was defined with a learning rate of 0.1, a max depth of 3, and 150 estimators. The hyperparameters were tuned using the scikit-learn GridSearch cross-validation tool.

The results (Figure 21) demonstrate that the NSR calculation is predominantly influenced by the lead variable, with significantly less relevance attributed to the other features. The correlation coefficient between NSR and Pb is 0.95, indicating that the current economic function used in the GMF could potentially be simplified to a single-feature function without a substantial loss in predictive capability. The greater contribution of Pb to the overall NSR uncertainty was further illustrated by plotting the point-scale uncertainty together (Figure 22). It becomes evident that NSR is highly dependent on lead, closely following its trend and magnitude.

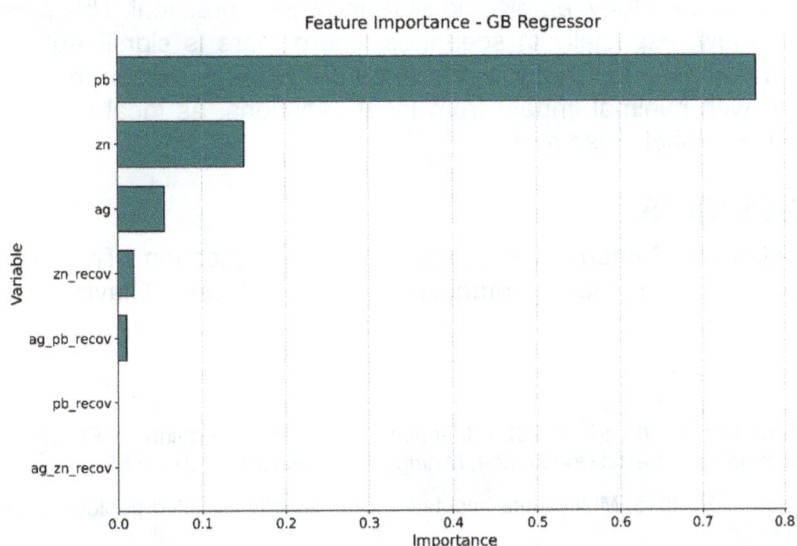

FIG 21 – Gradient Boosting feature importance. NSR is the independent variable, Pb, Zn, Ag and recovery, are the dependent ones.

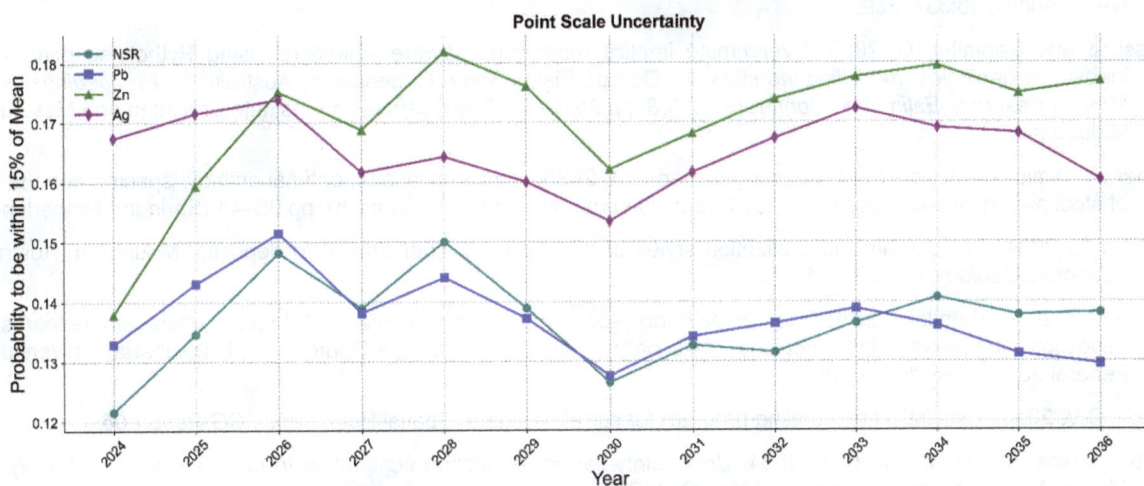

FIG 22 – NSR, Pb, Zn, and Ag Point-scale uncertainty represented as the probability for NSR being within 15 per cent of the mean.

CONCLUSION

This study is not intended by the author to be final or definitive for the George Fisher Mine. The methodology presented here is applicable to a wide range of uncertainty problems, providing information for better-informed decisions in geology, mine planning, and processing at GFM. The flexibility of changing variables and scale of uncertainty is one advantage of using simulated realisations over other probabilistic geostatistical methods. Simulation addresses the change of support, allowing us to upscale the point to match relevant volumes in the operation. This property permits transferring the primary point uncertainty to any scale and production time frame.

However, it is important to acknowledge that the NSR and recovery coefficients used in this analysis were assumed to be true. There is inherent uncertainty associated with these functions, which suggests that the actual uncertainty in profit could be greater than predicted here. The risk assessment did not include a sensitivity analysis of the economic functions, which should be the next step in this study. Evaluating the NSR and recovery equations with reconciliation exercises to confirm their accuracy and whether the NSR is primarily influenced by lead could offer valuable insights. Furthermore, recent internal studies in a similar geological context indicate that uncertainty in tonnes, grade, and metal is not highly sensitive to domaining uncertainty in this deposit style.

Given the broad scope of this study, employing SGS may be impractical. This presents an important drawback in the workflow, especially in scenarios where there is significant anisotropy variation across the simulated domains. In GFM the reference ore plane is reasonably uniform with respect to strike and plunge, with minimal impact from local variations, as local orientations only deviate slightly from the defined global anisotropy.

ACKNOWLEDGEMENTS

The author acknowledges Glencore for supporting this publication. Thanks also to the GFM geologists and engineers for their contributions and Professor Clayton for comments and recommendations.

REFERENCES

Acorn, T, Boisvert, J B and Leuangthong, O, 2020. Managing Geologic Uncertainty in Pit Shell Optimization Using a Heuristic Algorithm and Stochastic Dominance, *Mining, Metall, Explor,* 37:375–386.

Barnett, R M and Deutsch, C V, 2015. Multivariate imputation of unequally sampled geological variables, *Mathematical Geoscience,* 47(7):791–817.

Barnett, R M, 2017. Projection Pursuit Multivariate Transform, in *Geostatistics Lessons* (ed: J L Deutsch). Available from: <http://geostatisticslessons.com/lessons/lineardecorrelation>

Barnett, R M, Manchuk, J G and Deutsch, C V, 2014. Projection pursuit multivariate transformation, *Mathematical Geosciences,* 46:337–359.

Bertossi, L and Carvalho, D, 2023. Overcoming implicit modelling software limitations using Python scripting – an innovative geological modelling workflow for George Fisher Mine, Queensland, Australia, in *Proceedings of the Mineral Resource Estimation Conference 2023,* pp 251–265 (The Australasian Institute of Mining and Metallurgy: Melbourne).

Boucher, A, Dimitrakopoulos, R and Vargas-Guzman, J A, 2005. Joint simulations, optimal drillhole spacing and the role of stockpile, in *Geostatistics Banff 2004* (eds: O Leuangthong and C Deutsch), pp 35–44 (Springer: Netherlands).

Chapman, L, 2004. Geology and mineralization styles of the George Fisher Zn-Pb-Ag deposits, Mount Isa, Australia, *Economic Geology,* 99:223–255.

Da Silva, C Z, 2019. Metodologias de inserção de dados sob o mecanismo de falta MNAR para modelagem de teores em depósitos multivariados heterotópicos, Tese para obtenção do título de Doutora em Engenharia, Universidade Federal do Rio Grande Do Sul.

Deutsch, C V, 2004. A statistical resampling program for correlated data: Spatial Bootstrap, *CCG Report 06.*

Dimitrakopoulos, R and Ramazan, S, 2004. Uncertainty-based production scheduling in open pit mining, *Society For Mining, Metallurgy and Exploration,* 316:106–112.

Dowd, P A, 1994. Risk assessment in reserve estimation and open-pit planning, *Transactions of the Institution of Mining and Metallurgy, Section A: Mining Technology.*

Englund, E J and Heravi, N, 1993. Conditional simulation: Practical application for sampling design optimization, in *Proceedings of the Fourth International Geostatistics Congress* (ed: A Soares), pp 613–624 (Kluwer Academic Publishers).

Friedman, J H, 1987. Exploratory projection pursuit, *Journal of the American Statistical Association,* 82:249–266.

Friedman, J H, 2001. Greedy function approximation: A gradient boosting machine, *The Annals of Statistics,* 29(5):1189–1232. https://doi.org/10.1214/aos/1013203451

Gomes, C G, Boisvert, J and Deutsch, C V, 2022. Gaussian Mixture Models, in *Geostatistics Lessons* (ed: J L Deutsch), Available from: <http://www.geostatisticslessons.com/lessons/gmm>

Ho, T K, 1995. Random decision forests, in *Proceedings of the Third International Conference on Document Analysis and Recognition (ICDAR)*, pp 278–282.

Isaaks, E H, 1990. The application of Monte Carlo methods to the analysis of spatially correlated data, PhD thesis, Department of Applied Earth Sciences, Stanford University, Stanford.

Jelvez, E, Ortiz, J, Morales Varela, N, Askari-Nasab, H and Nelis, G, 2023. A Multi-Stage Methodology for Long-Term Open-Pit Mine Production Planning under Ore Grade Uncertainty, *Mathematics*, 11:3907. https://doi.org/10.3390/math11183907

Journel, A G, 1974. Geostatistics for conditional simulation of ore bodies, *Economic Geology*, 69(5):673–687.

Leuangthong, O, McLennan, J A and Deutsch, C V, 2004. Minimum acceptance criteria for geostatistical realizations, *Natural Resources Research*, 13(3):131–141.

Maleki, M, Madani, N and Jelvez, E, 2021. Geostatistical Algorithm Selection for Mineral Resources Assessment and its Impact on Open-pit Production Planning Considering Metal Grade Boundary Effect, *Nat Resour Res*, 30:4079–4094.

Martínez-Vargas, A, 2017. Optimizing grade-control drillhole spacing with conditional simulation, *Minería y Geología*, 33(1):1–12. Available from: <https://www.redalyc.org/articulo.oa?id=223549947001>

Matheron, G, 1973. The intrinsic random functions and their applications, *Advances in Applied Probability, Jerusalem*, 5(3):439–468.

Parker, H M and Dohm, C E, 2014. Evolution of Mineral Resources Classification, Amec lecture presentation. Available from: <https://www.crirsco.com/docs/H_Parker_Finex.pdf>

Rezvandehy, M, 2016. Geostatistical reservoir modelling with parameter uncertainty in presence of limited well data, PhD thesis, University of Alberta.

Rubin, D B, 1976. Inference and Missing Data, *Biometrika*, 63(3):581–592.

Silva, D S F and Deutsch, C V, 2015. Transformation for multivariate modelling using Gaussian mixtures with exhaustive secondary data, CCG Report 17.

Combining geological and grade uncertainties in drill hole spacing studies

C Gomes[1]

1. Senior Geostatistician, SLR Consulting, Edmonton AB, Canada.
   Email: cgomes@slrconsulting.com

ABSTRACT

Drill hole Spacing (DHS) Optimisations and Mineral Resource Classification Criteria derive from varying forms of modelling and quantification of risk within mineral resources. While grade traditionally takes precedence in such assessments, recent years have emphasised that uncertainty in mineral resources extends beyond grade alone. Despite this recognition, practitioners still lack a practical methodology to incorporate other sources of uncertainty effectively in such studies. Geological uncertainty, often linked directly to estimation domains, represents a significant additional source of variability. Domaining can be established based on various geological properties such as lithologies, alterations, structural controls or grade-thresholds, depending on the context of the mineral deposit.

This study advocates for a probabilistic modelling approach that integrates categorical and grade simulation methods to provide a more comprehensive assessment of uncertainty. Utilising Hierarchical Truncated Pluri-Gaussian (HTPG) for categorical simulation nested with Sequential Gaussian Simulation (SGS) for grades. Using a one-to-one nesting methodology, this technique generates a continuous realisation of grade for each simulated geological scenario. The HTPG-based categorical simulation approach encompasses sufficient parameters and steps to ensure robust domain modelling of different deposit types in terms of spatial orientation, anisotropy and variances per domain,

The results of this advanced technique are compared against traditional grade-only simulation from data sets of distinct mineral deposits, like gold and iron ore. The diverging outcomes for both techniques related to Optimal Drill hole Spacing definition, mineable panels and Resource Classification are exemplified. Certain deposits exhibit domains with higher uncertainty, while others demonstrate greater variability in grades. Nonetheless, the workflow integrating categories and grades offers a more comprehensive and realistic approach to risk management and should be applied provided expertise and processing resources are available.

INTRODUCTION

Modelling mineral resource uncertainty is a long-standing practice within the geoscience and mining industry (Chiles and Delfiner, 2012; Goovaerts, 2001). Usually aimed at establishing classification criteria, determining optimal drill hole spacing, or quantifying risk at a Life-of-Mine plan. The methods for modelling uncertainty as well as the understanding of which are the main sources of resource model uncertainty have evolved jointly over recent decades (Gómez-Hernández and Srivastava, 2021; Journel, 1994; Isaaks and Srivastava, 1989; Pyrcz and Deutsch, 2014).

Historically, grade uncertainty was the major aspect considered for uncertainty quantifications. Presently, it is known that along with grade there are other important sources of model uncertainty, ie the estimation (geological) domains, drilling spatial precision, sampling histogram, and variogram parameters (Deutsch, 2018; Deutsch, 2023). Regarding current modelling techniques, simulation-based methods are the most accepted tools in mineral resources. While continuous sequential Gaussian simulations (Deutsch and Journel, 1997) for grades have been well established and extensively used since the 1990s, a widely accepted categorical method for simulating geological domains has been a challenge for resource modellers up to the present date. Main techniques include Sequential Indicator Simulation (SIS) (Deutsch and Journel, 1997), Multiple Point Statistics (Strebelle, 2002), and most recently, Hierarchical Truncated Pluri-Gaussian Simulation (HTPG) (Silva and Deutsch, 2019; Velasquez Sanchez, 2023), Vein Simulation (Carvalho, 2018) and Machine Learning (ML) based methods like Generative Adversarial Networks (Azevedo et al, 2020).

The work of this paper presents the utilisation of HTPG for modelling domain uncertainty nested with SGS for grades on a 1:1 realisation set-up, where for each new domain realisation, a subsequent grade realisation is performed. This configuration is conceived to carry uncertainty evenly downstream on mineral resource models (Deutsch, 2018), in contrast to the most common simulation approach where the same static estimation domains are used repeatedly across all grade realisations. To further outline the impact of the HTPG methodology, a drill hole spacing (DHS) analysis is executed on two different case studies, comparing traditional 'grade-only' (G-O) simulations against 'domain+grade' (D+G) models and seeing how uncertainty at mineable panels and optimal DHS results differ between them.

SIMULATION MODELLING – CARRYING UNCERTAINTY DOWNSTREAM

The central premise of this work is that uncertainty exists in every component of a mineral resource model, including samples, assays, variogram parameters, geological interpretations, and more. Consequently, propagating this uncertainty throughout the modelling process is the appropriate approach when the goal is to quantify risk (Deutsch, 2023; Journel, 2018).

In traditional (G-O) simulation workflows (Figure 1), the focus is predominantly on exploring uncertainty arising from sample grades. However, deterministic estimation domains—whether based on geological interpretations or grade thresholds—do not contribute to the *modelled* risk. This omission results in a highly constrained representation of uncertainty, limiting the model's ability to capture the full range of potential variability.

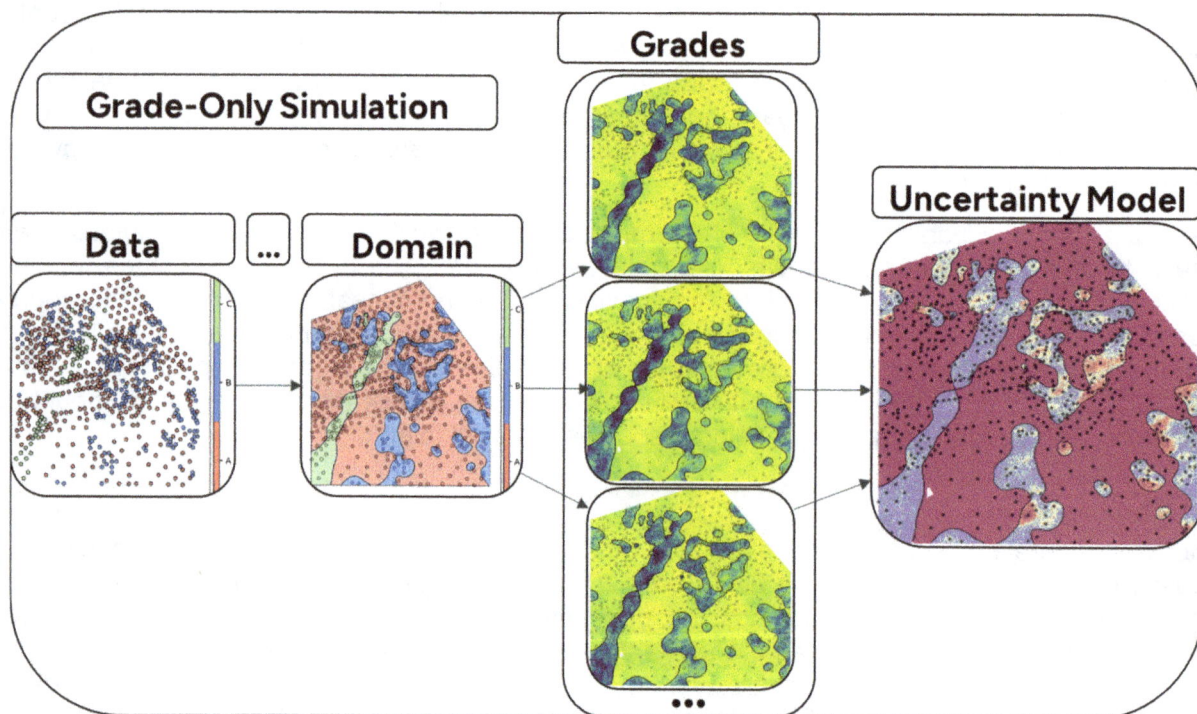

FIG 1 – Flow chart of traditional G-O simulation methodology where uncertainty is only explored through simulation of sample grades.

In contrast, a workflow in which domains are simulated nested with simulation of grades incorporates two important sources of uncertainty, geological interpretation and spatial distribution of grades away from the data location. Such a workflow enables uncertainty from two major sources to accumulate, leading to a more robust final uncertainty model. Other sources of uncertainty, such as the histogram of grades, can also easily be simulated through the spatial bootstrap technique, adding one more uncertainty aspect to the model. The flow chart of D+G simulation is presented as Figure 2, highlight how it differs from traditional in incorporating domain realisations on a 1:1 configuration with grade realisation, properly carrying uncertainty into the final probabilistic model.

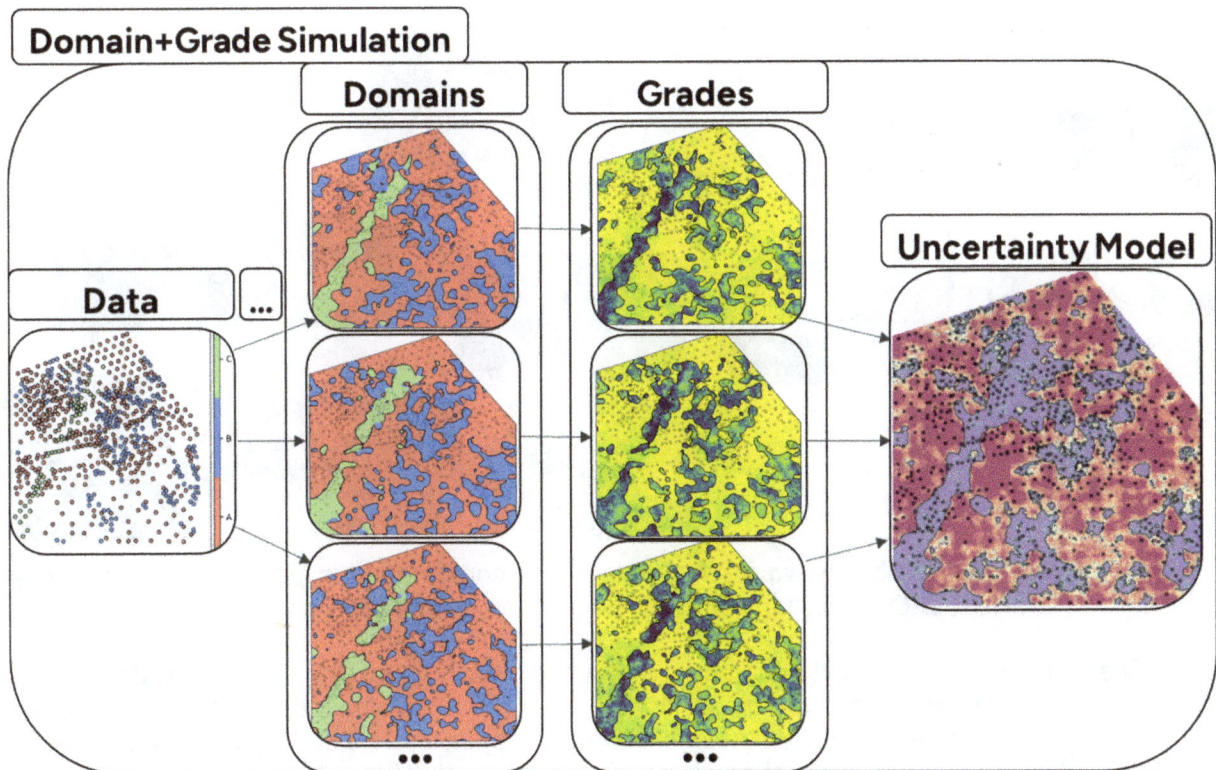

FIG 2 – Flow chart of D+G simulation, highlighting how geological uncertainty is nested with grade simulation on a 1:1 configuration, which carries uncertainty properly to the final uncertainty model.

Given the very well-researched status of continuous grade simulation with SGS, emphasis is placed on briefly presenting the HTPG technique for simulating geological domains prior to embarking on the application case studies. In both methods, locally varying anisotropy (LVA) fields (Boisvert and Deutsch, 2011) are fit and applied to either simulated or static domains for better interpolation of grades.

Categorical simulations of geological domains

HTPG is a methodology for the multivariate spatial modelling of categorical variables which uses the hierarchical truncated pluri-Gaussian approach (Silva and Deutsch, 2018). HTPG uses three rounds of variograms for modelling two or multiple categories, and preferentially benefits itself from Trend Modelling (Harding and Deustch, 2021) of indicated categories on its initial stage.

The three levels of variogram involved in HTPG categorical simulation are:

1. Variograms of indicators.
2. Variograms of residuals from indicator trend model.
3. Variograms of latent Gaussians deriving from truncation mask.

The HTPG simulation method is only briefly described in this paper, which focuses instead in showing its practical implementation and results in real cases rather than explaining its engines. The method is complex and involves multiple other advanced geostatistical tools in its implementation, such as Trend-Modelling, Multiple Data Imputation (Silva and Deutsch, 2018), Step-wise Conditional Transform (Leuangthong and Deutsch, 2003) and PPMT decorrelations (Barnett, Manchuk and Deutsch, 2014), which will not be detailed here. Interested readers are encouraged to follow the references for more fundamental information on all those methods.

The two main streams of simulating uncertainty, G-O simulation on static domains and D+G simulation, are now as part of a larger workflow encompassing both options to calculate risk and/or optimise drill hole spacing.

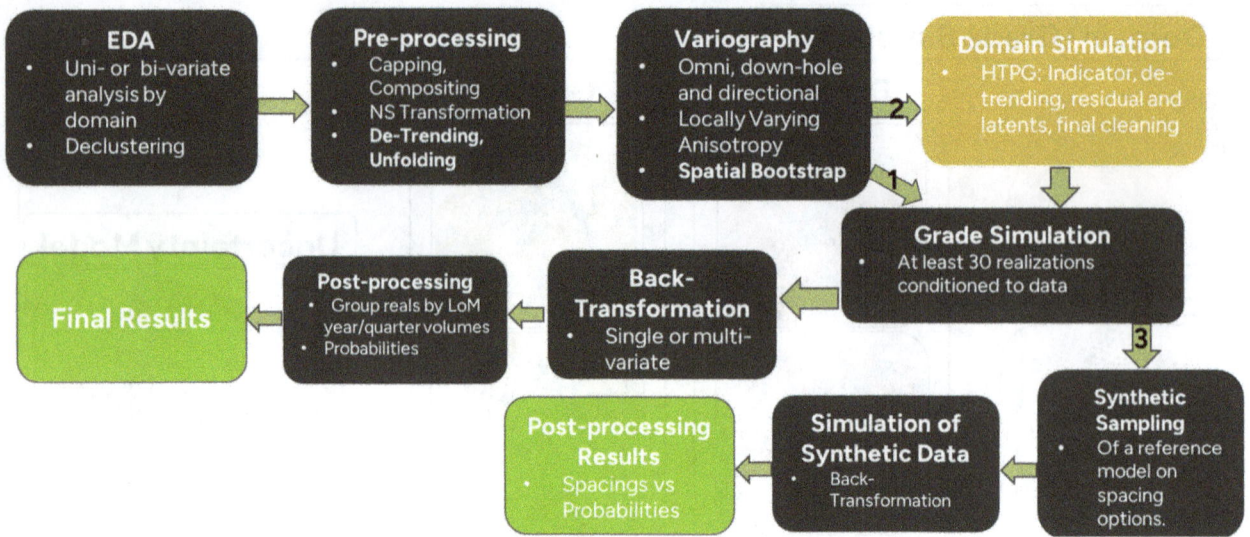

FIG 3 – Flow chart exhibiting two simulation methods compared plus the dill-hole spacings analysis: G-O simulation (1), D+G simulation (2), and DHS analysis (3).

The G-O simulation approach will only model the spatial continuity of the preprocessed variable of interest and model it constrained by the static domains. The D+G simulation will proceed towards building realisations of the indicated lithologies/domains, prior to simulating grades within each of the simulated domains, which will change for every new realisation of grade. The deterministic domains, usually built with implicit modelling tools, might be used as input on the domain simulation or not, depending on the deposit type.

The domain simulation through HTPG is broken down into more specific steps on Figure 4 in a simplified manner, as describing those steps in more detail is not the aim of this paper.

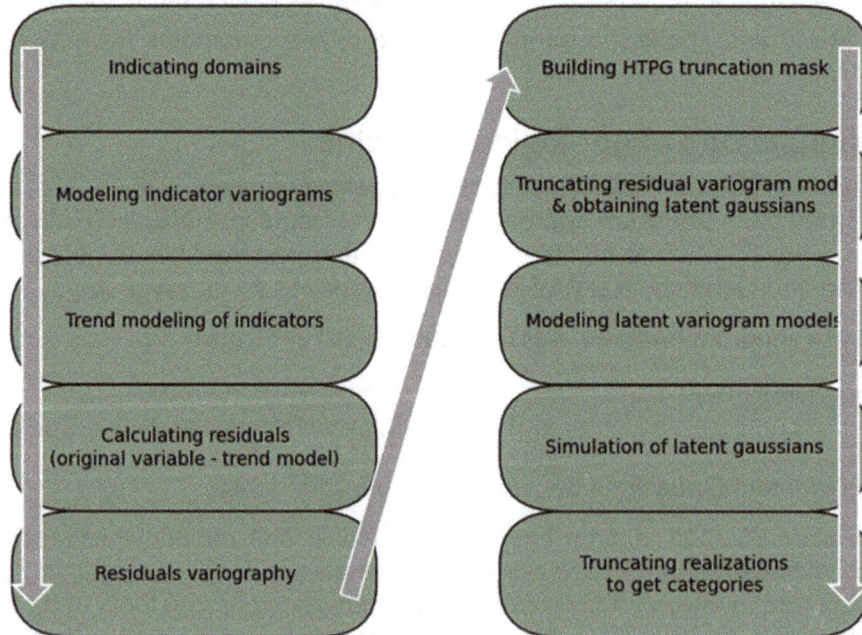

FIG 4 – Simplified sequence of steps involved in building a HTPG categorical model.

From the categorical realisation, grade realisations are built to then having final probabilistic models being post-processed and undergoing re-sampling/re-simulations, which are part of the Drill hole Spacing Analysis.

DRILL HOLE SPACING ANALYSIS

Drill hole spacing (DHS) analysis is a common practice in mineral resource modelling, aimed at determining optimal drill hole spacings based on a specified criterion. These criteria can include confidence intervals, profit, misclassification errors, and various other metrics. In this work, the chosen criterion is confidence intervals (Verly, Potolski and Parker, 2014).

DHS analysis involves synthetically re-sampling a simulated representation of a deposit at various selected spacings. Each re-sampled grid is then re-simulated or re-estimated, and its performance is evaluated against the selected metric, with results grouped by spacing intervals.

CASE STUDIES

The implementation and comparative results of uncertainty modelling using D+G simulated scenarios are presented for two distinct geological and operational contexts: an open pit massive oxide iron ore sedimentary deposit and an underground orogenic narrow-vein gold deposit. These two case studies, selected for their contrasting geologies, aim to provide insights into the suitability and impact of incorporating domain simulation into uncertainty studies across different types of mineral deposits.

This section focuses on explaining the context of each case and highlighting comparative results while being brief on implementation details given the extended nature of domain simulation works.

Case study 1 – iron ore open pit

Geological context

The iron ore case study consists of an undisclosed source of short-term drilling data of a pit bench (Figure 5). The geological setting is a massive sequence of weathered iron-rich layers (Lithology 'A' – which is the ore), occasionally containing hard fresh BIF boulders and bodies (Lithology 'B') which are controlled by complex interactions between different chemical components, mineralogy, surface and underground water, alteration, faulting and structural geology. Moreover, the sequence is cross-cut by occasional dykes and/or conforming sills of meta-basic intrusive rocks (Lithology 'C'). The lithological implicit model is also retrieved to serve as the domain for the G-O simulation approach, and is also shown on Figure 5.

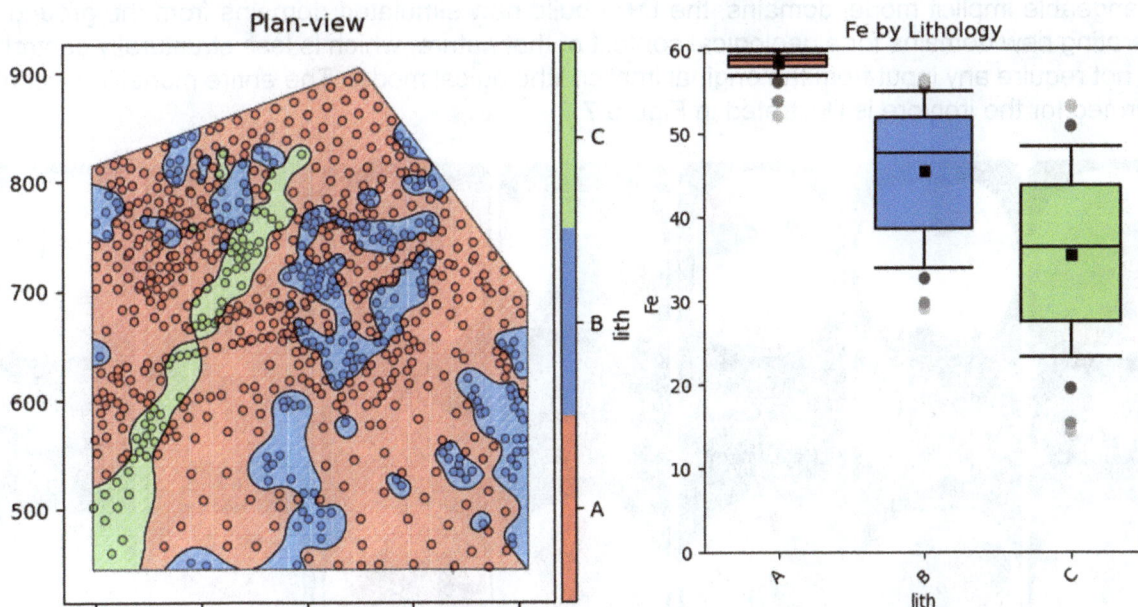

FIG 5 – Case study 1 drilling data and implicit model coloured by lithotype (on the left) and box-plot distributions of Fe by lithotype (on the right hand-side).

Interesting about this deposit is that while iron ore grades are mostly very high and homogeneous within the hematite throughout the mine, the problems in mining derive from the uncertainty on the locations of fresh BIF boulders, often not detected by the grade-control drilling (view Figure 6).

FIG 6 – Mining Face of a hard BIF pocket within weathered hematite from case study 1 deposit.

The iron ore case exhibits lithological boundaries not heavily structured in pre-determined directions or locations, especially the BIF hard pockets, which occur in multiple forms, sizes and orientations. This 'loose' type of geological boundaries favours HTPG categorical modelling, without the need for strong constraints to generate realistic realisations. Beyond that, lithological units are critical for processing efficiency and display hard boundaries with each other, meaning that grade continuities change abruptly at boundaries.

Categorical and grade simulation

While the G-O method will basically model Fe spatial continuity and simulate it *n* times within static unchangeable implicit model domains, the D+G build new simulated domains from the ground up. Generating new domains for a geological context of that nature, which is less structurally controlled, does not require any input from the original implicit lithological model. The entire modelling workflow performed for the iron ore is illustrated in Figure 7.

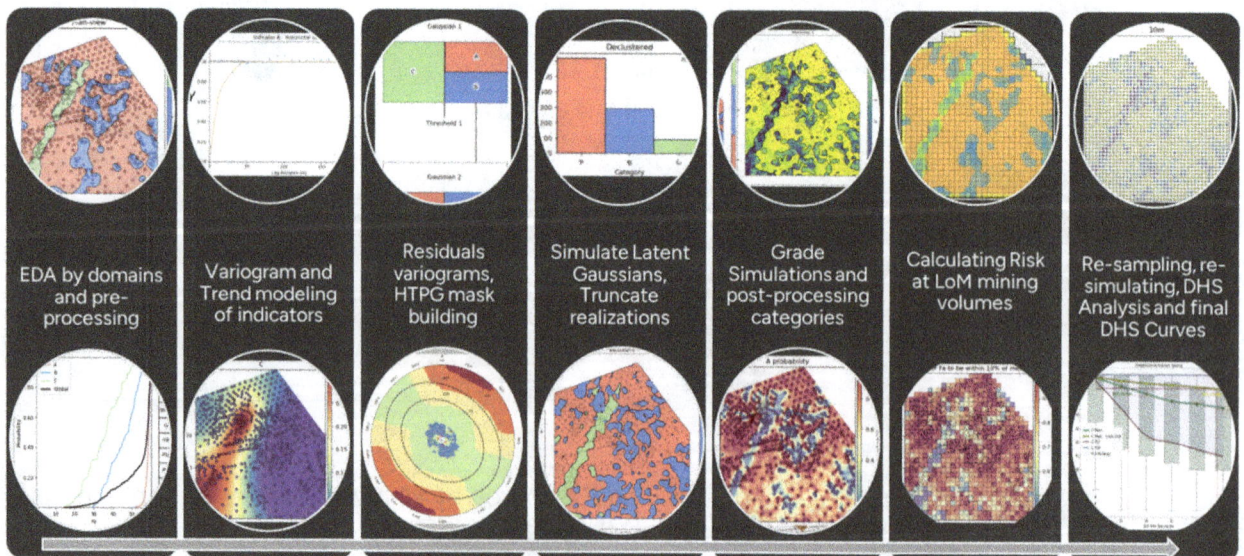

FIG 7 – Schematic sequence of processes involved in modelling and post-processing categorical realisations for the iron ore case study.

Each of the HTPG categorical realisations is unique, while equally honouring the domains at the data locations and matching the domains' global proportions. Validation of probabilistic models is just as crucial as it is for deterministic models. Although not shown here, multiple validation steps should be employed prior to extracting probabilities from simulated models, including but not limited to, swath-plots, histogram and variogram reproduction, 3D visual validations, accuracy plot and grade-tonnage curves.

Results

Three of the generated domain realisations nested with simulated grades of the iron ore deposit can be seen on the following Figure 8.

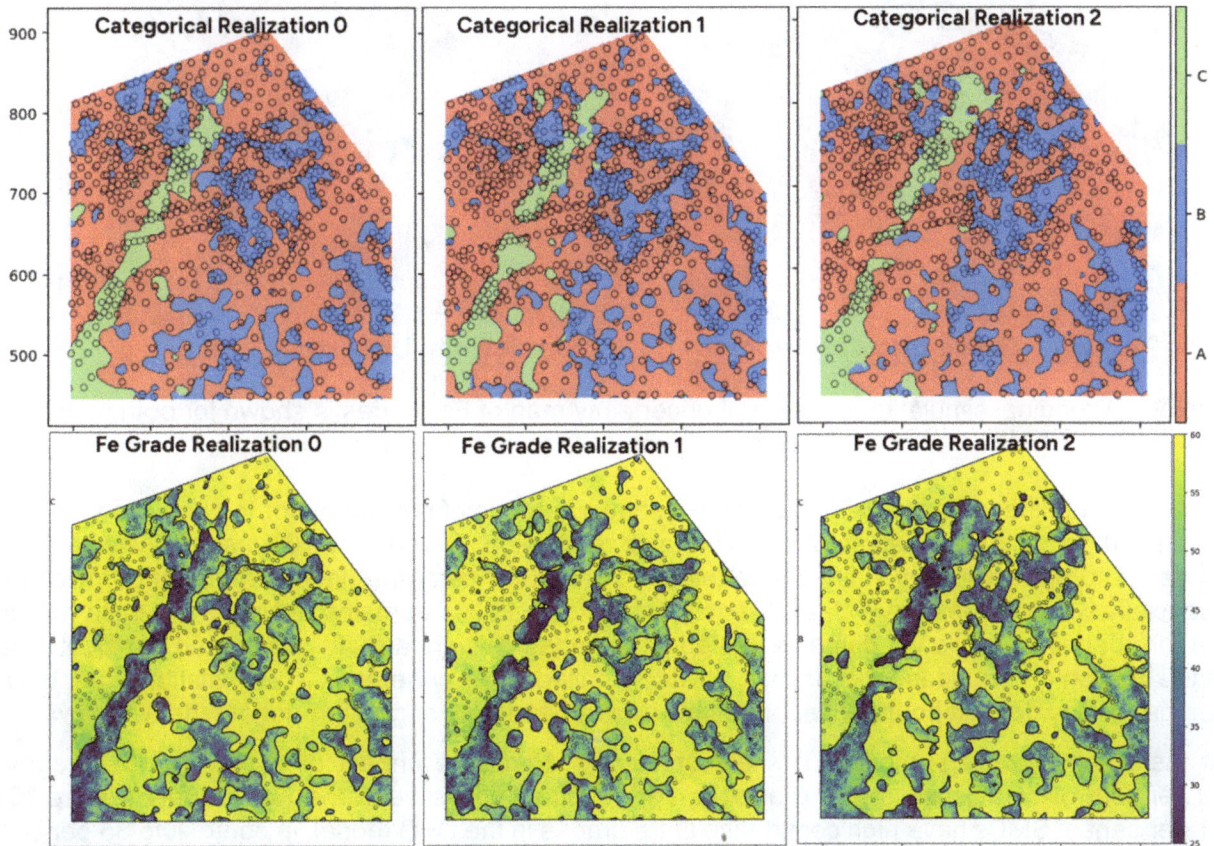

FIG 8 – Unique equiprobable HTPG categorical realisations nested to grade realisations of Fe.

Categorical realisations are globally consistent in assigning the same domain where samples are available, but exhibit significant local variability, particularly at domain contacts. This variability, inherent in each new realisation, allows uncertainty in lithological boundaries to propagate downstream within the modelling workflow. The uncertainty introduced by domains combines with the simulated grades, as evident in the comparison of post-processed final simulations derived from both methods (Figure 9).

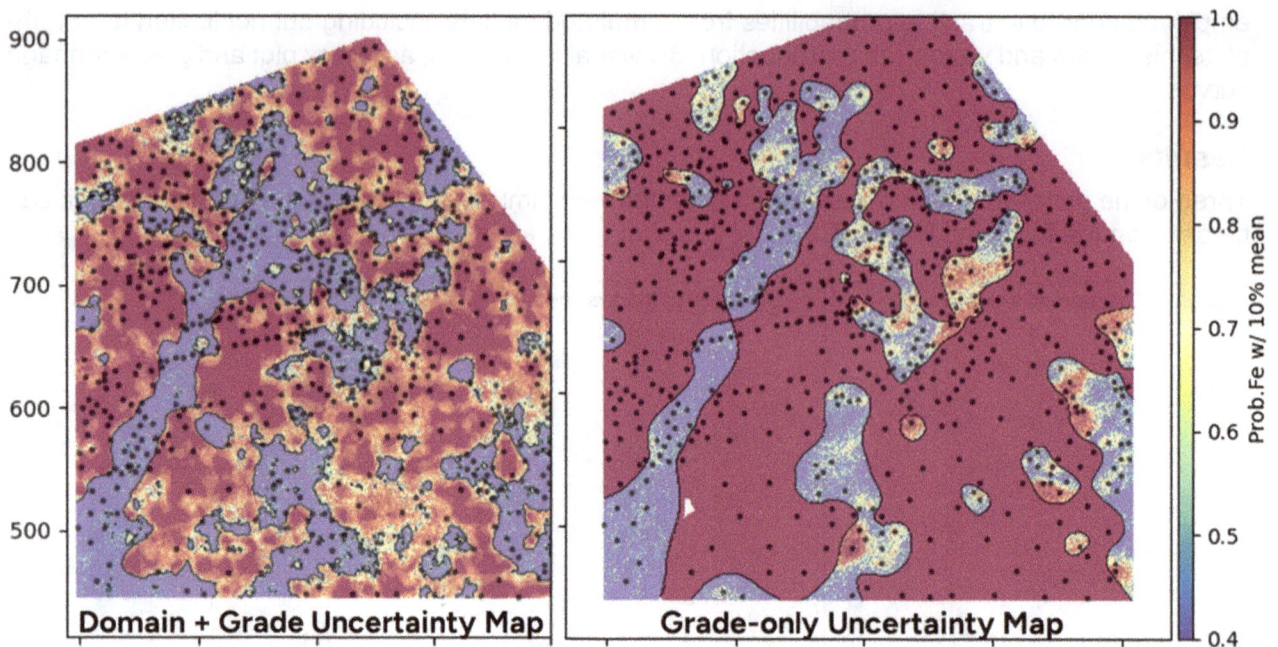

FIG 9 – At original simulation scale, final uncertainty map of Fe grades is shown for both methods: D+G on the left, and G-O on the right side.

The resultant difference in final uncertainty is shown in terms of probability for the iron grade to vary within ±10 per cent the mean across all simulated realisations (*'Prob.Few/10%mean'*). When nesting domain to grade realisations, final uncertainty on grade will carry the uncertainty from domain in it, especially when grades have clear distinct distributions across domains, which is the case for this iron ore deposit. Simulating domains nested with grades delivers a realistic scenario of risk at most of the lithological boundaries, especially at widely drilled areas, whereas G-O simulation is unable to see risk in any domain boundaries, once those have been unchanged throughout the workflow.

At the smallest-mining-unit (SMU) scale of 15 m × 15 m, the impact of mineable blocks above 90 per cent of confidence on Fe grades is very contrasting between the two models (Figure 10). 61 per cent of SMUs have high confidence in Fe grades in the D+G model, in contrast to 85 per cent from the G-O model. The uncertainty associated with sparsely drilled locations—whether within the weathered hematite domain or at the interfaces between hematite and either intrusive dykes or hard BIF—is entirely overlooked in the G-O approach. This omission results in an unrealistically optimistic outlook for probabilistic modelling of the deposit.

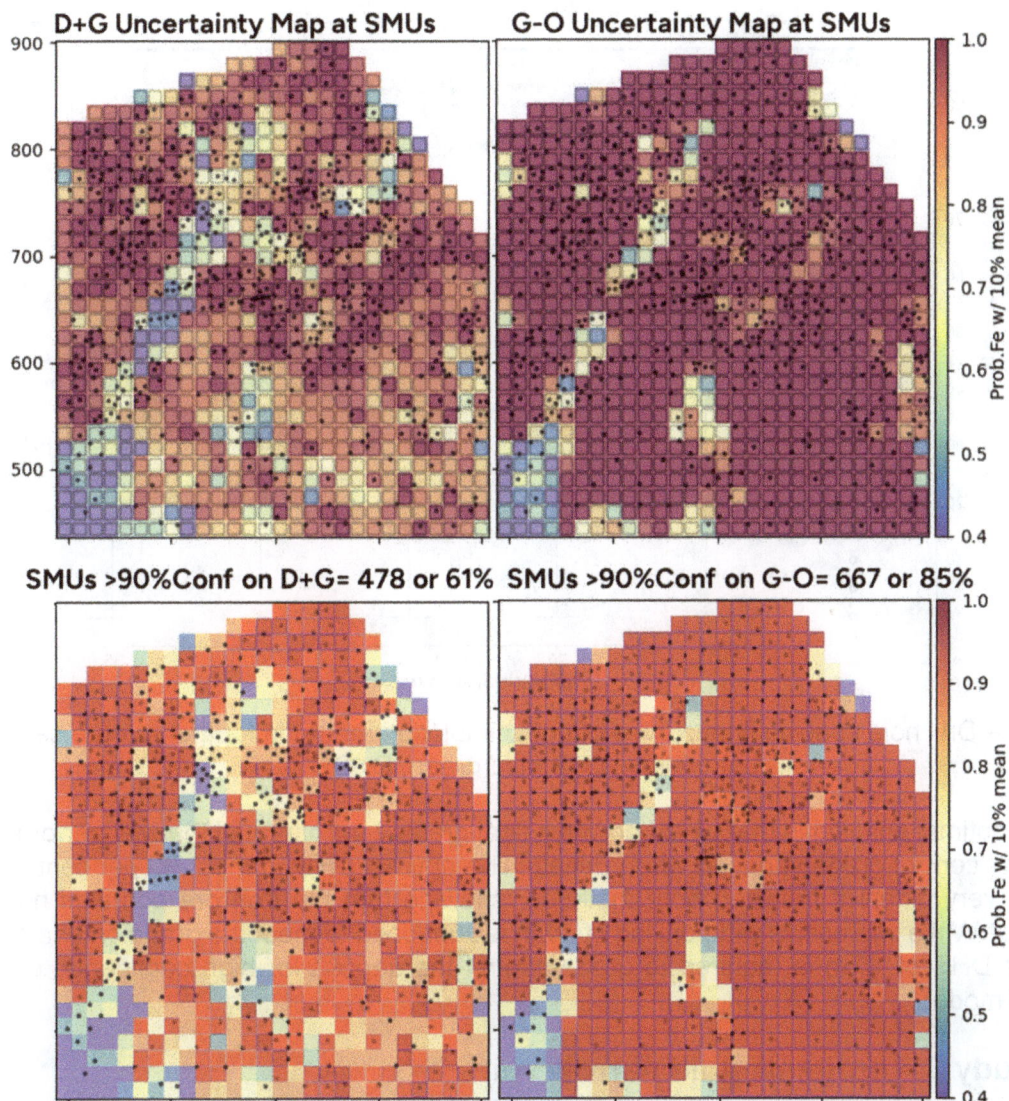

FIG 10 – Probability of Fe being within 10 per cent of the mean at the SMU blocks between both simulation methods. High confidence blocks (>=90 per cent) are 25 per cent less on D+G model than when simulating uncertainty from grades alone.

After re-sampling the area at certain DHS options and re-simulating the area for each spacing, a final DHS optimal curve is available (Figure 11).

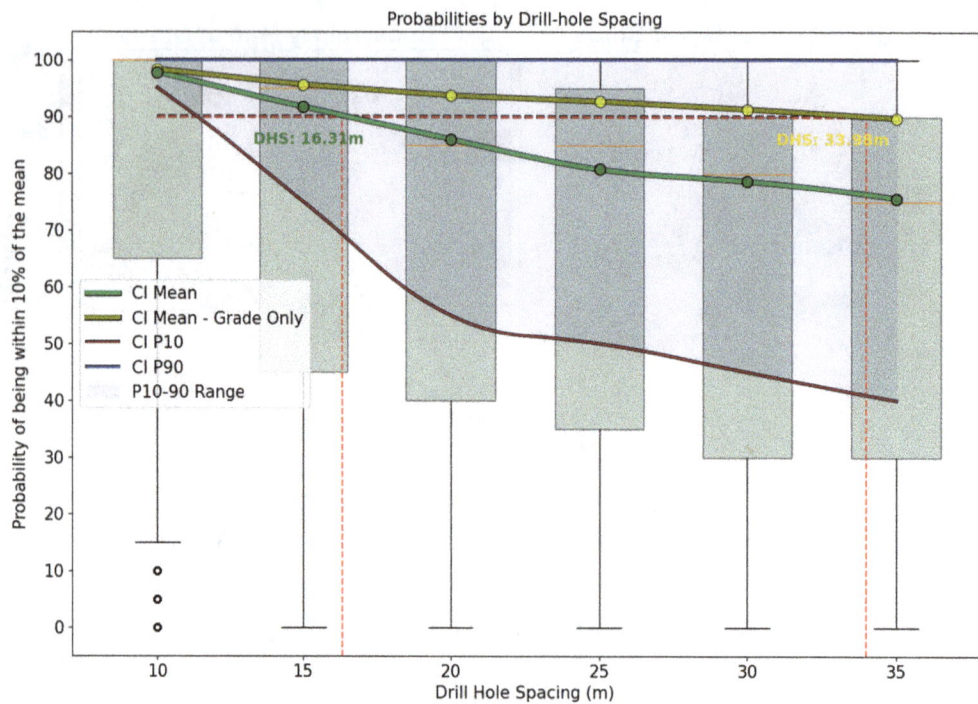

FIG 11 – Drill hole spacing final curves of both simulation methods for the iron ore case study (yellow: G-O, and green: D+G).

The DHS optimal spacing is that which grants a final grade-control model that will perform with less than 10 per cent of grade variation when mined. The optimal DHS to meet 90 per cent confidence interval is very different between both methods, showing that the iron ore domains hold a lot of uncertainty. While impressive wide drilling of 34 m is supposed to meet the grade criteria on the G-O method, a DHS of 16 m is required to grade-control the deposit for high confidence on the D+G simulated model.

Case study 2 – orogenic gold narrow vein

The second deposit in which the probabilistic models were built consists of a single narrow vein mineralised with gold in an orogenic context. It possesses a low-grade (LG) halo and a high-grade (HG) mineralisation shell in the inner parts of the vein (see Figure 12). The LG and HG domains are developed based on grade-threshold and make no use of geological information on their construction. The modelling workflow followed for this case study is very similar to the iron ore case, therefore focus is entirely given to results, aiming for conciseness.

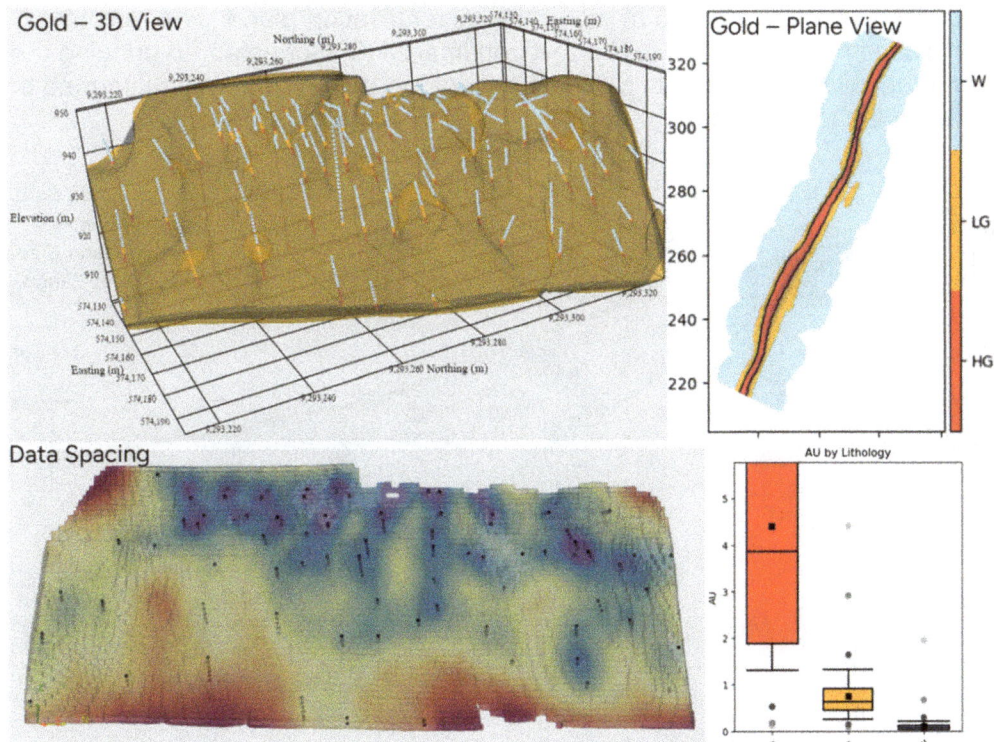

FIG 12 – Overview of gold case study, which involves two domains determined by gold-grade thresholds.

As the grade boxplot on Figure 12 suggests, there is significant gold-grade variability in the HG domain. Also, grade distributions are hard cut by domains with very few overlaps, since domains are built from grade thresholds. A few from the D+G simulated realisations are seen next in Figure 13:

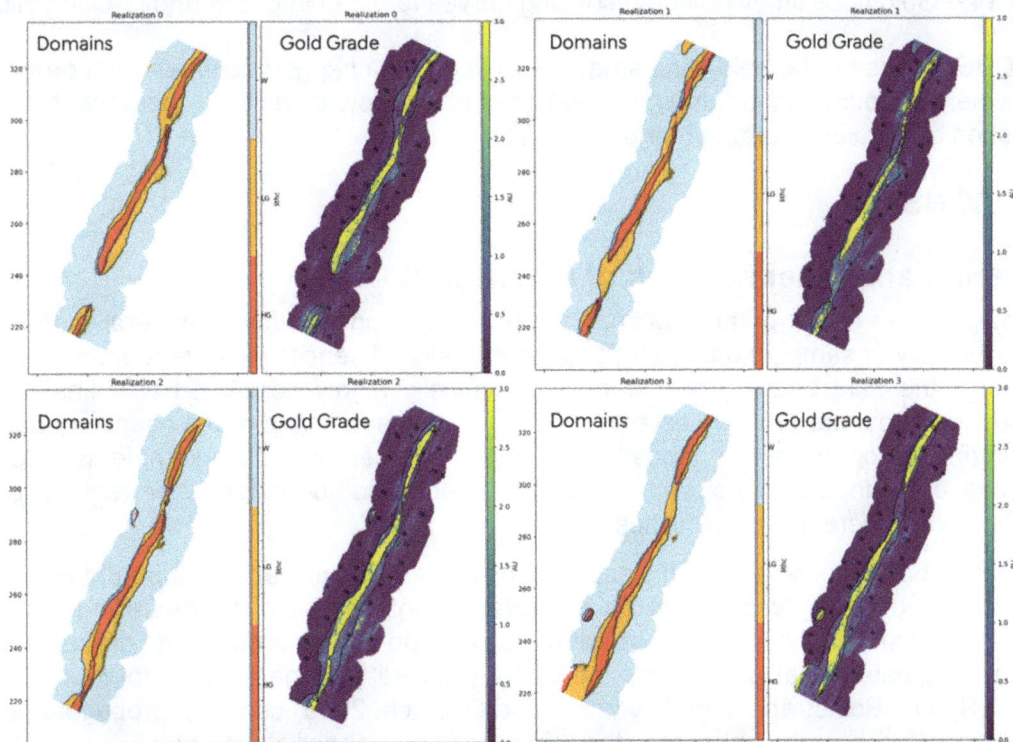

FIG 13 – Categorical domain realisations nested 1:1 with gold grade realisations of the gold case study.

The domain realisations follow a very similar configuration but are slightly different from each other. The volumes on which the grade probabilities are post-processed are synthetically created resembling UG stopes of 25 m × 25 m along the mineralisation plane. The resultant DHS optimal curves are exhibited next (Figure 14). The D+G optimal DHS suggested for bringing the deposit for a 90 per cent probability of being within 20 per cent of the gold grade mean informed by the model is 7.3 m, while for G-O it is 12 m.

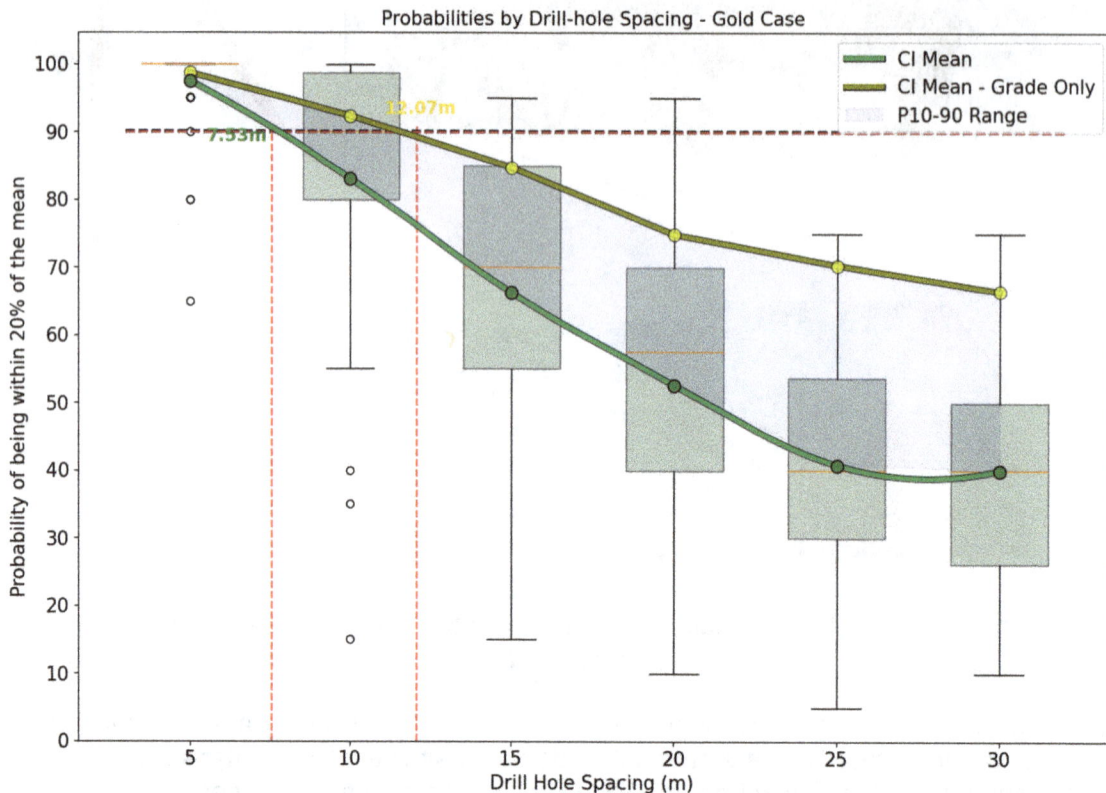

FIG 14 – Gold case study drill hole spacing curves for D+G in green and G-O in yellow.

The final DHS curves for the gold case study again suggest a big gap between both probabilistic methods, where accounting for domain simulation of the narrow gold veins contributes to a deeper representation of risk, leading to tighter drilling needs.

DISCUSSIONS

Data spacing and other sources of uncertainty

Data spacing is one of the primary factors contributing to uncertainty in mineral resources, as it reflects the density of sampled data, which fundamentally influences mineral resource evaluations. Consequently, the data spacing of a deposit is typically highly correlated with final uncertainty models, provided both are properly established. While data spacing is a root cause of uncertainty, it is not an intrinsic property of the deposit itself. Domain uncertainty, for example, is influenced by data spacing but can also persist in areas near geological boundaries or regions of intense deformation, even where drilling densities are high.

Beyond domain boundaries—which this paper highlights as a significant contributor to uncertainty—other sources of uncertainty can also be incorporated into probabilistic modelling workflows for comprehensive risk analysis. Key contributors include uncertainties in grade histograms, variograms, and categorical proportions, collectively called parameter uncertainties. Techniques such as the Spatial Bootstrap method (Vincent and Deutsch, 2019) can help propagate uncertainty from these elements, thereby enhancing the robustness of probabilistic models.

As Journel (1994) emphasises, uncertainty models based on prior decisions about randomisation are not designed to be entirely objective. Instead, they should be 'maximally charged with the data deemed relevant to the 'unknown' at hand'. This perspective underscores the importance of integrating all relevant uncertainties into the modelling process to produce realistic and practical assessments.

Different geological settings and uncertainty

The two case studies illustrate how geological settings significantly influence uncertainty distribution within a mineral deposit. The iron ore example is particularly notable due to its high, stable, and lithologically controlled Fe grades, especially within the ore zone (weathered hematite). This stability results in low risk when assessed solely through grades. However, the recurring risk of encountering hard BIF pockets and intrusive rocks is evident, a factor that can only be adequately addressed through dynamic domain simulation rather than static domain assumptions.

The gold case study highlights the versatility of domain simulation across diverse geological contexts. Modelling probabilistic realisations of a tabular deposit like the narrow gold veins requires a different set of techniques within the categorical modelling process to be accomplished. It also demonstrates that even domains derived from grades carry inherent uncertainty, which cannot be captured by simulating grades alone.

Implementation challenges and limitations

Implementing a D+G uncertainty workflow is significantly more computationally demanding than simulating grades on a fixed domain. Each categorical realisation pairs with a grade realisation, effectively doubling the number of simulations required. Additionally, defining an optimal DHS introduces further complexity. At the re-sampling and re-simulation stages, different domains must be generated for each spacing and each realisation, exponentially increasing the number of categorical simulations needed.

While D+G workflows are feasible and already implemented in specialised software, they remain computationally intensive. A resource modeller should carefully assess the geological and mining context to determine whether D+G assessments are critical. The key considerations include the geological characteristics of the deposit, metallurgical processing requirements and mining selectivity. Understanding these factors helps mitigate computational challenges while ensuring D+G workflows are applied where they provide the most value.

CONCLUSIONS

Modelling mineral resource uncertainty through a combination of domain and grade simulation has been explained and compared in two real data sets of distinct nature. Results provided insight on how final models of uncertainty and risk can greatly differ between a G-O and D+G analysis, depending on the geological context. Finally, it leads to the understanding that probabilistic models should rely on as many reliable sources of uncertainty, like grade, boundaries (domains) and other parameter uncertainties (ie histograms, variograms and proportions) as possible. Given that good quality data is available along with expertise, time and processing means, a multiple-sourced uncertainty model should always be built for better-informed decision-making on the risk of a mineral project.

ACKNOWLEDGEMENTS

The author gratefully acknowledges SLR Consulting Canada, especially Renan Lopes and Luke Evans, for supporting the idea and allowing the hours and resources that made such scientific research possible. We also thank Brandon Wilson and Resource Modelling Solutions, for their technical support on the first implementations of HTPG models in RMSP. Another thanks to Bruno Afonseca for the continued exchange of ideas regarding the contents of this work. Finally, the author also thanks Clayton Deutsch for his informal review of this manuscript.

REFERENCES

Azevedo, L, Paneiro, G, Santos, A and Soares, A, 2020. Generative adversarial network as a stochastic subsurface model reconstruction, *Computational Geosciences*, 24(4):1673–1692.

Barnett, R M, Manchuk, J G and Deutsch, C V, 2014. Projection pursuit multivariate transform, *Mathematical Geosciences*, 46:337–359.

Boisvert, J B and Deutsch, C V, 2011. Programs for kriging and sequential Gaussian simulation with locally varying anisotropy using non-Euclidean distances, Computers and Geosciences, 37(4):495–510.

Carvalho, D, 2018. Probabilistic resource modelling of vein deposits, Master's Thesis, University of Alberta, Edmonton, AB.

Chiles, J P and Delfiner, P, 2012. *Geostatistics: modelling spatial uncertainty,* vol 713 (John Wiley and Sons).

Deutsch, C and Journel, A, 1997. *Gslib geostatistical software library and user's guide,* second edition, 369 p (Oxford University Press: New York).

Deutsch, C, 2018. *All realisations all the time, Handbook of mathematical geosciences: 50 years of IAMG*, pp 131–142.

Deutsch, C, 2023. The Place of Geostatistical Simulation through the Life Cycle of a Mineral Deposit, *Minerals*, 13(11):1400.

Gómez-Hernández, J J and Srivastava, R M, 2021. One Step at a Time: The Origins of Sequential Simulation and Beyond, *Math Geosci*, 53:193–209. https://doi.org/10.1007/s11004-021-09926-0

Goovaerts, P, 2001. Geostatistical modelling of uncertainty in soil science, *Geoderma*, 103(1–2):3–26. https://doi.org/10.1016/S0016-7061(01)00067-2

Harding, B and Deutsch, C V, 2021. Trend modelling and modelling with a trend, *Geostatistics Lessons*, pp 1–3.

Isaaks, E and Srivastava, R, 1989. *An introduction to applied geostatistics*, Oxford University Press.

Journel, A G, 1994. Modelling uncertainty: some conceptual thoughts, *Geostatistics for the Next Century: An International Forum in Honour of Michel David's Contribution to Geostatistics*, pp 30–43 (Springer: Netherlands).

Journel, A, 2018. Roadblocks to the evaluation of ore reserves—the simulation overpass and putting more geology into numerical models of deposits, *Advances in applied strategic mine planning*, pp 47–55.

Leuangthong, O and Deutsch, C V, 2003. Stepwise conditional transformation for simulation of multiple variables, *Mathematical Geology*, 35:155–173.

Pyrcz, M and Deutsch, C, 2014. *Geostatistical reservoir modelling* (Oxford University Press: USA).

Silva, D and Deutsch, C, 2019. Multivariate categorical modelling with hierarchical truncated pluri-gaussian simulation, *Mathematical Geosciences*, 51(5):527–552.

Silva, D S and Deutsch, C V, 2018. Multivariate data imputation using Gaussian mixture models, *Spatial statistics*, 27:74–90.

Strebelle, S, 2002. Conditional simulation of complex geological structures using multiple-point statistics, *Mathematical Geology*, 34:1–21.

Velasquez Sanchez, H G, 2023. Truncation Trees in Hierarchical Truncated PluriGaussian Simulation, Master's Thesis, University of Alberta, Edmonton, AB.

Verly, G, Potolski, T and Parker, H, 2014. Assessing Uncertainty with Drill Hole Spacing Studies – Applications to Mineral Resources, in *Proceedings of the Orebody Modelling and Strategic Mine Planning Symposium*, pp 24–26.

Vincent, J D and Deutsch, C V, 2019. The Multivariate Spatial Bootstrap, *Geostatistics Lessons* (ed: J L Deutsch). Available from: <http://geostatisticslessons.com/lessons/spatialbootstrap>

Onto copper-gold deposit, Indonesia – evaluation of geological and grade uncertainty in a block cave operation

R C Hague[1]

1. Specialist Resource Evaluation Geologist, PT. Vale Eksplorasi Indonesia, Jakarta Selatan 12190, Indonesia. Email: richard.hague@vale.com

ABSTRACT

The Onto Copper-Gold deposit is a 3 billion tonnes (Bt) hybrid porphyry/high sulfidation deposit located in the Hu'u Contract of Work, Sumbawa Island, Indonesia. Drilling has defined a large tabular mineralised volume measuring at least 1.5 × 1 km with a vertical thickness of ≥ 1 km located 400–700 m below surface. The prefeasibility study was completed in December 2024 incorporating drilling results and various geological, mineral resource, geometallurgical, geotechnical, hydrothermal and mine engineering studies. Given the size, location and grade distribution, a large cave mining operation feeding a conventional ball mill – semi-autogenous grinding (SAG) – flotation circuit is being considered. The current resource estimate is 3.0 Bt @ 0.72 per cent Cu and 0.39 g/t Au in a combination of Indicated and Inferred resource categories defined by drilling on nominal 200 × 200 m and 400 × 400 m drilling densities respectively.

Simulation methods were used to provide a probabilistic framework to determine uncertainty in the estimated grades and the effect that this has upon resource classification within the planned cave operation due to uncertainty inherent in the current drill hole spacing – volumes within the deposit that are defined by low drilling densities contribute higher risk to the project's economic assessment than volumes defined by closer spaced drilling densities.

Resource estimation domains, representing underlying rock-type/alteration combinations, were simulated into the block model to capture geological variability. For each geological model, a set of Cu grades was simulated. Each simulated grade model captures both geological and grade variability that contribute to the overall uncertainty in the estimated grades due to drilling density. These simulations were subsequently used to objectively investigate uncertainty in the following items:

- Resource classification.
- Production Run of Mine (RoM) grade predictions throughout the mine schedule.
- Overall project economic metrics.
- Positioning of extraction levels.

These data will provide objective measures of risk, in terms of uncertainty, being carried by the project due to drilling density. A decision can be made whether the risk level is acceptable or if further drilling is warranted to reduce the risk. Simulation studies involving planned drilling programs to reduce geological and grade estimation variation will provide trade-off study information of drilling costs against risk reduction.

INTRODUCTION AND PROJECT DESCRIPTION

The Onto Copper-Gold Deposit is located on eastern Sumbawa Island in Indonesia within the larger Hu'u Contract of Work (CoW) (Figure 1) held by PT. Sumbawa Timur Mining (STM), a company owned 80 per cent by Eastern Star Resources Pty Ltd (a Vale Base Metals Limited (VBM) subsidiary) and 20 per cent by PT. Antam Tbk.

The Onto Deposit was discovered in 2013 and has undergone a significant amount of drill definition work culminating in the completion of the prefeasibility study (PFS) in December 2024.

Onto is a hybrid porphyry/high sulfidation deposit. Mineralisation is hosted within a multiphase porphyry intrusive complex into a diatreme sequence (polymictic breccia). A barren andesite dome sequence overlies the host rocks (Figure 2). The high-sulfidation event created a thick advanced argillic alteration lithocap of up to 1 km in thickness. The bulk (>90 per cent) of the mineralisation is contained within the lithocap consisting of quartz-alunite and quartz-pyrophyllite±diaspore advanced argillic alteration assemblages. Copper mineralisation is primarily covellite (CuS) with minor amounts of enargite (Cu_3AsS_4). The relative sulfide mineralisation abundance content is pyrite >> covellite > enargite > chalcocite > bornite.

FIG 1 – Deposit location.

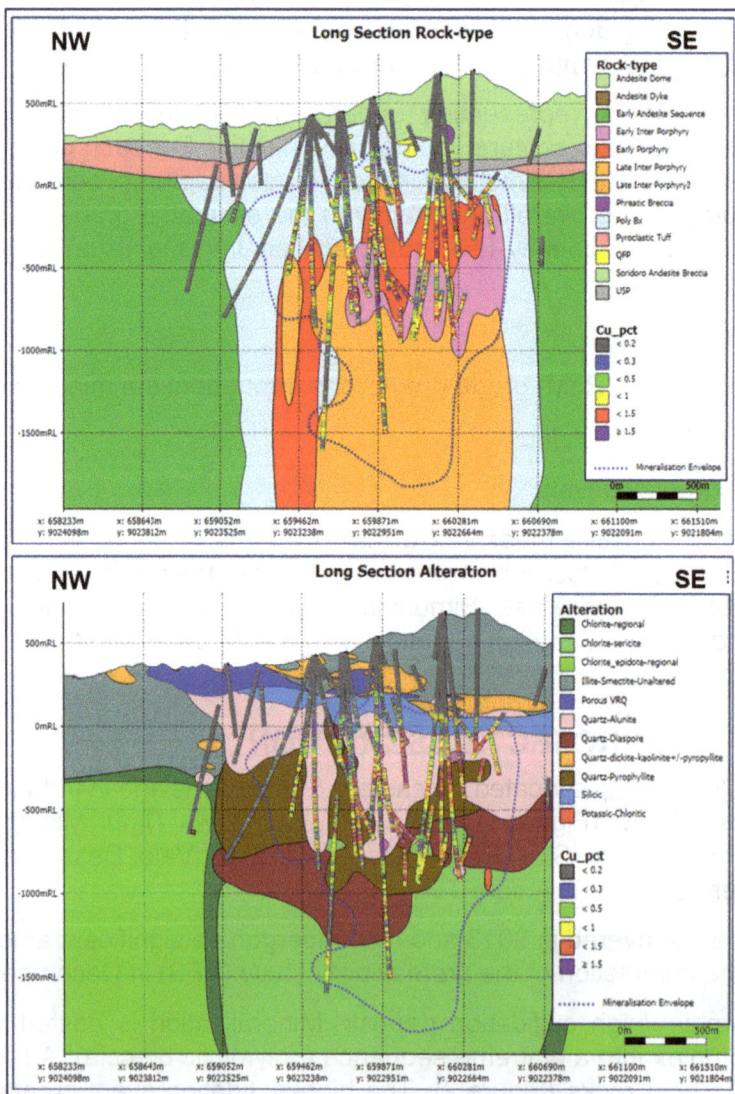

FIG 2 – Long-section showing geology (top rock-type, bottom alteration).

The two deepest holes in the project (about 2050 m) pass through the base of the lithocap and into primary alteration styles associated with the porphyry emplacement – mainly chlorite-sericitic alteration. Within this alteration style copper and gold grades are generally of lower tenor represented by chalcopyrite ($CuFeS_2$) and bornite (Cu_5FeS_4) minerals

The deposit also contains Au and Ag, however the project's economic value is driven by copper.

The Onto system is exceptionally young (<1 Ma) and there is no indication of post-mineralisation faulting offsetting or limiting mineralisation or the presence of late stage unmineralised intrusions stoping-out the mineralised rock mass. The high sulfidation event has enhanced grade tenor and grade continuity.

Drilling to date has defined a large, mineralised body measuring about 1.5 km along strike (NW-SE), about 1 km across strike (NE-SW) and at least 1.5 km vertically. The deposit is open to the SW and, significantly, at depth. The top of the mineralised mass is at about sea level (0 m El) about 400–700 m below surface.

This mineralised mass is planned to be exploited by a dual lift block cave operation with a maximum production rate of 48 Mt/a feeding a conventional milling, flotation, and smelting process. Cave lift heights of about 400 m (Lift-1) and 400–800 m (Lift-2) are planned. The current mine life is about 50 years, with operations planned to commence in the early 2030s.

The current resource was defined by a combination of nominal 200 × 200 m and 400 × 400 m spaced drilling (Table 1). All drilling to date was by triple diamond-coring methods with the bulk of the samples within the mineralised rock mass being HQ3 (40 per cent) or NQ3 (47 per cent) in size. Assay samples are two metres in length, with ½ core subjected to crushing to -2 mm and splitting using a Boyd combination crusher/rotary splitter, followed by pulverising to -75 µm in an Essa LM2 disc pulveriser. Assaying was by 30 g fire-assay/AAS finish for Au and a multi-element package of four acid digest/ICP OES finish for Cu, Ag, As, Fe, S, Mo, Pb and Zn and others.

TABLE 1

Resource database statistics.

Item	Value
No. drill holes	170
Meterage	171 800
Assay data	73 420
Bulk density data	22 103

A comprehensive quality control program is implemented in parallel with the sampling process, and includes duplicate samples collected at all comminution and mass reduction stages; insertion of matrix matched critical raw materials; field and analytical blanks; and mass and particle size determinations at each comminution stage. Dry bulk density was determined using oven dried samples and the calliper method with periodic water displacement checks.

The current resource estimate (Table 2) uses an ordinary kriging estimator for Cu, Au, Ag and bulk density and multiple indicator kriging for arsenic using four metre assay compositing. Four main estimation domains are used for grade estimation being the combination of two rock-types (polymictic breccia and porphyry) and the two main advance argillic (AA) alteration styles (quartz-alunite and 'basal AA', the latter being combined quartz-pyrophyllite and quartz-diaspore). The resource estimate is reported in accordance with the JORC Code (2012).

TABLE 2

Mineral resource estimate.

Resource category	Tonnage (Bt)	Grade			Contained metal	
		Cu (%)	Au (g/t)	As (ppm)	Cu (Mt)	Au (Moz)
Indicated	1.6	0.85	0.51	350	13.9	26.9
Inferred	1.4	0.57	0.24	230	7.8	10.6
Total	3.0	0.72	0.39	300	21.7	37.4

Three items were identified as the main drivers of grade uncertainty within the resource estimate:

1. Geological variability.

2. Assay variability.

3. The block-cave mining method.

Geological and assay variability are inherently related to drilling data density – the greater the drilling density the lower the variability within the estimated resource grade.

This paper presents a strategy to objectively evaluate variability in the estimated resource grade, and the risk to the project economic metrics due to the prevailing drilling density and mining method.

THE CHALLENGE

Since discovery, the most significant resource definition-related issue was providing evidence to Vale's internal Mineral Resource/Mineral Reserve (MRMR) Group to support the resource classification assigned to the resource estimate. This was an ongoing contentious issue between the Onto Project Team and the MRMR Group.

The Deposit has a unique set of geological conditions that positively impinge upon grade estimation and the required drilling data density for resource classification:

- The high sulfidation event has enhanced grade continuity compared to that observed in typical porphyry-only style deposits. This results in less drilling being needed to define the required grade continuity to meet the definition of the various JORC resource categories (JORC, 2012).

- The deposit is extremely young. Age dating (Burrows et al, 2020) indicate that the porphyries stocks intruded ~0.7 Ma, and the high sulfidation event ~40 ka – there has been insufficient time for late stage barren intrusions stoping the mineralisation and has limited the amount of post-mineralisation faulting to being negligible. The Deposit is a large intact body of mineralisation. There is no requirement for dense drilling to, for example, define barren intrusions or fault offsets.

- There is a strong correlation between Cu and Au, ie there are no gold-only parts to the mineralised mass that may require denser drilling to elucidate.

In addition, the mining method (caving) introduces a homogenisation process to the orebody before it is extracted for processing by virtue of the mass movement of rock through the cave, and that production at any scale is derived from a large number of drawpoints – ie daily production comes from a large number of drawpoints across the extraction level to ensure an even drawdown of material across the broken rock pile sitting in the cave (Figure 3).

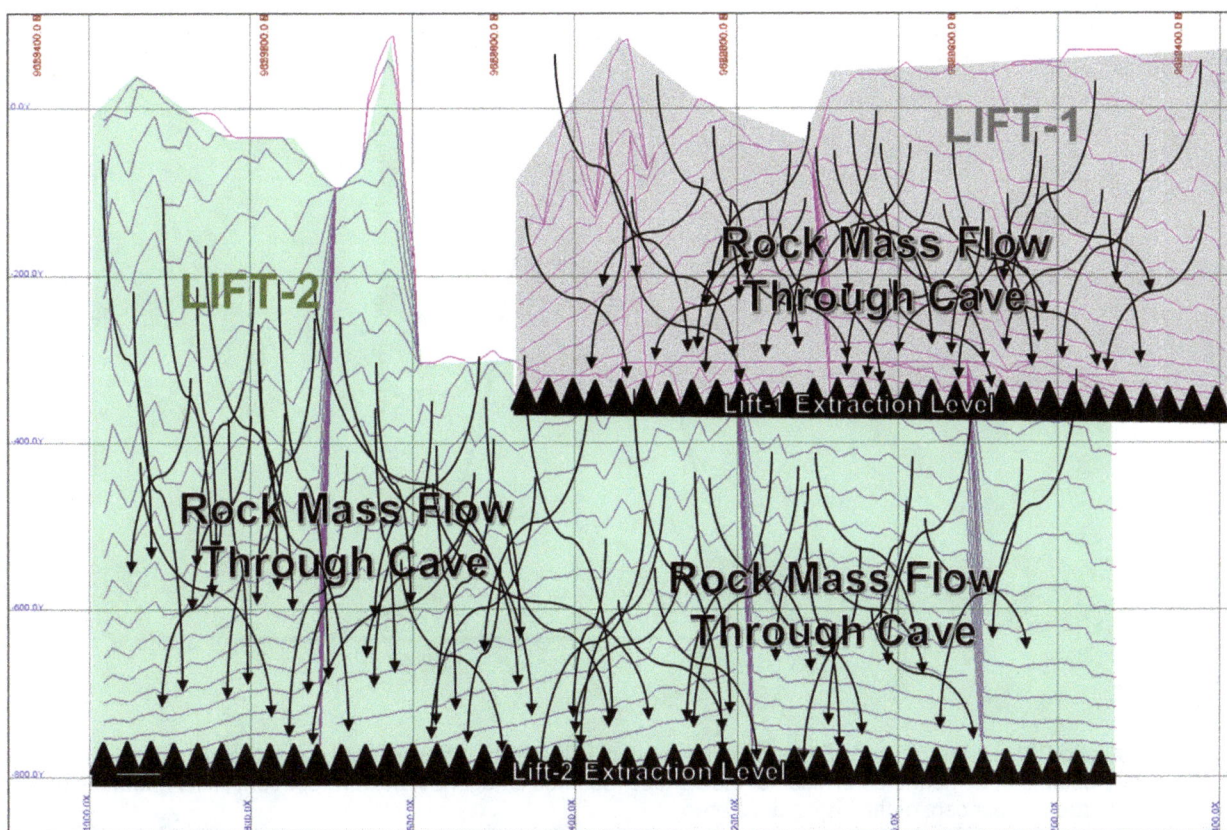

FIG 3 – Diagrammatic section showing the grade homogenisation affect due to chaotic rock mass flow in a cave operation coupled with daily production coming from numerous drawpoints across the extraction levels.

Since the earliest resource estimates and based upon the items described above and the proposed high production rates (24 Mt/a in Lift-1 and 48 Mt/a in Lift-2), the Project Team was confident that a 400 × 400 m drill spacing was sufficient to define Inferred Resources, and 200 × 200 m sufficient for Indicated. Internal guidelines do not require Measured Resource to be defined at Onto until the feasibility stage of evaluation.

To initially provide the MRMR Group with evidence to support the resource classification, a benchmarking exercise was undertaken. Unfortunately, this work indicated that in other porphyry/block-cave operations significantly denser drill hole spacings were used to define Inferred and Indicated resources, and this made it increasingly difficult to support the Project Team's view, despite the projects selected for benchmarking not particularly analogous to Onto.

This paper describes this journey.

OBJECTIVE DEFINITIONS OF RESOURCE CATEGORIES

In a presentation delivered by Dr Harry Parker (Parker and Dohm, 2014) reviewing the attempts by various workers since the 1970s to provide an objective definition of resource categories, Parker and Dohm concluded that the following objective, probabilistic definitions could be used (Figure 4):

- **Inferred**: Insufficient geological information to establish confidence levels.

- **Indicated**: ±15 per cent accuracy with 90 per cent confidence over annual production increment. Actual will be within 85 and 115 per cent of the estimate 90 per cent of the time. Annual production increments are typically used for Pre-feasibility and Feasibility cash flows. Can stand one year in 20 as being below 85 per cent of the estimate – normal business risk. If actual is less than 85 per cent, very often the mine will run a loss.

- **Measured**: ±15 per cent accuracy with 90 per cent confidence over a quarterly or monthly production increment. Actual will be within 85 and 115 per cent of the estimate 90 per cent of the time. Quarterly or monthly production increments are typically used for Operating Budget cash flows. If error is less than 15 per cent can usually rework the mine plan and prevent a loss.

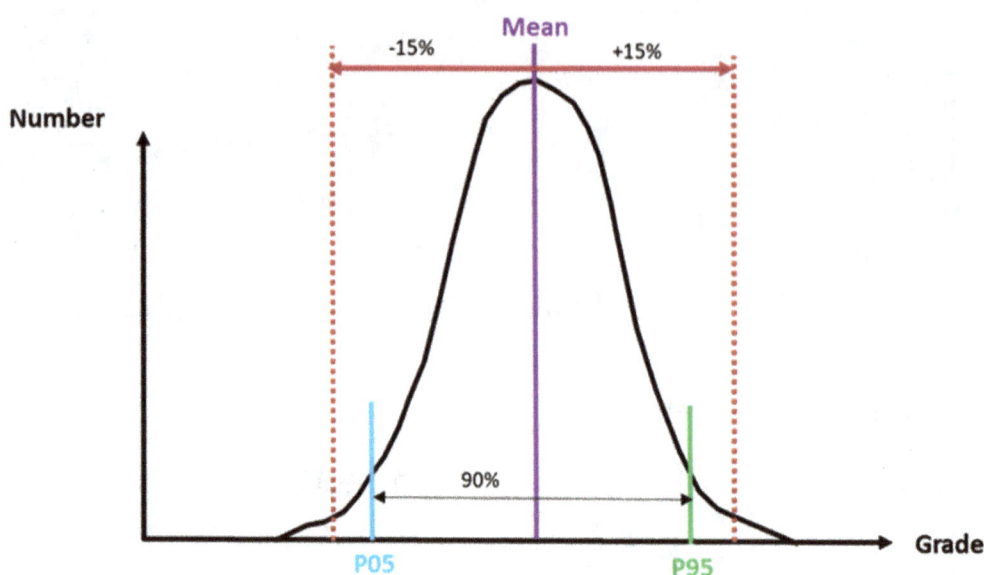

Notes:

- P05 and P95 are the 5th and 95th percentiles respectively which define 90 per cent of the histogram's area. The idea being that to meet the definition of the various resource categories, the ±15 per cent limits about the mean captures ≥90 per cent of the histogram's area.

- The histogram reflects the production rate required for the evaluation of the particular resource category – Measured: quarterly production rate; Indicated: annual production rate. This is related to the so-called 'volume-variance' relationship – ie for a given drill density, larger volumes (eg annual volumes) will be less variable than smaller volumes (eg quarterly volumes).

FIG 4 – Histogram showing probabilistic concepts of the 90:15 approach. Note while the histogram is shown as a normal distribution, it may not necessarily be so in reality.

This set of definitions is referred to as the '90:15 approach'. These definitions account for:

- Drilling density – denser drilling reduces grade and geological variation.

- Production tonnages – larger production volumes reduce grade variation.

It was these definitions of resource categories that the Project Team decided to pursue, with the Inferred category re-defined in a similar fashion as the other two categories: 'Inferred is ±15 per cent accuracy with 90 per cent confidence over the whole resource volume'.

KRIGING VARIANCE – DRILL HOLE SPACING ANALYSIS

The paper by Verly, Postolski and Parker (2014) provides a simple method and a set of equations that allows a 90:15 evaluation for a given drill hole density and volume based upon kriging variance (KV) (noting that KV is independent of actual grades, but dependent upon data configuration, the variogram model and block volume).

The method is simple:

- Set-up a block model with one block representing the size of interest (eg if evaluating the required drill hole spacing to meet an Indicated category, the block's volume will represent an annual mining volume) and a set of dummy assay data in drill holes at various drill hole spacings. Run kriging into the block and report KV for the various drilling densities. Calculate the estimation variance using the provided formula (Verly, Postolski and Parker, 2014):

$$90\% \; CI_{REL} = 1.645 \cdot \left(\frac{KV}{mean}\right)$$

(Refer to Verly, Postolski and Parker (2014) for more information and applicable equations for standardised variogram models.)

- Graph drilling density versus 90 per cent CI$_{REL}$ to determine the appropriate drill hole spacing that gives a 15 per cent error.

For Onto this showed that for a 36 Mt/a operation (the production rate being envisaged at the time of the analysis) a 200 m spacing would meet the 90:15 definition for Indicated. To meet the Measured definition for a 36 Mt/a operation (=9 MtpQ) this analysis indicated that a ~120 m drill hole spacing would meet the threshold (Figure 5).

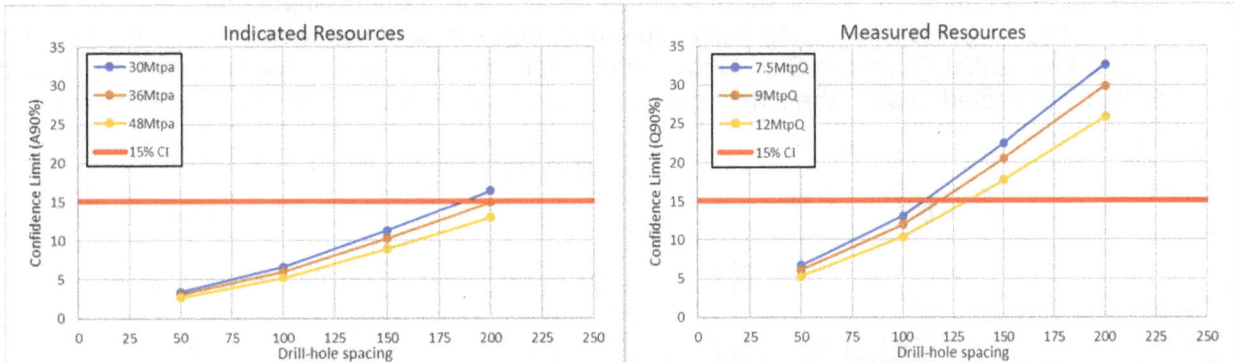

FIG 5 – Drill hole density evaluation to provide the required drilling density to meet the 90:15 definition of Indicated (left) and Measured (right) resource classification.

This analysis presented objective evidence that 200 × 200 m drill density would meet the 90:15 definition for Indicated resources.

However, it was recognised that this method had the following failings:

- The method assumes a normal distribution of possible outcomes. Although, perhaps, this is a reasonable assumption for a porphyry copper deposit.

- Relies upon KV, which does not account for actual grades. It would be expected that, for a given drill hole spacing, in those parts of the deposit where there is a mixture of high and low-grades the estimation variance would be higher than in those parts of the deposit with relatively uniform grades.

- It does not account for geological variation.

- This type of analysis does not account for the homogenisation process inherent in the caving mining method.

This resulted in the Project Team to investigate simulation methods to address these weaknesses.

CONDITIONAL SIMULATION

Simulating geology to capture geological variability is important as the grade estimated into a particular block is determined by the block's underlying geological conditions which were used to derive the estimation domain which in turn determines which assay data, variogram model, search routine, and any other differences in the kriging plan parameters between domains are used for grade estimation.

As mentioned above, at Onto, four estimation domains were used. These reflected the underlying rock-type/alteration styles. For simulating 'geological variation', this was simplified to simulating estimation domains.

Data sets

The data used for the simulations consisted of 4 m composite Cu grades and 2 m data flagged against the current estimation domain model. To evaluate the changes to the estimated grade uncertainty due to drilling density, three data sets representing the data available at various points in time through Onto's evaluation history were used (Figure 6):

1. 'DS1' – the drilling data set as of December 2020. This data set was used to develop the maiden publicly declared resource estimate.

2. 'DS2' – the drilling data set as of December 2023. This data set was used to develop the PFS resource block model and the most recent public resource estimate declaration.

3. 'DS3' – DS2 plus a future drilling program. During the upcoming feasibility study (FS) a drill program has been developed to convert the Indicated resources in Lift-1 to Measured, and consists of 35 holes for about 35 000 m. To provide the required inputs into the simulation processes, Cu grades were simulated into 4 m composite samples within the planned drilling by the Sequential Gaussian Simulation (SGS) method and 2 m samples within the planned drilling intersected against the existing estimation domain model to provide this information.

Item	Dataset		
	DS1	DS2	DS3
Planview (colour coded to Estimation Domain number)	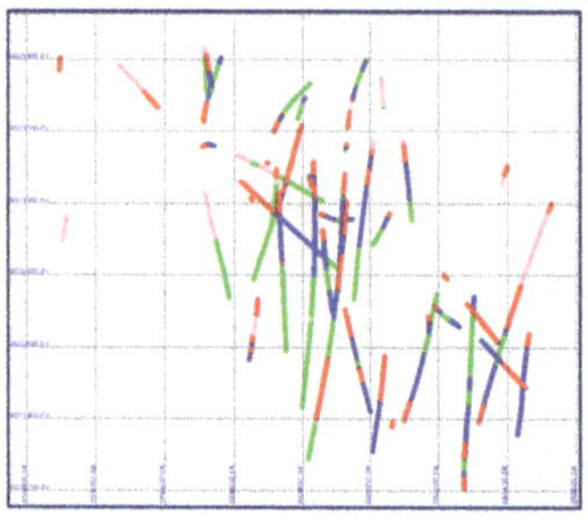	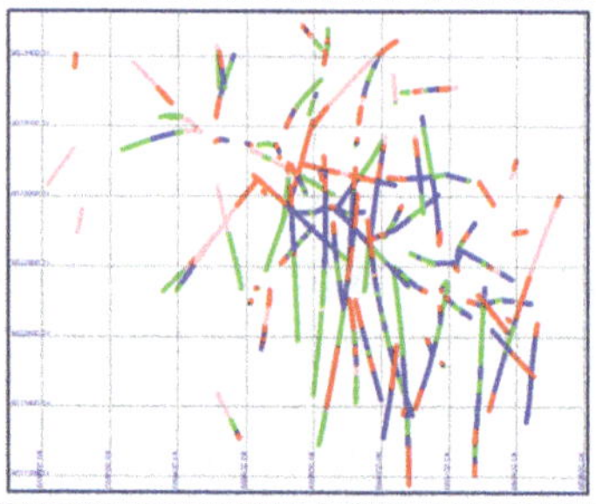	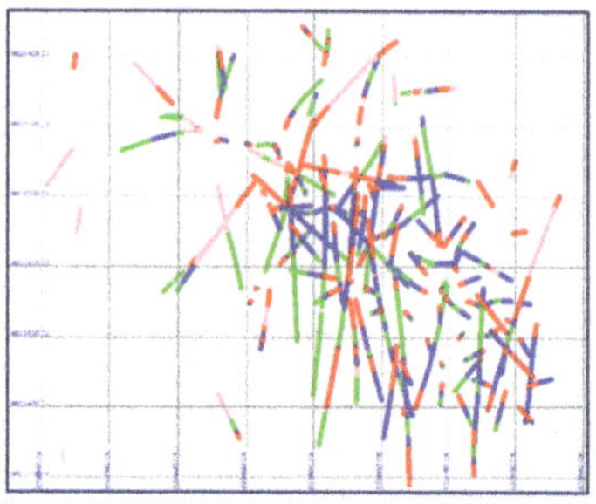
Section (colour coded to Estimation Domain number)	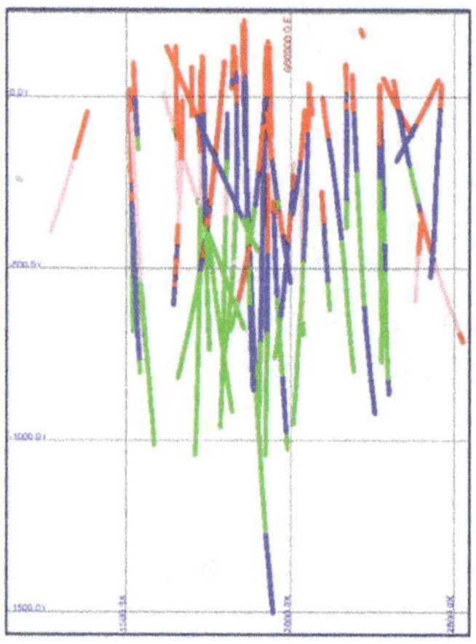	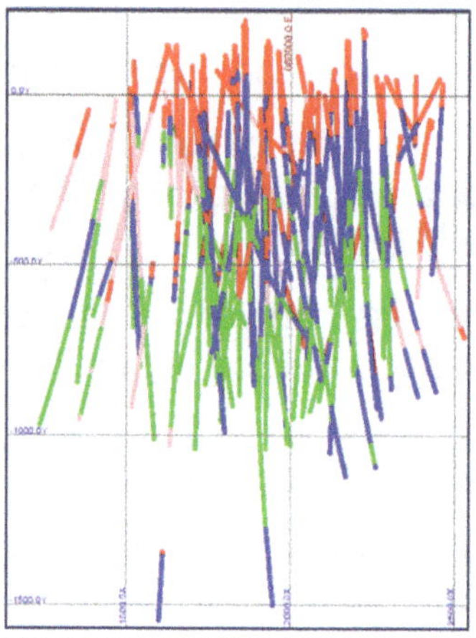	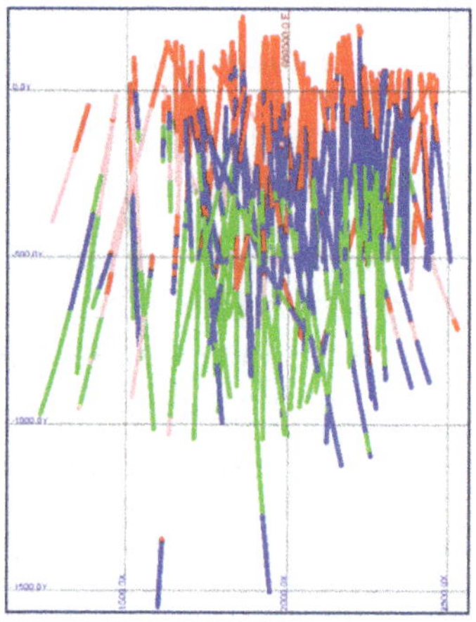
No. holes	43	88	123
No. 4 m Cu composites	7 060	14 193	19 098
No. estimation domain data	14 172	28 416	38 211

FIG 6 – Description of the three input data sets.

Thus, there is an increase of drilling density from DS1, to DS2 and finally into DS3.

Geological and grade simulation

To capture geological (estimation domain) variation the Sequential Indicator Simulation (SIS) method was used. Other methods of categorical simulation are available to retain any relative geological zonation (eg sedimentary facies) aspects, however this was considered unnecessary for Onto where, on an estimation domain basis, zonation is not particularly evident. The parent block model (45(N) 45(E) × 15 m(El) sized blocks) was directly populated with a domain number for each realisation.

One of the key items was to ensure that a suitable number of realisations were developed to capture the underlying categorical variability. Initially a simulated data set of 100 realisations were developed from DS2, and the simulated proportions for each estimation domain assessed. This indicated that 40 categorical realisations were sufficient to capture the underlying geological (estimation domain) variability.

For each input data set (DS1, DS2 and DS3) one simulated data set of 40 realisations were developed.

The Sequential Gaussian Simulation (SGS) method was used to simulate copper grades into each of the 40 geological realisations. Point support realisations on a 5 × 5 × 5 m grid were accumulated into the parent block size of 45(N) × 45(E) × 15 m(El) – there being 243 point simulated grades averaged into each parent block (=9 × 9 × 3).

Analysis of a large set of realisations derived from the DS2 data set, indicated that 30 Cu grade realisations were sufficient to capture the underlying Cu assay variability.

Thus, for each underlying input data sets 1200 realisations of Cu grade were developed (=40 geological realisations × 30 Cu grade simulations) for a total of 3600 realisations over the three input data sets – each block in the block model now had 3600 Cu (simulated) estimates to evaluate uncertainty in grade estimation reflecting geological and assay variability and data density.

Results

The 90:15 definitions for Indicated and Measured resource categories required evaluation within annual and quarterly production volumes and this analysis is presented in the next section. The adopted definition for inferred is an evaluation on a global basis presented in Figure 7.

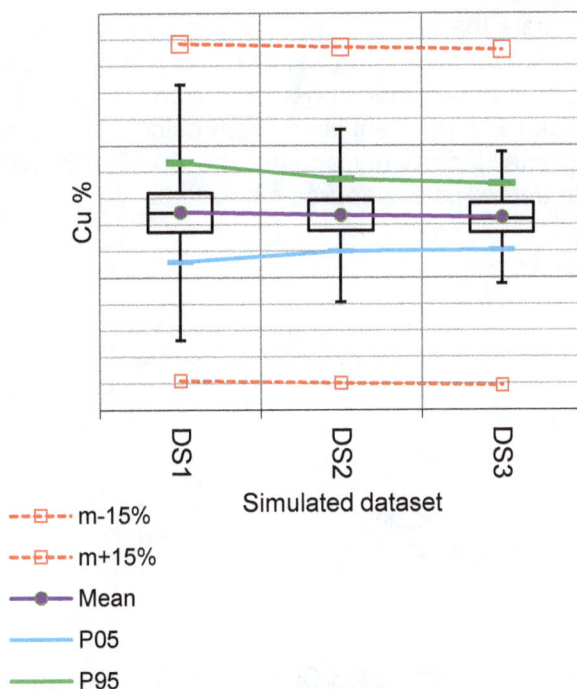

FIG 7 – Box and Whisker plot of the simulated data sets (each of 1200 realisations) derived from DS1, DS2 and DS3.

As mentioned above, to meet the various 90:15 definitions, the ± 15 per cent limits about the mean need to capture ≥90 per cent of the histogram's area (defined by the area between P05 and P95). As shown in Figure 7, in all simulated data sets derived from DS1, DS2 and DS3, the ± 15 per cent limits about the mean easily captures 90 per cent of the realisations and in fact captures 100 per cent of the realisations. On this basis the current drill hole spacing (DS2) sufficiently defines uncertainty to meet the Inferred 90:15 criterion, as did the DS1 data set.

ACCOUNTING FOR CAVING AND PRODUCTION VOLUMES

An interesting aspect of the block caving mining method is that this uniquely relies upon large scale rock mass movement through the cave before the material is extracted at drawpoints on the extraction level at the base of the Lift. Broken material may be flowing through the rock pile in the cave for several years before it reaches the extraction level. In addition, to reduce the likelihood of hang-ups of broken rock in the cave, it is important that extraction of the broken material comes from multiple drawpoints over the entire footprint of the Lift to evenly draw the rock pile downwards. At Onto the planned drawpoint spacing is 20 × 16 m.

This rock mass movement, coupled with production being derived over a large area on a daily basis, acts as a large homogenisation process – when a loader bucket of material is extracted from a particular drawpoint the material is not derived from one particular block, but represents parts of many blocks across the cave which have fallen from the cave back, onto the cave pile, and moved over several years through the rock pile, self-breaking akin to the breakage that occurs in a semi-autogenous grinding (SAG) mill, resulting in a homogenisation process.

Specifically in caving operations, specialist software is widely used to model rock mass flow and predict the resultant grades at each drawpoint over time.

For the purpose of capturing the above described homogenisation effect, it was considered important that the realisations be passed through the mine schedule. This would also capture the effects of changes in production rates through the mine life – ramping up to 24 Mt/a in Lift-1 and then to 48 Mt/a in Lift-2.

Results

The PFS produced a deterministic production schedule based upon the resource model. Since the object of this study is to evaluate the uncertainty in predicted production Cu grades, the production schedule was fixed, and each simulated data set (each of 1200 realisations) was passed through the same schedule. An example of summarising a simulated data set is shown in Figure 8 whereby the mean, P05, P95, and mean ±15 per cent limits were calculated from the 1200 realisations. This allows an assessment of grade uncertainty against the 90:15 definitions through the production profile, and determination of appropriate resource categories.

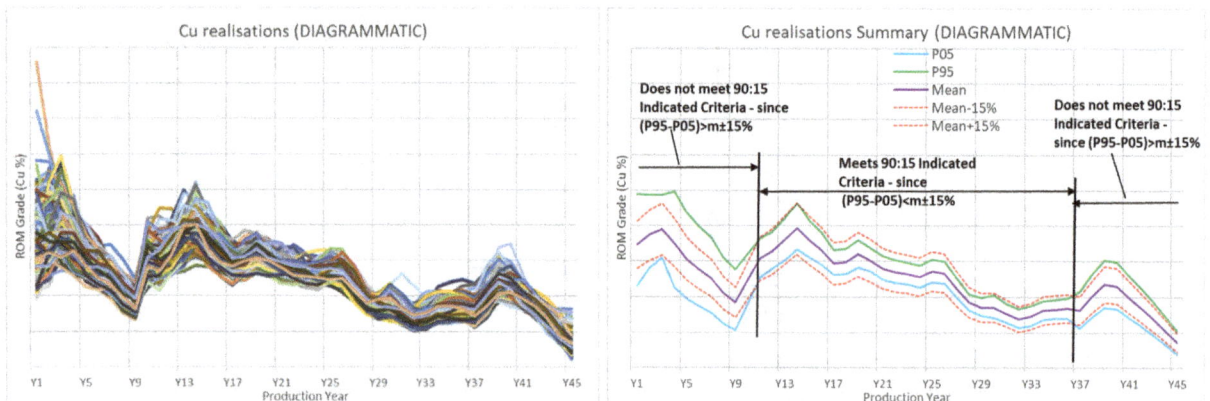

FIG 8 – Diagrammatic example of a simulated Cu data set (of 250 realisations) passed through an annualised mine schedule (left) and the summarising of the simulated data set to determine which parts of the production profile meet the 90:15 Indicated criterion (right).

The annual production graphs (Figure 9) show that for the existing drill hole spacing, the bulk of production meets the 90:15 Indicated criterion in all DS1, DS2 and DS3 data sets, excepting the beginning and end of the mine life where production tonnages are small. Indeed, according to this assessment, the data density within the DS1 data set were sufficient to define Indicated classification for a significant amount of the mine production.

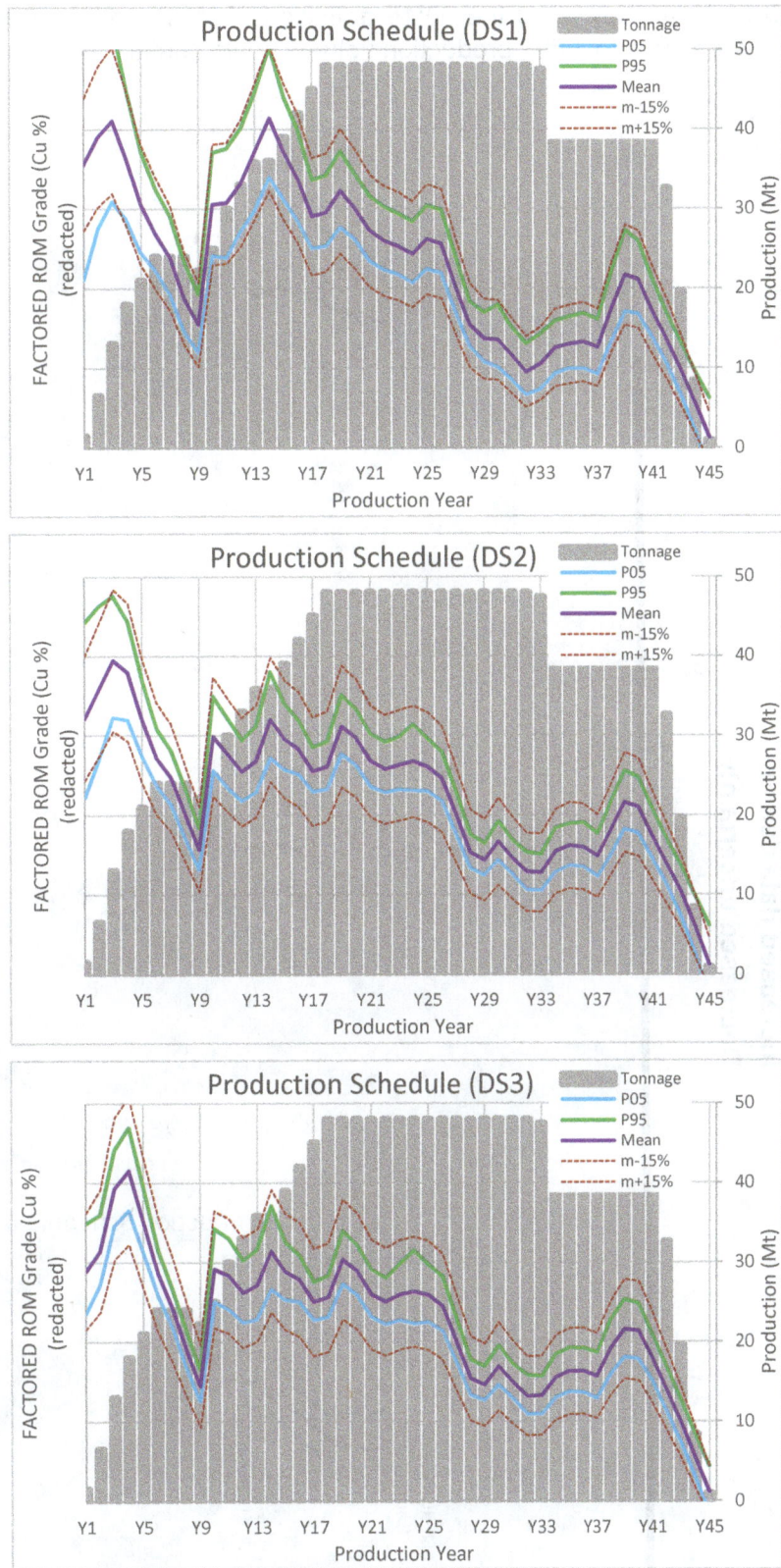

**Increased data density
Decreased uncertainty**

FIG 9 – Annual production schedule showing summary simulation results for DS1, DS2 and DS3 top to bottom respectively.

This provided the Project Team with an objective analysis showing that the prevailing drill hole spacing does meet the 90:15 Indicated criterion.

To investigate whether the proposed drilling program (DS3) would be successful in converting Lift-1 Indicated resources to Measured, an analysis of the simulation data sets by quarterly production within Lift-1 was made (Figure 10). This showed that at the current drill spacing production between years 3 to 12 already meet the 90:15 definition of Measured, however the planned drilling (DS3) would ensure the total Lift-1 production volume could confidently be placed in a Measured category.

Increased data density
Decreased uncertainty

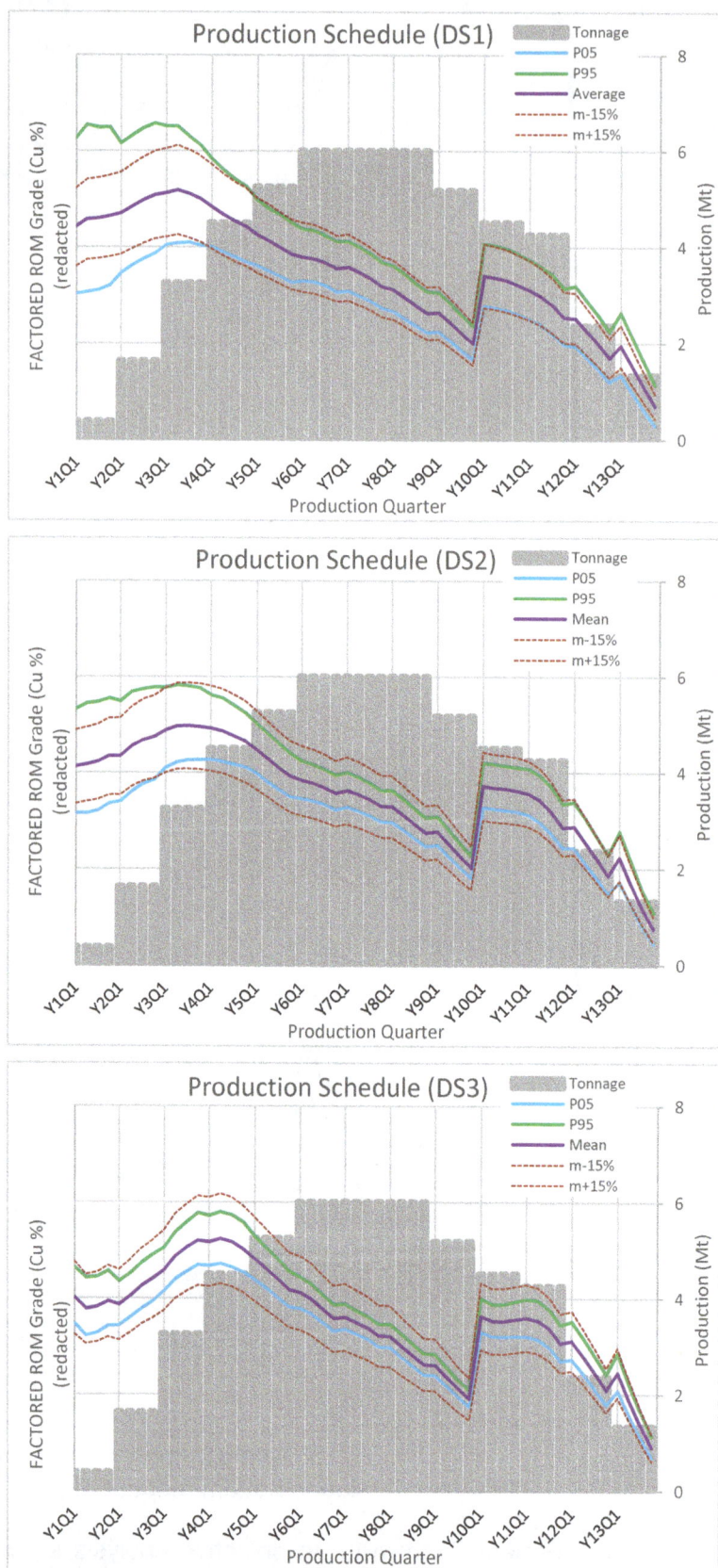

FIG 10 – Lift-1 quarterly production schedule showing summary simulation results for DS1, DS2 and DS3 top to bottom respectively.

DISCUSSION AND CONCLUSIONS

In any resource estimate there will be uncertainty in the grade estimate based upon the prevailing drill hole density. The uncertainty in the grade estimate, in turn, is affected by underlying variability of the assay data set and geological modelling components. Simplistically, the limiting cases of maximum and minimum uncertainty is a resource estimate based upon two drill holes versus an estimate based upon exhaustive drilling collared on (say) 1 m centres. The former case (two holes) would produce an estimate with a significant amount of uncertainty, while the latter (drill holes on a 1 m grid) would produce an estimate with little or no uncertainty.

A thought exercise to visualise the uncertainty inherent in a resource grade estimate due to drill hole spacing, as shown in the histogram in Figure 4, is to take two drill holes, and randomly drill these into the deposit and use the results to determine a grade estimate. If this process was repeated (say) 100 times and all grade estimates were plotted on a histogram the variation (or uncertainty) within all two-hole estimates would be extremely large. Conversely, undertaking the same exercise but with our 1 m spaced drilling, would result in a histogram of all possible resource grade estimates with very little variation (uncertainty) about the average. A similar argument can be made about geological uncertainty – the lower the drilling density the more uncertainty is inherent in the geological model; the higher the drilling density, the lower the uncertainty in the geological model. If there is a significant correlation between geology and assay grades, then the uncertainty in the geological model itself contributes to the uncertainty in the resultant resource grade estimate.

Generally, resource classification is largely determined by data density, and the role of the resource geologist in assigning resource classification, is to determine if the prevailing drill hole spacing is enough to meet the various 90:15 criteria required to assign Measured, Indicated, and Inferred resource categories.

The aim of this paper was to provide a narrative of how this was achieved for the Onto mineral resource. Simply, simulation methods were used to model the variability in the underlying geological and assay data and the manifestation of uncertainty in the final resource grade estimate at the prevailing drill data density. The simulated uncertainty was then used to determine if the 90:15 Indicated and Inferred resource classification criteria were met. For Onto, a proposed block cave operation, the homogenisation of grades inherent in this mining method, was considered significant enough to warrant an additional layer of model processing prior to evaluation against the resource classification criteria. In more conventional mining methods (eg longwall stoping, open cut mining) this would be an unnecessary step.

This work culminated in providing objective evidence that a nominal 200 × 200 m drilling density would be sufficient to define Indicated resources, while a nominal 400 × 400 m drilling density would be sufficient to define Inferred Resources.

The simulation of grades into a planned drill program aimed at converting Lift-1 Indicated resources to Measured (DS3) provided a data set to evaluate whether the 90:15 Measured threshold is met by this drilling, or if additional drilling is required. This type of analysis provides a mechanism for trade-off studies by quantifying both the cost of drilling and the amount of risk reduction achieved.

This study required the expenditure of a considerable amount of time mainly due to developing the required number of realisations and the passing of these through the mine schedule. The required simulation software is readily available, and it is anticipated that with ever increasing computation speed of computer hardware that over time these types of studies can be completed quickly; become an essential component of a well-executed resource evaluation study; and require small budgets to complete.

Additional studies

Additional studies, not presented in detail, are described below, to provide commentary on other uses of quantified uncertainty.

Project economics

In addition to the prediction of the uncertainty of estimated Cu grade throughout the production profile, each simulation, of 1200 realisations, provided the ability to calculate a probabilistic measure around the project economic metrics, eg net present value (NPV) and internal rate of return (IRR), such that Management could be presented not only with the likely NPV and IRR (the median value) but also with the P05 and P95 limiting values representing the 90 per cent Confidence Level. Furthermore, the economic metrics provided by the simulation derived from DS3 (future drilling), provides a cost estimate (ie the cost of the planned drilling) to reduce the Confidence Intervals by a defined amount, being the difference between P05 and P95 of the simulation from DS3 (future drilling) and the simulation from DS2 (existing drilling). This probabilistic approach provided Management with an objective measure of the uncertainty inherent in Project Economic metrics related to the current drill hole spacing, and a cost estimate to reduce the uncertainty by a determined amount.

Hill of value

As part of the PFS, a study was undertaken to determine the optimal elevations of the two extraction levels to maximise NPV and IRR by calculating both metrics for a series of lift elevation combinations (Figure 11).

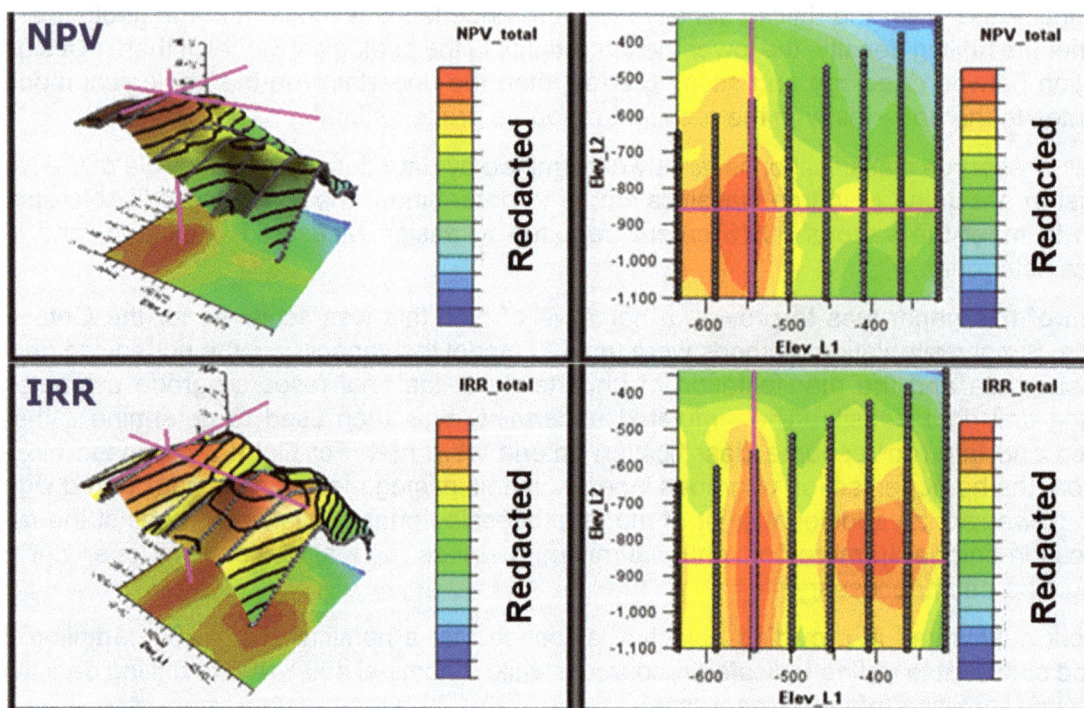

FIG 11 – Hill of Value to determine the optimal Lift-1 and Lift-2 elevations to maximise NPV (top) and IRR (bottom).

As a further piece of work, it would be of interest to pass all realisations though this process to investigate the effect of uncertainty – due to the prevailing drill hole density – has upon this analysis, and whether the drill hole density is sufficient to confidently identify these optimal elevations.

Geometallurgy

In parallel to the resource block model, the Onto project carries a geometallurgical model of various metallurgical recovery and comminution variables. Copper metallurgical recovery and concentrate grade are defined by a series of equations involving a geological rock-type indicator, and other grade variables. Simulated rock-types coupled with co-simulation methods for grade variables could be used to investigate variability in metallurgical recovery due to geological and grade variability and the effect this has on project economics and whether the variability meets corporate expectations.

ACKNOWLEDGEMENTS

In the mid-1990s, Neil Schofield (FSSI Consultants (Australia) Pty Ltd) first introduced the author to various methods for evaluating and exploiting uncertainty to achieve optimal outcomes. These discussions over several years were instrumental in framing and culminated in this work.

Several months were spent developing the 3600 realisations – no small task – and Jan Graham, Ian Glacken and Dr Oscar Rondon (Datamine) along with Pak Ignas Meak (PT. Vale Eksplorasi Indonesia) all helped develop the realisations.

Pak Eddy Samosir (Vale Exploration Pty Ltd) spent several weeks passing all 3600 realisations through the mine schedule and presented the results in a simple to use format.

Pak Swandi Manurung and Osmario Neto (Vale Exploration Pty Ltd) provided ongoing management support to undertake this work.

VBM and STM are acknowledged for providing permission to publish this work.

REFERENCES

Burrows, D R, Rennison, M, Burt, D and Davies, R, 2020. The Onto Cu-Au discovery, Eastern Sumbawa, Indonesia: A large, middle Pleistocene lithocap-hosted high-sulfidation covellite-pyrite porphyry deposit, *Economic Geology*, 115:1385–1412.

JORC, 2012. Australasian Code for Reporting of Exploration Results, Mineral Resources and Ore Reserves (The JORC Code) [online]. Available from: <http://www.jorc.org> (The Joint Ore Reserves Committee of The Australasian Institute of Mining and Metallurgy, Australian Institute of Geoscientists and Minerals Council of Australia).

Parker, H M and Dohm, C E, 2014. Evolution of mineral resource classification from 1980 to 2014 and current best practice, the Julius Wernher Lecture presented at Finex'14, London.

Verly, G, Postolski, T and Parker, H M, 2014. Assessing uncertainty with drill hole spacing studies – applications to mineral resources, in *Proceedings of the Orebody Modelling and Strategic Mine Planning Symposium 2014*, pp 109–118 (The Australasian Institute of Mining and Metallurgy: Melbourne).

Consuming *in situ* uncertainty into mine planning processes – from reserve calculation to dynamic decision-making

I Minniakhmetov[1] and J Serjeantson[2]

1. Principal Geoscientists, Resource Centre of Excellence, BHP, Perth WA 6000.
   Email: ilnur.minniakhmetov@bhp.com
2. Senior Engineer Mining Production Scheduling, BHP, Perth WA 6000.
   Email: julia.serjeantson@bhp.com

ABSTRACT

The 3D resource model provides the foundation for decision-making in mining operations. This model represents our understanding of the *in situ* resource and serves as the cornerstone for critical aspects such as mine planning, geotechnical analysis, water management, financial assessments, and eventual closure strategies. However, these 3D models are constructed based on imperfect information and therefore contain uncertainties.

Conditional simulations are known to be the best technique for quantifying uncertainty in complex non-linear systems. However, they have not found much application in the mining industry due to the inability to integrate them into mine planning processes.

This work advances the state of 'consuming' geological scenarios from simple reserve calculations and box-and-whisker plots of monthly tons into an actual emulation of the processes and decisions that will be made when real variability and uncertainty impact operations and planning.

The importance of *in situ* uncertainty quantification for mine planning is demonstrated through a real iron ore deposit consisting of four pits feeding into a single crusher with multiple stockpiles of different quality materials. This work focuses on two-year budget mine planning, particularly on scheduling and operational aspects of mining.

First, 100 realisations for each pit have been generated using the latest advancements in geostatistics. Each realisation is then used as an *in situ* reality to emulate the mine planning process and operations. The initial resource model is developed using conventional estimation process. Every month, new grade control information is collected (sampled from a realisation) and used to update the short-term model. The updated model is used to re-optimise scheduling decisions and material flow. This process is repeated for all 24 months for each realisation and analysed in terms of the implications of uncertainty and variability for additional rehandling costs, bottlenecks, product quality, and potential delays due to insufficient materials in stocks.

INTRODUCTION

The 3D resource model is the cornerstone of decision-making in mining operations, encapsulating our understanding of *in situ* resources. This model supports critical processes, including mine planning, geotechnical analysis, water management, financial evaluations, and closure strategies. It functions as a shared framework that aligns efforts across diverse teams such as Operations, Geoscience, Resource Engineering, Exploration, and Portfolio Strategy and Development.

However, it is crucial to recognise that these 3D models are constructed from incomplete and imperfect data, introducing inherent uncertainties. These uncertainties arise from limitations in data availability, quality, and interpretation. Quantifying such uncertainties involves calculating probabilities and confidence levels in data, models, and decisions — forming the foundation of a risk-based decision-making approach. High-quality 3D models, combined with quantified uncertainty, are vital for optimising mining operations and managing associated risks effectively.

The resolution and uncertainty of 3D models vary depending on the planning and decision time frame — ranging from life-of-mine (LOM) strategies to daily operations. For instance, a ±10 m uncertainty in the ore-waste boundary may be acceptable for long-term strategic decisions like crusher placement but becomes critical for daily scheduling. This study focuses on the two-year budget (2YB) production scheduling process, which offers a well-defined time frame for evaluating geological uncertainties without the complicating influence of macroeconomic factors, such as commodity price fluctuations, regulatory changes, or technology shifts. Unlike LOM strategies, where external factors can overshadow geological variability, the 2YB time frame allows a more targeted analysis of uncertainties, while its financial impact is higher compared to daily or weekly decisions.

Traditionally, 3D resource models have been used in mine planning without explicitly accounting for uncertainty. Over the past two decades, advancements in decision-making under uncertainty and risk analysis have emerged (Ramazan and Dimitrakopoulos, 2012; Goodfellow and Dimitrakopoulos, 2016). These studies often rely on conditional simulations or Monte Carlo realisations (Journel and Alabert, 1990), substituting a single resource model with multiple scenarios representing possible resource outcomes. For example, conditional simulations may provide a range of *in situ* grade profiles within a given time frame (see Figure 1).

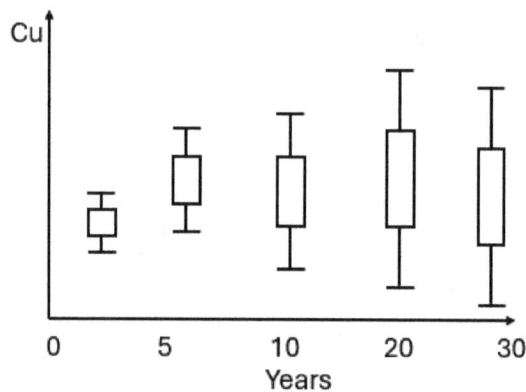

FIG 1 – Cu grade profile uncertainty over life-of-mine. Box-plots are created by running realisations through a fixed schedule.

While these studies are valuable, they have notable limitations. For instance, uncertainties depicted through box-whisker plots of *in situ* uncertainties often fail to translate into tangible business or financial implications. Consider a scenario where the overall tonnages and grades within a 2YB volume remain consistent across realisations, but monthly high-grade ore tonnages vary by ±10 per cent. If adequate stockpiles and fleet resources are available, such variability might not pose a material risk to business outcomes. Additionally, many studies assume that the mine plan will be executed as initially designed, neglecting dynamic decision-making or adjustments based on new data.

In this study, we go beyond simulating *in situ* resource scenarios by emulating mine planning processes and operational decisions. This approach integrates geological uncertainties into dynamic decision-making, effectively converting uncertainty into actionable business risks. In an ideal scenario, this would culminate in the development of a probabilistic digital twin of mining operations, encompassing:

- Probabilistic digital twin of *in situ* resources: Using conditional simulations to model resource uncertainty.

- Digital twin of data collection: Modelling when and how infill drilling, geophysics, and sampling are conducted.

- Digital twin of modelling processes: Simulating how and when resource models are updated based on new data.

- Digital twin of planning processes: Representing how mine design and scheduling are conducted and dynamically updated in response to new models.

- Digital twin of operations: Simulating material handling processes, including blasting, hauling, crushing, and processing, while accounting for delays or critical events (eg wall failures).

- Digital twin of decisions: Capturing how decisions evolve in response to changing inputs and conditions.

While fully achieving this vision requires substantial effort, this paper focuses on the first four components, particularly within the context of scheduling processes in mine planning.

The paper is structured as follows: First, the conventional approach to mine planning is discussed in detail. Next, the proposed methodology is introduced. A case study involving a real-world iron ore deposit is then presented, followed by a discussion of the results. Finally, the paper concludes by summarising key findings and their implications. (Note: All numerical data, including drill hole spacings, have been scaled due to sensitivity considerations.)

DESCRIPTION OF PLANNING CYCLE

The production scheduling process for the two-year budget (2YB) plan begins with the importing of a short-term model. This model is based on the most recent data, including exploration, infill, and grade control drill hole samples. Inputs from the long-term or five-year plan — such as pit designs, tonnage and product quality targets, equipment specifications (crushers, trucks, diggers), and mine set-up constraints (eg stockpile policies, strip ratios, slope angles) — are integrated. Additional guidance is provided by relevant teams such as marketing and supply chain.

The core objective of production scheduling is to determine the optimal timing and destinations for mining activities. These activities include drilling and blasting, excavation, haulage, and processing, all while adhering to constraints (eg slope angles, block precedence rules) and meeting product targets (eg tonnage, lump-to-fines ratio, alumina and silica limits, Fe grades, and moisture levels).

The primary outputs of this process are:

- Detailed 3D production schedule: Mining sequence represented as 3D shapes/solids for each month.

- Tabular activity summary: Each row in the table specifies the activity, start and end times, source and destination, equipment utilisation (eg truck usage, cycle time), and material details (eg tonnes, volumes, head grades, lump/fines split, material handling index, and moisture levels).

The projected ore tonnage and quality from the 2YB plan are locked into the business budget, forming a commitment that guides operations.

The 2YB plan serves as the foundation for subsequent monthly, weekly, and daily plans, as well as for operational execution. As mining progresses, new grade control information is collected, triggering updates to the short-term resource model. This, in turn, necessitates updates to the production schedule. For simplicity, this study assumes that grade control data becomes available at monthly intervals. The process of updating the schedule is depicted in Figure 2.

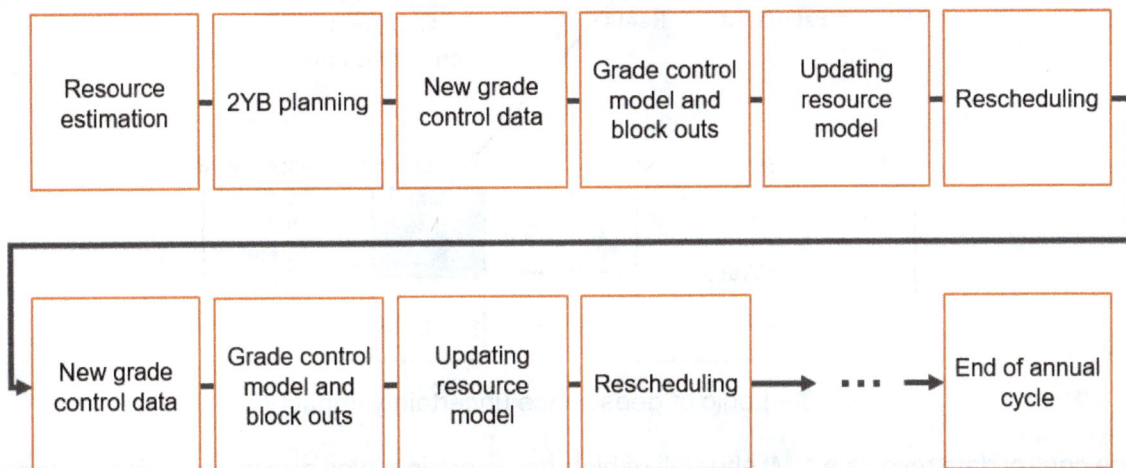

FIG 2 – 2YB mine planning cycle process.

The update process begins with incorporating new grade control samples into the short-term resource model. Parameters such as variogram ranges, domain definitions, and neighbourhood search rules are assumed to remain constant. Block-out polygons are then created for different material types (eg waste, low-grade ore, blend-grade ore, high-grade ore) based on predefined rules. These polygons are integrated back into the short-term model and subsequently passed to the scheduling team.

The scheduler then reruns the optimisation process, adhering to the same constraints and striving to meet the crusher tonnage commitments for each month. Updates to the grade control data primarily affect the ore-waste boundary, which in turn influences the availability of ore and waste at the open pit's active faces for the upcoming month.

Because the capacity of trucks and diggers is limited, monthly tonnages from *in situ* material are designed to closely match the original plan. Any surplus ore is redirected to stockpiles, while shortfalls are supplemented from stockpiles, potentially resulting in additional rehandling and productive movements.

An example of this process is shown in Figure 3 for a single month, assuming:

- Total material movement in the pit: 10 million tonnes (Mt).
- Crusher target: 3 Mt.
- Scheduled ex-pit ore tonnage: 4 Mt (2 Mt to crusher, 2 Mt to stockpiles).
- Scheduled stockpiles-to-crusher tonnage: 1 Mt.

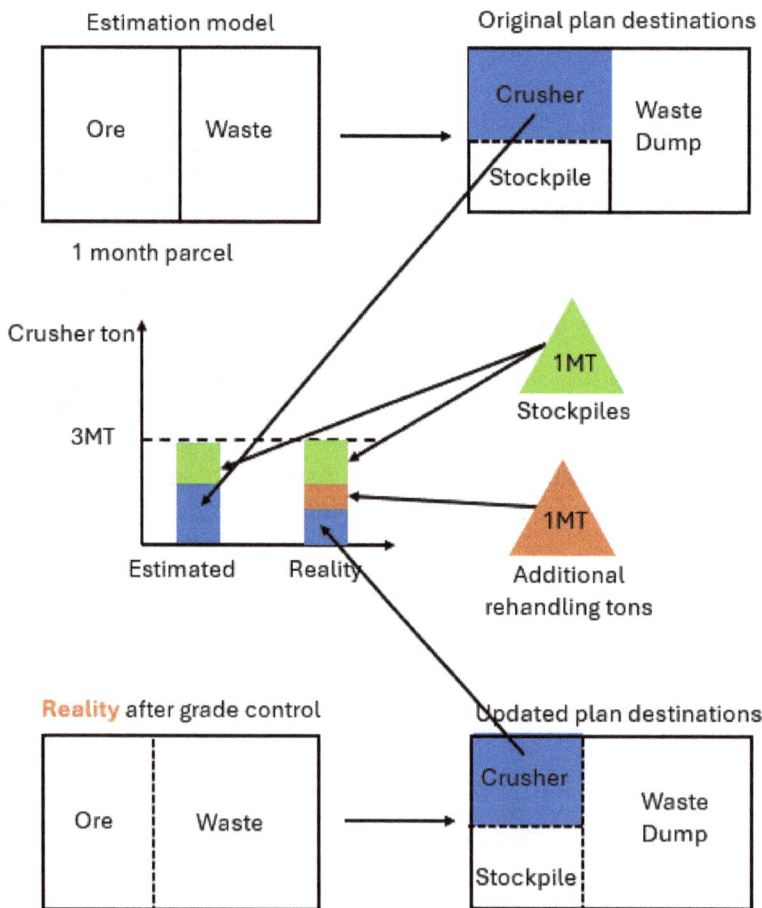

FIG 3 – Logic of geoscience uncertainty impact.

If grade control data reveals a 1 Mt shortfall of high-grade ore in the pit due to geological uncertainty, the operation must still move 10 Mt from the pit. To compensate crusher deficit, an additional 1 Mt must be retrieved from stockpiles, resulting in 1 Mt of extra productive movement.

This additional movement could simply incur higher costs or, if the fleet becomes a bottleneck, result in production delays. In such cases, operations might leave waste material in the pit to prioritise ore extraction; however, this waste will need to be removed later to access underlying ore.

PROPOSED APPROACH

This work aims to emulate the mine planning and production scheduling process while incorporating a probabilistic perspective. It provides insights into iron ore product volumes and quality, along with potential business risks such as failure to meet targets, delays, and unplanned costs. The proposed workflow is illustrated in Figure 4 and operates monthly for each geological scenario.

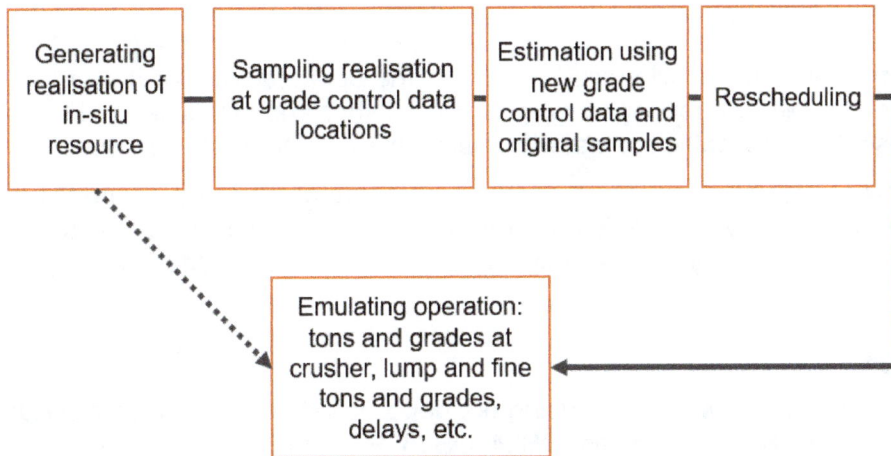

FIG 4 – Proposed approach.

Resource emulation

A critical component of this workflow is the accurate representation of *in situ* properties, which are necessary for simulating grade control sampling and evaluating ore loss and dilution. The industry standard for *in situ* resource simulation is the use of conditional simulations.

In this work, we simulate grades for Fe, Al_2O_3, SiO_2, P, and LOI at a high resolution of 2.5 m × 2.5 m × 4 m per block. This resolution is essential for quantifying ore loss and dilution caused by imperfect grade control sampling at a typical spacing of ~20 m × 10 m. Geological uncertainty is the sole focus, and the workflow is executed independently for each geological scenario (also referred to as a realisation).

Grade control sampling

For this study, grade control samples are assumed to be collected monthly. These samples are used to update the short-term resource model for the subsequent month. Sampling is simulated by extracting data from the geological realisation at 20 m × 10 m spaced locations for the solid corresponding to the upcoming month.

Short-term modelling

New grade control samples are incorporated into the data set, and a grade control model is generated using ordinary kriging. This updated model reevaluates all geometallurgical properties, such as lump-to-fines ratios, moisture content, and material classifications (eg waste, low-grade ore, blend-grade ore, and high-grade ore), according to predefined rules and formulas.

The revised 3D resource model is then imported into the production scheduling software.

Production scheduling

The production schedule is updated using the revised resource model, while maintaining the same constraints and product quality targets (eg Fe grades, impurity limits) as the original schedule. The primary outputs include:

- Updated block destinations: Adjustments to where mined blocks are sent (eg crusher or stockpile).

- Stockpile movements: Modifications to stockpile usage to meet crusher tonnage targets.

In this study, it is assumed that the monthly solids (representing the physical mining sequence) remain unchanged by the update process. As a result, only the material destinations and associated stockpile movements are affected.

This approach enables the integration of geological uncertainty into the decision-making process, offering a dynamic and probabilistic view of risks to production and business outcomes.

Execution

It is assumed that operations will strictly adhere to the updated plan for the upcoming month. As density is not simulated, the tonnage of ore at the crusher and stockpiles will match the updated schedule, unless there is insufficient material in stockpiles or the pit to meet the target.

Given the inherent uncertainty in grade control model, the grades and geometallurgical attributes (such as lump and fines) will be different in *in situ* realisation and the updated schedule. These differences are used to assess ore loss, dilution, and the expected product quality to be delivered to customers.

CASE STUDY

The proposed methodology was tested using ore deposits within BHP's Western Australia Iron Ore (WAIO) Newman Hub. Three orebodies (OB24, OB25_P1, and OB32) were selected based on their varying degrees of geological complexity. This selection allowed us to analyse the impact of uncertainty on the risks associated with a typical mine plan, which sources ore from multiple locations to feed stockpiles and crushers.

The analysis simulated conditions as of November 2019, using data available at that time. Data collected post-2019 was used for reconciliation purposes, ensuring that differences observed were solely due to uncertainties and limitations in current mine planning processes.

The value chain studied consisted of four pits, 33 stockpiles, three crushers, and two ore handling plants (OHPs) (Figure 5). However, since the OHPs were fed from adjacent deposits, the analysis was limited to crusher tonnage and grade. Future studies will expand to include all deposits, crushers, and OHPs contributing to fine and lump products (eg Newman Fines). This would require creating conditional simulation models for an additional five pits.

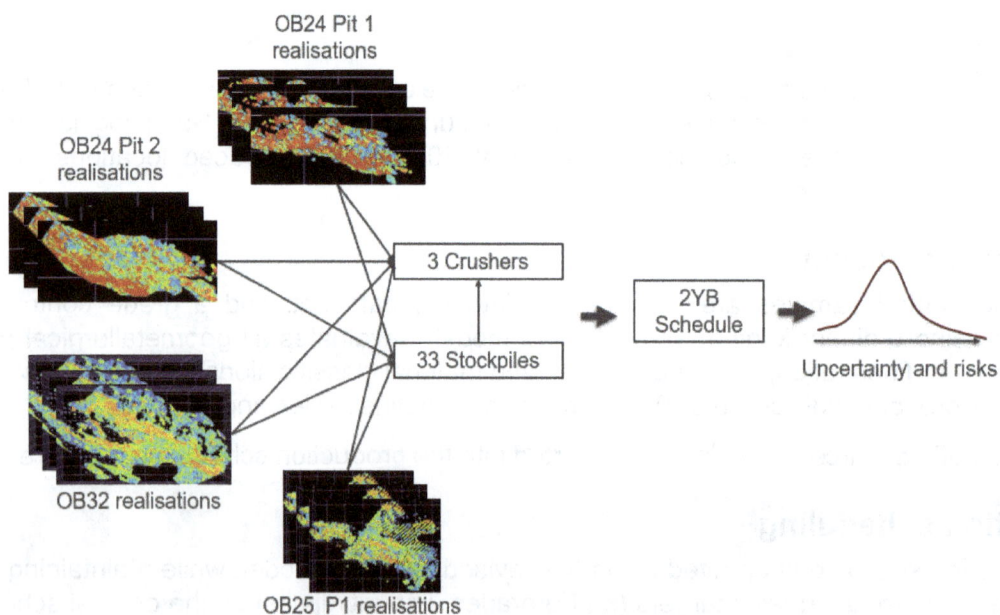

FIG 5 – Case study set-up.

Conditional simulations

The study modelled four pits at a resolution of 2.5 × 2.5 × 4 m, representing approximately 50 million blocks. This was based on around 2000 drill hole samples and 100 000 blasthole samples. Key variables modelled included Fe, SiO_2, Al_2O_3, LOI, and P.

The advanced workflow, detailed in Figure 6 and fully described in Williams and Minniakhmetov (2025); began with exploratory steps to define the simulation domains and variables' continuity. For modelling iron ore grades, three factors—stratigraphy, mineralisation, and weathering—were critical in domain definition. While stratigraphy was considered well-defined, mineralisation (mineralised versus non-mineralised) and weathering boundaries (fresh, transitional, and weathered) were modelled using hierarchical truncated pluri-Gaussian simulation (HTPGS). These variables exhibited high non-stationarity, requiring spatial domaining and local proportions or trends. In total, 100 realisations of mineralisation and weathering boundaries were generated.

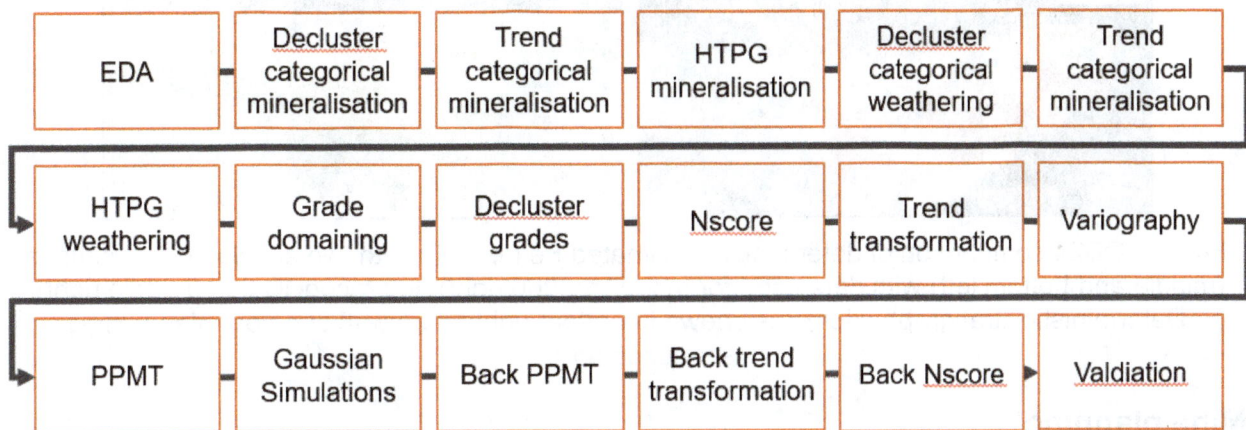

FIG 6 – Modelling workflow.

Grade domains for each block and sample were defined based on exploratory data analysis and the realised values of mineralisation and weathering variables, resulting in 100 realisations of grade domains. A kriging-based declustering approach was used to calculate weights, which were stored for use in subsequent transformations and validations.

Grades were transformed into Gaussian variables using normal score transformations. Due to non-stationary mean, variance, and local distributions, a non-stationary kernel density estimation (KDE) approach was applied to derive stationary residuals (Minniakhmetov, 2023). Then, variography has been performed to quantify spatial correlations, and the projection pursuit multivariate transform (PPMT) approach (Barnett, Manchuk and Deutsch, 2014) was employed to decorrelate the data while preserving complex multivariate relationships. Following transformations, turning bands simulation (Journel, 1974) was used to generate continuous variable realisations, which were back-transformed in reverse order, starting with back-PPMT.

A comparison of the short-term model (used for the two-year budget plan) with two of the 100 realisations highlights the limitations of traditional estimation models (Figure 7). These models often overstate grade continuity due to smoothing effects, which have significant implications discussed in the results section.

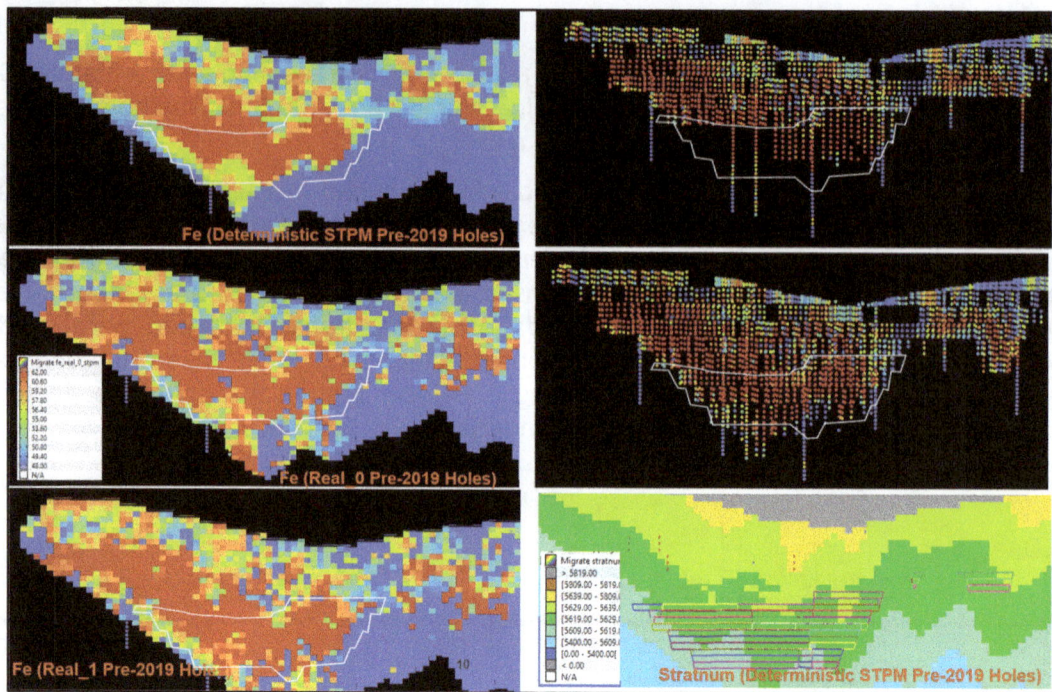

FIG 7 – OB24 comparison of deterministic estimated Fe (top left) with two simulated realisations (middle and bottom left) with data used for modelling (top right) and reconciliation (middle right). Deterministic stratigraphy domains shown in bottom right figure with two-year plan mining sequence solids.

Mine planning

To evaluate the impact of geological uncertainty on production outcomes, we developed a production scheduling framework guided by key objectives and constraints. This framework ensured that the scheduling process adhered to realistic operational conditions while addressing the variability introduced by stochastic grade models.

Key assumptions for scheduling

1. Sequence re-optimisation:

   o The mine plan assumes minimal re-optimisation of the mining sequence during the 24 month study period. This means that the material designated for extraction remains consistent throughout the planning horizon, reducing the complexity of sequence adjustments while isolating the effects of new grade data.

2. Monthly rescheduling:

   o Only the month impacted by updated grade control information is rescheduled. The full mine plan is not re-optimised monthly to avoid compounding the computational effort. However, this simplification introduces potential impacts on long-term outcomes, which may be revisited in future studies with a more dynamic re-optimisation approach.

3. Grade control:

   o Grade control data is updated monthly, revealing the grades for the entire month. This ensures that monthly scheduling reflects the most accurate representation of *in situ* material properties.

4. Block grouping by material type:

   o Blocks (10 m × m 10 × m 4 m) are categorised based on material types, such as high-grades (HG1–HG6), low-grade (LG), beneficiation-grade (BG), and waste. Once grade control data is collected, no additional adjustments are made to block groupings, ensuring consistency in material classification.

Production scheduling process

1. Objective-driven scheduling:

   The scheduling algorithm prioritises the following objectives in order of importance:

   o Maximise crusher feed tonnage: The primary goal is to ensure consistent ore delivery to the crushers to maintain production targets.

   o Achieve target rehandling ratios: A secondary goal is to manage stockpile rehandling operations effectively. The rehandling ratio is constrained to an upper limit while maintaining a minimum threshold of 0.2 to reflect realistic stockpile movement.

   o Stockpile composition: Efforts are made to align the material type and quality in stockpiles with operational requirements, ensuring that crusher feed meets grade and blend targets.

2. Constraints in scheduling:

   The production schedule is constrained by several operational and geological factors:

   o Grade quality: Scheduled material must meet grade constraints, ensuring crusher feed remains within specified quality parameters for iron content and impurities (eg SiO_2, Al_2O_3).

   o Slope stability: Pit designs and mining sequences must adhere to slope constraints to ensure safe operations and long-term pit wall stability.

   o Material movement: Monthly schedules must account for the physical limitations of mining equipment and haulage operations.

3. Monthly schedule execution:

   Each month, updated grade control data informs the allocation of blocks to the crushers, stockpiles, or waste dumps. The scheduling process ensures that the extracted material satisfies the crusher's tonnage and grade requirements. Blocks are dynamically assigned to stockpiles or crushers based on their material type, while waste material is directed to designated dumps.

4. Rehandling operations:

   Rehandling stockpile material is an integral part of the schedule. Material is retrieved from stockpiles to supplement crusher feed when the *in situ* material cannot meet tonnage or grade targets. The schedule balances rehandling operations to minimise excessive movement while maintaining operational flexibility.

This production scheduling approach integrates geological uncertainty into monthly mine planning by leveraging conditional simulation models. By iteratively incorporating new grade control data, the schedule adapts to variations in material availability and quality, ensuring operational targets are met within the defined constraints. The results section further explores the implications of these scheduling decisions on production outcomes.

Results

Using the proposed methodology, the production schedule was updated for each realisation monthly. Tons and grades at the crushers were assessed for each scenario.

The left-hand graphs in Figure 8 illustrate the total tons processed at the crushers. The right-hand graphs display the monthly stockpile balance. The blue line represents the short-term resource model (STPM) used for the 2-Year Budget (2YB) schedule; black dots indicate the fixed tonnage targets committed in the budget; red violin plots show the distribution of tons across 100 simulation scenarios. Top graphs represent the base case, while bottom graphs show the aspirational plan, designed to deliver 15 per cent more ore to the market.

According to the graphs the base plan can handle the variability and uncertainty by utilising stockpiles. However, if the management decides to implement the aspirational plan the operation will be exposed to significant risks. Stockpiles are likely to be depleted by the end of the first year, leaving insufficient material to feed the crushers. This creates a cascading 'domino effect,' impacting the second year's production targets and resulting in substantial operational challenges.

It should be noted that the STPM model (blue line) projects an overly optimistic view, suggesting achievable plans with a stable stockpile balance, even for the aspirational plan. However, the violin plots highlight the risks due to geological uncertainty, which are not captured in the STPM model.

Figure 9 examines the Fe grade and impurities (Al_2O_3, SiO_2, P) at the crusher feed: black lines indicate upper and lower specification limits; the blue line reflects STPM model predictions, while the violin plots represent the uncertainty from simulations.

Even for the base case, simulations reveal that Fe grades consistently fall short of the target due to dilution effects. Similar patterns are observed for impurities, where critical specifications for smelting processes are frequently exceeded, potentially impacting product quality and customer satisfaction. The dilution effect results in dual losses: lower Fe grades and higher impurity levels, both of which are detrimental to product quality and marketability.

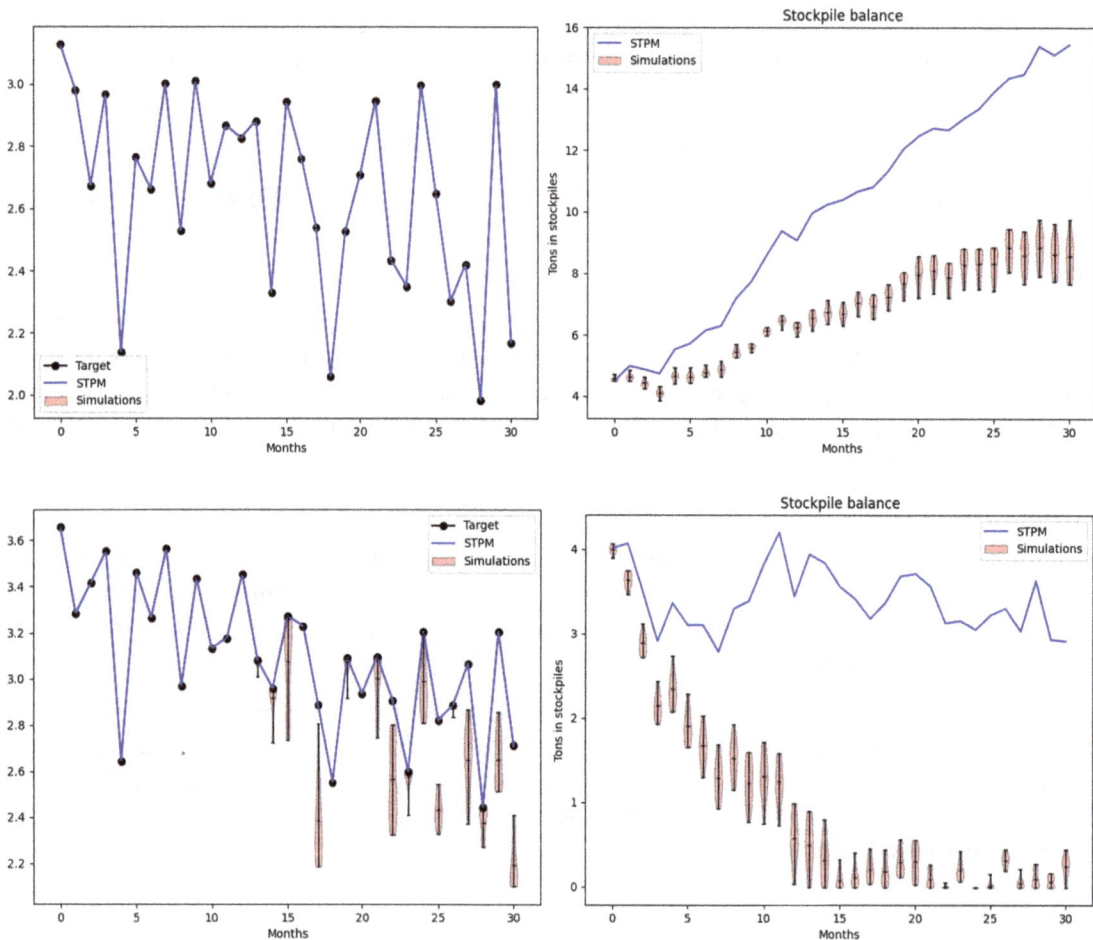

FIG 8 – Left graphs: the total tons processed at the crushers. The right-hand graphs display the monthly stockpile balance. Top graphs are for the base case and bottom graphs for 15 per cent bigger throughput. The blue line represents the STPM used for the 2-Year Budget (2YB) schedule; black dots indicate the fixed tonnage targets committed in the budget; red violin plots show the distribution of tons across 100 simulation scenarios.

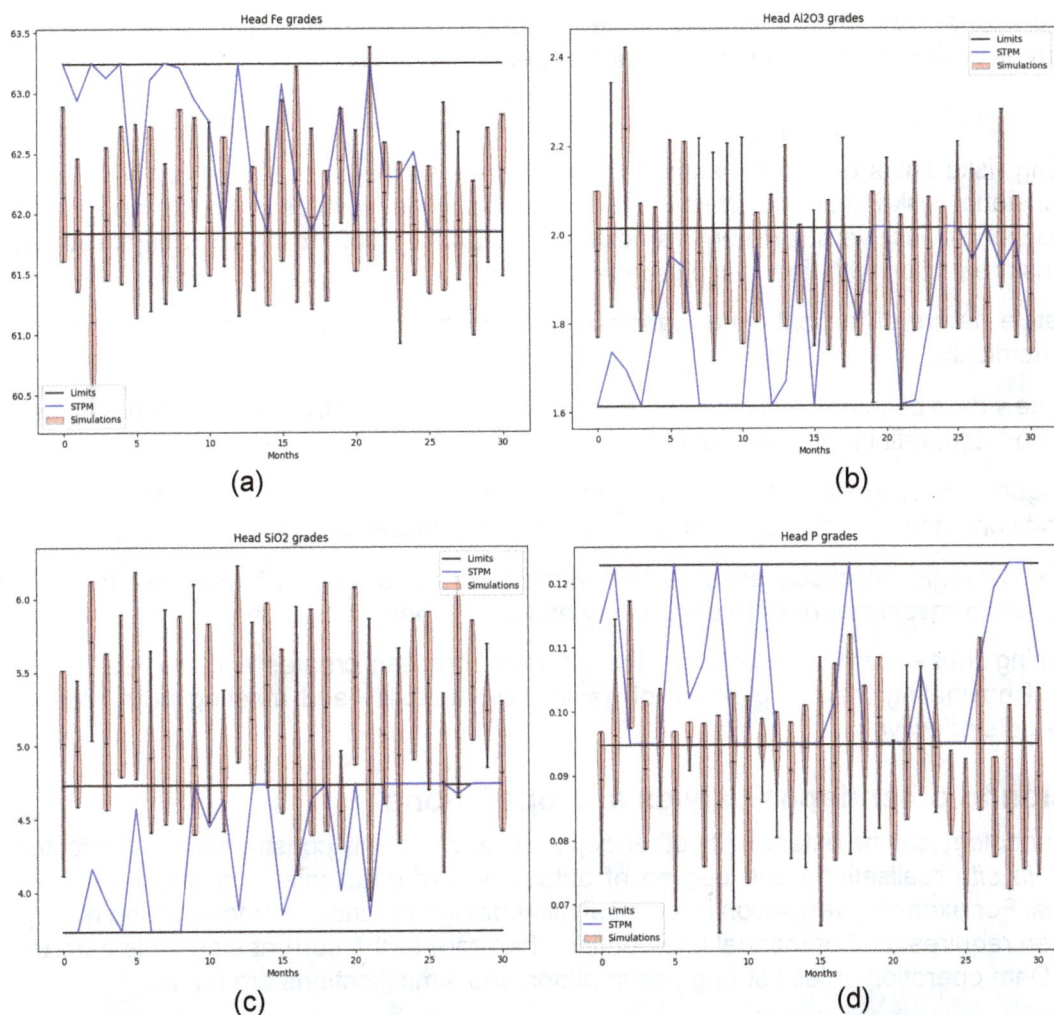

FIG 9 – Grades at crushers feed for base case. black lines indicate upper and lower specification limits; the blue line reflects STPM model predictions, while the violin plots represent the uncertainty from simulations: (a) Fe grades, (b) alumina grades, (c) silica grades, (d) phosphorus grades.

DISCUSSION

Risk mitigation

The current business practice relies heavily on STPM model predictions (blue lines in the figures). This approach does not account for risks arising from geological uncertainty, which are only evident when examining the red violin plots. Traditionally, businesses attempt to account for these risks through historical performance adjustments; for example, as applying dilution factors.

The proposed methodology enables a more robust, quantified approach to risk assessment and mitigation:

1. Stockpiling policies: adjusting stockpile policies, such as increasing buffer stocks or changing stockpile compositions, can help mitigate risks.

2. Sequencing adjustments: re-sequencing mining operations to prioritise high-confidence areas could reduce variability in crusher feed.

3. Rehandling ratios: optimising rehandling ratios can provide flexibility to address material shortages during critical periods.

4. Additional drilling: infill drilling in key areas can reduce geological uncertainty, improving grade estimates and reducing the probability of deviations.

5. Operational investments: increasing fleet capacity or creating additional stockpiles, provided there is sufficient lead time, can enhance operational resilience.

Dynamic decision-making

Quantifying risks shifts decision-making from traditional single-axis metrics (eg NPV or ROI) to a two-dimensional Risk-Return perspective. Mitigation activities, such as those outlined above, can be evaluated using a Risk versus Return framework (Markowitz, 1952) to identify optimal strategies that balance risk reduction with financial performance.

The iterative nature of tactical mine planning, updated quarterly or monthly, allows businesses to adapt dynamically:

- As new data becomes available (eg from grade control or infill drilling), uncertainty models can be updated, refining the view of risks.

- Mitigation measures can be re-assessed and adjusted in real time based on changing conditions and time frames, such as lead times for drilling or fleet deployment.

- This dynamic approach ensures that decisions are always informed by the latest data, improving responsiveness to uncertainty and operational challenges.

Incorporating these methodologies into the planning process creates a more resilient operation capable of navigating the complexities of geological variability and meeting both production and quality targets effectively.

Extension to other deposit styles and operations

This methodology can be extended to other deposits, and it is only constrained by two factors: ability to model *in situ* realisations and degree of automation of data collection and mine optimisation processes. For example, ventilation is a critical mine design process for underground mine planning, and it often requires a lot of manual intervention, therefore at the current stage it is hard to apply to Olympic Dam operation unless strong assumptions and simplifications are made.

CONCLUSIONS

This work presents a novel approach to integrating geological uncertainty into mine planning and scheduling processes by emulating the dynamic decisions made in response to updated grade control data. The results demonstrate that while traditional short-term resource models (STPM) provide a static view of production schedules, incorporating probabilistic scenarios allows for a more comprehensive risk assessment. This enables businesses to move from deterministic decision-making to a risk-informed framework, where both variability and potential outcomes are quantified.

The case study on WAIO's Newman Hub highlights significant findings:

- Stockpile risks: for the aspirational plan aiming to deliver 15 per cent more ore, there is a substantial risk of stockpile depletion within the first year. This would result in operational disruptions and a cascading impact on the second year's production. Conversely, the base plan provides sufficient buffers to mitigate uncertainties, ensuring stable crusher feed.

- Product quality risks: even under the base plan, deviations in Fe grades and impurities such as Al_2O_3 and SiO_2 were observed, indicating potential challenges in meeting product quality specifications. These deviations, driven by geological dilution, have dual impacts—lower Fe grades and compromised product quality, both of which are critical to customer requirements.

- Risk visibility: traditional approaches relying solely on STPM fail to reveal the variability and risks arising from geological uncertainties. The proposed methodology, leveraging simulations, provides actionable insights by translating geological variability into tangible risks such as unmet production targets, unplanned costs, and quality issues.

- Dynamic planning opportunities: the methodology enables adaptive, dynamic decision-making by quantifying risks and updating plans iteratively as new grade control data becomes available. This process not only accounts for uncertainties but also allows for timely interventions, such as additional stockpiling, rehandling, or infill drilling.

Implications for mining operations

This research underscores the importance of integrating probabilistic modelling into mine planning. By transitioning from a deterministic to a probabilistic framework, businesses can:

- Enhance risk mitigation: quantified risks allow for tailored mitigation strategies, including changes in stockpiling policies, sequencing adjustments, or strategic investments in additional equipment and drilling.

- Optimise decision-making: a risk-return optimisation framework enables management to evaluate various mitigation strategies holistically, identifying the most efficient trade-offs between cost and risk reduction.

- Drive operational resilience: dynamic updates to plans ensure that operations remain agile and responsive to changing conditions, reducing the likelihood of disruptions and ensuring consistency in production and product quality.

Future directions

Future work could expand the scope of the analysis to include all deposits, crushers, and ore handling plants contributing to product supply chains. Developing fully integrated probabilistic digital twins encompassing resource, modelling, planning, and operational decision-making processes represents the next frontier for mining operations.

In conclusion, this methodology provides a robust foundation for modern mine planning, transforming uncertainty from a risk into an opportunity for strategic optimisation. By adopting these advanced practices, businesses can achieve greater operational efficiency, product quality assurance, and financial resilience in an increasingly uncertain mining environment.

ACKNOWLEDGEMENTS

The authors would like to acknowledge the support provided by Juan Mai and Will Patton from WAIO Resource Engineering and Geoscience teams, Craig Williams and Ranjan Saha from BMA Geoscience Team, Glen Elliot from Resource Centre of Excellence and the critical support from the leadership team of the Resource Centre of Excellence at BHP: Daniel Kahler, Kery Turnock, Cam McCuaig, and Tim O'Connor. Their collaboration, contributions, and support have been instrumental in the successful completion of this work.

REFERENCES

Barnett, R M, Manchuk, J G and Deutsch, C V, 2014. Projection Pursuit Multivariate Transform, *Math Geosci*, 46:337–359.

Goodfellow, R C and Dimitrakopoulos, R, 2016. Global optimisation of open pit mining complexes with uncertainty, *Applied Soft Computing*, 40:292–304. https://doi.org/10.1016/j.asoc.2015.11.038

Journel, A G and Alabert, F, 1990. New method for reservoir mapping, *Journal of Petroleum Technology*, 42(2):212–218.

Journel, A G, 1974. Geostatistics for Conditional Simulation of Ore Bodies, *Economic Geology*, 69:673–687.

Markowitz, H, 1952. Portfolio selection, *The Journal of Finance*, 7(1):77. https://doi.org/10.2307/2975974

Minniakhmetov, I, 2023. Local Transformations using Kernel Density Estimation for simultaneous detrending and Gaussian anamorphosis' in IAMG 2023, 22st Annual Conference of the International Association for Mathematical Geosciences, Norway.

Ramazan, S and Dimitrakopoulos, R, 2012. Production scheduling with uncertain supply: A new solution to the open pit mining problem, *Optimisation and Engineering*, 14(2):361–380. https://doi.org/10.1007/s11081-012-9186-2

Williams, C and Minniakhmetov, I, 2025. In-situ uncertainty quantification and implications for risk mitigation – a case study from WAIO Newman Hub, in *Proceedings of the Mineral Resource Estimation Conference 2025*, pp 333–356 (The Australasian Institute of Mining and Metallurgy: Melbourne).

In situ uncertainty quantification and implications for risk mitigation – a case study from WAIO Newman Hub

C Williams[1] and I Minniakhmetov[2]

1. Principal Resource Modelling, BHP Mitsubishi Alliance, Brisbane Qld 4000.
   Email: craig.williams1@bhp.com
2. Principal Global Modelling and Data, Resource Centre of Excellence, BHP, Perth WA 6000.
   Email: ilnur.minniakhmetov@bhp.com

ABSTRACT

Conditional simulation is widely accepted in the mining industry as the best way to quantify uncertainty in tonnage and grade. Uncertainty quantification needs to be evaluated against the potential negative consequences of uncertainty, so that effective risk mitigation strategies can be identified. However, conditional simulation models need to be carefully constructed with robust validations, otherwise risk mitigation strategies based on sub-optimal simulations, may be ineffective.

A conditional simulation case study is presented, with simulation of both categorical and continuous variables, for multiple orebodies from BHP WAIO's Newman Hub in Western Australia, using BHP's SBRE conditional simulation platform. The orebodies considered display a range of geological complexity, both in terms of the structural setting and in the presence of deleterious elements and internal waste.

SBRE conditional simulation workflows were initially constructed using exploration drilling data only, followed by validation of the simulations using blastholes. After successful validation, workflows were used to run the final simulation models for each orebody, using both exploration and blastholes, drilled prior to December 2019.

An existing deterministic estimation model, constructed using data acquired prior to December 2019, together with the simulation models constructed for this case study, were then cross validated against 'reality', based on closely spaced blastholes drilled over a 31-month reconciliation period post-November 2019. Cross validation results show the value of the simulation models over the deterministic ones, in terms of the true range of potential grade values being represented in the simulation models. The use of accuracy plots to validate simulation models is discussed.

Final simulation models were used to model downstream blending options from multiple orebodies with different levels of geological complexity and uncertainty. Simulation models highlight the risk to planned crusher tonnes, with an increase in stockpile tonnes for higher production rates required to mitigate this risk.

INTRODUCTION

Conditional simulation is widely recognised as a best practice for quantifying geological uncertainty (Journel, 1974). In this study, we present a case study aimed at assessing *in situ* uncertainty across both categorical domain variables and continuous grade variables (Fe, SiO_2, Al_2O_3, LOI, and P) through the development of conditional simulation models for several iron ore deposits within BHP's Western Australia Iron Ore (WAIO) Newman Hub. Three orebodies (OB24, OB25_P1, and OB32) were selected to represent varying degrees of geological complexity, allowing for an evaluation of uncertainty impacts on a typical mine plan that integrates ore from multiple sources to supply stockpiles and crushers. The simulations were conducted using BHP's Simulation-Based Resource Evaluation platform (SBRE), which serves as the standard platform for designing, testing, and executing conditional simulation workflows across BHP assets. The SBRE framework adopts a modular workflow approach, enabling iterative model development and incorporating robust validation processes to optimise workflows for routine deployment of stochastic models in uncertainty quantification and risk mitigation studies (Figure 1).

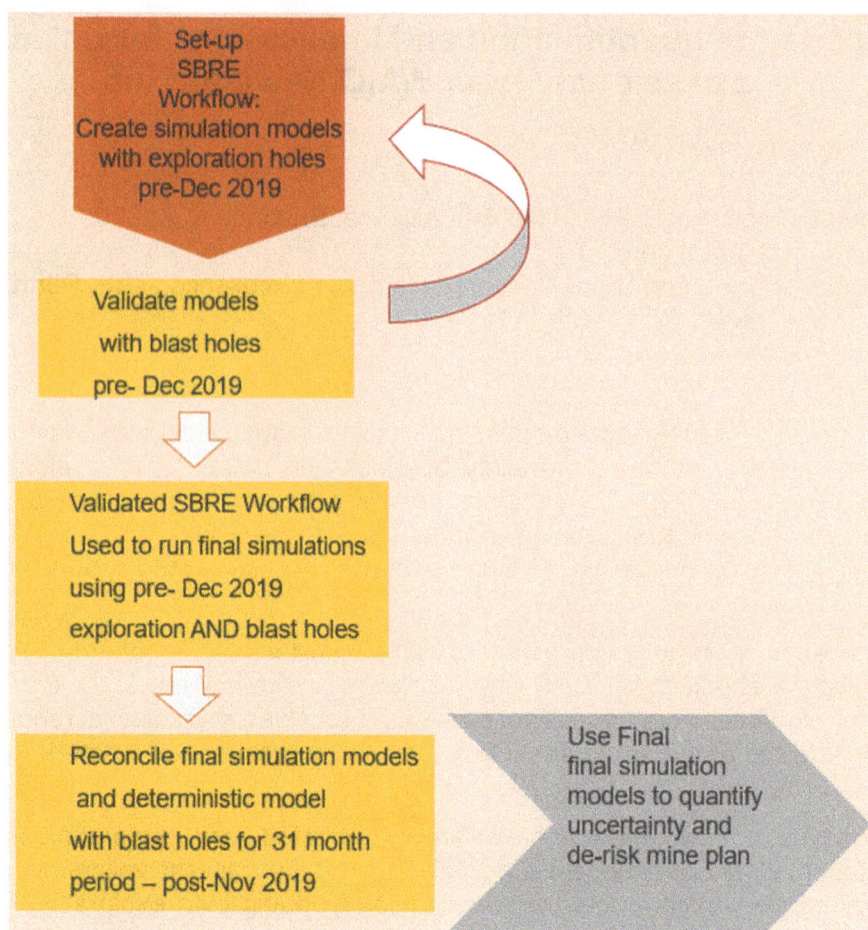

FIG 1 – SBRE workflow development process flow diagram.

As highlighted by Journel (1974), conditional simulations, unlike kriged estimates, are not subject to smoothing and are therefore more suitable for quantifying uncertainty. However, simulated models, as with all modelling techniques, are prone to errors and strong assumptions, ie the global distribution of grades in realisations should reproduce the global distribution of grades in the samples. Therefore, declustering weights can significantly affect global statistics which are unknown. For example, in this study we observed that blasthole information significantly affected mean and other statistics of the global Fe distribution, which were not known at the stage of modelling. Next, the choice of trend parameters affects uncertainty represented by the realisations. For example, very detailed trend (small ranges of smoothing window) will create realisations which are slightly fluctuating around trend and misrepresent uncertainty. Therefore, validation of simulation models is critical to ensuring robust models are used for subsequent uncertainty analysis and risk mitigation strategies.

To evaluate the accuracy and precision of conditional simulation models, Deutsch (1997) introduced the use of accuracy plots. These plots compare modelled values to a set of true values, typically obtained through cross-validation techniques such as leave-one-out cross-validation. An accuracy plot displays the proportion of true values (y-axis) falling within a given probability interval against the modelled values (x-axis), centred on the median value. Figure 2 illustrates several examples of accuracy plots from Deutsch (1997). In the first accuracy plot (Figure 2, top row left), perfect agreement between the model and reality is observed, with the plot closely following the 45° line and complete overlap between the model and true value histograms. This scenario indicates both high accuracy and high precision. In the second plot, where the modelled distribution spans from -0.5 to 1.5, the accuracy plot lies above the 45° line, indicating high accuracy but low precision, as the modelled values exhibit a broader spread compared to true values. Conversely, the third plot shows an accuracy plot below the 45° line, indicating low accuracy, as the proportion of true values within the probability intervals is consistently lower than the modelled values. Although the modelled values in this case are more tightly clustered (indicating higher precision), this scenario represents an oversmoothed model that fails to capture the true distribution of uncertainty.

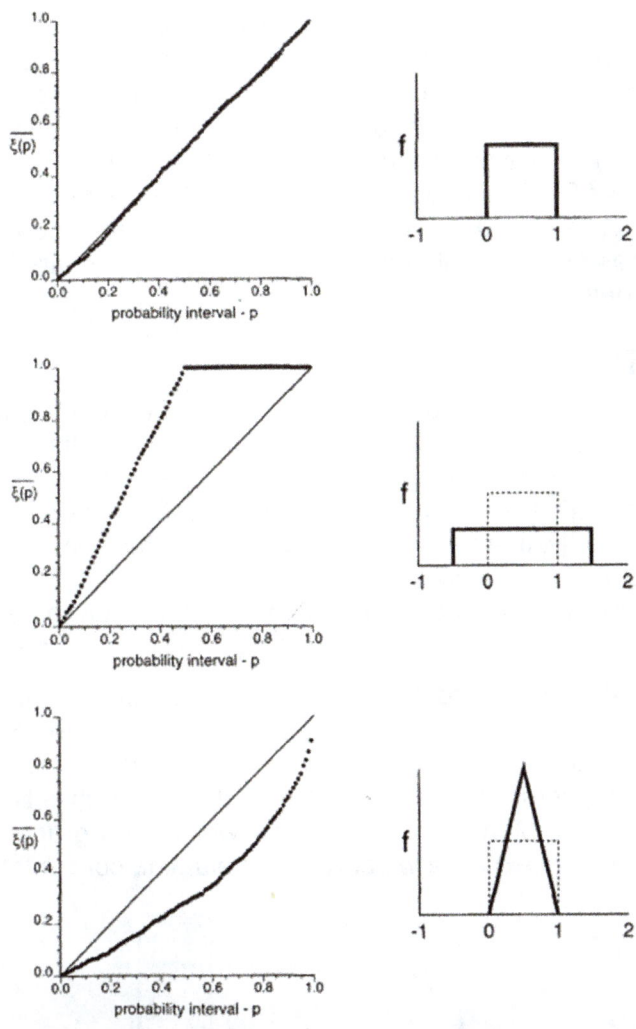

FIG 2 – Accuracy plot examples from Deutsch (1997).

Accuracy plots have been effectively used to validate and reconcile conditional simulation models in this study. The paper provides firstly a brief description of the geological setting of the Newman Hub, focusing on the contrasting geological complexity seen in the case of the two largest orebodies, OB24 and OB32. The process of domaining and descriptive statistics of the continuous variables for each stratigraphic domain is also described.

The development of the SBRE conditional simulation workflow is described, focusing on the iterative approach which uses validation against blasthole data to optimise the workflow. The summarised SBRE conditional simulation workflow is shown below in Figure 3.

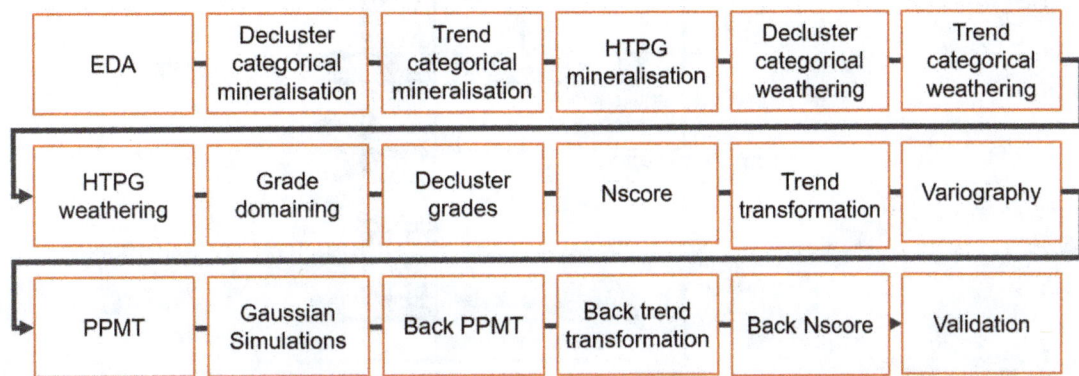

FIG 3 – General SBRE conditional simulation workflow.

The initial EDA step includes data import. During the initial workflow optimisation step, pre-Dec 2019 exploration holes were imported for the simulation and blastholes drilled over the same period were used for validation. Data used for modelling and validation was cut-off at the end of November 2019 so that following successful validation of workflows, final simulations generated using all pre-Dec 2019 data, inclusive of blasthole data, could then be used in a retrospective study looking at the impact of uncertainty on short-term mine planning for a 31-month period post-Nov 2019. This involved firstly reconciliation of the simulation models and existing deterministic estimation model, built using the same pre-Dec 2019 data, against blasthole data for a 31-month period post-Nov 2019. Secondly, the 100 realisations from each of the three final simulation models were put through both a base case and an increased mining rate mine plan and results compared to that obtained from the existing deterministic models.

GEOLOGICAL SETTING

The mineralised iron formation units modelled for OB24 and OB25_P1 form part of the Proterozoic aged Brockman Iron Formation. The immediate footwall shales of the Mount McRae Shale unit, which contains very sparse iron mineralisation, have also been modelled. Stratigraphically above the Mount McRae shales there are two mineralised zones within the Brockman Iron Formation, as shown below in Figure 4. Firstly, there is the Dales Gorge Member, which contains mineralised sub-units D1 to D4. Overlying this are the lower two mineralised units of the Joffre Member (J1 and J2). In between these two main mineralised zones are the mainly non-mineralised shales of the Mt Whaleback Shale Member.

The mineralised iron formation units modelled in OB32 form part of the Marra Mamba Iron Formation (Figure 4). Again, there are two main mineralised zones, a lower grade zone in the lower portion of the Marra Mamba Iron Formation, comprising the MacLeod Member and the Nammuldi Member, immediately overlain by a higher-grade zone comprising three sub-units N1 to N3, within the Mount Newman Member. Overall, the Marra Mamba ore is lower in iron grade than the Brockman Iron Formation mineralised units and exhibit a wider range in alumina concentration.

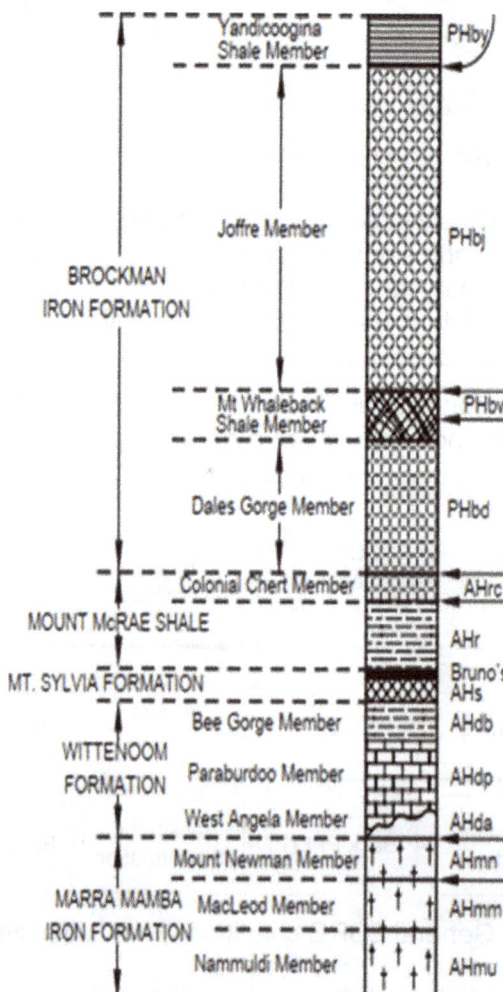

FIG 4 – Hamersley area stratigraphic column (source BHP).

The location of the three or bodies modelled within the Newman area is shown in Figure 5. OB32 lies just to the north of the major east–west striking regional Homestead Fault, which appears to be associated with the intense overturned recumbent folding, with an east–west striking fold axis, seen in the case of OB32 (Figure 6). In contrast, OB24 and OB25_P1 are much less intensely structurally deformed, with most of the mineralisation in the case of OB24 seen in a broad east–west striking synclinal trough, moving into a more tightly folded anticline to the north (Figure 6).

Two oxidation zones are recognised across all three orebodies, weathered or hard cap ore at the surface with transitional ore below. The hard cap ore is a surficial ferricrete zone formed by weathering, ranging between 10 m and 30 m thick, which is draped over the underlying bedded transitional ore. The underlying supergene enriched transitional ores form the high-grade portion of the orebody. Limited intersections of fresh primary banded iron formation may be found in deeper holes but are rare and do not form part of the mineralised portion of the deposits.

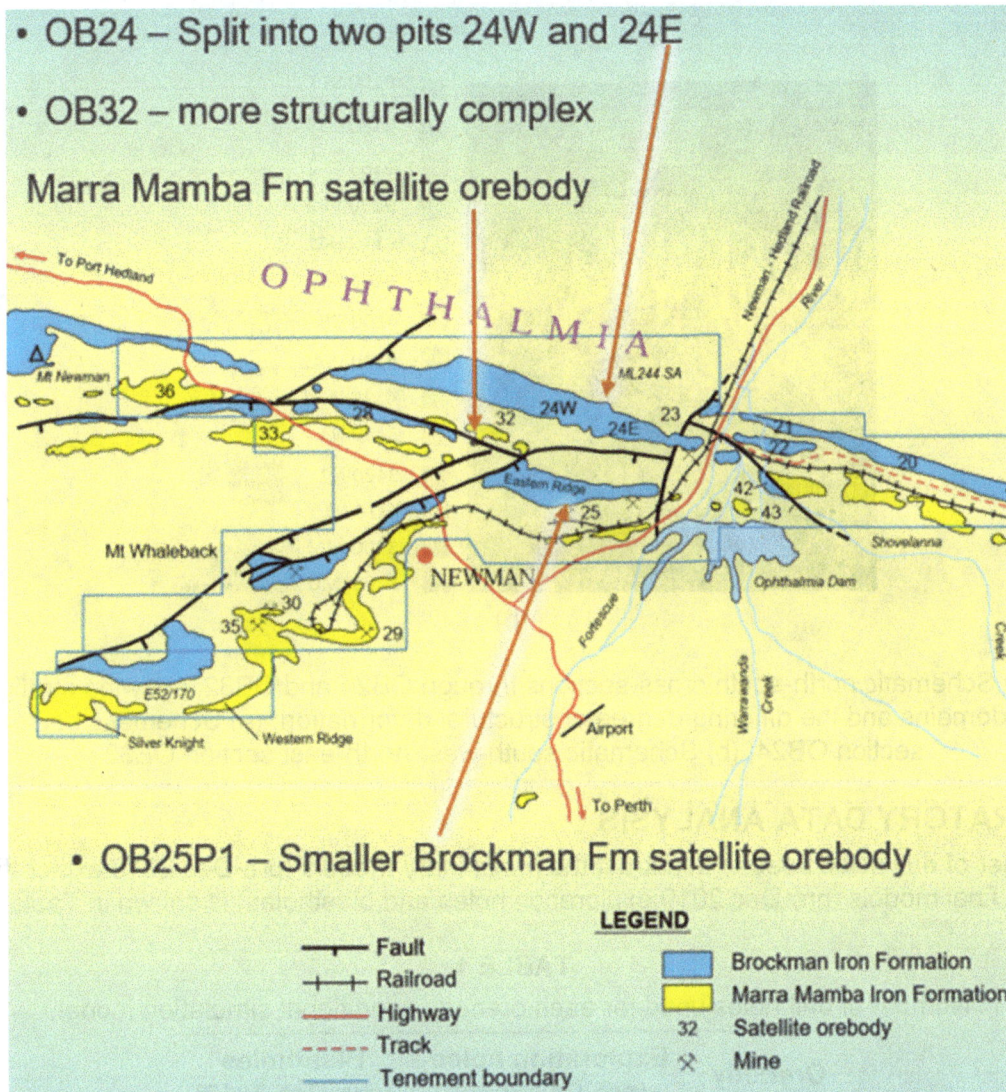

FIG 5 – Simplified geological map of the Newman area (source BHP).

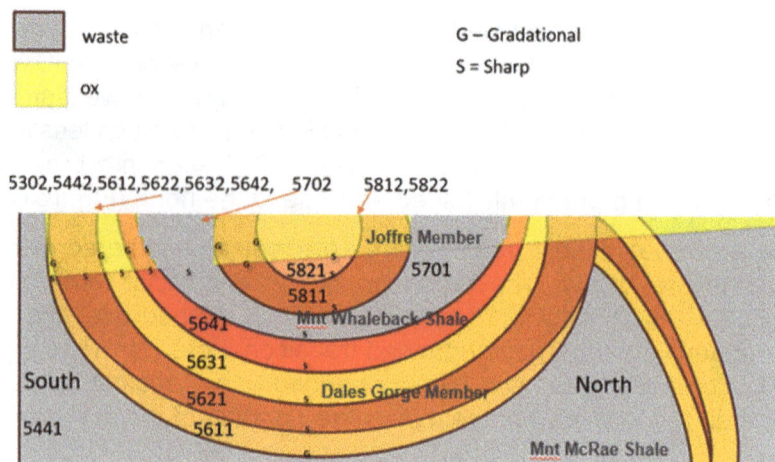

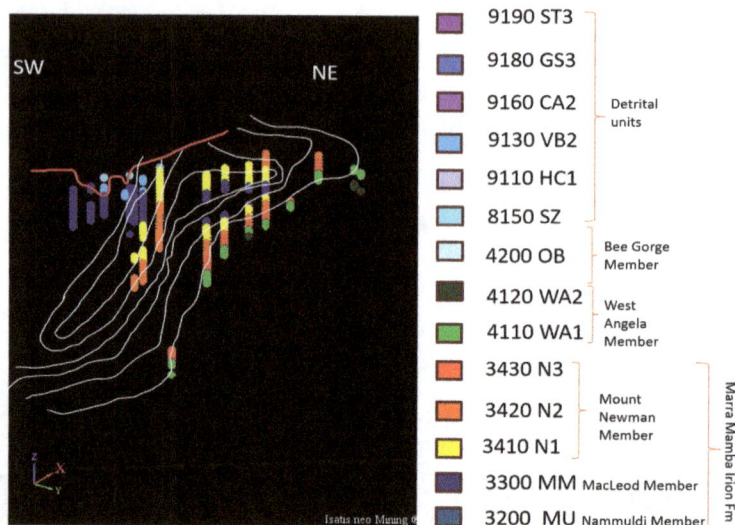

FIG 6 – Schematic north-south cross-sections through OB24 and OB32, showing stratigraphic grade domains and the differing degree of structural deformation: (a) Schematic south–north section OB24; (b) Schematic south-west-north-east section OB32.

EXPLORATORY DATA ANALYSIS

The number of drill holes used to construct the preliminary models (pre-Dec 2019 exploration holes only), and final models (pre-Dec 2019 exploration holes and blastholes) is shown in Table 1.

TABLE 1

Number of drill holes used for each orebody conditional simulation model.

Orebody	Exploration holes (pre-Dec 2019)	Blastholes (pre-Dec 2019)
OB24	963	74 390
OB32	617	13 565
OB25_P1	130	15 358

The stratigraphic categorical variable 'stratn' was combined with the weathering categorical variable to derive a combined domain categorical variable 'Comp_stratn' where the first three digits denote the stratigraphic unit and the last digit denotes the oxidation the state of weathered/hard cap (2) or transitional (1). For geological context, schematic south to north cross-sections through the two main orebodies in this study (OB24 and OB32) showing the stratigraphic position of the mineralised units discussed above and the differing degree of folding, are provided in Figure 6. For OB24, the position of the hard cap units relative to the parent transitional units is shown. For OB32, the scale is too small to show the weathered layer, but it follows the unconformity (red line) between the older Proterozoic sediments and the Cenozoic detrital units above.

Contact analysis was performed to identify which stratigraphic units should be grouped together for modelling purposes, based on the presence of hard or soft iron grade contacts between the Comp_stratn units. Results for OB24 support the modelling of the Dales Gorge Member transitional zone as four separate units (5611, 5621, 5631 and 5641 also referred to as Dales Gorge D1 to D4) and the Joffre Member transitional zone as two separate units (5811 and 5821, also referred to as Joffre J1 and J2) separated by the Mount Whaleback shales (5701). In contrast however there are gradational contacts between weathered Dales Gorge and Joffre member oxidised units, supporting the grouping of the weathered Dales Gorge and Joffre member units as two combined units (Ox1 and Ox3); separated by weathered Mount Whaleback shales (Ox2=5702) as shown in Figure 6.

In the case of OB32, there is a gradational contact between the Nammuldi Member and the Macleod Member transitional zone units, so this was modelled as a single unit termed the Lower Marra Mamba. Similarly, there are gradational contacts between the three N1-N3 sub-units of the Mount Newman Member in the transitional zone and this was modelled as a single unit termed the Upper Marra Mamba. On the basis of a gradational iron grade contact between the two West Angela transitional units 4111 and 4121, these were modelled as a single unit. The uppermost Proterozoic unit within the transitional zone in the drilling data, the Bee Gorge Member, was also modelled separately. All oxidised zone units were modelled as a single oxidised horizon for OB32, due to gradational contacts between all of them.

In addition to the stratigraphic and oxidation state domain coding discussed above, an additional categorical variable 'mintype' was created based on an iron grade cut-off of 48 per cent. Samples with greater than 48 per cent Fe were classified as ore and those with 48 per cent or less Fe were classified as waste.

Waste samples were filtered out of the database prior to calculating the global descriptive statistics for ore samples of the continuous variables, grouped by the Comp_stratn domain variable. Only Fe and Al_2O_3 are shown in Tables 2 and 3, for OB24 and OB32 respectively, as these two continuous variables comprise the main grade variable (Fe) and a key deleterious element variable (Al_2O_3).

TABLE 2

OB24 Selected continuous variable statistics for samples classed as ore (>48 per cent Fe).

Comp_stratn Domain	Number Fe samples	mean	min	max	Number Al_2O_3 samples	mean	min	max
5611	14428	57.64	48.01	68.58	14428	3.58	0.28	10.89
5612	1035	53.91	48.03	65.91	1035	3.95	0.36	9.03
5621	44848	62.81	48.01	68.93	44848	1.57	0.10	10.14
5622	9416	58.76	48.01	67.80	9416	2.46	0.25	9.06
5631	31604	58.77	48.01	68.62	31604	3.19	0.11	11.21
5632	22266	56.26	48.01	67.68	22266	3.41	0.24	10.17
5641	21363	61.29	48.01	69.10	21363	1.96	0.10	10.98
5642	20511	57.27	48.01	68.43	20511	3.20	0.21	9.99
5701	745	56.91	48.01	67.93	745	4.02	0.28	13.73
5702	1299	54.31	48.01	67.49	1299	4.14	0.37	10.61
5811	1996	57.99	48.04	67.65	1996	3.64	0.31	11.54
5812	4292	56.65	48.01	66.84	4292	3.58	0.46	9.00
5821	418	60.40	49.66	66.42	418	2.18	0.47	6.48
5822	1470	58.57	48.05	65.66	1470	2.55	0.42	7.72

TABLE 3

OB32 Selected continuous variable statistics for samples classed as ore (>48 per cent Fe).

Comp_stratn Domain	Number Fe samples	mean	min	max	Number Al$_2$O$_3$ samples	mean	min	max
3201	161	57.45	48.44	62.98	161	3.22	0.73	9.09
3202	110	55.65	48.85	62.67	110	3.11	1.12	8.34
3301	2659	57.81	48.05	66.32	2659	2.80	0.33	9.86
3302	2426	55.47	48.07	65.85	2426	3.10	0.81	9.08
3411	5096	60.88	48.15	68.25	5096	1.63	0.20	9.58
3412	3985	55.96	48.01	65.40	3985	2.65	0.45	15.05
3421	3680	60.15	48.09	66.69	3680	1.87	0.17	9.25
3422	4015	55.75	48.02	65.04	4015	2.64	0.67	11.80
3431	2839	60.37	48.02	66.81	2839	1.57	0.16	12.87
3432	3306	56.38	48.01	65.30	3306	2.85	0.53	12.46
4111	1488	56.85	48.03	66.66	1488	2.77	0.28	11.65
4112	613	56.12	48.08	65.07	613	3.58	0.92	9.00
4121	271	54.52	48.04	66.19	271	3.93	0.66	8.64
4122	66	54.84	48.43	62.24	66	4.36	1.30	11.20
4201	103	55.16	48.02	63.51	103	4.58	1.11	10.90
4202	110	54.48	48.05	63.09	110	5.75	1.99	14.18
8150	181	54.38	48.08	63.16	181	4.31	1.01	10.13
9110	56	59.12	48.44	65.97	56	3.90	1.17	13.32
9130	416	53.66	48.03	64.52	416	4.55	0.86	10.90
9160	5	56.38	52.55	60.05	5	4.29	3.59	4.93
9180	184	54.63	48.04	64.28	184	6.37	1.59	14.80
9190	440	54.41	48.05	64.40	440	5.42	1.25	12.33

Figure 7 shows boxplots of Fe and Al$_2$O$_3$ grouped by Comp_stratn, for OB24 and OB32. Transitional zone domains, generally have higher Fe and lower Al$_2$O$_3$ than corresponding weathered zone domains. Within the transitional zone domains, for OB24, the lowermost Dales Gorge D1 unit (Comp_stratn = 5611) and the third D3 unit (Comp_stratn = 5631) have lower Fe and higher Al$_2$O$_3$ than the adjacent D2 and D4 units (Comp_stratn = 5621 and 5641). In terms of the two main mineralised zones for OB32, the Lower Marra Mamba (Comp_stratn = 3201 and 3301) has lower Fe and higher Al$_2$O$_3$ than the Upper Marra Mamba (Comp_stratn = 3411, 3421 and 3431).

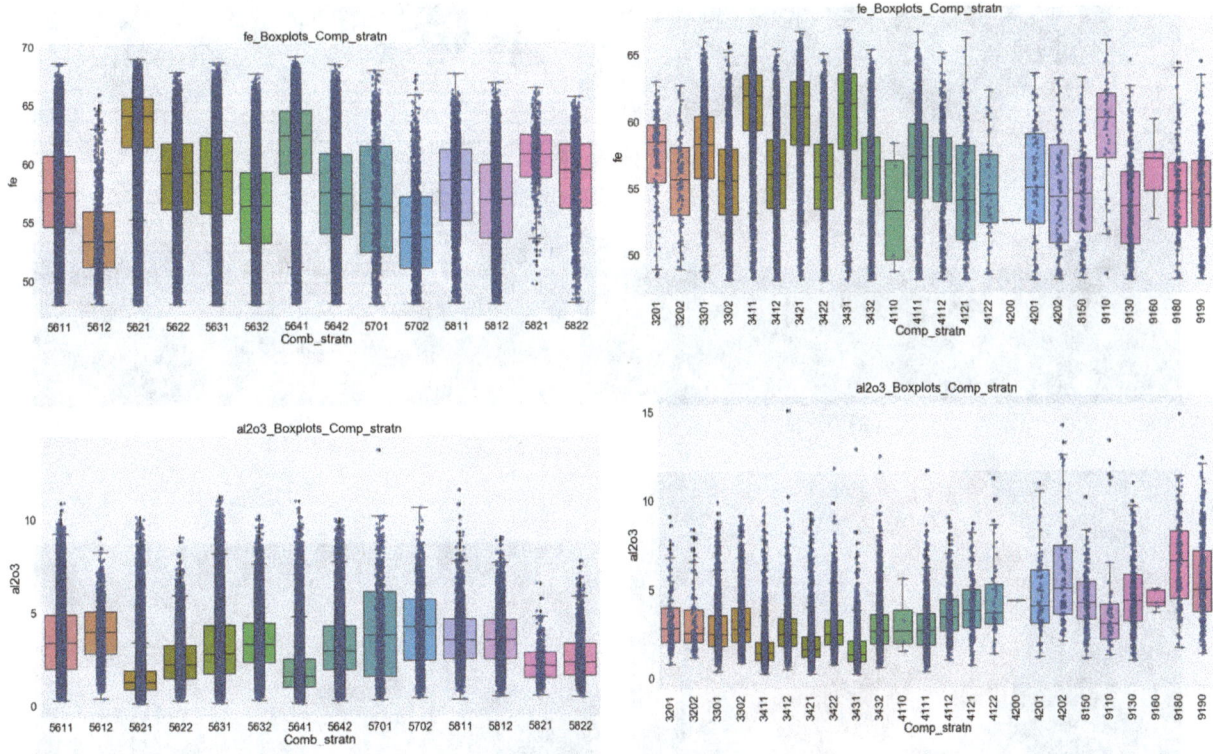

FIG 7 – Box and Strip plots for Fe and Al$_2$O$_3$ distributions per domain for OB24 (left) and OB32 (right).

CONDITIONAL SIMULATION WORKFLOW

Categorical conditional simulation

Following data import, truncated Gaussian simulation (TG) was used to simulate the mintype domain variable, where mintype = 1 is mineralised (>48 per cent Fe) and mintype = 0 is waste. The advantage of being able to run conditional simulation workflows iteratively followed by validation of each run, is demonstrated in Figure 8 which shows that initial runs were unsuccessful in reproducing the variability of waste samples seen in the pre-Dec 2019 blasthole data used for validation. Review of the data showed that a change in the continuity of waste samples occurs in the north, controlled by the tightly folded anticline. Following domaining of the mintype categorical variable into north and south domains, much better validation of the simulated model with pre-Dec 2019 blasthole data was achieved.

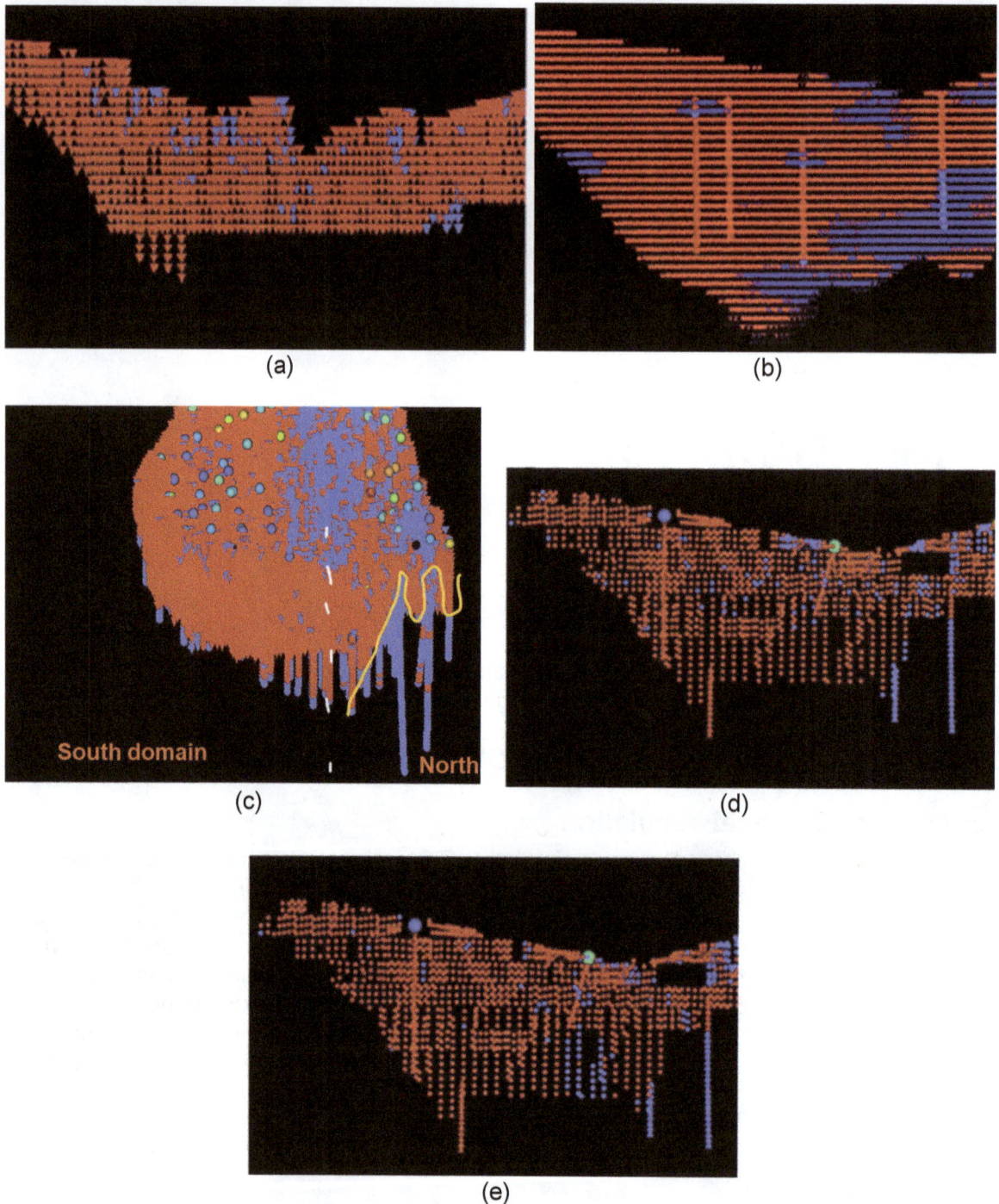

FIG 8 – OB24 Mintype domain simulation: (a) Section showing pre-Dec 2019 blastholes, colour-coded by mintype waste = blue, ore = red. (b) Same section, initial categorical simulation run, one realisation, simulated onto grid nodes. (c) Oblique view looking west of pre-Dec 2019 exploration and blastholes colour-coded by mintype, note – tightly folded anticline in the north causes change in ore/waste continuity. (d) Different section showing exploration and blastholes colour-coded by mintype; exploration holes have collar colour-coded by continuity factor. (e) Same section as (d), final mintype categorical simulation colour-coded by mintype, single realisation.

Following waste and ore domain simulation, the weathered/hard cap and transitional domains were simulated, also using TG. TG is known to be less effective simulating tabular domains, but from Figure 9 it can be seen that a typical realisation from OB24 shows good agreement with the exploration holes used for the simulation and the blastholes (closer spaced, shallower holes) used to validate.

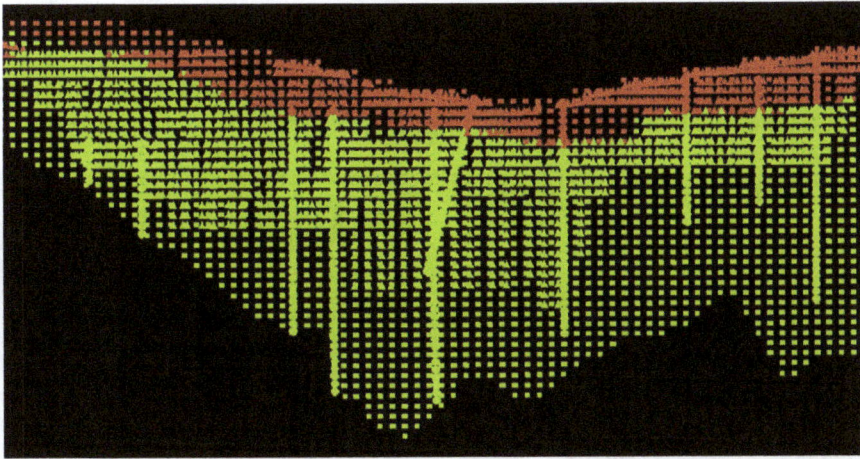

FIG 9 – OB24 ox zone simulation, showing exploration and blastholes versus one realisation.

In the case of OB32 however, the simulated hard cap zone (pink in Figure 10) is more variable with small, isolated patches enclosed by transitional zone, which are geologically not possible. This is likely due to the much more variable nature of the unconformable surface, which is cut by detrital channels, under which the weathered hard cap is found. Although the actual location of the isolated islands of hard cap are not geologically possible, they are an expression of the variability of the surface and allow for the uncertainty of likely yet undiscovered detrital channels to be accounted for.

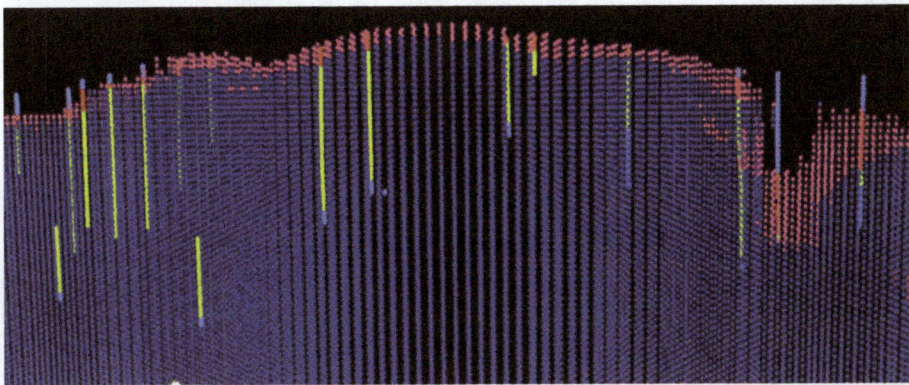

FIG 10 – OB32 ox zone simulation, showing exploration holes versus one realisation.

Continuous variable conditional simulation

The sample data and grid nodes are domained into grade domains identified based on contact analysis groupings discussed above and based on the mintype and weathering categorical variables. In the case of the grid nodes, the categorical simulation realisations following waste and weathering categorical simulation were used to create a set of realisations of domained grid nodes for continuous variable simulation. Due to time constraints and the likely significant increase in complexity, stratigraphic domains were not simulated using categorical simulation. The deterministic stratigraphic boundaries contained in the estimation model were used to domain the grid nodes into stratigraphic categories using the 'stratn' variable.

Domained sample data was then transformed by declustering, normal scores and detrending operations (to account for non-stationarity), following which variography was conducted on the transformed data. Following variography, the data was transformed into completely uncorrelated continuous variables using the projection pursuit multivariate transform (PPMT) approach (Barnett, Manchuk and Deutsch, 2014). PPMT allows for complex multivariate relationships in the data to be preserved through independent simulation of the completely decorrelated transformed variables. As shown by Barnett, Manchuk and Deutsch (2014), PPMT does not change the spatial structure of the data, allowing PPMT to be done post-variography. Continuous variable simulation was then done using turning bands (Journel, 1974), followed by back transformations in reverse order, starting with back-PPMT.

Simulation realisations for Fe and Al_2O_3 for the final models built using both exploration and blasthole data pre-Dec 2019, are shown in Figures 11 and 12 for OB24 and in Figures 13 and 14 for OB32. All the above figures also show the deterministic estimation model and the data used for simulation and reconciliation, as well as the deterministic stratigraphic units, for comparison.

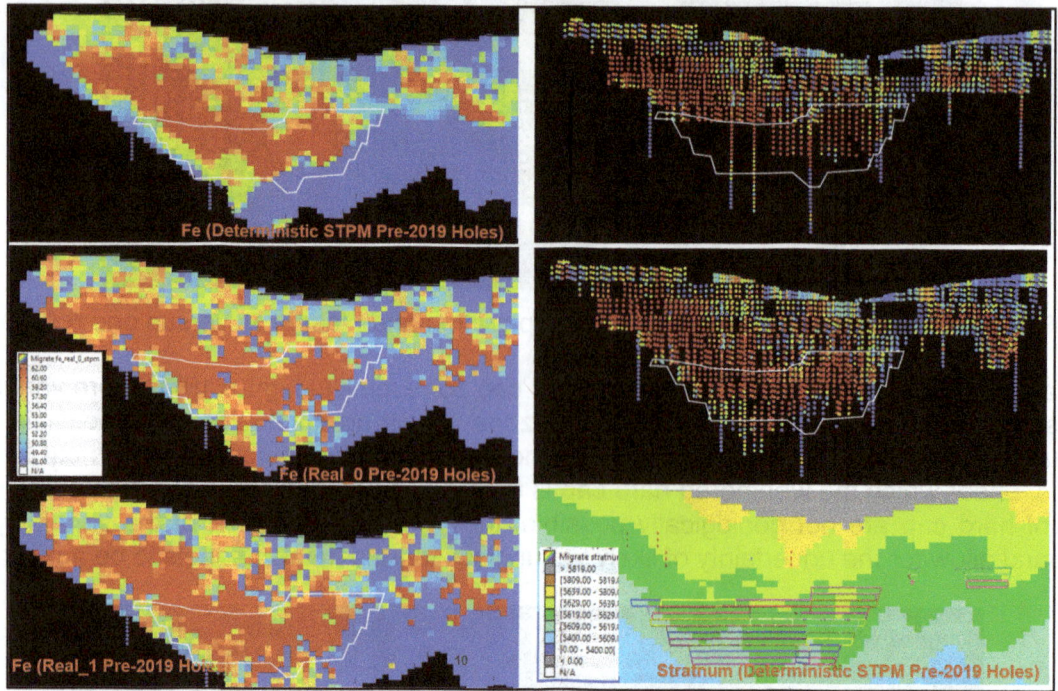

FIG 11 – OB24 comparison of deterministic estimated Fe (top left) with two simulated realisations (middle and bottom left) with data used for modelling (top right) and reconciliation (middle right). Deterministic stratigraphic domains shown in bottom right figure with two-year planned mining sequence solids.

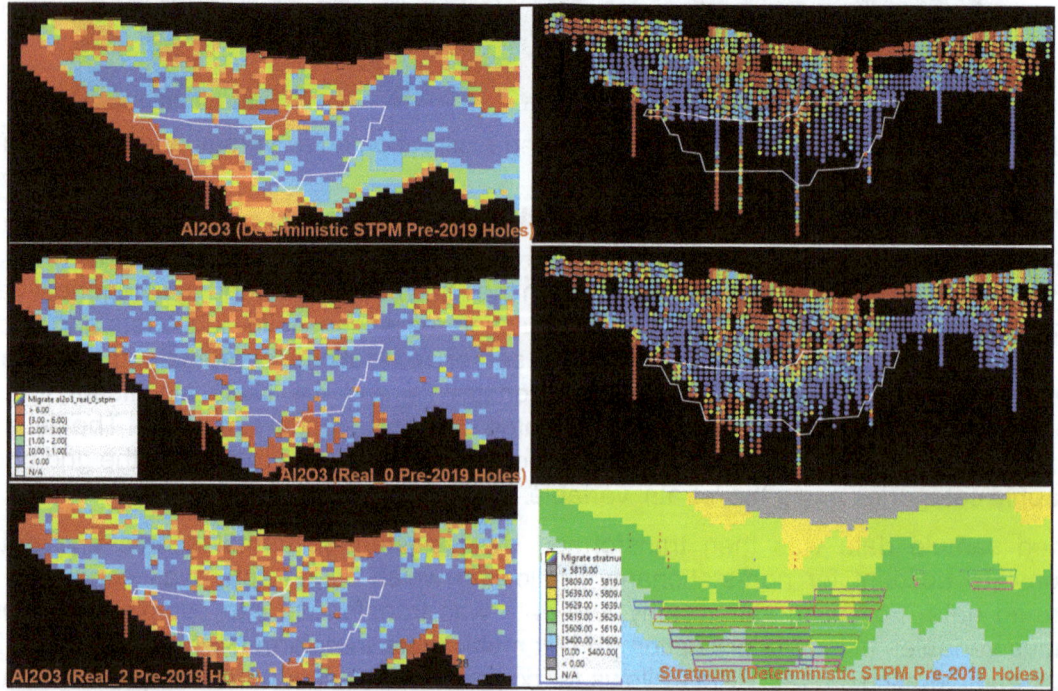

FIG 12 – OB24 comparison of deterministic estimated Al_2O_3 (top left) with two simulated realisations (middle and bottom left) with data used for modelling (top right) and reconciliation (middle right). Deterministic stratigraphic domains shown in bottom right figure with two-year planned mining sequence solids.

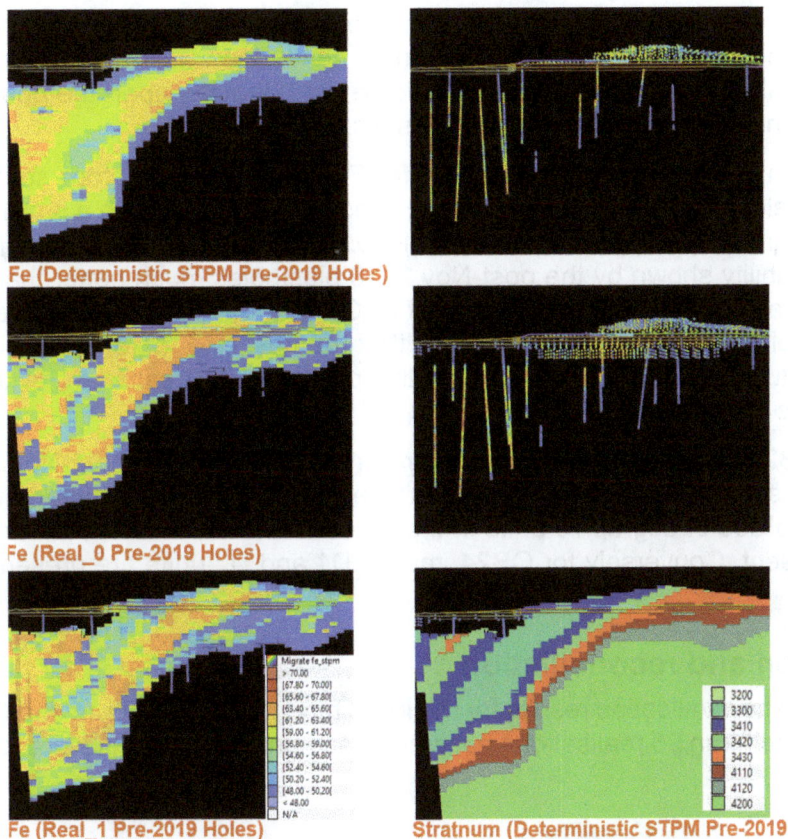

FIG 13 – OB32 comparison of deterministic estimated Fe (top left) with two simulated realisations (middle and bottom left) with data used for modelling (top right) and reconciliation (middle right). Deterministic stratigraphic domains shown (bottom right).

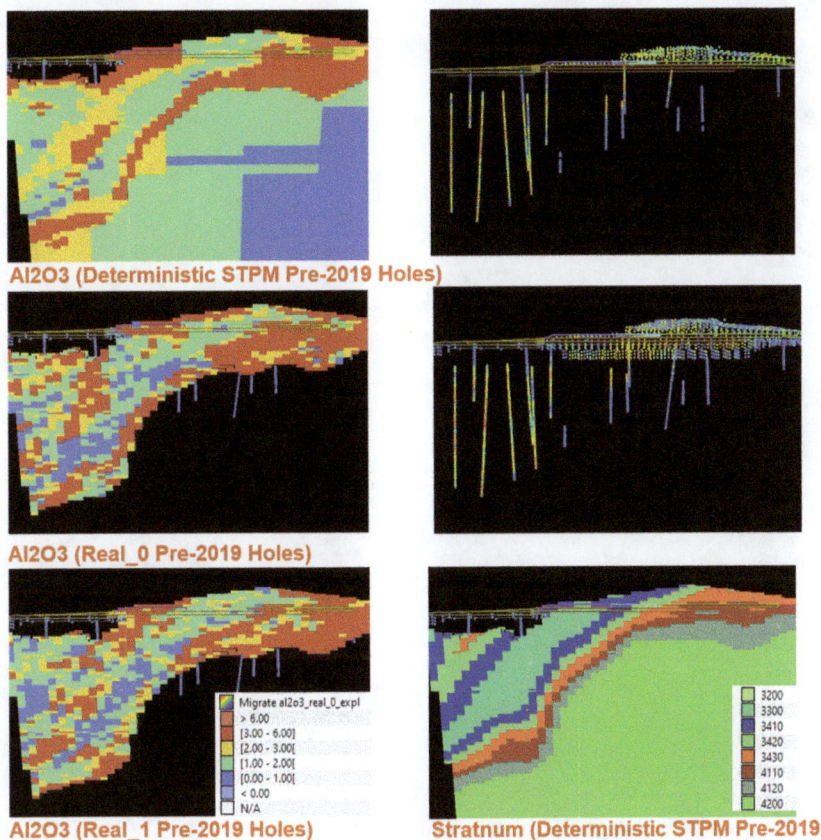

FIG 14 – OB32 comparison of deterministic estimated Al₂O₃ (top left) with two simulated realisations (middle and bottom left) with data used for modelling (top right) and reconciliation (middle right). Deterministic stratigraphic domains shown (bottom right).

For OB24, the area of the model reconciled using post-Nov 2019 blastholes is shown by the white pit-shell outline. The data used for modelling (top right) does not adequately sample this void compared to the data used for reconciliation (middle right) leaving open the possibility for the new information to potentially impact reconciliation results.

In the area through which the section was cut for Figures 11 and 12, four stratigraphic units are present with the Dales Gorge D1 at the base, followed by D2 and D3 with a small amount of D4 just visible in the core of the syncline, near to the surface. In the case of Fe, the simulations capture the real Fe grade variability shown by the post-Nov 2019 blastholes much better than the deterministic estimation model, as is expected. In the case of Al_2O_3, the realisations are more variable than the deterministic model as expected, but it appears that the post-Nov 2019 blastholes have additional variability not captured by the simulations, as the pre-Dec 2019 sampling of this area does not adequately capture this variability.

In the case of OB32 in Figures 13 and 14, the area reconciled against post-Nov 2019 blastholes is a lot more evenly sampled with the pre-Dec 2019 exploration and blasthole data. There are also more of the mineralised stratigraphic units in the reconciled void, with the Lower and Upper Marra Mamba being present. Conversely for OB24, mainly D1 and D2 units are present in the reconciled void.

Model validation and reconciliation

Figures 15 and 16 show validation results for the initial models built for OB24 and OB32 respectively. The initial models both ran 25 realisations each and used pre-Dec 2019 exploration data to build the models.

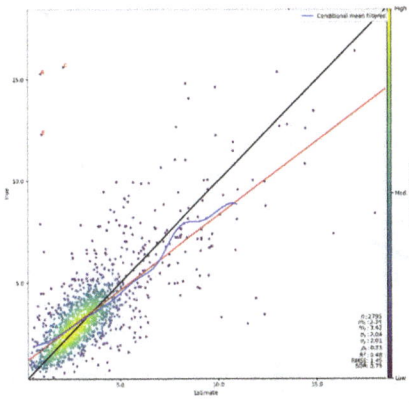
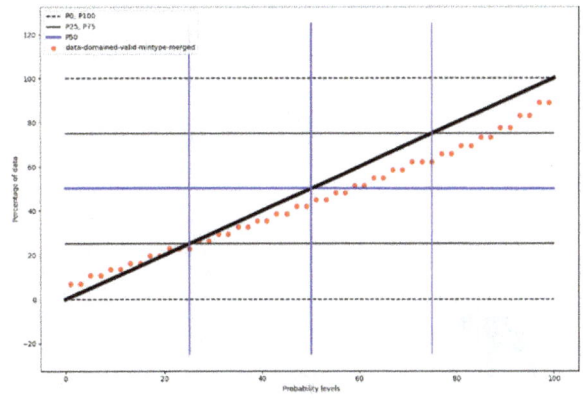

Al₂O₃

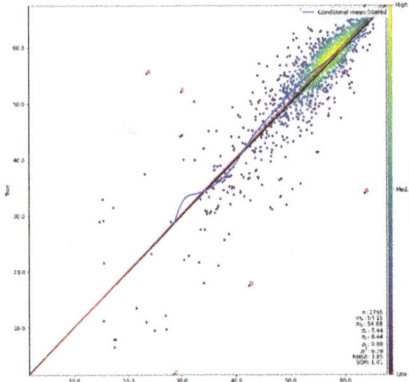
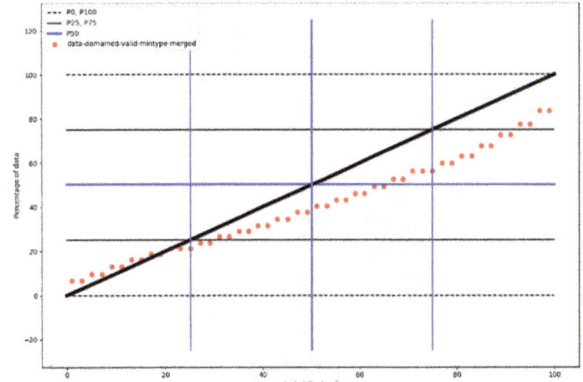

Fe

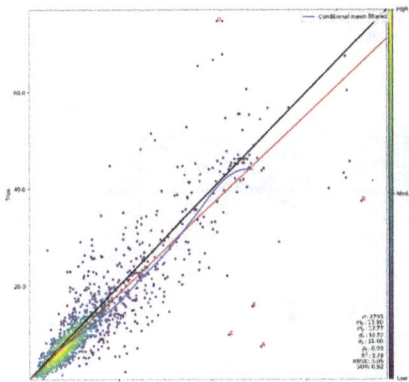
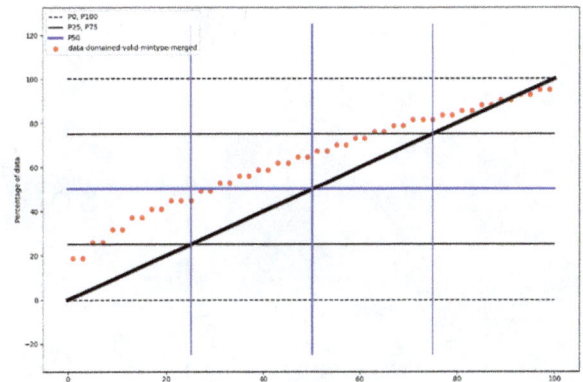

SiO₂

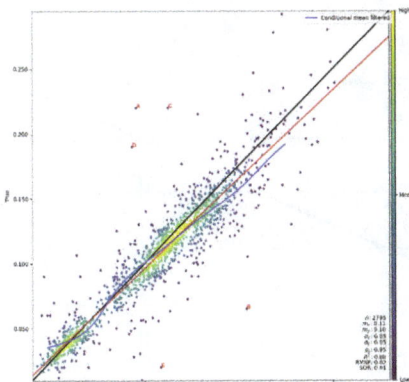
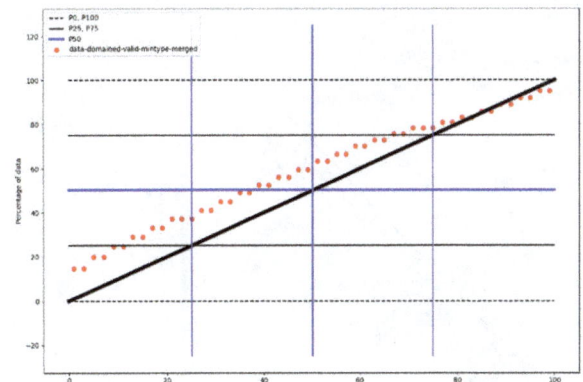

P

FIG 15 – OB24 25 simulations initial model validated against pre-Dec 2019 blastholes (scatter plots left and accuracy plots right).

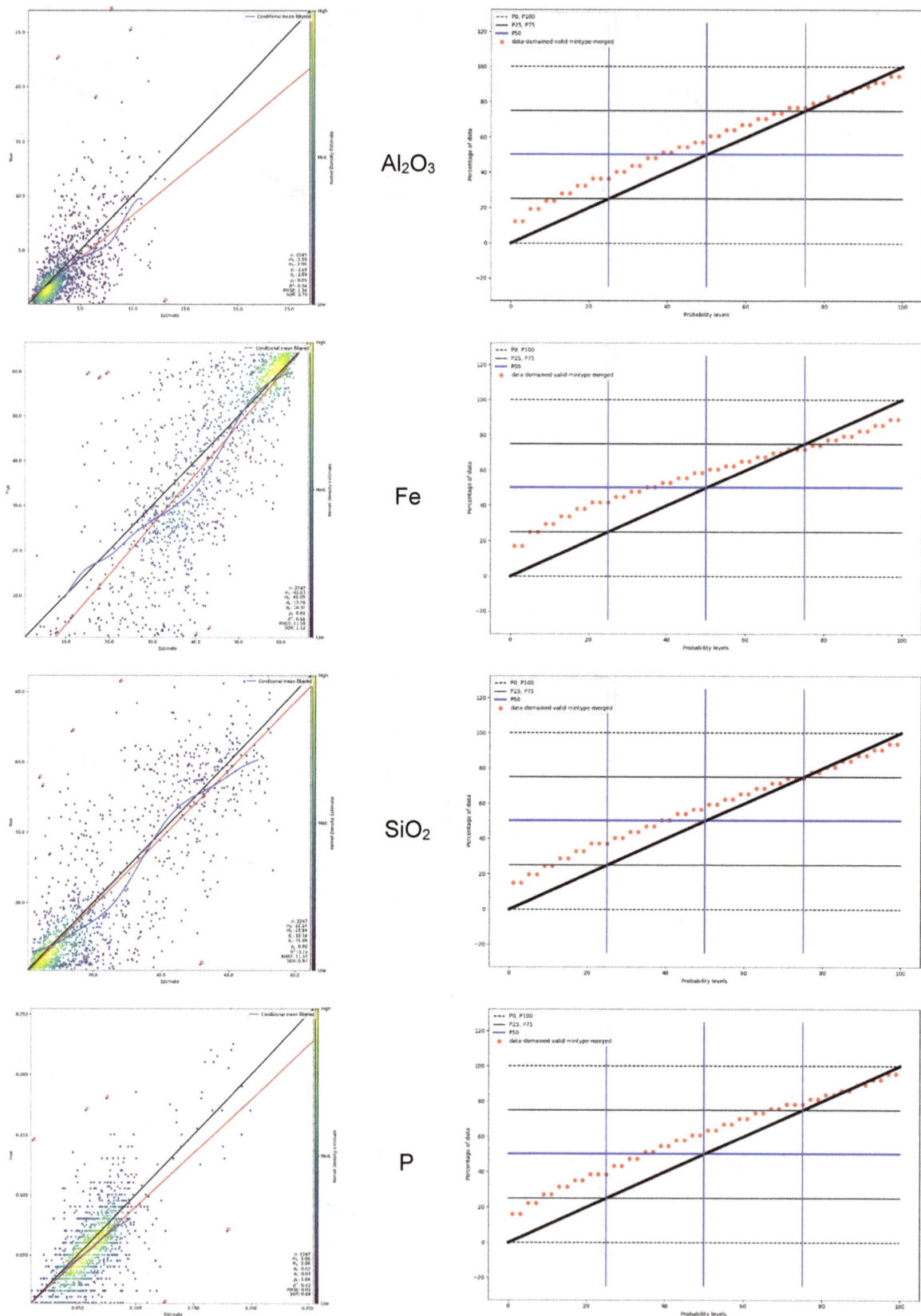

FIG 16 – OB32 25 simulations initial model validated against subset of 30 per cent exploration holes (scatter plots left and accuracy plots right).

In the case of OB24, pre-Dec 2019 blastholes were used as the validation data on the y axis in the scatter plots (left) and accuracy plots (right). Generally, these validation plots are considered acceptable, with some minor deviations in the simulated data seen in the case of the Al_2O_3 scatter plot slope of regression being less than 45° due to some extreme outliers and the SiO_2 accuracy plot showing poor precision at low probability intervals. SiO_2 is strongly impacted by the categorical simulation of ore and waste which may be the cause of the slightly lower level of precision in the SiO_2 simulations compared to the other continuous variables.

For OB32, due to the limited spatial extent of the pre-Dec 2019 blastholes, as mining of this deposit has only commenced relatively recently, a 30 per cent subset of pre-Dec 2019 exploration data was used to validate the initial 25 simulation model. Figure 16 shows generally good validation results; however, the high degree of scatter in the Fe scatter plot and associated lower precision in the accuracy plot is notable when compared with OB24 results. This is considered to reflect the higher-grade variability for OB32 which is impacting the model.

Figure 17 shows scatter plots and accuracy plots for the reconciliation of the final 100 simulation OB24 model against 31 months of post-Nov 2019 blasthole data. The main economic variable, Fe, reconciles well and SiO_2 reconciliation is acceptable, barring lower precision noted previously. However, Al_2O_3 and P do not show good reconciliation with Al_2O_3 plotting significantly below the 45° line in the accuracy plot and P conversely plotting significantly above the 45° line in the accuracy plot. Notably in both these variables, there is also a strong bias in the respective scatter plots, not seen for the other two variables which reconcile well.

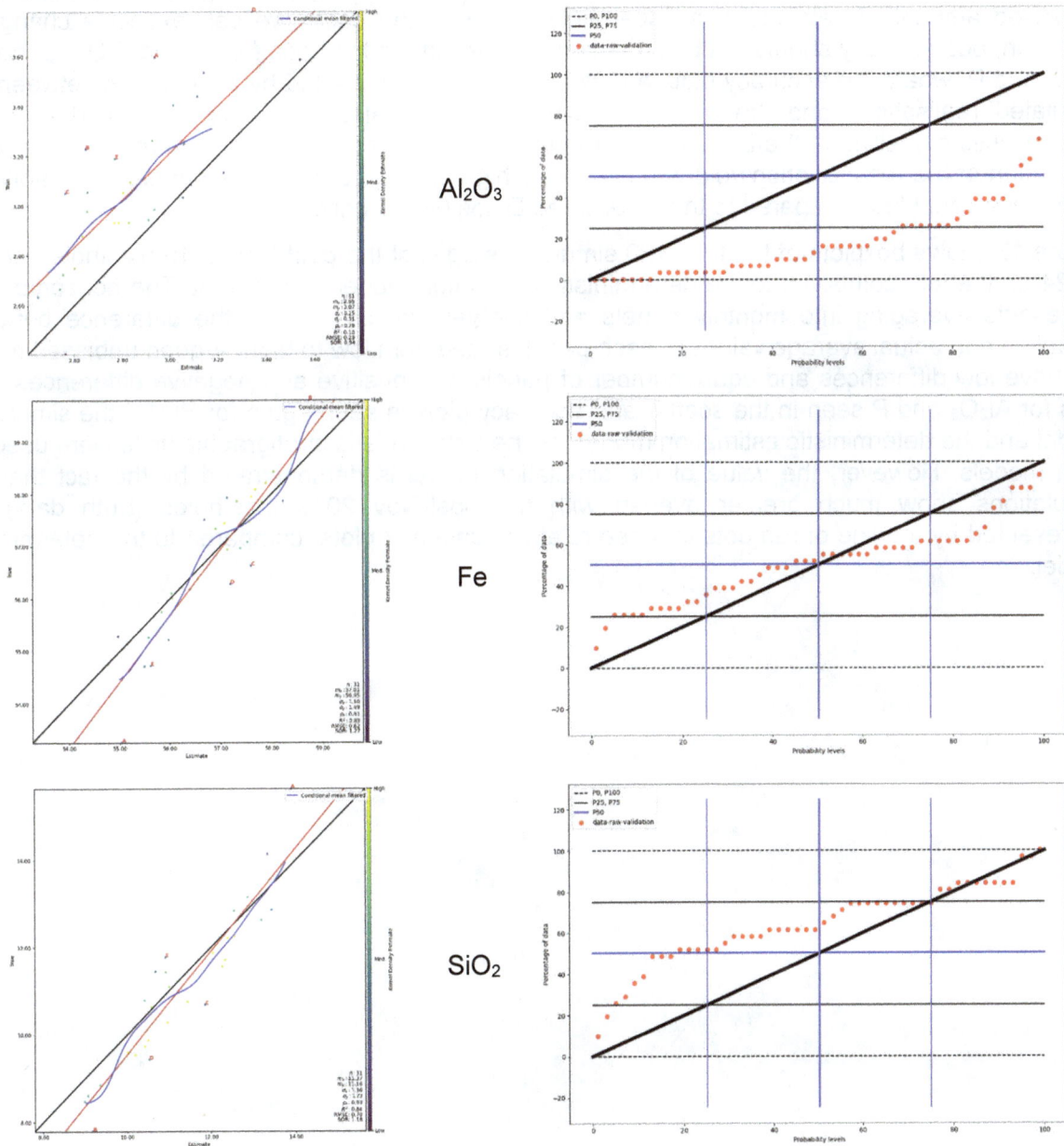

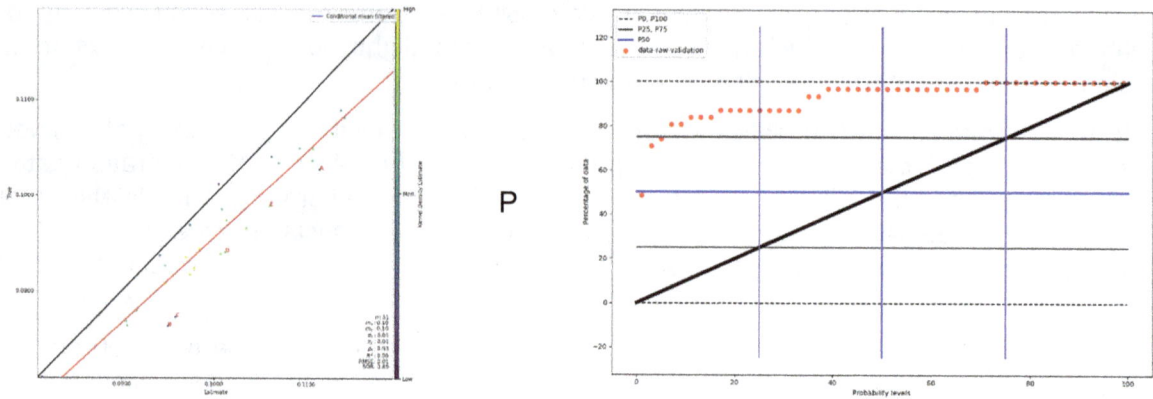

FIG 17 – OB24 100 simulations final model reconciliation results (scatter plots left and accuracy plots right).

As discussed previously, simulations are sensitive to initial assumptions about global distributions observed in samples. If new data (blasthole samples) do not change the global distribution, the increased amount of data reduces uncertainty. If this is the case, we can expect a change in precision, but accuracy should not be impacted. This is indeed the case for Fe and SiO_2 but not for Al_2O_3 and P, where the accuracy plots are clearly negatively impacted by a clear bias between the simulated realisations and the post-Nov 2019 blasthole data. This bias is likely due to the deterministic modelling of the D1 Dales Gorge unit (Comp_stratn = 5611) which makes up a large proportion of the reconciliation void and is notably high in Al_2O_3 and P with increased variability in both these variables, compared to the other three Dales Gorge units.

Figure 18 shows boxplots of the final 100 simulations against the post-Nov 2019 blasthole data for OB24 on the left, compared to the deterministic estimation model on the right. The horizontal axis represents averaging into monthly panels and the vertical axis shows the difference between modelled and actual average values in each panel, sorted from low to high. A good unbiased model will have low differences and equal number of panels with positive and negative differences. The bias for Al_2O_3 and P seen in the scatter and accuracy plots is seen again for BOTH the simulation model and the deterministic estimation model, as the deterministic stratigraphic units were used for both models. However, the value of the simulation model is demonstrated by the fact that the simulations show much greater overlap with the post-Nov 2019 blastholes (truth data), as represented by the line of red dots in these relative scale box plots, compared to the deterministic model.

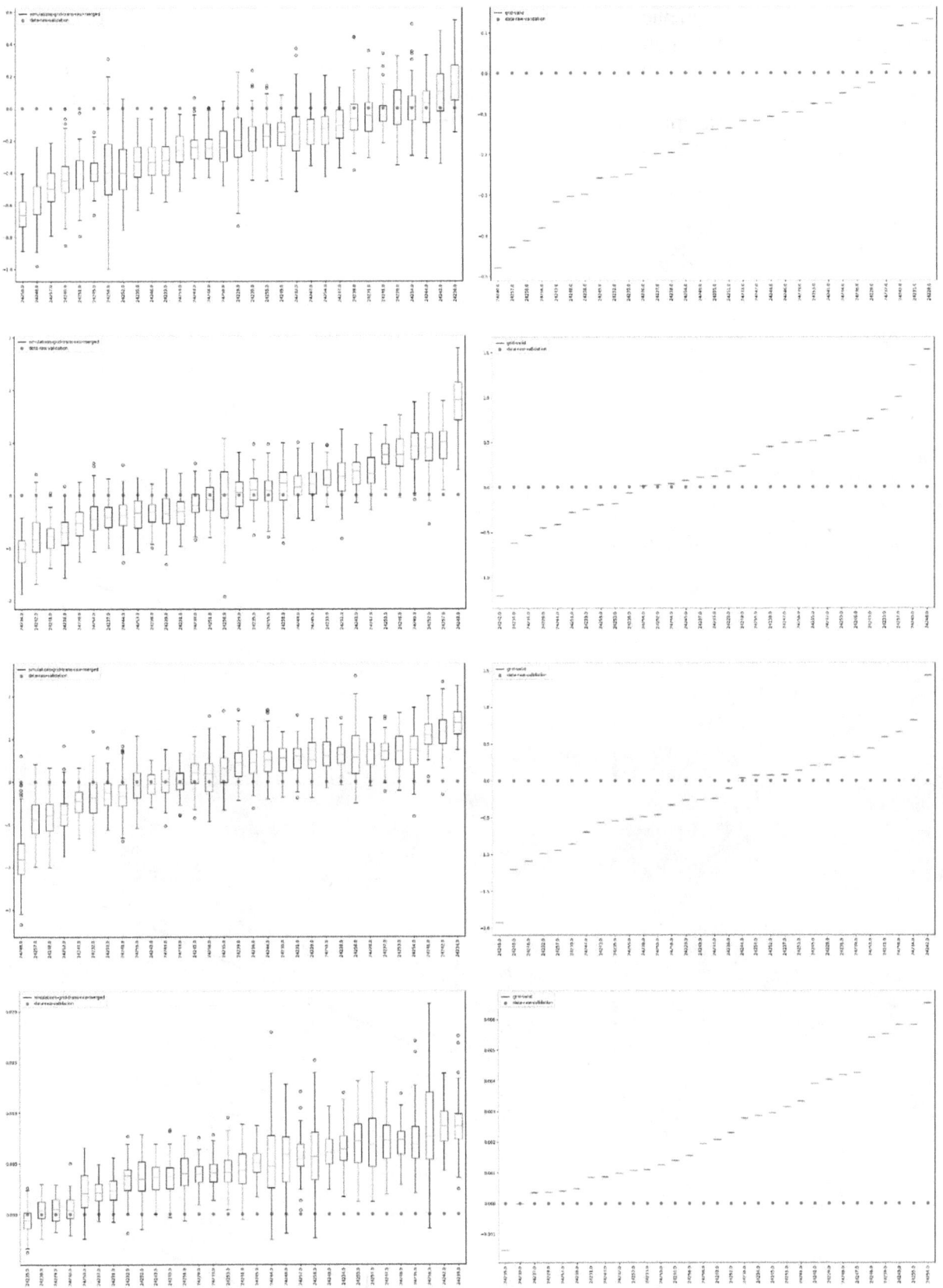

FIG 18 – OB24 final model (left) and deterministic estimation model (right) reconciliation results in order Al_2O_3, Fe, SiO_2 and P from top to bottom.

Figures 19 and 20 show scatter plots, accuracy plots and boxplots for the reconciliation of the final 100 simulation OB32 model against 31 months of post-Nov 2019 blasthole data. Overall reconciliation of the OB32 final simulation model is very good, apart from the lower precision for SiO_2 in the accuracy plot, noted previously. There is a general drop in precision for all continuous variables in the accuracy plots in Figure 19, due to the new information in the post-Nov 2019 blastholes (truth data). Precision of the simulations is adversely affected by the fact that the new information which forms part of the truth data has changed the local statistics of the data, as compared to the wider spaced data used to build the models. Scatter plots in Figure 19 show that the simulations provide unbiased estimates, which manifests in good accuracy plots, compared to that seen for the equivalent accuracy plots for OB24 (Figure 17). This points to more representative deterministic modelling of the mineralised stratigraphic units in the case of OB32 compared to OB24.

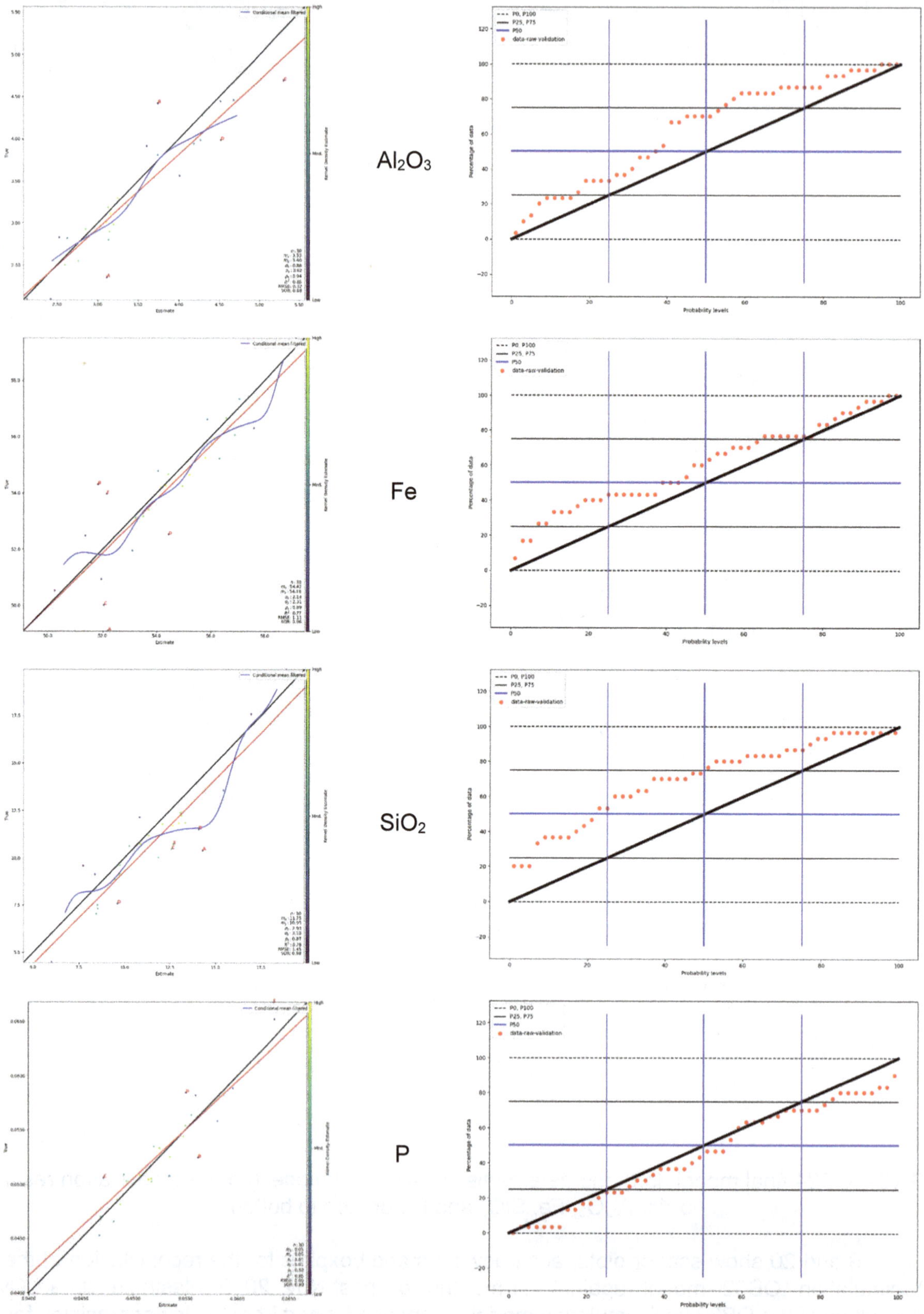

FIG 19 – OB32 100 simulations final model reconciliation results (scatter plots left and accuracy plots right).

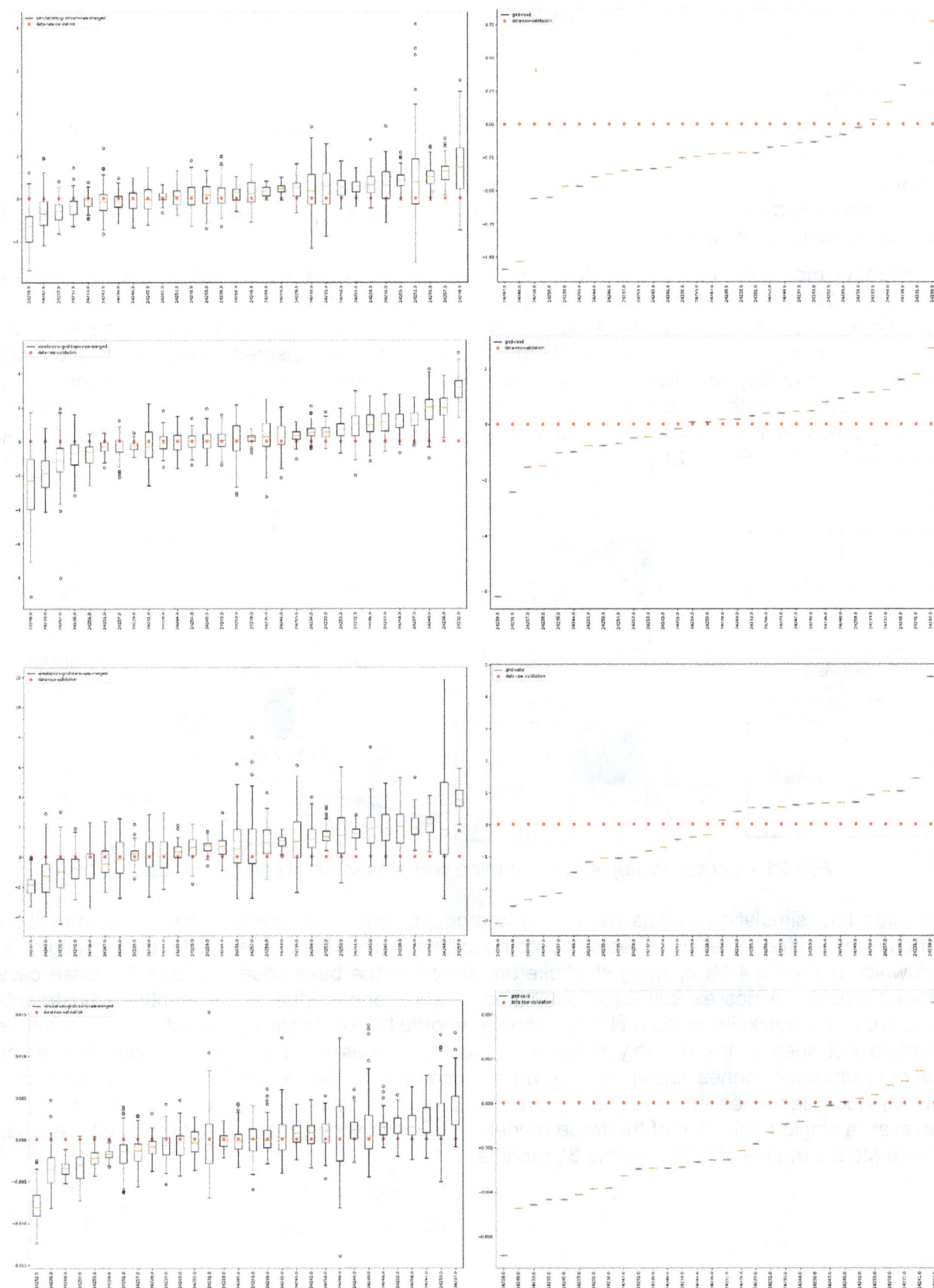

FIG 20 – OB32 final model (left) and deterministic estimation model (right) reconciliation results in order Al$_2$O$_3$, Fe, SiO$_2$ and P from top to bottom.

The OB32 boxplots for the simulations shown in Figure 20 also show that the simulations are unbiased, however there is bias seen in the deterministic model boxplots in the case of Al$_2$O$_3$ and P. The value of the simulation model is again demonstrated by the fact that the simulations show much greater overlap with the post-Nov 2019 blastholes truth data and additionally the simulations for all the grade variables are unbiased compared to the truth data, which is not the case for the deterministic model for OB32.

Risk analysis

The final 100 simulation models for OB24, OB32 and OB25_P1 all reconciled well against the post-Nov 2019 blasthole data, apart from the stratigraphic domain based bias seen in OB24 for Al_2O_3 and P. Notwithstanding this bias, seen in the deterministic model used for short-term planning as well, the final simulation models for all three deposits are considered to be adequate for the purpose of *in situ* grade uncertainty quantification.

The negative material impact of uncertainty is known as risk. There is clear risk in meeting targeted Fe, Al_2O_3, SiO_2 and P grade thresholds, as evidenced by the variance in the boxplots from the final simulations compared to the reality of the post-Nov 2019 blasthole data, which has been set to zero in the relative box plots shown in Figures 18 and 20. However, the expected impact of loss of product tonnes or out of quality specification shipments is not seen during actual mining, as the mine plan is not static. Every month, dig plans are adjusted based on grade control data and any shortfall in product tonnes at the correct specifications is made up from the stockpile, as shown by the rehandling tonnes in Figure 21.

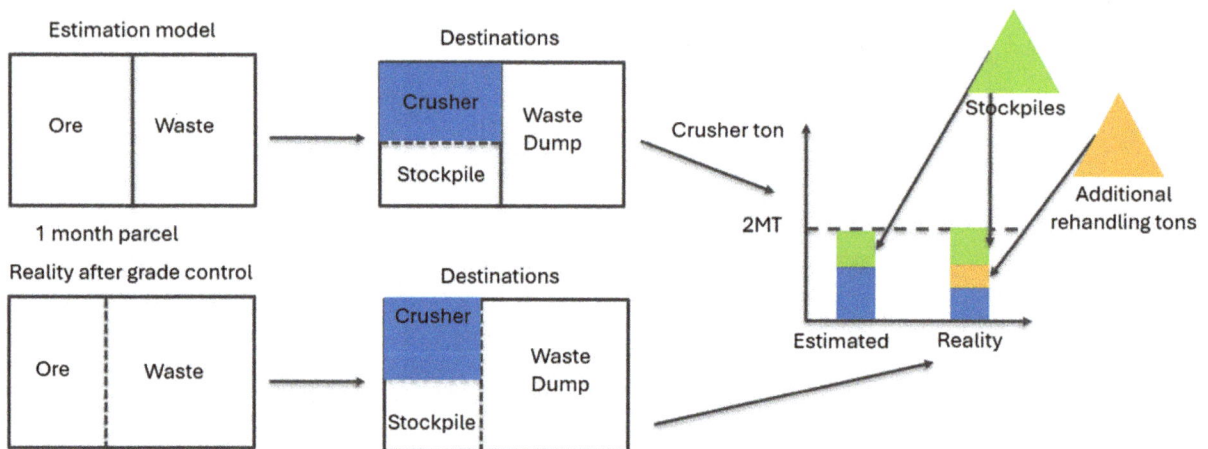

FIG 21 – Iteratively adjustment of mine plan based on grade control data.

The three final simulation models were used to model downstream blending options from multiple orebodies with different levels of geological complexity and uncertainty. Results for the existing mine plan which utilises a 4 Mt opening stockpile are shown in the base case in Figure 22. Base case planned crusher tonnes exactly equal simulated crusher tonnes, due to the monthly adjustments made from the stockpile to correct any monthly shortfall. The negative impact of uncertainty is therefore not seen in the monthly delivery of crusher tonnes compared to the plan, but in the changing stockpile tonnes shown on the right. According to the deterministic estimation model (STPM) stockpile tonnes increase linearly over the 31-month mining period from 4 Mt to over 14 Mt. However, a single realisation of the three modelled deposits shows that the final stockpile reaches only 10 Mt, a variance of 4 Mt over the 31 months.

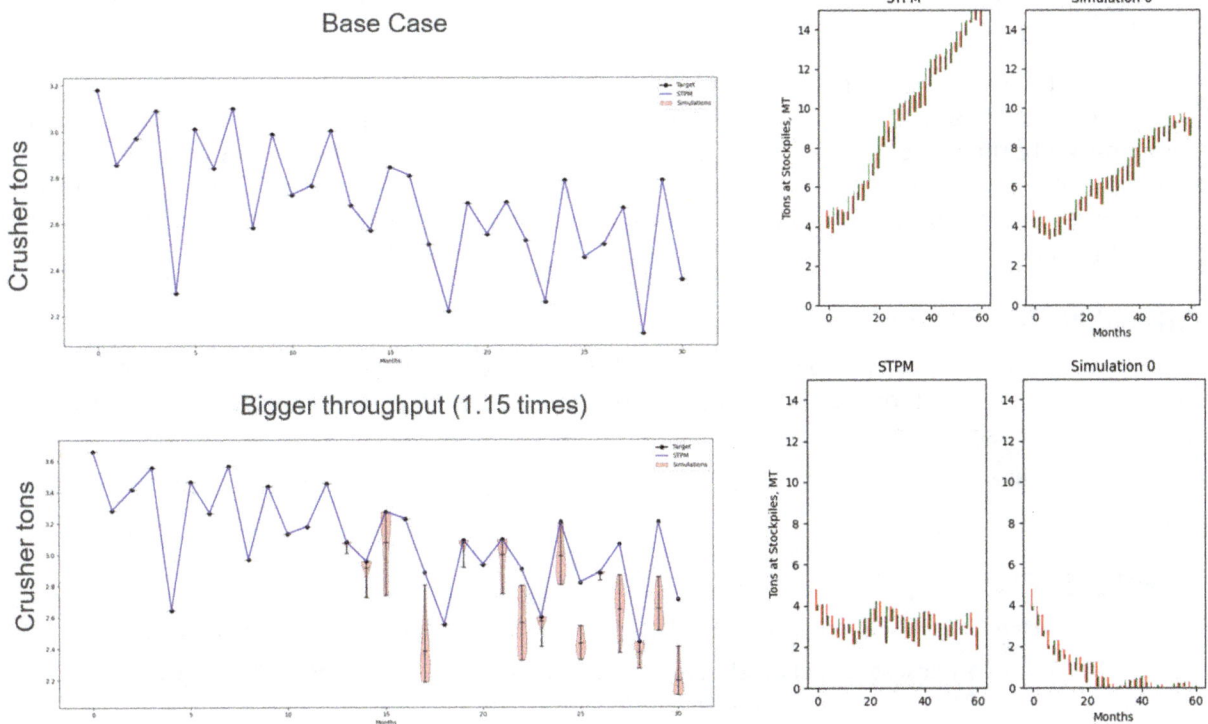

FIG 22 – Simulations show negative impact of uncertainty with increased production rate.

If the planned mining rate is increased from the base case to 1.15 times the base case, then the starting 4 Mt stockpile is shown to be inadequate to smooth out crusher tonnes variance, resulting in significant shortfall in simulated crusher tonnes compared to planned tonnes in the mine plan based on the deterministic STPM model.

Therefore, the simulation models highlight the risk to planned crusher tonnes, with an increase in stockpile tonnes required for higher production rates required to mitigate this risk.

CONCLUSIONS

BHP's SBRE conditional simulation platform has been used successfully to create simulation models

of three orebodies from BHP's WAIO Newman iron ore mining hub. These three orebodies were selected as they display a range in geological complexity, related both to relative stratigraphic position within the Proterozoic Hamersly Group and the degree of structural deformation.

Simulation models were validated against pre-Dec 2019 blastholes and reconciled against post-Nov 2019 blastholes, to ensure they are fit for purpose. The SBRE workflow approach allowed for iterative adjustments to be made quickly and effectively during workflow optimisation. Accuracy plots proved to be very useful in identifying any accuracy or precision problems with the simulation models prior to final models being used for uncertainty quantification purposes.

Issues identified following reconciliation over a 31-month period against the post-Nov 2019 blastholes are a decrease in precision in the simulated models compared to blastholes, which is due to new information and some local bias between simulated models and the post-Nov 2019 blastholes, due mainly to deterministic modelling of stratigraphic units. This bias in Al_2O_3 and P is only seen for OB24 and is not considered critical since the same issue impacts the deterministic STMP model used for short-term mine planning as well. Notwithstanding this bias, simulation models show a clear risk in meeting targeted Fe, Al_2O_3, SiO_2 and P grade thresholds, as evidenced by the variance in the boxplots from the final simulations compared to the reality of the post-Nov 2019 blasthole data.

The final simulation models for the three deposits were used to model downstream blending options from multiple orebodies with different levels of geological complexity and uncertainty. Results for the existing base case mine plan which utilises a 4 Mt opening stockpile show that planned crusher tonnes exactly equal simulated crusher tonnes, due to the monthly adjustments made from the stockpile to correct any monthly shortfall.

If the planned mining rate is increased from the base case to 1.15 times the base case, then the starting 4 Mt stockpile is shown to be inadequate to smooth out crusher tonnes variance, resulting in significant shortfall in simulated crusher tonnes compared to planned tonnes in the mine plan based on the deterministic STPM model. Therefore, the simulation models highlight the risk to planned crusher tonnes, with an increase in stockpile tonnes required for higher production rates to mitigate this risk.

ACKNOWLEDGEMENTS

The authors would like to acknowledge several WAIO staff members for productive discussions around the WAIO mine planning methods and the simulation results, which helped to frame our ideas around the conclusions for this study.

We would also like to thank BHP for allowing us the time to work on this study and to publish the results.

REFERENCES

Barnett, R M, Manchuk, J G and Deutsch, C V, 2014. Projection Pursuit Multivariate Transform, *Math Geosci*, 46:337–359.

Deutsch, C V, 1997. Direct Assessment of Local Accuracy and Precision, in *Geostatistics Wollongong '96 – Proceedings of the Fifth International Geostatistics Congress* (eds: E Y Baafi and N A Schofield), pp 115–125.

Journel, A G, 1974. Geostatistics for Conditional Simulation of Orebodies, *Economic Geology*, 69:673–687.

Novel stratigraphy simulation – a case study for ESG and mine planning application

B Wilson[1], S Sanchez[2], D Carvalho[3] and S Horan[4]

1. Project Geostatistician, GeologicAI, Edmonton, Canada. Email: bwilson@geologicai.com
2. Geostatistician, GeologicAI, Edmonton, Canada. Email: ssanchez@geologicai.com
3. Principal Geologist, GeologicAI, Brisbane Qld 4000. Email: dcarvalho@geologicai.com
4. Principal Geologist, GeologicAI, Toronto, Canada. Email: shoran@geologicai.com

ABSTRACT

Characterisation of the quantity and location of potentially acid-forming (PAF) material and the embedded tonnage and grade uncertainty is critical to determine and mitigate the impact of handling procedures on operations, project evaluation and ESG compliance and responsibilities. An uncertainty analysis case study was conducted for an open pit operation to characterise the variation of PAF volume on-site.

The study details the simulation of sulfur within simulated stratigraphy, integrating domain uncertainty into the result. The relevant subsurface comprises a series of folded pyritic shale strata, intercalated with breccia units. The presence of folding complicated the continuous and categorical simulations as the intended workflow relied on the implementation of an unfolding transformation. However, the relative thickness of the target unit combined with the strength of folding led to artifacts in the unfolding result.

This behaviour led to the development and implementation of a novel stratigraphy simulation workflow that facilitates the reproduction of true strata thicknesses in cartesian space. The workflow, Sequential Surface Offset Simulation (SSOS), proceeds by calculating and simulating the normal distance between a designated reference surface and intersections with the target (simulated) surface. The simulated normal distance is truncated against a gridded reference thickness to generate a contact realisation, and surface realisations are extracted where the contact is equal to zero, effectively applying an equality constraint. This process is repeated for each target surface, using previously simulated surfaces as the reference where available.

The integrated SSOS and sulfur simulation workflow is post-processed to calculate PAF tonnage and uncertainty. When compared to deterministic modelling, there is a clear difference in global results, and a notable increase in the possible PAF extents. Comparing to simulation of sulfur within deterministic domains shows the impact of modelling domain uncertainty with the SSOS and highlights greater PAF potential at the top and bottom of the package. This improved understanding of the behaviour of PAF material is being integrated into mine planning, drill hole planning and waste dump management.

INTRODUCTION

Stratigraphic deposits provide significant and essential sources of metals and energy resources. Key types include sedimentary-exhalative (SEDEX) (also defined as clastic-Dominated (CD)), magmatic sulfide, lateritic, and coal and organic-related deposits. Examples of these deposits include Mount Isa (Cu-Pb-Zn-Ag), Central African Copperbelt (Cu-Co), Kupferschiefer (Cu-Ag), Bushveld Complex (PGE), Goro (Ni), Weipa (Al), Bowen Basin (Coal), Athabasca Oil Sands (Oil), and many others. Given their critical role in resource supply, accurate mineral resource estimates (MRE) are essential for effectively defining, developing, and operating these deposits.

In the mining industry, MRE workflows typically focus on defining the geometry, grade, chemical and physical characteristics of mineralised bodies. This focus is self-evident, as the term 'mineral resource estimation' inherently emphasises the quantification of resources. This prioritisation on resources is fundamental, as demonstrating the economic viability of a deposit is a prerequisite for developing a mining project. Similarly, in an operating mine, accurate resource estimates are critical to meeting economic targets, ensuring the operation remains profitable, sustainable, and aligned with long-term goals. Most MREs focus on single estimates and models without clear understanding of uncertainty.

Waste characterisation, while often secondary to resource estimation, plays a crucial complementary role in mining operations. Waste characterisation ensures the sustainable management of the materials left behind. Understanding the chemical and physical characteristics of waste helps identify risks, guiding the design of stable waste rock dumps and tailings storage facilities. Modelling waste properties also aids in identifying recoverable by-products, supports compliance with regulatory requirements, and facilitates economic modelling.

Deposits often contain sulfides in both ore and waste zones, requiring additional infrastructure and specific handling procedures to mitigate potential acid mine drainage (AMD). Proper management of potentially acid-forming (PAF) material is essential for sustainable mining practices and meeting ESG obligations. To support mine planning, modelling PAF tonnage is crucial and can be achieved through methods such as domain modelling or estimating and simulating sulfur content.

Building on the need for comprehensive waste characterisation and uncertainty quantification, this paper presents a novel technique to simulate stratigraphic geometry and evaluate the tonnage uncertainty of PAF and non-acid-forming (NAF) materials within waste zones of stratigraphic deposits. By understanding the spatial distribution of grades and geometric uncertainties, effective waste management strategies can be developed to mitigate environmental risks like AMD and optimise storage designs and drilling plans. While this approach is demonstrated for PAF/NAF characterisation, it is equally applicable to mineralised orebodies or any geological model within a stratigraphic framework.

CURRENT METHODS

Modelling stratigraphic units is a non-trivial problem but quite recurrent in the mining industry. Models can be either deterministic, where a single representative interpretation of the domains is generated, or probabilistic, in which the geometry and grades are simulated for hundreds of realisations, quantifying their associated uncertainty.

Deterministic

Explicit and implicit modelling techniques currently dominate geological modelling in the mining industry. Explicit modelling is a traditional method relying on manual digitisation and interpretation to define orebody geometry. It involves creating 2D sections and linking them into 3D wireframes, allowing detailed incorporation of geological insights. Implicit modelling automates wireframe creation using distance functions, offering faster, repeatable results. It uses data as lithological contacts and structural measurements, typically with radial basis functions (RBF) or kriging (Cowan *et al*, 2003; Hillier *et al*, 2014).

Both methodologies are used in stratigraphic modelling with specific attention points in each case. Generating continuous tabular stratigraphic units with explicit modelling is challenging due to difficulties in maintaining thickness and avoiding wireframe intersections in between drill hole sections and intercepts, as well as tying 2D polylines into a 3D volume. Updating wireframes is laborious, especially when faulting is involved. In contrast, implicit techniques facilitate the quick generation of wireframes while honouring thicknesses and avoiding intersections. Considerable manual input can be used in this technique as well.

For each method, grade modelling is most commonly performed using linear techniques, such as kriging or other interpolation methods. Deterministic modelling approaches, by design, do not provide insight into the uncertainty or variability of geometries and grades.

Simulation

Categorical simulation is growing in popularity as it facilitates the integration of domain uncertainty throughout a modelling workflow. Common domain simulation methods such as hierarchical truncated pluri-Gaussian (HTPG) (Silva and Deutsch, 2019) and indicator simulation (Deutsch, 2006) have limitations that often lead to poor results when used to simulate continuous, extensive units, as is the case in sediment hosted deposits. Ensuring geologically consistent results that respect domain contacts can be particularly difficult with these methods, as there are no inherent controls to restrict domain occurrence to fall strictly in continuous units. Alternatively, the position of domain contacts can be simulated directly and used to generate realisations of domain uncertainty.

Uncertainty in the position of contacts between stacked stratigraphic sequences can be modelled with co-simulation of unit thicknesses. While independent simulation of the units could be considered, this would not reproduce the relationships between units that may have correlated thickening or thinning due to the nature of the underlying geological processes.

Simulation of a single tabular structure follows the general framework developed by Carvalho (2018):

1. Identify intersections with the top and bottom surfaces of each unit.

2. Rotate the frame of reference so the unit thickness is vertical.

3. Simulate position of the first surface (top or bottom) in the rotated reference.

4. Calculate thickness data associated with each simulated surface realisation.

5. Simulate thickness and convert it to second surface position.

6. Back rotate to cartesian space.

The location of pinch-outs may also be simulated as a part of this process using HTPG on a 2D grid in the plane of the unit. This workflow is extended to the simulation of multiple contacts by simulating each thickness sequentially, treating the exhaustive set of previously simulated thicknesses as a super-secondary variable for use in intrinsic collocated co-kriging.

In some circumstances, the assumptions underlying co-simulation of sequential contact surfaces may be violated, and the method cannot be used. Additional data preprocessing can be considered in some instances to allow simulation in a space where the method is more appropriate. For instance, in highly folded structures, the thickness simulation may not be representative as the orientation of the thickness is not consistent across the unit. Unfolding procedures may be sufficient to transform to a consistent frame of reference and allow for simulation to proceed. However, in some instances with complex folding or a thick set of strata that cannot be unfolded consistently with existing tools, contact simulation cannot be reasonably undertaken.

Due to the issues with applying traditional techniques, an alternate workflow was conceived, leveraging the flexibility of a Python-based workflow to explore and develop a new surface simulation method mid project stream. The ultimate result was the development of the sequential surface offset simulation (SSOS), a novel workflow for the simulation of contacts within complex stratigraphic sequences.

SEQUENTIAL SURFACE OFFSET SIMULATION

SSOS is a surface simulation workflow that is developed as a modification and extension of single-vein simulation (Carvalho, 2018). The workflow was implemented in Python using Resource Modelling Solutions Platform (RMSP). While true vein thicknesses are simulated in traditional single vein simulation, requiring transformation to a suitable frame of reference, SSOS instead considers a field of normal thicknesses in cartesian space.

SSOS consists of sequential simulation of individual domain contact positions. The process can be broken down into two main components: the simulation of a surface from a given reference, and the process to extend to multiple surfaces.

Single surface simulation

The base surface simulation procedure is based on the idea of simulating a contact as the normal thickness from a reference surface. The workflow consists of:

- Select a reference surface for thickness calculations.

- Calculate the distance between the reference surface and contact surface intersects.

- Simulate thickness, conditioning to contact intersects.
- Truncate simulated thickness realisations against distance to the reference surface.

Figure 1 shows a schematic view defining terminology for single surface simulation.

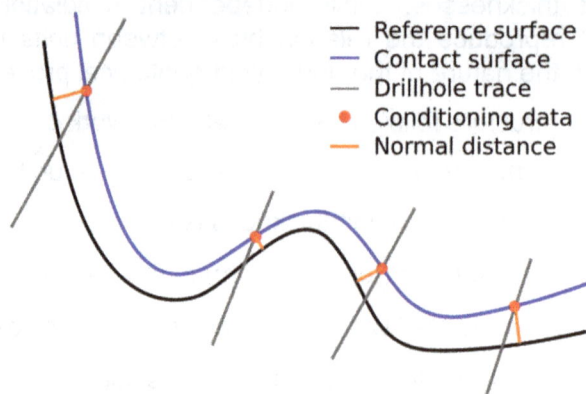

FIG 1 – Schematic cross-section view defining surface simulation terminology.

There are several implementation details that must be considered to ensure a reasonable outcome. First, the selection of the reference surface is critical to the results away from conditioning data, as simulated thickness is truncated based on distance to the reference. Normal thickness simulation follows geostatistical best practices, including considerations for data transformations in the presence of sparse data, trend calculation and removal (Qu, 2018), and simulation with locally varying anisotropy, where appropriate. Declustering and trend calculations are conducted on a plan projection of the simulation grid. Data are also projected to 2D for these operations. This mitigates the potential for vertical intersect position to unduly influence the declustering weights and trend values.

Results may be post-processed to ensure expected continuity and conditioning data are reproduced by using a filter-snap transformation, where the resulting thicknesses are smoothed using a Gaussian filter and corrected to match conditioning with an inverse distance estimate. The simulated surface realisations are used to flag gridded domain realisations for continuous variable simulation.

Sequential simulation

Extending single surface simulation to a full SSOS workflow consists of adding an initialisation stage and using previously simulated surfaces as the reference for the simulation of subsequent contacts. Initialisation consists of:

- Select a deterministic reference surface and stochastic seed contact.
- Simulate selected seed contact.
- Simulate additional surfaces moving progressively away from stochastic seed surface.

Figure 2 shows the SSOS process, including initialisation.

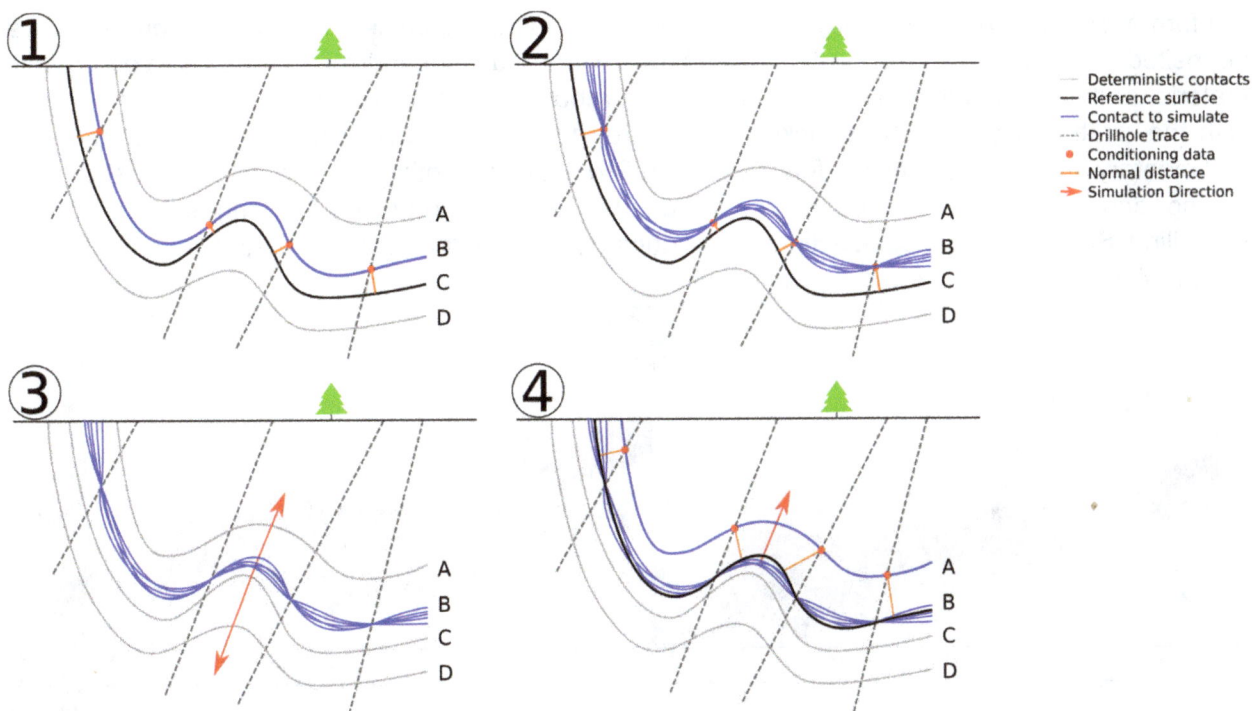

FIG 2 – Initialisation of SSOS – (1) selection of reference and seed surfaces, (2) simulation of stochastic seed contact, (3) simulation of surfaces directionally from stochastic seed, (4) simulation proceeds using each simulated surface as a reference for a single target surface realisation.

Initialisation of the SSOS is conducted to allow for the simulation of all contacts within the stratigraphic package, including the initial deterministic reference. The contact that is simulated using the deterministic reference is deemed the stochastic seed, as it forms the basis for all subsequent contact simulations, including the simulation of the initial deterministic reference contact.

The selection of a deterministic reference and stochastic seed has a direct effect on the results for all simulated surfaces, particularly away from conditioning. In highly folded structures, the surfaces should be selected where folding is strongest as the folding is attenuated with each subsequent simulated surface, potentially leading to understated folding in the final simulated realisations. Once SSOS is initialised, contacts are simulated sequentially in spatial order. If the reference is in the middle of the package, contacts are simulated independently upwards and downwards before they are combined.

After the full set of surfaces are simulated, the results are used to flag gridded domain model realisations. Flagging order is critical to ensure results are geologically consistent, respecting pinch-outs and overprinting of domains. In general, flagging proceeds in the opposite order of simulation. This results in any locations in a simulated contact that cross the associated reference being treated as a pinch-out. That is, when simulating a surface that lies above a given reference, any instances where the simulated surface passes below the reference represent a pinch out of the associated domain.

CASE STUDY

Background

Due to confidentiality agreements with our client, the case study excludes the full scope of work, detailed geological descriptions, images, specific locations, implications, and final uncertainty results and actions. Instead, an overview of the challenge, geological setting, and selected results is provided, supplemented by anonymised values and figures.

The deposit lies within a major sedimentary basin formed in a tectonically stable environment, characterised by thick sequences of marine sedimentary units that created favourable conditions for ore formation. Its fine-grained, organic-rich host rocks, including black shales and dolomites, were deposited in a restricted, deep-water environment with reducing conditions that preserved organic material and promoted mineral precipitation. The orebody, laterally extensive and stratiform, aligns with the stratigraphy of its host rocks and features fine laminations of sulfides interbedded with sedimentary material. Structural features, including faulting (both pre- and post-mineralisation), folding, and brecciation, further influence the deposit, with folding posing challenges for geological modelling. Some vertical cross-sections, highlighting the deterministic stratigraphic waste model, are shown oi Figure 3.

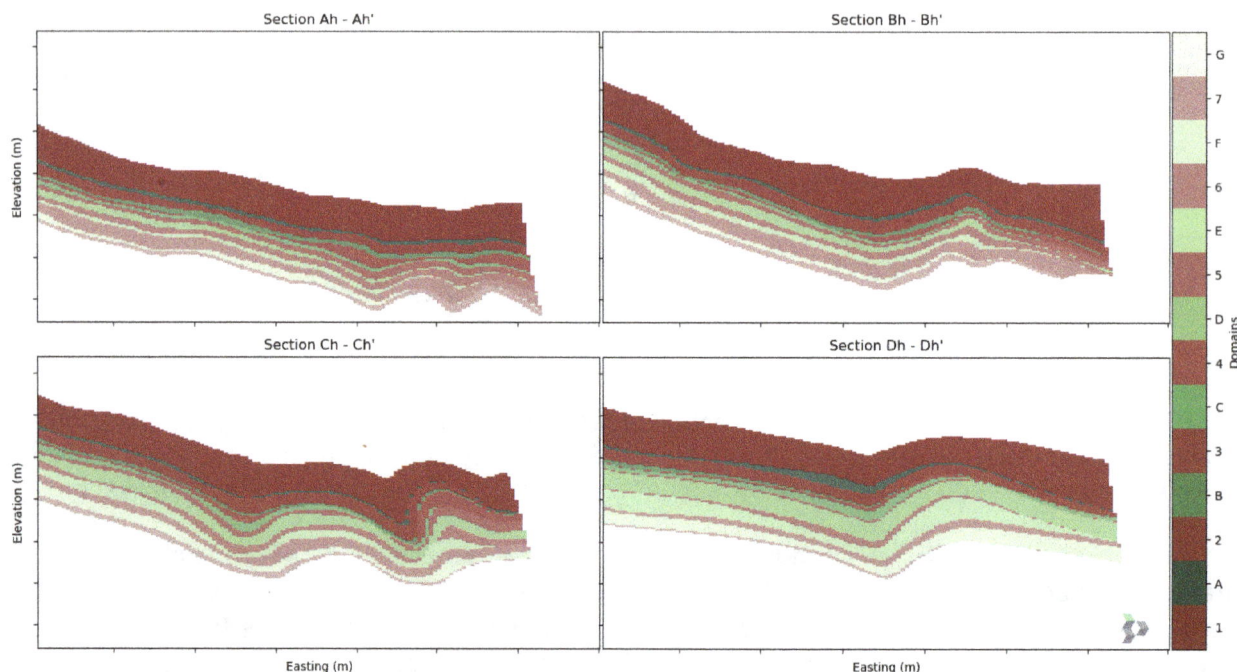

FIG 3 – Cross-sections of the waste stratigraphic units of the deposit.

The presence of PAF material is primarily observed within certain waste units of the deposit. Due to its significance for environmental and health considerations, the client sought to assess the uncertainty associated with the tonnages and spatial distribution of this material. Understanding this uncertainty was critical for informed decision-making, directly supporting drill hole planning and strategic mine planning.

Scope and data

The original methodology to characterise PAF content in the deposit consisted of:

- creating a grade shell containing samples above PAF cut-off
- slicing the volume only considering shale layers
- perform ordinary Kriging to estimate sulfur and iron inside and outside the sliced grade shell.

A downside of this methodology is that it assumes that all shale layers have a similar average sulfur grade, which is not the case. Thus, samples from layers with low sulfur grade will affect layers with high sulfur and *vice versa*, smoothing the model. Furthermore, by relying on an estimation technique, the method does not provide a measure of uncertainty. Alternatively, each layer could be considered as a domain and estimated independently, but this would still fail to characterise uncertainty in either the sulfur grade or domain boundaries. Plans to mitigate PAF that do not consider the underlying uncertainty in PAF quantity and position represent a potential operational risk, as, for example, flexibility to increase waste disposal footprints may be limited if excess PAF is encountered. These risks are addressed through a simulation study that integrates domain and grade uncertainty to quantify the uncertainty in PAF material.

The proposed methodology consists of:

- domain simulation using the SSOS methodology
- grade simulation within simulated domains.

Drill holes are flagged with stratigraphic wireframes, composited, and declustered. A point scale grid is buffered from the original PAF volume to allow for simulation. Stratigraphic contacts are simulated using SSOS and categorical realisations are flagged for grade simulation. Grade variables are detrended and decorrelated before they are simulated independently by domain and then combined. The grade simulation model is post-processed to calculate PAF material uncertainty, first upscaling to the relevant SMU scale before evaluating the PAF content in each realisation by bench. The resulting distributions of PAF material are summarised and compared to the deterministic model.

Exploratory data analysis

The sedimentary unit contains high sulfur shale domains intercalated with low sulfur breccias. The difference in sulfur content between the shale and breccia strata is apparent when comparing the distributions (Figure 4). The difference in grade between domains within each subgroup is also evident, supporting independent simulation by domain. As sulfur and iron are highly correlated (Figure 5), a projection pursuit multivariate transform (PPMT) (Barnett, Manchuk and Deutsch, 2014) is applied prior to independent simulation to ensure that this relationship is reproduced. There is no capping nor soft boundaries applied to the data.

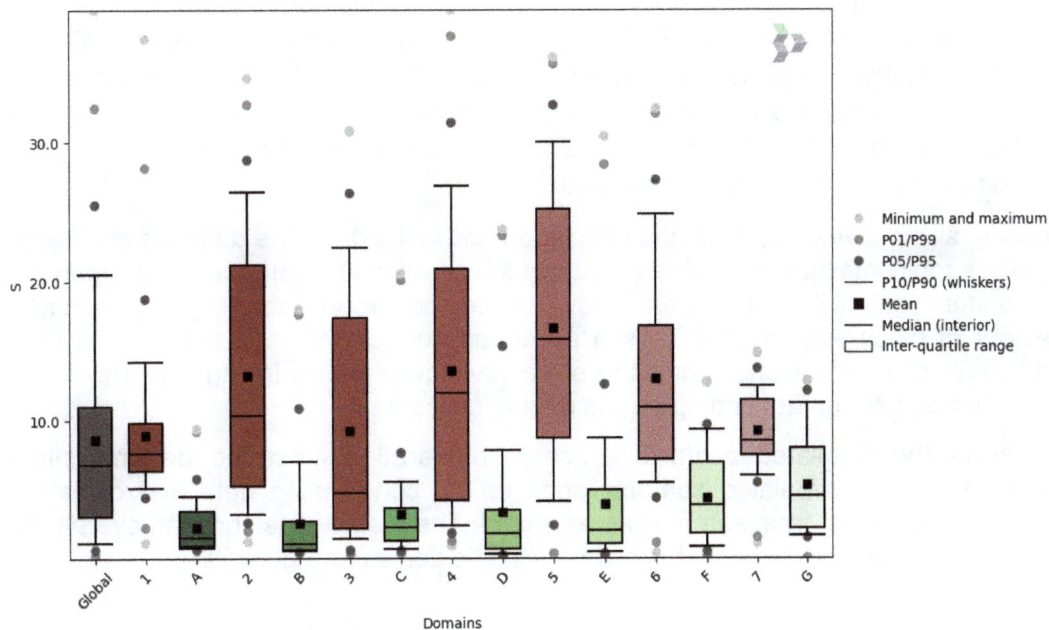

FIG 4 – Box plot of sulfur per domain.

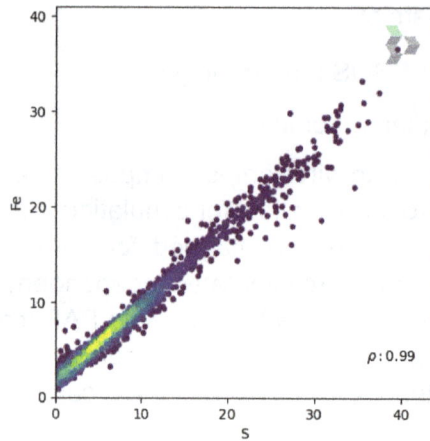

FIG 5 – Global correlation between S and Fe.

Domain simulation

SSOS is applied for domain simulation. The selection of the first reference surface is a key step in SSOS since the reference surface has a downstream impact on subsequent simulated surfaces geometry. While a simple sequence from top to bottom seems the most practical, contacts at the centre of the package have stronger folding. The fold is progressively less intense in units closer to surface, being barely noticeable at the top surface of the simulated package. Simulating from the top surface would distort the central layers by smoothing the fold. Consequently, the deterministic reference surface is chosen in the centre of the package at the contact between domains 3 and C to maintain fold geometry throughout the unit. As shown in Figure 2, the seed contact is initialised at the contact between domains C and 4, forming the basis for simulation of subsequent contacts upwards and downwards from this point. The contacts are then simulated sequentially from the centre to the top and from the centre to the bottom.

After obtaining all simulated contacts, the simulation grid is flagged. The domains are flagged below their respective top contacts if the nodes are above the seed contact surface and flagged above their respective bottom contact if the nodes are below the seed contact surface. The flagging order reflects the behaviour of pinch-outs in SSOS, as a simulated surface that crosses over its reference is assumed to have pinched out and should therefore have lower priority in flagging. The flagging order can be modified and set-up depending on the geological context.

Figure 6 shows the simulated contacts (traces) compared against the deterministic domains (background grid). The simulated contacts snap to the conditioning data and deviate from the deterministic model further from samples as expected. The realisations show an overall reasonable proportion reproduction compared with the deterministic domains (Figure 7).

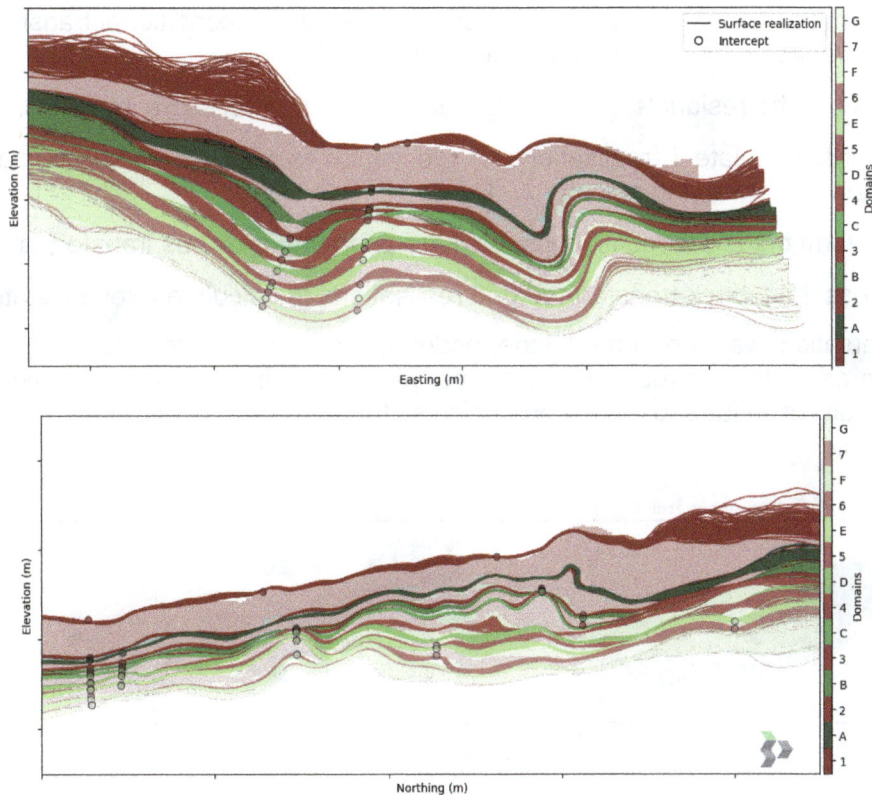

FIG 6 – Cross (top) and longitudinal (bottom) sections of the deterministic grid and simulated contacts.

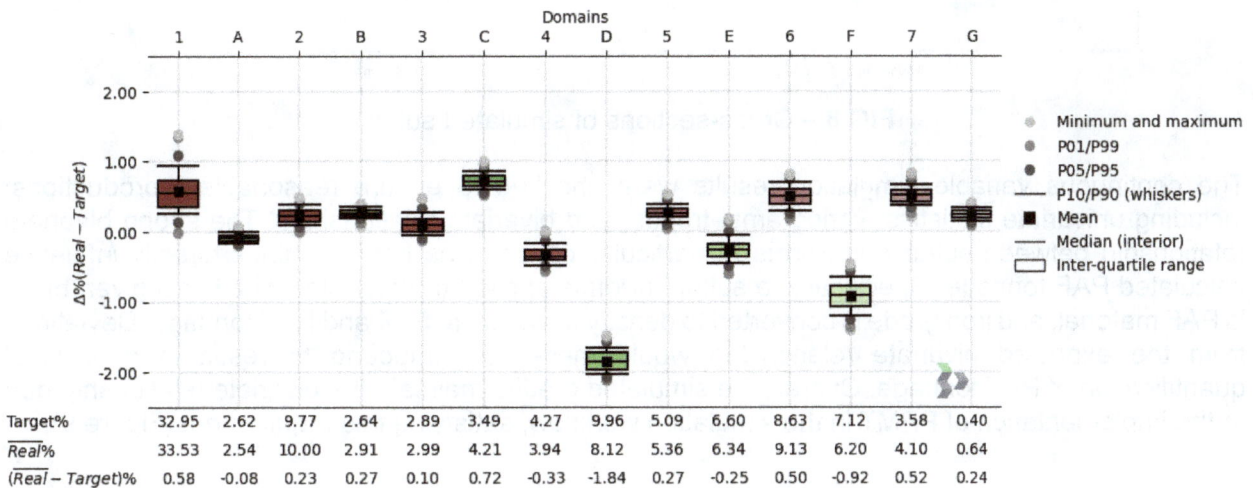

	1	A	2	B	3	C	4	D	5	E	6	F	7	G
Target%	32.95	2.62	9.77	2.64	2.89	3.49	4.27	9.96	5.09	6.60	8.63	7.12	3.58	0.40
$\overline{Real}$%	33.53	2.54	10.00	2.91	2.99	4.21	3.94	8.12	5.36	6.34	9.13	6.20	4.10	0.64
$(\overline{Real} - Target)$%	0.58	-0.08	0.23	0.27	0.10	0.72	-0.33	-1.84	0.27	-0.25	0.50	-0.92	0.52	0.24

FIG 7 – Absolute difference between simulated and deterministic domain proportions.

Grade simulation

Both sulfur and iron must be simulated to calculate PAF tonnage. Sulfur is used to determine if a block is PAF and iron is used to calculate the density of each block. Due to presence of trends in the data and the correlation between sulfur and iron, a detrending and decorrelation simulation workflow was followed:

- Normal score transform the variables.
- Independently detrend the variables:
  - Model the trend for the normal scores of each variable.
  - Fit a Gaussian mixture model (GMM) to each of the normal score variable-trend pairs.

- o Transform the normal score variables using the stepwise conditional transformation (SCT), yielding variables that are independent of the trend.
- Apply PPMT to the residuals, generating independent normal score variables.
- Independently simulate L realisations of the variables within domain realisations in a 1:1 fashion, using local varying anisotropy (LVA).
- Back-transform the L grid realisations, back-transforming previous transformation steps.
- Post-process the L back-transformed grid realisations to calculate relevant statistics.

Conditional simulation was performed independently on each domain for a given categorical realisation, then each domain realisation was merged into a single model. Viewed in section, the non-linear continuity of sulfur and restriction of high sulfur material to pyritic shale simulated domains is clear (Figure 8).

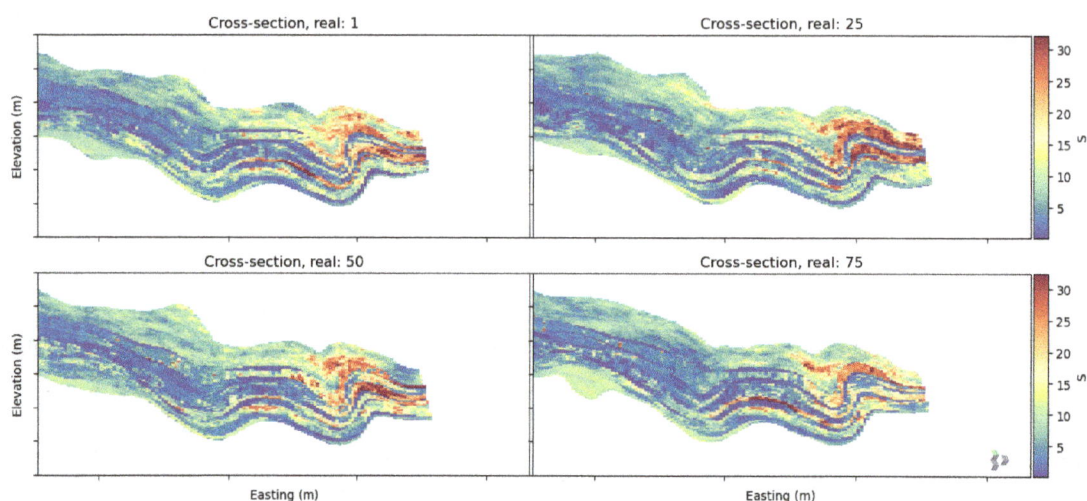

FIG 8 – Cross-sections of simulated sulfur.

The continuous variable simulation results were checked to ensure reasonable reproductions, including univariate statistics, variograms, trends, and bivariate relationships. The strong bivariate relationship between sulfur and iron is of particular concern, as both variables directly influence calculated PAF tonnage. Specifically, a sulfur cut-off is applied to determine whether a given block is PAF material, and iron grade is converted to density to calculate PAF and NAF tonnage. Deviations from the expected bivariate relationship would therefore compound to result in poor local quantification of PAF tonnage. Overall, the simulated grades maintain the bivariate relationship due to the implementation of PPMT in the simulation workflow, satisfying this requirement (Figure 9).

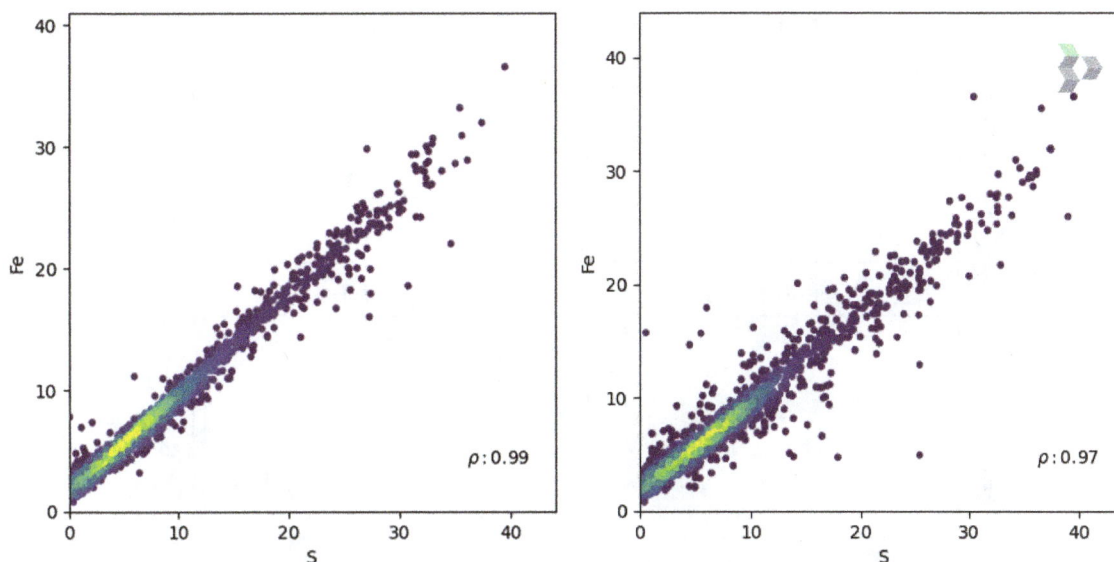

FIG 9 – Sulfur-iron bivariate scatterplot for data (left) and one simulated realisation (right), showing reasonable reproduction.

Post-processing

Post processing was conducted to convert simulated sulfur and iron into PAF tonnages for use in planning. The point-scale grade realisations are block averaged to an SMU scale that reflects the selectivity of waste stripping. SMU scale blocks on each realisation are then compared to a sulfur cut-off to determine whether the contained material constitutes PAF or NAF material, generating 100 realisations of waste material properties. An e-type model of sulfur is also generated to allow for a direct comparison with the previous sulfur estimate.

When compared with the original kriged sulfur model, the impact of simulation is evident at an SMU scale (Figure 10). Most notably, the folding observed in the stratigraphic sequence is highly visible, as are the high-sulfur shale units and intercalated low-sulfur breccias. The kriged model, conversely, has greater smoothing with limited areal influence of high sulfur samples.

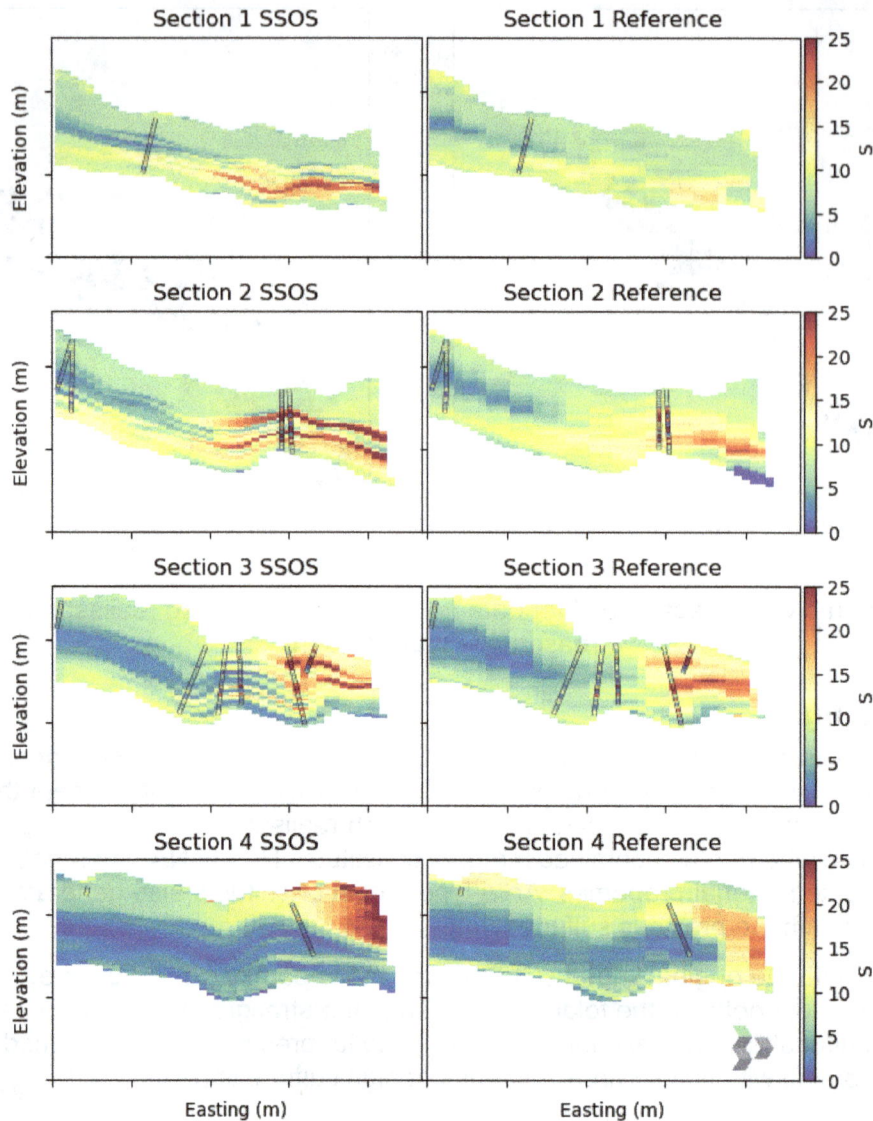

FIG 10 – E-Type vertical sections comparing the simulated sulfur e-type estimate (left) to the kriged model (right).

Binary categorical realisations that indicate whether a block is PAF or NAF are generated by applying the sulfur cut-off at SMU scale. Averaging these realisations generates a probabilistic summary of PAF content. All blocks contain a value between 0 and 1 which reflects the proportion of realisations where that block was considered PAF. This proportion can be interpreted as the probability that the block will contain PAF. Visualising the resulting probability in map view (Figure 11) shows the PAF tonnage statistics within the planned pit volume were summarised globally by bench to characterise the impact of the simulation study on expected PAF content relative to the deterministic results, and to summarise the modelled uncertainty spatially. Resulting tonnages are anonymised for presentation in this paper. Figure 12 shows the PAF tonnage broken down by bench. PAF material is constrained to 15 consecutive benches within the simulated stratigraphic sequence and displays some cyclicity with depth. The highest quantities of PAF are expected in benches 12 and 13, with benches 6–10 having moderate PAF content. The greatest uncertainty also occurs in these phases of waste stripping. This is a substantive change relative to the deterministic model, where PAF is estimated to be lower than the simulated average for benches 1–4 and 9–14. The most significant differences are on benches 8 and 12. In the deterministic model, bench 8 has the highest deterministic PAF tonnage at about 55 000 t, while the simulated model shows 26 000 t, on average. Bench 12, on the other hand, has under 10 000 t of PAF in the estimated model, and an average of 41 000 t based on the simulation.

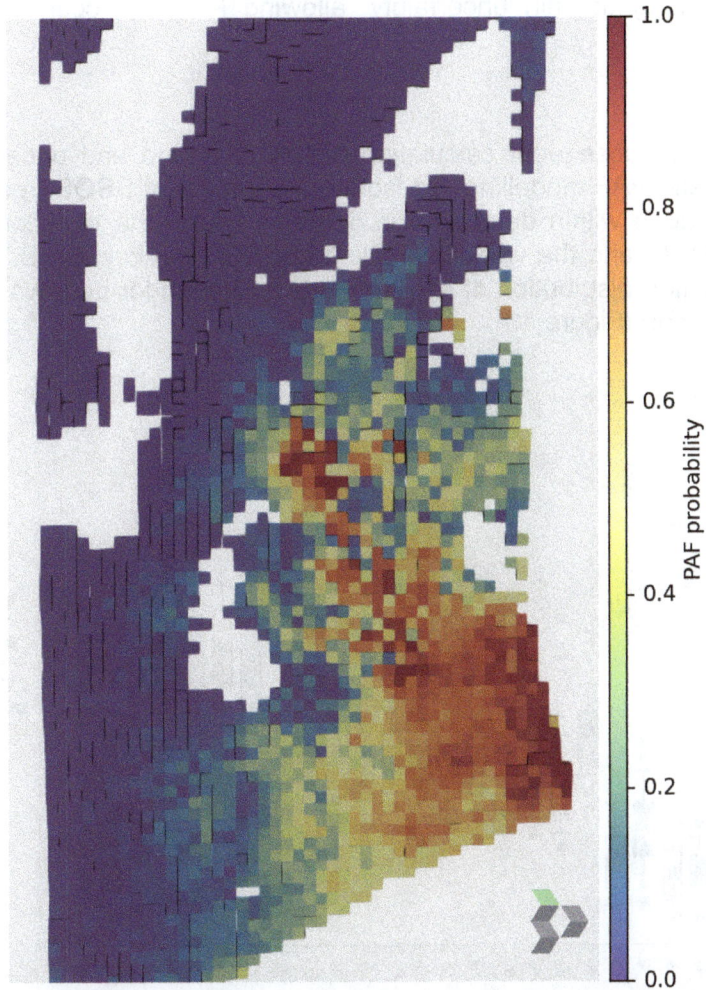

FIG 11 – Map view of PAF probability for selected domain.

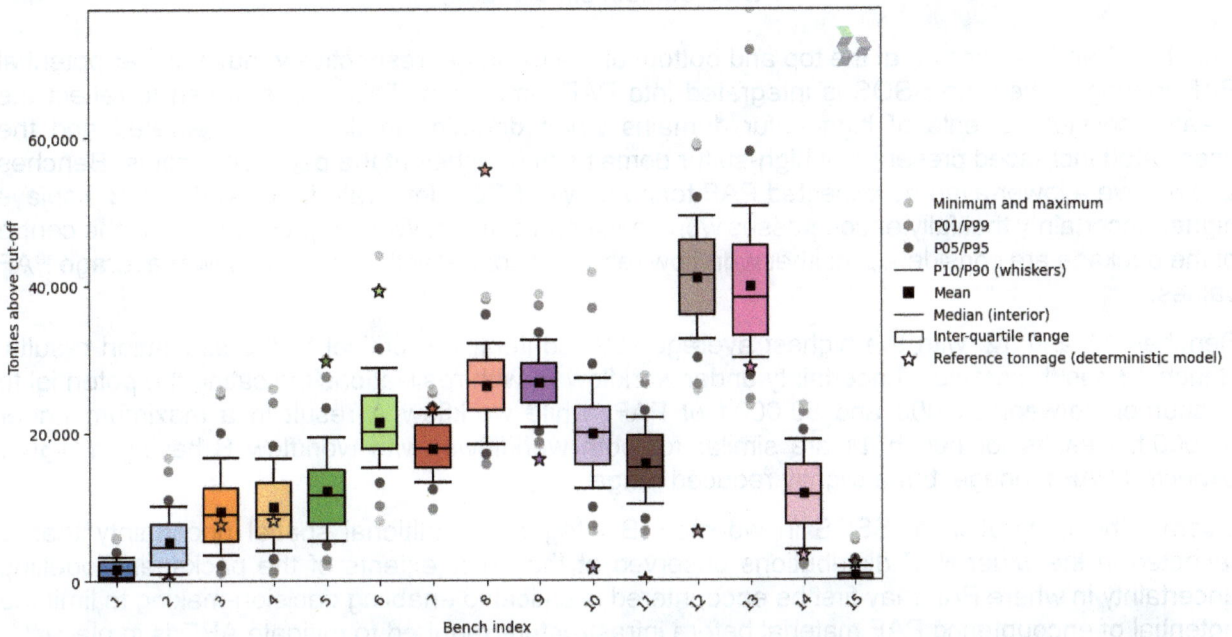

FIG 12 – Box plot showing PAF content and uncertainty within the planned pit volume by bench.

The difference in model tonnages is attributed to a combination of factors, including the smoothing inherent to estimation, the LVA applied in the simulation aligning the sulfur grades with the host units, and the SSOS integrating domain uncertainty, allowing PAF to occur over a wider range of elevations.

IMPACT OF SSOS

Simulation of sulfur and subsequent calculation of PAF tonnage and uncertainty shows a clear benefit over deterministic PAF modelling. To isolate the impact of SSOS, results are compared to simulation of sulfur grades within deterministic domains. Here, the workflow excluding SSOS is referred to as workflow A, and the workflow including SSOS is workflow B. A series of box plots representing the simulated distribution of PAF tonnage by bench for both workflows are generated to facilitate the comparison (Figure 13).

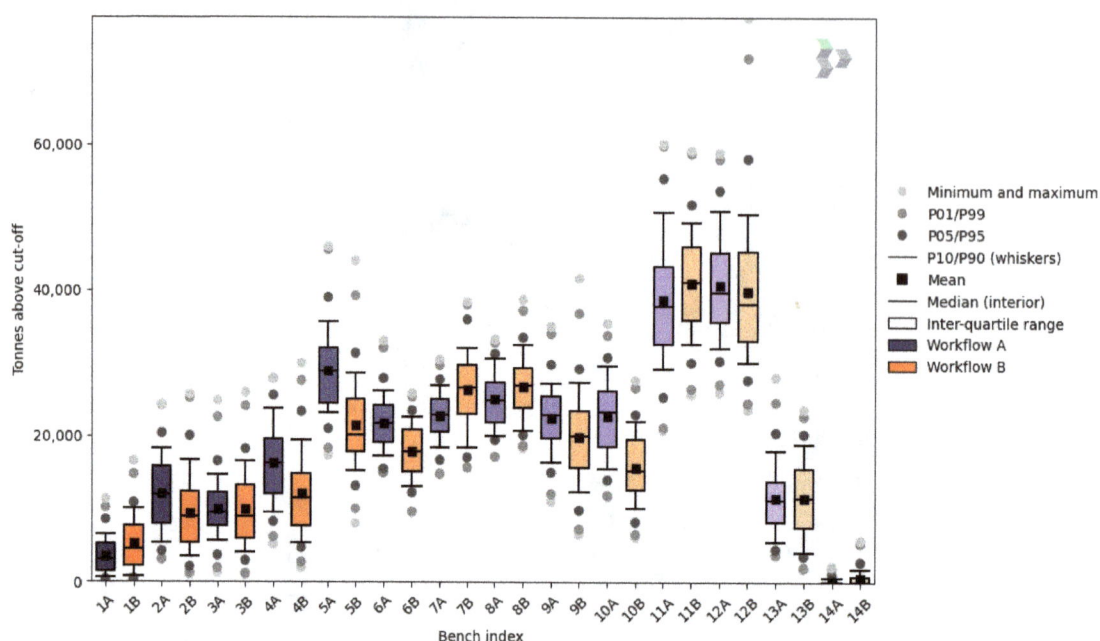

FIG 13 – Boxplots comparing PAF tonnage distributions from simulation without (A) and with (B) SSOS domain simulation.

Benches 1 and 14, located at the top and bottom of the package, respectively, have higher potential PAF tonnage when the SSOS is integrated into PAF simulation. This is interpreted to reflect the greater possible extents of high-sulfur domains when domain simulation is integrated and the associated increased presence of high-sulfur domains in benches at the package extents. Benches 2 to 4 have a lower average expected PAF tonnage with SSOS integrated, but workflow B displays higher uncertainty that fully encompasses workflow A realisations. When benches towards the centre of the package are considered, neither workflow results in consistently higher nor lower average PAF values.

Benches 11 and 12, with the highest average PAF content, are critical to the simulation results. Bench 12 has the greatest uncertainty under workflow B, with realisations indicating the potential to encounter between 20 000 and 80 000 t of PAF, while workflow A result in a maximum under 60 000 t. Results for bench 11 are similar for both workflows, with workflow B having a higher expected PAF tonnage, but a slightly reduced range.

Overall, the integration of SSOS in workflow B integrates additional spatial uncertainty that is reflected in the wider PAF distributions observed at the outer extents of the package. Modelling uncertainty in where PAF may first be encountered is critical to enabling decision-making to limit the potential of encountering PAF material before infrastructure required to mitigate ARD is in place.

CONCLUSIONS

Management of PAF material is critical to facilitating operational planning, financial risk management, and ensure integration of ESG considerations. A case study is presented showing the benefits simulating PAF material tonnage. The simulation includes the uncertainty of domain contact positions using a novel surface simulation tool, SSOS. This tool facilitates surface modelling where current techniques are not suitable due to high domain continuity and the presence of strong folding that cannot be addressed with unfolding routines.

The integration of SSOS and continuous grade modelling in a state-of-the-art geostatistical workflow for multivariate simulation with a trend and locally varying anisotropy results in a model that respects geological contacts and their uncertainty, and the non-stationary grade continuity within the modelled units. The simulated results will inform future planning efforts to ensure effective management of PAF material. They will guide drill hole planning to improve understanding of the material's geometry and grade distribution, while also aiding in the development of infrastructure capable of accommodating the anticipated volumes.

Due to the novelty of the SSOS and its development as a practical tool under time constraints, there is room for further development that could lead to improved results. First, there is a known relationship between the thickness of units within a stratigraphic package due to the thickness-sum constraint. The generally negative correlation of unit thickness is not currently incorporated into the independent surface simulation implemented in SOSS. While a method to determine an exhaustive super-secondary for use in co-simulation was not obvious at the time of the initial implementation, further development target the implementation of intrinsic collocated co-simulation to ensure these relationships are reproduced. The current independent simulation may result in excessive uncertainty away from conditioning data, particularly for contacts distant from the initial reference. Second, the implementation details of the SSOS merit further study to ensure it is a repeatable process that is easily transferrable to other deposits and applications, including in ore. Finally, the correct method to transition from simulated contacts to gridded domain realisations is unclear, especially where the geological environment leads to complex interactions that may conflict with the current flagging scheme.

REFERENCES

Barnett, R M, Manchuk, J G and Deutsch, C V, 2014. Projection pursuit multivariate transform, *Mathematical Geosciences*, 46:337–359.

Carvalho, D, 2018. Probabilistic resource modeling of vein deposits, MSc thesis (unpublished), University of Alberta, Canada.

Cowan, E J, Beatson, R K, Ross, H J, Fright, W R, McLennan, T J, Evans, T R, Carr, J C, Lane, R G, Bright, D V, Gillman, A J and Oshust, P A, 2003, November. Practical implicit geological modelling, in *Proceedings of the 5th International Mining Geology Conference*, pp 89–99 (The Australasian Institute of Mining and Metallurgy: Melbourne).

Deutsch, C V, 2006. A sequential indicator simulation program for categorical variables with point and block data: BlockSIS, *Computers and Geosciences*, 32(10):1669–1681.

Hillier, M J, Schetselaar, E M, de Kemp, E A and Perron, G, 2014. Three-dimensional modelling of geological surfaces using generalised interpolation with radial basis functions, *Mathematical Geosciences*, 46:931–953.

Qu, J, 2018. A practical framework to characterise non-stationary regionalised variables, PhD thesis (unpublished), University of Alberta, Canada.

Silva, D and Deutsch, C, 2019. Multivariate categorical modeling with hierarchical truncated pluri-gaussian simulation, *Mathematical Geosciences*, 51(5):527–552.

Author index

Afonseca, B C	269	Mattingley-Scott, M	75
Ambach, D	243	Maxwell, K	19
Assibey-Bonsu, W	209	McDivitt, J A	147
Barrios, A	119	Minniakhmetov, I	319, 333
Barton, C	3	Nwaila, G	101
Benthen, G	227	Pari, J	133
Bertossi, L	171	Ramos, E	185
Carvalho, D	171, 357	Rondón, O	209
Cavelius, C	87	Roux, M	35, 45
Clark, J M	35, 45, 133, 227	Samson, M	45, 227
Conn, D	227	Sanchez, S	357
Coombes, J	75	Sen, S	185
Dalrymple, J	3	Serjeantson, J	319
Dangin, J	87	Sims, D A	57
Fiddes, C	35, 45	Snider, L	133
Gomes, C	119, 289	Soodi Shoar, P	87
Greene, D	147	Stefoni, L	243
Hadavand, M	147	Vandelle, M	45
Hague, R C	303	Wilhelmsen, A	243
Hart, E	147	Williams, A	45
Harvey, J	87	Williams, C	333
Hodkiewicz, P	75	Wilson, B	147, 357
Horan, S	357	Wilson, T	87
Le Cornu, C	45, 133, 227	Wilson, V	119
Machukera, J	185	Zhang, S	101
Matthews, M	185		

www.ingramcontent.com/pod-product-compliance
Lightning Source LLC
Chambersburg PA
CBHW061104210326
41597CB00021B/3969